Control of Foodborne Microorganisms

FOOD SCIENCE AND TECHNOLOGY

A Series of Monographs, Textbooks, and Reference Books

1. Flavor Research: Principles and Techniques, *R. Teranishi, I. Hornstein, P. Issenberg, and E. L. Wick*
2. Principles of Enzymology for the Food Sciences, *John R. Whitaker*
3. Low-Temperature Preservation of Foods and Living Matter, *Owen R. Fennema, William D. Powrie, and Elmer H. Marth*
4. Principles of Food Science
   Part I: Food Chemistry, *edited by Owen R. Fennema*
   Part II: Physical Methods of Food Preservation, *Marcus Karel, Owen R. Fennema, and Daryl B. Lund*
5. Food Emulsions, *edited by Stig E. Friberg*
6. Nutritional and Safety Aspects of Food Processing, *edited by Steven R. Tannenbaum*
7. Flavor Research: Recent Advances, *edited by R. Teranishi, Robert A. Flath, and Hiroshi Sugisawa*
8. Computer-Aided Techniques in Food Technology, *edited by Israel Saguy*
9. Handbook of Tropical Foods, *edited by Harvey T. Chan*
10. Antimicrobials in Foods, *edited by Alfred Larry Branen and P. Michael Davidson*

11. Food Constituents and Food Residues: Their Chromatographic Determination, *edited by James F. Lawrence*
12. Aspartame: Physiology and Biochemistry, *edited by Lewis D. Stegink and L. J. Filer, Jr.*
13. Handbook of Vitamins: Nutritional, Biochemical, and Clinical Aspects, *edited by Lawrence J. Machlin*
14. Starch Conversion Technology, *edited by G. M. A. van Beynum and J. A. Roels*
15. Food Chemistry: Second Edition, Revised and Expanded, *edited by Owen R. Fennema*
16. Sensory Evaluation of Food: Statistical Methods and Procedures, *Michael O'Mahony*
17. Alternative Sweeteners, *edited by Lyn O'Brien Nabors and Robert C. Gelardi*
18. Citrus Fruits and Their Products: Analysis and Technology, *S. V. Ting and Russell L. Rouseff*
19. Engineering Properties of Foods, *edited by M. A. Rao and S. S. H. Rizvi*
20. Umami: A Basic Taste, *edited by Yojiro Kawamura and Morley R. Kare*
21. Food Biotechnology, *edited by Dietrich Knorr*
22. Food Texture: Instrumental and Sensory Measurement, *edited by Howard R. Moskowitz*
23. Seafoods and Fish Oils in Human Health and Disease, *John E. Kinsella*
24. Postharvest Physiology of Vegetables, *edited by J. Weichmann*
25. Handbook of Dietary Fiber: An Applied Approach, *Mark L. Dreher*
26. Food Toxicology, Parts A and B, *Jose M. Concon*
27. Modern Carbohydrate Chemistry, *Roger W. Binkley*
28. Trace Minerals in Foods, *edited by Kenneth T. Smith*
29. Protein Quality and the Effects of Processing, *edited by R. Dixon Phillips and John W. Finley*
30. Adulteration of Fruit Juice Beverages, *edited by Steven Nagy, John A. Attaway, and Martha E. Rhodes*
31. Foodborne Bacterial Pathogens, *edited by Michael P. Doyle*
32. Legumes: Chemistry, Technology, and Human Nutrition, *edited by Ruth H. Matthews*
33. Industrialization of Indigenous Fermented Foods, *edited by Keith H. Steinkraus*
34. International Food Regulation Handbook: Policy • Science • Law, *edited by Roger D. Middlekauff and Philippe Shubik*
35. Food Additives, *edited by A. Larry Branen, P. Michael Davidson, and Seppo Salminen*
36. Safety of Irradiated Foods, *J. F. Diehl*
37. Omega-3 Fatty Acids in Health and Disease, *edited by Robert S. Lees and Marcus Karel*
38. Food Emulsions: Second Edition, Revised and Expanded, *edited by Kåre Larsson and Stig E. Friberg*
39. Seafood: Effects of Technology on Nutrition, *George M. Pigott and Barbee W. Tucker*

40. Handbook of Vitamins: Second Edition, Revised and Expanded, *edited by Lawrence J. Machlin*
41. Handbook of Cereal Science and Technology, *Klaus J. Lorenz and Karel Kulp*
42. Food Processing Operations and Scale-Up, *Kenneth J. Valentas, Leon Levine, and J. Peter Clark*
43. Fish Quality Control by Computer Vision, *edited by L. F. Pau and R. Olafsson*
44. Volatile Compounds in Foods and Beverages, *edited by Henk Maarse*
45. Instrumental Methods for Quality Assurance in Foods, *edited by Daniel Y. C. Fung and Richard F. Matthews*
46. *Listeria*, Listeriosis, and Food Safety, *Elliot T. Ryser and Elmer H. Marth*
47. Acesulfame-K, *edited by D. G. Mayer and F. H. Kemper*
48. Alternative Sweeteners: Second Edition, Revised and Expanded, *edited by Lyn O'Brien Nabors and Robert C. Gelardi*
49. Food Extrusion Science and Technology, *edited by Jozef L. Kokini, Chi-Tang Ho, and Mukund V. Karwe*
50. Surimi Technology, *edited by Tyre C. Lanier and Chong M. Lee*
51. Handbook of Food Engineering, *edited by Dennis R. Heldman and Daryl B. Lund*
52. Food Analysis by HPLC, *edited by Leo M. L. Nollet*
53. Fatty Acids in Foods and Their Health Implications, *edited by Ching Kuang Chow*
54. *Clostridium botulinum*: Ecology and Control in Foods, *edited by Andreas H. W. Hauschild and Karen L. Dodds*
55. Cereals in Breadmaking: A Molecular Colloidal Approach, *Ann-Charlotte Eliasson and Kåre Larsson*
56. Low-Calorie Foods Handbook, *edited by Aaron M. Altschul*
57. Antimicrobials in Foods: Second Edition, Revised and Expanded, *edited by P. Michael Davidson and Alfred Larry Branen*
58. Lactic Acid Bacteria, *edited by Seppo Salminen and Atte von Wright*
59. Rice Science and Technology, *edited by Wayne E. Marshall and James I. Wadsworth*
60. Food Biosensor Analysis, *edited by Gabriele Wagner and George G. Guilbault*
61. Principles of Enzymology for the Food Sciences: Second Edition, *John R. Whitaker*
62. Carbohydrate Polyesters as Fat Substitutes, *edited by Casimir C. Akoh and Barry G. Swanson*
63. Engineering Properties of Foods: Second Edition, Revised and Expanded, *edited by M. A. Rao and S. S. H. Rizvi*
64. Handbook of Brewing, *edited by William A. Hardwick*
65. Analyzing Food for Nutrition Labeling and Hazardous Contaminants, *edited by Ike J. Jeon and William G. Ikins*
66. Ingredient Interactions: Effects on Food Quality, *edited by Anilkumar G. Gaonkar*
67. Food Polysaccharides and Their Applications, *edited by Alistair M. Stephen*

68. Safety of Irradiated Foods: Second Edition, Revised and Expanded, *J. F. Diehl*
69. Nutrition Labeling Handbook, *edited by Ralph Shapiro*
70. Handbook of Fruit Science and Technology: Production, Composition, Storage, and Processing, *edited by D. K. Salunkhe and S. S. Kadam*
71. Food Antioxidants: Technological, Toxicological, and Health Perspectives, *edited by D. L. Madhavi, S. S. Deshpande, and D. K. Salunkhe*
72. Freezing Effects on Food Quality, *edited by Lester E. Jeremiah*
73. Handbook of Indigenous Fermented Foods: Second Edition, Revised and Expanded, *edited by Keith H. Steinkraus*
74. Carbohydrates in Food, *edited by Ann-Charlotte Eliasson*
75. Baked Goods Freshness: Technology, Evaluation, and Inhibition of Staling, *edited by Ronald E. Hebeda and Henry F. Zobel*
76. Food Chemistry: Third Edition, *edited by Owen R. Fennema*
77. Handbook of Food Analysis: Volumes 1 and 2, *edited by Leo M. L. Nollet*
78. Computerized Control Systems in the Food Industry, *edited by Gauri S. Mittal*
79. Techniques for Analyzing Food Aroma, *edited by Ray Marsili*
80. Food Proteins and Their Applications, *edited by Srinivasan Damodaran and Alain Paraf*
81. Food Emulsions: Third Edition, Revised and Expanded, *edited by Stig E. Friberg and Kåre Larsson*
82. Nonthermal Preservation of Foods, *Gustavo V. Barbosa-Cánovas, Usha R. Pothakamury, Enrique Palou, and Barry G. Swanson*
83. Milk and Dairy Product Technology, *Edgar Spreer*
84. Applied Dairy Microbiology, *edited by Elmer H. Marth and James L. Steele*
85. Lactic Acid Bacteria: Microbiology and Functional Aspects: Second Edition, Revised and Expanded, *edited by Seppo Salminen and Atte von Wright*
86. Handbook of Vegetable Science and Technology: Production, Composition, Storage, and Processing, *edited by D. K. Salunkhe and S. S. Kadam*
87. Polysaccharide Association Structures in Food, *edited by Reginald H. Walter*
88. Food Lipids: Chemistry, Nutrition, and Biotechnology, *edited by Casimir C. Akoh and David B. Min*
89. Spice Science and Technology, *Kenji Hirasa and Mitsuo Takemasa*
90. Dairy Technology: Principles of Milk Properties and Processes, *P. Walstra, T. J. Geurts, A. Noomen, A. Jellema, and M. A. J. S. van Boekel*
91. Coloring of Food, Drugs, and Cosmetics, *Gisbert Otterstätter*
92. *Listeria*, Listeriosis, and Food Safety: Second Edition, Revised and Expanded, *edited by Elliot T. Ryser and Elmer H. Marth*
93. Complex Carbohydrates in Foods, *edited by Susan Sungsoo Cho, Leon Prosky, and Mark Dreher*
94. Handbook of Food Preservation, *edited by M. Shafiur Rahman*

95. International Food Safety Handbook: Science, International Regulation, and Control, *edited by Kees van der Heijden, Maged Younes, Lawrence Fishbein, and Sanford Miller*
96. Fatty Acids in Foods and Their Health Implications: Second Edition, Revised and Expanded, *edited by Ching Kuang Chow*
97. Seafood Enzymes: Utilization and Influence on Postharvest Seafood Quality, *edited by Norman F. Haard and Benjamin K. Simpson*
98. Safe Handling of Foods, edited by Jeffrey M. Farber and Ewen C. D. Todd
99. Handbook of Cereal Science and Technology: Second Edition, Revised and Expanded, *edited by Karel Kulp and Joseph G. Ponte, Jr.*
100. Food Analysis by HPLC: Second Edition, Revised and Expanded, *edited by Leo M. L. Nollet*
101. Surimi and Surimi Seafood, *edited by Jae W. Park*
102. Drug Residues in Foods: Pharmacology, Food Safety, and Analysis, *Nickos A. Botsoglou and Dimitrios J. Fletouris*
103. Seafood and Freshwater Toxins: Pharmacology, Physiology, and Detection, *edited by Luis M. Botana*
104. Handbook of Nutrition and Diet, *Babasaheb B. Desai*
105. Nondestructive Food Evaluation: Techniques to Analyze Properties and Quality, *edited by Sundaram Gunasekaran*
106. Green Tea: Health Benefits and Applications, *Yukihiko Hara*
107. Food Processing Operations Modeling: Design and Analysis, *edited by Joseph Irudayaraj*
108. Wine Microbiology: Science and Technology, *Claudio Delfini and Joseph V. Formica*
109. Handbook of Microwave Technology for Food Applications, *edited by Ashim K. Datta and Ramaswamy C. Anantheswaran*
110. Applied Dairy Microbiology: Second Edition, Revised and Expanded, *edited by Elmer H. Marth and James L. Steele*
111. Transport Properties of Foods, *George D. Saravacos and Zacharias B. Maroulis*
112. Alternative Sweeteners: Third Edition, Revised and Expanded, *edited by Lyn O'Brien Nabors*
113. Handbook of Dietary Fiber, *edited by Susan Sungsoo Cho and Mark L. Dreher*
114. Control of Foodborne Microorganisms, *edited by Vijay K. Juneja and John N. Sofos*

Additional Volumes in Preparation

Food Additives: Second Edition, Revised and Expanded, *edited by Alfred Larry Branen, P. Michael Davidson, Seppo Salminen, and John H. Thorngate, III*

Flavor, Fragrance, and Odor Analysis, *edited by Ray Marsili*

Characterization of Cereals and Flours: Properties, Analysis, and Applications, *edited by Gönül Kaletunç and Kenneth J. Breslauer*

Postharvest Physiology and Pathology of Vegetables: Second Edition, Revised and Expanded, *edited by Jerry A. Bartz and Jeffrey K. Brecht*

Food Lipids: Chemistry, Nutrition, and Biotechnology: Second Edition, Revised and Expanded, *edited by Casimir C. Akoh and David B. Min*

Handbook of Food Enzymology, *edited by John R. Whitaker, A. G. J. Voragen, and Dominic Wong*

Handbook of Food Toxicology: Second Edition, Revised and Expanded, *S. S. Deshpande, K. Salunkhe, and Jose M. Concon*

Control of Foodborne Microorganisms

Edited by

VIJAY K. JUNEJA

U.S. Department of Agriculture
Wyndmoor, Pennsylvania

JOHN N. SOFOS

Colorado State University
Fort Collins, Colorado

CRC Press
Taylor & Francis Group
Boca Raton London New York

CRC Press is an imprint of the
Taylor & Francis Group, an **informa** business

First published 2002 by Marcel Dekker, Inc.

Published 2019 by CRC Press
Taylor & Francis Group
6000 Broken Sound Parkway NW, Suite 300
Boca Raton, FL 33487-2742

First issued in paperback 2019

ISBN 13: 978-0-367-45512-5 (pbk)
ISBN 13: 978-0-8247-0573-2 (hbk)

Visit the Taylor & Francis Web site at
http://www.taylorandfrancis.com

and the CRC Press Web site at
http://www.crcpress.com

Preface

The need for better control of microbial contamination in foods is paramount. Foods previously thought not to be involved in foodborne diseases or believed to be infrequent sources of foodborne diseases have recently been associated with outbreaks or sporadic episodes of illness, some of which have been fatal. Our understanding of foodborne pathogens and of the types of microorganisms that have been linked with documented outbreaks of illness has dramatically increased in the past two decades. These food safety concerns are magnified because of consumer preferences for minimally processed foods that offer convenience in availability and preparation. Thus, it is necessary for scientists and regulators in the industry to reconsider the approach to food preservation and pathogen control to enhance food safety.

Every effort was made to make this book as comprehensive and timely as possible. *Control of Foodborne Microorganisms* is written at a level that presupposes a general background in microbiology, which is needed to understand the basic mechanisms of microbial control or inactivation. The book covers microbial control by traditional techniques, such as by heat, cold, and chemicals, as well as new and novel approaches to microbial inactivation by high pressure, pulsed electric fields, and hurdle technologies. Chapters are arranged in a logical, concise sequence and presented in an easy-to-follow format by internationally renowned experts in their fields.

It is necessary for the food industry and regulatory agencies to have personnel who are knowledgeable about methodologies applied in the control or inactivation of microorganisms that may be present in foods. Until now, such information has been available only in diverse sources that are not always readily available. Accordingly, this book should be of special benefit to individuals who have little or no opportunity for additional classroom training, and it is a valuable text for those who are directly or indirectly involved in the production, processing, distribution, and serving of food; in the control of hazards and spoilage of food products; and in doing research on microbial control or inactivation, which includes individuals in academic, industrial, and government institutions, food consultants, and food lobbyists.

We are grateful to all our contributors for their relentless efforts. The credit for making this book a reality goes to them. We hope that the book will help identify new approaches to controlling foodborne pathogens and significantly contribute to decreasing the incidence of foodborne disease outbreaks.

Vijay K. Juneja
John N. Sofos

Contents

Contributors

Gary R. Acuff Department of Animal Science, Texas A&M University, College Station, Texas

Gustavo V. Barbosa-Cánovas Department of Biological Systems Engineering, Washington State University, Pullman, Washington

Robert L. Buchanan Center for Food Safety and Applied Nutrition, U.S. Food and Drug Administration, Washington, D.C.

Alejandro Castillo University of Guadalajara, Guadalajara, Mexico

Michael L. Chikindas Department of Food Science, Rutgers University, New Brunswick, New Jersey

Stephanie Clark Department of Food Science and Human Nutrition, Washington State University, Pullman, Washington

Martin B. Cole Food Science Australia, North Ryde, New South Wales, Australia

P. Michael Davidson Department of Food Science and Technology, University of Tennessee, Knoxville, Tennessee

James S. Dickson Department of Microbiology, Iowa State University, Ames, Iowa

Stephanie Doores Department of Food Science, The Pennsylvania State University, University Park, Pennsylvania

Colin O. Gill Lacombe Research Centre, Agriculture and Agri-Food Canada, Lacombe, Alberta, Canada

Kathleen A. Glass Department of Food Microbiology and Toxicology, Food Research Institute, University of Wisconsin–Madison, Madison, Wisconsin

Marsha H. Golden Eastern Regional Research Center, Agricultural Research Service, U.S. Department of Agriculture, Wyndmoor, Pennsylvania

Grahame W. Gould* Microbiology Department, Unilever Research Laboratory, Bedford, England

Gerhard J. Haas School of Natural Sciences, Fairleigh Dickinson University, Teaneck, New Jersey

Margaret D. Hardin Sara Lee Foods, Cordova, Tennessee

Federico M. Harte Department of Biological Systems Engineering, Washington State University, Pullman, Washington

Dallas G. Hoover Department of Animal and Food Sciences, University of Delaware, Newark, Delaware

Joseph H. Hotchkiss Department of Food Science, Cornell University, Ithaca, New York

Eric A. Johnson Department of Food Microbiology and Toxicology, Food Research Institute, University of Wisconsin–Madison, Madison, Wisconsin

Vijay K. Juneja Eastern Regional Research Center, Agricultural Research Service, U.S. Department of Agriculture, Wyndmoor, Pennsylvania

Lothar Leistner† International Food Consultant, Kulmbach, Germany

* Retired.

† Formerly of the Institute for Microbiology, Toxicology, and Histology at the Federal Centre of Meat Research, Kulmbach, Germany.

Christopher R. Loss Department of Food Science, Cornell University, Ithaca, New York

Paul D. Matthews Department of Biological Sciences, Lehman College of the City University of New York, Bronx, New York

Aubrey F. Mendonca Department of Food Science and Human Nutrition, Iowa State University, Ames, Iowa

Juan Fernández Molina Department of Biological Systems Engineering, Washington State University, Pullman, Washington

Thomas J. Montville Department of Food Science, Rutgers University, New Brunswick, New Jersey

Sevugan Palaniappan New Technologies and Commercialization, The Minute Maid Company, Houston, Texas

Terry A. Roberts Consultant, Reading, England

M. Fernanda San Martín Department of Biological Systems Engineering, Washington State University, Pullman, Washington

Sudhir K. Sastry Department of Food, Agricultural, and Biological Engineering, The Ohio State University, Columbus, Ohio

O. Peter Snyder, Jr. Hospitality Institute of Technology and Management, St. Paul, Minnesota

John N. Sofos Department of Animal Sciences, Colorado State University, Fort Collins, Colorado

Barry G. Swanson Department of Food Science and Human Nutrition, Washington State University, Pullman, Washington

Gaurav Tewari ESL Global Inc., San Antonio, Texas

Richard C. Whiting Center for Food Safety and Applied Nutrition, U.S. Food and Drug Administration, Washington, D.C.

Control of Foodborne Microorganisms

Control of Foodborne Microorganisms

1

Microbial Control in Foods: Needs and Concerns

John N. Sofos
Colorado State University
Fort Collins, Colorado

I. INTRODUCTION

Food safety has been at the forefront of societal concerns in recent years, and the major emphasis it has received is very likely to continue in many parts of the world and into the future. The complexity of food safety issues has increased, as has the number of emerging pathogenic microorganisms, some of which are resistant to antibiotics or to traditional preservation methods, and which sometimes cause illness with low infectious doses (1,2). In addition to microorganism-associated concerns, societal changes including changing consumer food preferences, lack of adequate food handling education, increases in human populations at-risk for foodborne illness, complex food distribution patterns, increased international trade, and better methods of testing for microbial detection also emphasize the complexity of the safety concerns and the need for development of strategies to address these challenges. In general, food safety is at the forefront of our societal concerns, and it should continue to be there for a number of years to come, because food safety problems and challenges that need to be addressed may actually increase. It is estimated that foodborne diseases cause approximately 76 million illnesses, 325,000 hospitalizations, and 5,000 deaths in the United States each year (3). Known pathogens cause 14 million illnesses, 60,000 hospitalizations, and 1,800 deaths, while unknown agents are involved in 62 million cases, 265,000 hospitalizations, and 3,200 deaths. Only a small portion of foodborne illness episodes are reported and investigated annually, and only a small portion of the reported foodborne illnesses are resolved (3,4). It is therefore important to emphasize development and application of control processes for microorganisms with the objective of improving the safety of our food supply.

II. EMERGING AND ADAPTING PATHOGENS

A number of "new," "emerging," "reemerging" or "evolving" pathogenic microorganisms have been associated with documented foodborne illness episodes in the past 15–25 years, and their number appears to be increasing. Among these are *Listeria monocytogenes*, *Campylobacter jejuni*, *Escherichia coli* O157:H7 and other enterohemorrhagic *E. coli* serotypes, *Salmonella* serotypes Enteritidis and Typhimurium DT104, *Yersinia enterocolitica*, *Cyptosporidium parvum*, *Cyclospora cayetanensis*, and various viral hazards (1,2,5). Factors that may contribute to pathogen emergence can be classified as biological, environmental, food-related, societal, and consumer-associated.

Biological factors may be related to changes in genotypes leading to enhanced survival, resistance, or virulence. However, improvements in microbial detection methods and increased emphasis on sampling and analysis may also lead to isolation and confirmation of the role of suspected or currently unknown pathogens in food safety problems. Truly emerging pathogens are those that have developed new virulence genes or resistance to standard therapeutic doses of drugs (5). A known human pathogen, however, may be designated as an emerging foodborne pathogen when there is evidence of transmission and cause of illness through consumption of contaminated food. Pathogens may also be classified as emerging when they become involved in outbreaks in countries other than those in which they have caused problems in the past, while a pathogen may be involved in illness following a period of hiatus and be classified as reemerging. A pathogen may be considered new or newly emerging if no evidence of past existence or involvement in foodborne illness exists (5). As interest in and scrutiny of microbial foodborne illness episodes continue to increase, epidemiological efforts become more systematic, new and improved methods for pathogen detection are developed, and microorganisms continue to adapt, additional microorganisms may be recognized as important in food safety. As these risks become known, control approaches should be developed in order to enhance food safety.

Intensified research in recent years indicates continuous adaptation and development of resistance by pathogenic microorganisms to antibiotics and to the traditional food preservation barriers of low pH, heat, cold temperatures, dry or low–water activity environments, and chemical preservatives. Furthermore, there is evidence of strains of pathogens with enhanced ability for survival in their hosts, low infective doses, and increased virulence, sometimes after exposure to common environmental stresses (6–13). These findings indicate that the microbial ecology of our food supply is undergoing changes that go along with the modernization taking place in our food-processing and distribution industries and the transformation of our society (1,2,14). Environmental factors associated with variations in geographic location and climate, as well as natural stresses, may induce biological changes and lead to new pathogens or enhanced virulence.

Food-related factors that may lead to pathogen emergence, increased resistance, or enhanced virulence include changes in food-production and harvesting practices, modifications in food-processing operations, new marketing practices, modern food preparation methods, and development of new food products to address consumer preferences and lifestyles (14). These changes may lead to new stresses exerted on microorganisms and removal or reduction of existing food preservation hurdle intensities, leading to increased pathogen resistance or virulence as well as failure of traditional food-processing hurdles to assure food safety.

III. FOODS AS VEHICLES OF FOODBORNE ILLNESS

As the number of pathogenic microorganisms documented as being transmitted through food has increased, so has the number of foods involved in foodborne illness. A variety of food items not previously associated with confirmed foodborne illness episodes have been linked with transmission of microbial foodborne illness in recent years, including a variety of foods of plant origin (15). Examples of food vehicles and associated pathogens include *E. coli* O157:H7 and other hemorrhagic *E. coli* serotypes from ground beef, fruit juices, alfalfa, radish and other types of sprouts, jerky, mayonnaise, watermelon, other produce, and dry fermented meats; *Salmonella* from ice cream, cantaloupes, watermelon, potatoes, alfalfa sprouts, and produce; *Salmonella* Enteritidis from eggs, originating in the ovaries of the hen before the eggshell is formed; *Shigella* from produce; *Y. enterocolitica* from chitterlings and tofu; *Vibrio vulnificus* from oysters; *L. monocytogenes* from milk, cheeses, coleslaw, hot dogs, and luncheon meats; *Clostridium botulinum* from potato salad, garlic sauce, sauteed onions, eggplant, bean dip, clam chowder, and olives; *C. parvum* from water and fresh-pressed apple juice; *C. cayetanensis* from raspberries and basil; hepatitis A virus from strawberries; and Norwalk-like virus from oysters, salads, and frostings. Additional information on outbreaks, clusters of foodborne illness, and the results of FoodNet Active Surveillance Network can be found in publications of the Centers for Disease Control and Prevention (CDC, at www.cdc.gov) such as the *Morbidity and Mortality Weekly Report* (MMWR) and other reports.

In general, there appears to be an increase in the number of foodborne illness outbreaks or clusters, an increase in the types of foods involved in foodborne illness, and an increase in the biological agents involved in foodborne illness and in their virulence and ability to survive in adverse, but sublethal, environments. This makes it necessary to reconsider our approach to food preservation for pathogen control in order to meet these new challenges and to enhance food safety.

IV. SOCIETAL AND CONSUMER CHANGES

The developments summarized above have increased interest in food safety among scientists, health agencies, and regulatory officials, but they have also been noticed by public interest groups, news-reporting media, and, consequently, a segment of the consuming public. The increased publicity has led to public awareness, concern, and more interest in food safety issues. This has also increased pressure on the private and public sectors to accelerate efforts that may lead to enhanced food safety.

The changing society and demographics may also be playing a role in pathogen emergence and may be contributing to our current food safety concerns. Society has undergone major changes in population numbers, household profiles, food preferences and expectations, lifestyles, life expectancy, and educational experiences. The number of people involved in agriculture and direct food production has decreased dramatically as our total population has increased and become more urban. The composition of our households has changed in ways that have led to changes in lifestyles and associated food preferences, food-handling practices, and expectations of or demands on our food supply (14). Consumers eat more meals away from home, the number of "take-home" meals has increased, and the use of preprepared or prepackaged meals, salads, and other food items that need minimal preparation and offer convenience, has increased. More consumers prefer or follow special diets, present-day consumers are exposed to limited education relative to proper food-handling practices, and an increasing number of consumers prefer minimally processed foods with low fat, reduced salt and other additives, fresh-like properties, and a long shelf life. Increasing numbers of consumers expect preservative-free and safe but mildly preserved foods with extended shelf life that are convenient to use (14,16–18). Some of these preferences may conflict and adversely affect food safety. For example, a lower fat content in a food is usually associated with higher moisture, which leads to dilution and further reduction of the lower level of salt and other additives. This further dilutes the preservative contribution of additives in a product that may also be minimally processed. These consumer preferences may lead to food safety risks, which become challenges to be addressed by those involved in assuring the safety of our food supply.

Food safety risks become even greater and more acute for consumers who are more sensitive to microbial infection. The number of consumers that are immunosuppressed, chronically ill, and of advanced age has increased. Our aging population includes more immunosuppressed and chronically ill persons, population segments that are more sensitive to foodborne illnesses and their consequences. In summary, consumer-associated factors affecting food safety are linked with changes in demographics, consumer expectations, eating habits, and lack of adequate and proper food-handling education. A number of chronic sequelae may be the result of foodborne infections, including ankylosing spondylitis, arthropathies, renal disease, cardiac and neurological disorders, and nutritional and other malabsorptive disorders (incapacitating diarrhea) (19). All this necessitates new strategies to provide safe foods for these increasing sensitive populations.

The increased urbanization of our population, the centralization of food production and processing, and the globalization of the food industry may also be important in pathogen emergence and resistance development. Furthermore, improved medical diagnostics, socioeconomic developments, news media scrutiny, and consumer advocate interest also enhance the concept of a food safety crisis. However, and as indicated, increases in host susceptibility and numbers of at-risk populations are well documented (5) and need to be considered as food products are being developed by the industry.

V. RESEARCH NEEDS TO CONTROL FOODBORNE PATHOGENS

The importance of food safety, especially of microbial food safety, has been emphasized in recent years through activities and initiatives undertaken by regulatory and public health agencies. Such initiatives include the application of new regulations, based on the principles of hazard analysis critical control point (HACCP), for the inspection of meat-, poultry-, and seafood-processing operations, and the United States National Food Safety Initiative (20–22). Control of pathogens to reduce risks should rely on science that generates knowledge about the interrelationship of foods, the environment, microbes, and humans based on quantitative risk assessment and predictive models (23–27). The knowledge base generated by research is necessary for regulatory decision making, industry solutions to food safety problems, and public education. Furthermore, this process allows training of future scientists for the public and private sector of society. The new and increased food safety concerns of our society have created new challenges and opportunities that need to be addressed. The response to food safety issues should be based on scientific advances accomplished through fundamental, basic, and applied research, as well as technology transfer and outreach public education.

Recent federal budgets have included increased funding to address food safety issues, with emphasis placed on epidemiology, research on procedures to control hazards, implementation of HACCP programs, risk-assessment activities, and public education on food safety (22). With the establishment of government standards for pathogen reduction in certain foods (20), there is also a need for effective methods to detect microorganisms, prevent contamination, and control pathogens. Various organizations, agencies, and scientific societies have formulated research needs and agendas to be followed for the management of food safety risks. A Food Safety Research Agenda was developed in 1996 for the Food Safety and Inspection Service by a Food Safety Research Working Group, which included research scientists representing several agencies. Research needs have also been formulated by the National Food Safety Initiative and groups such as the American Academy of Microbiology (28). Common themes of these research agendas are that risk assessment should be used to characterize and quantitate risks associated with foodborne hazards and their control and that better control procedures should be developed and implemented for pathogens of concern such as *E. coli* O157:H7, other enterohemorrhagic *E. coli* (EHEC) and Shiga-like toxin-producing *E. coli* (STEC), *S.* Enteritidis and *S.* Typhimurium DT104, *Campylobacter*, and *L. monocytogenes*.

Research priority areas in need of development of knowledge to be applied in the control of pathogens in foods include microbial ecology, microbial control processes, microbial resistance, and microbial methodology for detection of pathogens. In addition, the contribution of human factors to foodborne illness needs to be determined in order to develop food-handler and consumer education programs to assist with achievement of food safety objectives. Knowledge of pathogen ecology throughout the food chain can be instrumental in determining risk factors and niches and in developing intervention strategies for their control. Pathogen-control procedures are needed to assure food safety by preventing, reducing, inhibiting, inactivating, or eliminating pathogens. Use of predictive models can allow determination of the contribution of traditional, novel, physical, chemical, and biological methods to control pathogens. Studies should address various aspects, including microbial adhesion, penetration, removal, biofilms, inactivation, resistance, injury, inhibition, competition, culturability, superdormancy, mechanisms, and microbial and environmental interactions. A concern that needs to be addressed is the observation that pathogenic bacteria develop resistance to traditional preservative barriers such as low pH, cold, heat, water activity, disinfectants, and preservatives, as well as antibiotics. Resistance to antibiotics as well as traditional food-preservation hurdles is an important area that should be researched in terms of microbial strain ecology, factors, conditions and mechanisms of resistance development, microbial interactions, and approaches for control. Furthermore, the application of microbial control methods should be evaluated in association with their influence on practicality of application, quality and acceptability of treated foods, economics of application, and risks associated with potential subsequent contamination or product abuse.

Specific research targets and objectives should be based on application of quantitative risk assessment to determine intervention points with major impact on food safety enhancement (23–27,29,30). For example, among questions that may need to be addressed when designing research objectives are: Is it possible to, and how can we, produce "pathogen-free" raw/fresh foods? To what extent can testing assure us of a "pathogen-free" product? Will pathogens become a problem if decontamination increases shelf life? If decontamination is successful, can/should the intensity of subsequent preservation processes be reduced? How do new approaches to controlling microorganisms affect injury, resistance, and detection of pathogens? What are the mechanistic effects of the treatments? As new approaches to food safety reduce known risks, could other, unknown risks arise? Is there a false sense of security being created? Is complete elimination of all pathogens from foods necessary and feasible?

Another example is the subject of microbial testing and its role in food safety assurance efforts. A plethora of improved pathogen-detection procedures are being developed and marketed by the private sector. Such methods are needed to monitor contamination and its sources, to establish microbial ecology patterns, and to develop, validate, and verify HACCP principles and critical limits. Testing should not be relied upon as a food safety assurance process, but it is needed in the proper development of controls and their application and in the investigation of foodborne illness episodes (31). Levels or frequency of contamination are needed for use in development of risk-assessment models. Therefore, methodologies to address recovery, identification, and enumeration or quantitation of pathogens, their toxic metabolites, or acceptable indicators are needed. The methods should be sufficient to recover injured or stressed microorganisms, and potential interference by food components, properties of the foods, or other microorganisms should be avoided. In addition to potential matrix inhibition problems, research should address proper sampling, sample-handling, and sample-preparation techniques. Procedures are needed to remove, isolate, concentrate/enrich/amplify low levels of pathogenic cells or components to be detected by the rapid methods in various foods. For development and application of molecular detection methods there is a need to know the ecology, physiology, and pathogenesis of the target microorganism. Methods will also be needed for application in the field or plant or at distribution or retail outlets and in the form of biosensors.

VI. APPROACHES TO CONTROLLING PATHOGENS IN FOODS

The increased reports of foodborne illness of recent years have intensified public pressure for action to address the various challenges and to enhance the safety of our food supply. As indicated, when we consider approaches to meet these challenges, we need to ask questions such as do we need to apply microbial inactivation steps to all foods? If yes, which is the applicable and feasible method? Will application of a method affect the quality of the food and how? Could application of a new or an existing food-processing and preservation method to a food lead to the creation of any unpredicted problems or risks?

Traditional barriers used in food preservation that inactivate or inhibit microbial pathogens include heat, irradiation, low temperatures, natural fermentation, or direct acidification, reduced water activity, modified atmospheres, and use of chemical antimicrobial agents. These hurdles, singly or in combination, work effectively to ensure the safety of foods from pathogens that need to reach high numbers and to produce toxins (e.g., *Staphylococcus aureus*, *Clostridium botulinum*) in order to lead to human illness (32). The traditional hurdles that inhibit growth of microorganisms are effective against pathogens that act through high infectious doses, especially in healthy individuals, while heat and irradiation may be used to eliminate pathogens from foods, including those active at low infectious doses.

In recent years, however, our food safety concerns have been focused mostly on pathogens, such as *E. coli* O157:H7, certain *Salmonella* serotypes, parasites, and viruses, that may lead to infection through a small number of organisms present in foods without the need for proliferation before consumption or on those, such as *L. monocytogenes*, that are able to proliferate under refrigeration. It also appears that certain bacteria have developed potent virulence factors, and it is speculated that their environment may have contributed to their increased virulence through exposure to sublethal stresses and activation of resistance mechanisms. These concerns are especially important when they are associated with persons susceptible to low infectious doses.

As indicated, pathogenic bacteria develop systems that assist them to survive and adapt to environmental stresses such as heat, cold, or acid shock (7,8,12). Food environments are generally stressful for bacteria because most nutrients are in the form of complex substrates, free moisture levels may be restricted, presence of acids and other chemicals—some of which may be microbial metabolites—may be at stressful levels, and there is competition from other microorganisms present. To survive and cope with such stresses, microorganisms often undergo physiological changes that allow them to adapt to environmental conditions. Common food-associated stresses include heat or cold, acidity or alkalinity, osmolarity, oxidative stress, lack of a single nutrient such as iron, or general starvation. Exposure to one stress and activation of protective genes often leads to cross-protection against other stresses. According to Archer (6), the stresses that exist in foods, either naturally or through application of food-preservation processes, have a major effect on gene expression in bacterial pathogens, and some of the genes are associated with bacterial virulence. The question has been asked whether stress associated with food-preservation systems could potentiate virulence (6,10).

In addition, a number of novel processes have been developed and investigated with the objective of being applied for microbial control in foods. Novel and less traditional processes investigated, proposed, or to some degree applied to foods for microbial control include ionizing radiation, microwaves, increased hydrostatic pressure, pulsed electric fields, magnetic fields, ohmic heating, high-intensity light, use of chemicals such as gases, naturally occurring antimicrobials, and biopreservatives (33,34). Some of these treatments have been proposed for use as combinations in multiple hurdle systems or are applied to foods such as raw meat, poultry, and produce to reduce contamination without rendering the products ready for consumption. Application of some of these processes is still in the exploratory and developmental stages, while some have gained regulatory approval and are being introduced in HACCP plans and in the marketplace for consumer evaluation and acceptance (21).

It should be noted that not all of these processes are applicable to all foods, which is also the case with traditional food-preservation methods. Some are impractical to use in certain products, while others may damage the quality of some foods. As indicated above, the application of these processes should be based on results of well-designed and executed research as well as proper technological developments for specific foods. Application of any process, traditional or novel, or combination of processes with the objective of rendering a food ready and safe for consumption should be designed with the objective of inactivating a target number of pathogenic microorganisms of concern in order to meet established food safety objectives that are based on proper risk-assessment exercises (23–27).

Since some preservation procedures (e.g., chemical additives) at the intensity or concentration used in foods act by inhibiting growth, instead of inactivating microorganisms, their contribution may be more useful against pathogens that form toxins in foods or need to reach high numbers to cause foodborne illness, especially in healthy persons. However, to protect consumers at risk for foodborne illness or against microbes with low infectious doses, there is a need for complete inactivation and avoidance of recontamination of foods during processing, distribution, and preparation for consumption. Less than lethal doses or treatment intensities may be acceptable for use in decontamination applications in raw products that will be further processed before consumption. Reduction of the microbial load in these products should lower the probability of failure of subsequent processes such as the cooking of foods for consumption. Likewise, food safety can be improved by applying good agricultural, manufacturing, and hygienic practices throughout the food chain. Archer (6) stated that use of preservative systems is necessary, as is application of sanitation practices to maintain the lowest possible microbial numbers in foods. However, it is now established that foods may be contaminated with some pathogens that exhibit virulence at very low infective doses. As we continue employing traditional or novel food-preservation processes, we must be aware that they may be potentiating cross-protection, virulence, or unpredictable adaptive genetic changes in virulence genes. The relationship between stress and enhanced virulence needs to be studied. There may be a need to develop and apply processes that do not just inhibit but destroy or inactivate pathogens in order to assure the safety of susceptible individuals from pathogens acting with low infectious doses (6).

REFERENCES

1. J Lederberg. Infectious diseases as an evolutionary paradigm. Emerg Infect Dis 3: 417–423, 1997.
2. RV Tauxe. Emerging foodborne diseases: an evolving public health challenge. Emerg Infect Dis 3:425–434, 1997.
3. PS Mead, L Slutsker, V Dietz, LF McCaig, JS Bresee, C Shapiro, PM Griffin, RV Tauxe. Food-related illness and death in the United States. Emerg Infect Dis 5:607–625, 1999.
4. NH Bean, JS Goulding, MT Daniels, FJ Angulo. Surveillance for foodborne disease outbreaks—United States, 1988–1992. J Food Prot 60:1265–1286, 1997.
5. JG Morris, Jr., M Potter. Emergence of new pathogens as a function of changes in host susceptibility. Emerg Infect Dis 3:435–441, 1997.
6. DL Archer. Preservation microbiology and safety: evidence that stress enhances virulence and triggers adaptive mutations. Trends Food Sci Technol 7:91–95, 1996.
7. T Abee, JA Wouters. Microbial stress response in minimal processing. Int J Food Microbiol 50:65–91, 1999.
8. CK Bower, MA Daeschel. Resistance responses of microorganisms in food environments. Int J Food Microbiol 50:33–44, 1999.
9. D Ferber. Superbugs on the hoof. Science 288:792–794, 2000.
10. S Knochel, G Gould. Preservation microbiology and safety: quo vadis? Trends Food Sci Technol 6:127–131, 1995.
11. PS McManus. Antibiotic use and microbial resistance in plant agriculture. Am Soc Microbiol News 66:448–449, 2000.
12. JJ Sheridan, DA McDowell. Factors affecting the emergence of pathogens on foods. Meat Sci 49(suppl 1):S151–S167, 1998.
13. L Tollefson, MA Miller. Antibiotic use in food animals: controlling the human health impact. J AOAC Int 83:245–254, 2000.
14. SJ Knabel. Foodborne illness: role of home handling practices, scientific status summary. Food Technol 49(4):119–131, 1995.
15. LR Beuchat. Pathogenic microorganisms associated with produce. J Food Prot 59: 204–216, 1996.
16. S Brul, P Coote. Preservative agents in foods, mode of action and microbial resistance mechanisms. Int J Food Microbiol 50:1–17, 1999.

17. JE Collins. Impact of changing consumer lifestyles on the emergence/reemergence of foodborne pathogens. Emerg Infect Dis 3:471–479, 1997.
18. DL Zink. The impact of consumer demands and trends on food processing. Emerg Infect Dis 3:467–469, 1997.
19. JA Lindsay. Chronic sequelae of foodborne diseases. Emerg Infect Dis 3:443–452, 1997.
20. Food Safety and Inspection Service (FSIS). Pathogen reduction; hazard analysis critical control point (HACCP) systems: final rule. 9 CFR Part 304 et al. Fed Reg 61(144):38805–38989, 1996.
21. National Advisory Committee on Microbiological Criteria for Foods. Hazard analysis and critical control point principles and application guidelines. J Food Prot 61: 1246–1259, 1998.
22. National Food Safety Initiative. www.FoodSafety.gov, 2000.
23. RL Buchanan, RC Whiting. Risk assessment: a means for linking HACCP planning and public health. J Food Prot 61:1531–1534, 1998.
24. MH Cassin, GM Paoli, AM Lammerding. Simulation modeling for microbial risk assessment. J Food Prot 61:1560–1566, 1998.
25. AM Lammerding, GM Paoli. Quantitative risk assessment: an emerging tool for emerging foodborne pathogens. Emerg Infect Dis 3:483–487, 1997.
26. National Advisory Committee on Microbiological Criteria for Foods, Represented by Robert Buchanan, Chair for the Subcommittee on Risk Assessment. Principles of risk assessment for illness caused by foodborne biological agents. J Food Prot 61:1071–1074, 1998.
27. International Commission on Microbiological Specifications for Foods (ICMSF) Working Group on Microbial Risk Assessment. Potential application of risk assessment techniques to microbiological issues related to international trade in food and food products. J Food Prot 61:1075–1086, 1998.
28. S Doores. Food Safety: Current Status and Future Needs. American Academy of Microbiology Colloquium Report, Washington, DC, 1999.
29. RL Buchanan. Identifying and controlling foodborne pathogens: research needs. Emerg Infect Dis 3:517–521, 1997.
30. International Commission on Microbiological Specifications for Foods. Principles for the establishment of microbiological food safety objectives and related control measures. Food Control 9:379–384, 1998.
31. American Meat Science Association. The role of microbiological testing in food safety systems; the scientific perspective. American Meat Science Association, Kansas City, MO, 1999.
32. L Leistner. Principles and applications of hurdle technology. In: GW Gould, ed. New Methods of Food Preservation. Blackie Academic and Professional, New York, 1995, pp. 1–21.
33. Kinetics of microbial inactivation for alternative food processing technologies. U.S. Food and Drug Administration, Center for Food Safety and Applied Nutrition, Washington, DC, 2000.
34. JN Sofos, LR Beuchat, PM Davidson, EA Johnson. Naturally occurring antimicrobials in food: interpretive summary. Reg Toxicol Pharmacol 28:71–72, 1999.

2
Thermal Inactivation of Microorganisms

Vijay K. Juneja
U.S. Department of Agriculture
Wyndmoor, Pennsylvania

I. INTRODUCTION

Food preservation is designed to enhance or protect food safety while maintaining the organoleptic attributes of food. Inactivating or inhibiting the growth of undesirable microorganisms is very important for the successful and acceptable preservation of food. While a large number of preservation processes are at the disposition of food processors, the use of adequate heat treatment to destroy pathogenic and spoilage microorganisms is one of the most effective food-preservation processes in use today and has been used for centuries. Heat treatment designed to achieve a specific lethality for foodborne pathogens is a critical control point in food processing and is fundamentally important to assure the shelf life and microbiological safety of thermally processed foods. A key to optimization of the heating step is defining the target pathogen's heat resistance. While overestimating the heat resistance negatively impacts the product quality by altering the organoleptic attributes and nutritional qualities of a food, underestimating increases the likelihood that the contaminating pathogen will persist after heat treatment or cooking. Inadequate heat treatment or undercooking is an important contributing factor in food-poisoning outbreaks (1).

Mention of brand or firm name does not constitute an endorsement by the U.S. Department of Agriculture over others of a similar nature not mentioned.

The extent of heat treatment applied to foods falls into two categories:

1. Low-heat processing or pasteurization (named after Louis Pasteur) refers to the use of relatively mild heat treatment and is widely accepted as an effective means of destroying all non–spore-forming pathogenic microorganisms and significantly reducing the number of natural spoilage microflora, thereby extending the shelf life of pasteurized products. Pasteurization of milk has been performed for decades and is achieved by heating at 145°F (63°C) for 30 minutes or 161°F (72°C) for 15 seconds or equivalent treatments. These time-temperature combinations are sufficient to destroy the most heat-resistant target pathogens in milk, i.e., *Mycobacterium tuberculosis* (organism causing tuberculosis) and *Coxiella burnetti* (Q fever pathogen) and are designated as low-temperature, long-time (LTLT) and high-temperature, short-time (HTST) methods, respectively. The shelf life of pasteurized products depends on the type of food in addition to conditions of pasteurization and subsequent storage (2). Bacterial spores and some heat-resistant enzymes are not destroyed by the pasteurization process and limit the shelf life of the product.
2. High-heat processing or sterilization (in-container) refers to the complete destruction of microorganisms. *Bacillus stearothermophilus* spores, being extremely heat resistant, are often used to evaluate commercial sterilization. This is achieved by application of heat at very high temperatures for a short time to render food free of viable microorganisms that are of public health significance or that are capable of growing in the food at the temperature at which the food is likely to be held (under normal nonrefrigerated storage conditions) during distribution and storage. These products are termed as ''commercially sterile.'' However, this term is not correct because sterility implies the absence of living organisms. But these shelf-stable and microbiologically safe products may contain a low number of dormant bacterial spores. There are two limitations to the above ''in-container'' sterilization processes: first, products heat and cool at a relatively slow rate; second, final processing temperatures depend on the internal pressure generated (3). While thermal sterilization has been widely used in the food industry for over 200 years because of its proven reliability, ultra–high temperature (UHT) processing, which combines continuous flow thermal processing with aseptic packaging, has been introduced as an alternative sterilization process. Thus, the alternative term ''aseptic processing'' is also used. By UHT processing, quality of some products is improved because heating and cooling occurs at a faster rate and higher temperatures can be achieved by removing the pressure con-

straints (3). Milk heated to 150°C for 2–3 seconds can be stored at room temperature and has a shelf life of 3 months.

II. INACTIVATION KINETICS PARAMETERS

The higher the initial microbial population in a food, the longer is the processing/heating time at a given temperature required to achieve a specific lethality of microorganisms. Accordingly, the thermal process is designed based on the expected microbial load in the raw product. As such, the heat resistance of bacteria is described by two parameters: D- and z-values. The D-value is the time at a particular temperature necessary to destroy 90% of the viable cells or spores of a specific organism. It is a measure of the death rate or the heat sensitivity of the organism. The z-value is the change in heating temperature needed to change the D-value by 90% (1 log cycle), i.e., z-value provides information on the relative resistance of an organism at different destructive temperatures. D- and z-values are used for designing heat-processing requirements for desirable destruction of microorganisms in a particular food.

Generally, the rate of destruction of bacteria follows first-order kinetics, i.e., when a microbial population is heated at a specific temperature, the cells die at a constant rate. Log number of survivors decline in a linear manner with time (4,5) (Fig. 1). This traditional first-order kinetics model of thermal inactivation forms the basis of calculations used in thermal processing and has served the food industry and regulatory agencies for decades. However, this approach assumes that all of the cells or spores in a population have identical heat resistance, and it is merely the chance of a quantum of heat impacting a heat-sensitive target in a cell or spore that determines the death rate (6). Significant and systematic deviations from classical semi-logarithmic linear declines in the log numbers with time have been frequently observed, even when precise attention is paid to methodology (5,7,8). Such deviations are of two forms: (a) a shoulder or a lag period, i.e., time periods when the bacterial populations remain at the inoculation level; (b) a tailing, i.e., a subpopulation of more resistant bacteria that decline at a slower rate (Fig. 1).

The thermal inactivation data cannot be accounted for by experimental artifacts, and there is presently no satisfactory, unifying explanation for the variability in thermal death kinetics. Hansen and Riemann (9) suggested that the deviations in linear survival curves result from a cell population heterogenous in heat resistance, i.e., due to variability in heat resistance within a population. The ''shoulder effect'' observed may be attributed to the poor heat transfer through the heating menstruum or may be due to an initial requirement for the bacterial cells to sustain sufficient injury before the first-order inactivation kinetics in the log number of survivors with time. The shoulder also may be attributed to a

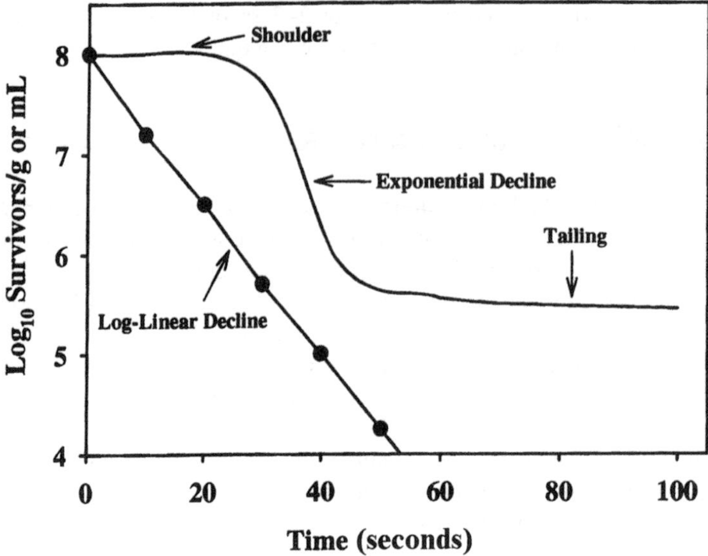

Figure 1 Thermal inactivation of microorganisms. The straight line represents the traditional first-order kinetics of microbial death, i.e., log number of survivors declining in a linear manner with time. The curved line represents microbial death showing a shoulder, a linear decline, and a tailing that are frequently observed in practice.

requirement for more than one damaging event or the need to activate the spores to make them more susceptible to thermal destruction. The ''tailing effect'' may be due to clumping of small number of cells in the heating menstruum resulting in their protection and increased resistance to the lethal effect of heat (4,9). Cerf (10) proffered two concepts for tailing:

1. A vitalistic mechanism that assumed a variability of heat resistances within a population, i.e., there may be a distribution of heat resistances within a population of cells that constitutes a suspension of apparently identical cells. The relative heat resistance is a genetic trait for an individual cell, and a population of cells presumably formed a normal distribution of cells.
2. A mechanistic concept, which assumed that resistance was the result of physiological and biochemical changes occurring in the cells. Under specific environmental conditions or at different times of its life cycle, an individual cell has varying degrees of heat resistance. A small population of cells is liable to be in a resistant state at the time of heating. Nonlinear survival curves may also be an artifact of the heating or

enumeration technique or nonuniform heat treatment. Also, cells acquire heat resistance as a result of sublethal heating. Such curves are increasingly observed due to the consumer demand for "fresh" and ready-to-eat products such as cook/chill, *sous-vide* foods, etc. Processing of such foodstuffs implies a mild increase in temperature, for which deviations from linearity in the survival curve are frequently observed (11).

Therefore, consideration of non–first-order inactivation kinetics is important in the safe application of milder heat processes or those relying on the combined effects of other factors such as pH, water activity, etc.

Typically, the traditional log-linear death model or a linear regression approach has often been employed to analyze thermal inactivation data despite a poor fit. This has likely resulted in false estimates of heat resistance values for nonlinear survival curves. Accordingly, attempts have been made to explain these deviations (nonlinear survivor curves) by various theories, and several alternate models have been developed to account for this behavior (12–14) and have been excellently reviewed by Whiting (15). Examples of the primary models used to describe survivor curves of microorganisms are given in Table 1. Abrahm et al. (16) hypothesized that the initial shoulder before an exponential decline resulted from the requirement for dormant spores to be activated before being destroyed by heat. Both activation and inactivation were first-order processes, and the first step was the limiting process. Later researchers advanced a population dynamic theory for thermal inactivation of spores. According to this theory, the initial decrease or increase in spore populations was due to a combination of first-order processes for the rapid inactivation of less heat-resistant spores followed by a period of activation of remaining spores to a more heat sensitive state and finally inactivation of remaining spores (17–20). Sapru et al. (21) used these concepts to model dormant spores being either inactivated or activated by heat in ultra–high-temperature sterilization. The temperature dependence of each of the three parameters followed the Arrhenius equation, and the model successfully predicted spore inactivation during variable heating regimes.

III. METHODS FOR DETERMINATION OF HEAT RESISTANCE

A. Traditional Methods

Existing methods for thermal inactivation of microorganisms include TDT (thermal death time) tube, TDT pouch (nylon), TDT can, flask, thermoresistometer, and capillary tube methods (2). These methods suffer from many disadvantages, including time-consuming operations, appreciable heating and cooling lags,

Table 1 Examples of the Primary Models Used to Describe Inactivation Curves of Microorganisms

Survival curves	Mathematical description (equation)
Exponential (first-order kinetic)	$N = Noe^{-kt}$ or $N = No10^{-t/D}$ $K = \frac{2 \cdot 303}{D}$
Linear model[a] (211)	$Y = No\ 0 < tT_L$
Logistic equations	$Y = No - (1/D)(t - T_L)\ t > T_L$
One slope (212)	$\log(M_t/M_o) = 2/[1 + e^{(kt)}]$
Biphasic (two slopes)[b] (212)	$\log(M_t/M_o) = \log(\{2F_l/[1 + \exp(k_1 t)]\} + \{2(1 - F_l)/[1 + \exp(k_2 t)]\})$
Initial lag followed by one-phase killing (one slope)[c] (212)	$\log(M_t/M_o) = \log[1 + \exp(-kt_{1/2})] - \log[1 + \exp\{k(t - t_{1/2})\}]$
Initial lag followed by two-phase killing (two slopes)[d] (211)	$\log(M/M_o) = \log[F_l(1 + \exp(-k_1 t_l))/(1 + \exp(k_1(t - t_l)))] + \log[(1 - F_l)(1 + \exp(-k_2 t_l))/(1 + \exp(k_2(t - t_l)))]$
Activated and dormant species[e] (213)	$n = (n_{oa} + n_{od})\exp\left[\frac{-t}{\theta_l}\right] - n_{od}\exp\left[\frac{-t}{\theta_{ai}}\right]$
Gompertz[f] (202)	$\log(n) = \log(n_o) + a\exp[-\exp(b + ct)] - a\exp[-\exp(b)]$
Vitalistic model[g] (8)	$\log_{10}(\text{viable cell number/ml}) = \alpha + \frac{\omega - \alpha}{1 + \exp\left[\frac{4\sigma(\tau - \log_{10}\text{time})}{\omega - \alpha}\right]}$

[a] T_2 is the lag time prior to initiation of inactivation.
[b] Where F_l was the fraction of population in the major group, k_1 was the inactivation rate parameter for the major population, and k_2 was the inactivation rate parameter for the subpopulation.
[c] Where $t_{1/2}$ was the time for $M = (M_o/2)$, a measure of the lag time.
[d] Where t_l was the lag period, F_l the fraction of cells in the major population, and k_1 and k_2 the respective rate parameters ($D = 2.3/k$). When the subpopulation did not exist, the fraction of cells in the subpopulation was set to an insignificantly low value. If the shoulder was not present, t_l was set to 0.0 and the model became a nearly straight line.
[e] n_{oa}, n_{od} are initial population sizes of activated spores and dormant spores, respectively; θ_i, time constant for inactivation; θ_{ai}, combined time constant for inactivation and activation.
[f] a, b, and c are fit parameters.
[g] Where α is the upper asymptote (log cfu/mL), ω is the lower asymptote (log cfu/mL), σ is the maximum slope of the death curve, log cfu/mL against log time, and τ is the log time at which maximum slope is reached.

splashing of contents, flocculation, high initial cost, and hazards of contamination during subculturing (2). Additionally, using traditional methods, the published literature on the heat resistance of certain organisms, such as *Listeria monocytogenes*, is conflicting. Using the holding technique of pasteurization in screw-capped test tubes placed in a water bath (61.7°C, 35 min), survival occurred when *L. monocytogenes* population levels exceeded 3 log cfu/mL (22). In contrast, using sealed borosilicate glass tubes, *L. monocytogenes* was unable to survive pasteurization (23). However, later investigation by Donnelly et al. (24) proved how important the methodology is in determining thermal inactivation. The authors compared the heat resistance of *L. monocytogenes* using the sealed tube and test tube methods and concluded that survival of the organism at pasteurization temperatures depended on the method used to inactivate cells and is not a biological phenomenon. Using the sealed-tube method of inactivation, *L. monocytogenes* was easily inactivated at pasteurization temperatures. However, when an identical cell population was heated using the test tube inactivation method, survival of the pathogen was observed, regardless of the heating temperature (62, 72, 82, or 92°C). The authors stated that the condensate and splashed cells could collect in the cap of the test tube above the level of the water bath and drip back into the heating menstruum. Thus, the tubes will have various levels of survivors depending on the amount of condensation in the cap. Also, the authors indicated that cells could coat the walls of the test tubes above the level of water bath; the only cell population exposed to the inactivation temperatures would be that which is below the level of water in the water bath.

B. Submerged-Coil Heating Apparatus

This is a novel and convenient tool for investigating the thermal inactivation kinetics of foodborne spoilage and pathogenic microorganisms (Fig. 2). It is com-

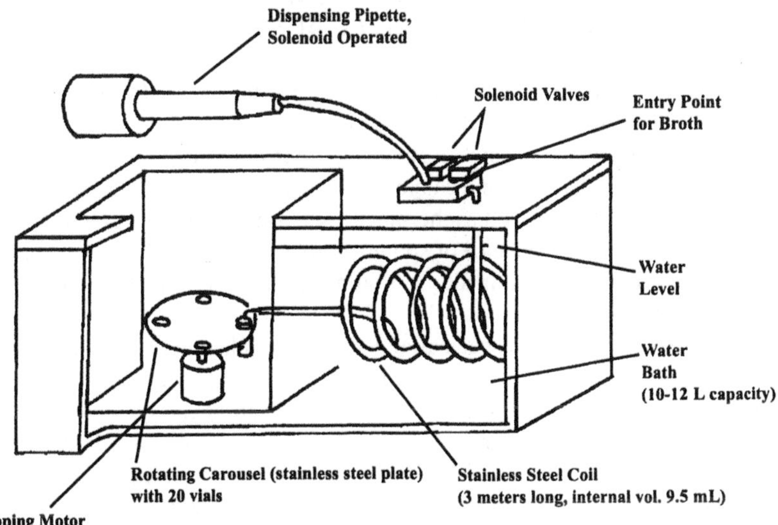

Figure 2 Submerged-coil heating apparatus.

prised of a stainless steel coil, fully submerged in a thermostatically controlled water bath, which allows microbial suspensions to be rapidly heated between 20 and 90°C. Since the equilibrium or come-up time of the samples is negligible, no bacterial inactivation occurs during this period. An electronically controlled displacement mechanism controls the sampling intervals and the number of samples to be dispensed. This makes the equipment fully automated. Various sampling times can be set at intervals ranging from 6 to 999 seconds.

C. Operation of the Apparatus

Bolus microbial suspensions (≤9.5 mL) are loaded rapidly (<1 s) via a solenoid valve system using a disposable syringe into the preheated coil. The come-up time is <1 second. Thereafter, several measured quantities of sterile water are introduced into the charging end of the tube, displacing the heated broth through the coil and discharging sample at the terminal end. These are collected in a series of vials carried on the carousel. The sample delivery needle is automatically flushed twice with hot culture prior to sample displacement to ensure no carry-over of the previous sample. The results obtained using this apparatus were highly reproducible. Thus, the heat resistance of microorganisms in a variety of liquid foods can be quantified accurately. The only disadvantage is that the apparatus cannot be used for assessing the heat resistance of microorganisms in solid foods.

IV. MECHANISM OF HEAT INACTIVATION OR BACTERICIDAL OR SPORICIDAL ACTIVITY

The resistance of bacterial cells and spores to heat has been a focus of study for decades. The literature on heat-resistance mechanisms includes excellent reviews (25–34). Heat resistance in spore-formers has been correlated with DNA content (35), dipicolinic acid (DPA), and calcium chelates with DNA (36), mineralization (37), dehydration (38), as well as thermal adaptation (39). Gerhardt and Marquis (26) reported that heat resistance of spores is attributed primarily to three physiochemical determinants that impact the protoplast: thermal adaptation, mineralization, and dehydration.

Thermal adaptation has been assumed to be an inherent or intrinsic molecular component that is genetically determined. In general, the spores of thermophilic species are inherently more resistant that those of mesophiles, which, in turn, are more resistant than spores of psychrophilic species (40). Furthermore, spores of a particular species or strain produced at higher temperature are more heat resistant than those prepared at optimum or low temperatures (26,39,41,42), and so there appears to be an extrinsic element imposed on the genetic element. Sporulation at higher temperatures inducing additional heat resistance may be attributed to the reduced water content in spores produced at higher temperatures (38). High sporulation temperatures have been argued to increase the mineralization of spores (39). The correlation between sporulation temperature and heat resistance has been investigated (43,44). Similarly, bacterial cells grown at higher temperatures are known to be more heat-resistant than those grown at lower temperatures (44,46).

Mineralization of the protoplast can have a significant effect on spore heat resistance; the heat resistance of spores is directly related to the level and type of divalent cation. In general, the higher the level of divalent cation, the higher the heat resistance of spores. However, the effect appears to be lower at higher temperatures (47). Setlow (48) reported on the analysis of spores of several microbial species from which all mineral ions were removed by acid titration and then restored by backtitration with a mineral hydrooxide. This work gave the following order of increasing spore heat resistance with different cations: $H^+ < Na^+ < K^+ < Mg^{2+} < Mn^{2+} < Ca^{2+}$ = untreated. Spores of *Clostridium botulinum* containing an increased iron content in the sporulation medium were more sensitive to heat compared to the spores with normal iron levels or spores with increased zinc or manganese. Spores formed with added manganese or zinc were better able to repair heat-induced injuries than spores with added Fe or Cu (49). While mineralization of spores in the later stages of sporulation is associated with an uptake of DPA, the latter is not thought to be necessary for attaining heat resistance, but there is growing evidence that it may have a role in retaining the resistance (26). Beaman and Gerhardt (38) reported that the increases in mineralization of spores is accompanied by reductions in protoplast water content, and at least a part of the increased heat resistance associated with mineralization is due to dehydration.

Dehydration of the protoplast restricts the mobility of heat-labile components of the spore core, i.e., vital macromolecules such as proteins, RNA, and DNA, and renders them less sensitive to irreversible thermal denaturation or even prevents denaturation (27). Gerhardt and Marquis (26) suggested that the dehydration of the protoplast is the only determinant necessary for the heat resistance of spores. Once protoplast dehydration is obtained or the resistant state is attained, its maintenance is clearly a function of an intact peptidoglycan cortex, but not a coat or exosporium (26). It must be noted that although dehydration is the only property necessary and sufficient in itself to impart heat resistance, it is enhanced by mineralization (especially by calcification) and thermal adaptation.

Spores contain a number of unique core-located, acid-soluble, basic proteins. *Bacillus subtilis* spores genetically deficient for synthesis of the α and β small acid-soluble spore proteins (α/β-type SASP) as a result of mutations in genes coding for major proteins of this type have been found to be more sensitive to heat than SASP sufficient spores (50–52). The D-values of *B. subtilis* spores lacking α/β-type SASP (termed as α-β spores) were 10- to 20-fold lower than those of wild-type spores (53). Inactivation of *B. subtilis* α-β spores is associated with induction of high level of mutations and significant DNA single-strand breakage compared to wild-type spores, in which killing by heat is associated with no significant mutagenesis and many fewer single-strand breaks (51,53). The precise cause of the DNA damage and mutagenesis by heat in α-β spores is not known, but Fairhead et al. (53) suggested that the heat treatment causes DNA depurination followed by strand cleavage. The authors demonstrated that the binding of α/β-type SASP to DNA in vitro reduces the rate of DNA depurination at least 20-fold. Thus, depurination is an attractive explanation for the DNA damage caused by heat treatment of α-β spores. Later, Setlow and Setlow (54) provided direct evidence for this process in vivo by analyzing the level of abasic sites in DNA from heat-inactivated α-β spores directly. Findings of this study suggested that a major mechanism responsible for the heat killing of α-β- (but not wild-type) spores is DNA depurination followed by strand breakage at the resultant abasic site.

Several studies have documented correlations between bacterial membrane fatty acid composition and heat resistance of the organism (55–58). Hansen and Skadhange (59) reported that the reduced heat resistance of cells grown at low temperatures may be due to an increase in the concentration of unsaturated fatty acids in the cytoplasmic membrane, which increases membrane fluidity and reduces viscosity, thereby decreasing thermotolerance. *Vibrio parahaemolyticus* grown at different temperatures and at different sodium chloride concentrations showed alterations in membrane fatty acid profile that correlated with an altered response to subsequent heat treatment (55). Juneja et al. (58) demonstrated an altered heat resistance along with altered membrane fatty acid profiles in *L. monocytogenes* grown in different acidic environments at different temperatures.

Dry heat lethal efficacy is less than that of wet heat treatment and requires higher temperatures and longer times to yield the same killing effect. Ernest (60) suggested that dry heat causes death by the destructive oxidation of cell components. Hashimoto et al. (61) proposed that the heat killing of spores is primarily due to physical and chemical alterations, which interfere with the absorption of water into the core during germination.

V. TARGETS OF HEAT DAMAGE

Heat is believed to be uniformly distributed in a cell, resulting in damage to only the most sensitive molecules within that cell (62). Potential targets of heat damage have been implicated with associations to various pathogen viabilities. These include proteins and enzymes, cellular membranes, as well as nucleic acids (63). Deficiency in DNA repair appears to cause spores to be more heat sensitive (64).

Hanlin et al. (64) suggest that heat-induced DNA damage, if repairable, may not result in inactivation of the spores, and the death of spores with damaged DNA may be due to the inability of spores to germinate and outgrow. Melting of spore DNA may occur at temperatures above 90°C and is reflected in differential calorimetric scans of spores as an endothermic peak at 90–91°C (65,66). In contrast, spore DNA is protected by small acid soluble proteins that bind tightly and specifically to the A form of DNA reducing the rate of depurination in vitro by at least 20-fold (67).

An increasing amount of evidence suggests that ribosome damage and degradation is the cause of cell death following thermal stress (68,69). Ribosome denaturation occurs in the same temperature region as thermal inactivation. Numerous investigators have used differential scanning calorimetry (DSC) to examine thermal transitions as indicators of potential sites of cellular injury (66,70). DSC has proven to be an effective technique in measuring changes in protein denaturation temperatures with corresponding changes in denaturation enthalpy (71). It was suggested that half the enthalpy of ribosome denaturation is associated with protein denaturation and/or disruption of higher-order interaction (72). Anderson et al. (70) reported that measurement of viability loss in the differential scanning calorimeter gave good correlation between cell death and the first major thermogram peak; the peaks observed in the thermogram of the ribosome cell fraction correspond to the major peak in whole cells. Allwood and Russell (73) observed a direct correlation between loss of RNA and heat-induced loss of viability of *Staphylococcus aureus* at temperatures up to 50°C. Magnesium is known to have a stabilizing effect on ribosomes. In a study involving mild heating of *S. aureus*, Hoa et al. (74) reported that heating results in membrane damage, leading to the loss of Mg^{2+} ions and destabilization of the ribosomes. In fact, depletion of Mg^{2+} leads to 70S ribosome dissociation into 30S and 50S subunits, ribonuclease activation, and finally destruction of 30S subunits (75–77). Earlier studies including a number of bacterial species showed that the 30S ribosomal subunit is specifically destroyed during heat treatment, while the 50S ribosomal subunit appears to be stable, and that 16S rRNA is the prime target of degradation in the heat-injured cells while 23S rRNA appears to be unaffected (78). Miller and Ordal (79) examined the rRNA profiles of cells at various times during heat injury at 47°C. The degradation of rRNA and ribosomal subunits occurs differently during heat injury; the 16S and 30S subunits are affected more readily following heating for 5 minutes, and the 23S and 50S subunits are degraded more slowly, disappearing after 30 minutes of heat treatment.

Stephens and Jones (80) proposed that the protection of the 30S subunit is a critical mechanism for increased thermotolerance. In their study, the osmotic and heat shock–induced increased thermotolerance response of *L. monocytogenes* was concurrent with increased thermal stability of the 30S ribosomal subunit, as measured by differential scanning calorimetry. The authors proposed that the stabilization of the subunits occurred through cellular dehydration leading to an increase in the internal solute concentration, including Mg^{2+} ions, which may contribute to tighter coupled particles of the 30S subunits. Tolker-Nielsen and Molin (81) reported that heat lethality of *S.* Typhimurium coincides with a significant reduction in the cellular content of 16S ribosomal RNA, thereby suggesting that the degradation of ribosomal RNA is a direct cause of cell death. This conclusion is based on the findings of carbon-starved and magnesium-supplemented cells, which survive heat treatment much better and which also maintain stable levels of ribosomal RNA.

Proteins and enzymes are also considered to be potential sites responsible for heat lethality. It has been postulated that water that is in close contact with the proteins inside the cell could be a factor determining the cell's inactivation. As the cell is heated, water molecules begin to vibrate, and this vibration causes the disulfide and hydrogen bonds in the surrounding proteins to weaken and break, altering the final three-dimensional configuration and possibly preventing the protein from functioning (82). The crucial protein that is the rate-limiting, primary target in heat killing is unknown, but the current belief is that membrane proteins may be denatured by heat initially because of peripheral locations followed by the denaturation of crucial proteins within ribosomes (65). There is evidence that catalase and superoxide dismutase (SOD) may be sensitive to heating. These enzymes detoxify oxygen radicals like superoxide and hydrogen peroxide, which form spontaneously in the presence of oxygen and, if undisturbed, can result in death of the cells as a result of lipid peroxidation and membrane damage (83). Amin and Olson (84) found that staphylococcal catalase activity decreased 10- to 20-fold faster at 54.4°C than at 37.8°C, and Dallmier and Martin (85) found that the specific activity of SOD produced by *L. monocytogenes* decreased quickly when heated to 55°C. In a study on the heat stability of *Bacillus cereus* enzymes within spores and in extracts, Warth (86) observed a range of sensitivities for spore enzymes and concluded that the enzymes in extracts of spores were inactivated at temperatures ranging from 24 to 46°C lower than those needed to inactivate the same enzymes within the intact spores. Membrane-bound ATPase has been associated with heat resistance/sensitivity of microorganisms. Coote et al. (87) suggested that ATPases are essential for the basal heat resistance of the cell to cope with elevated temperatures. Nevertheless, thermotolerance induced by sublethal heating is a mechanism independent of ATPase activity.

Flowers and Adams (88) suggested that the cell membrane is the site of thermal injury of spores subjected to mild or sublethal heating; membrane damage consequently increases sensitivity to environmental stresses. When spores are lethally heated, damage to the membrane permeability barrier results in the release of intracellular constituents and there is a temperature-dependent progressive loss of calcium and DPA (89–91). The death of spores proceeds faster than the release of DPA (65). When vegetative cells are heated, there is a rapid efflux of ions, amino acids, and low molecular weight nucleic acid components, thereby suggesting that interference with the semipermeability of membranes is a common consequence of heating (5).

A. Heat-Shock Response

While thermal processing guidelines are generally adequate for destruction of pathogens in foods, there may be conditions when the microorganism could become more heat resistant. Such conditions include environmental stresses occurring prior to cooking such as sublethal heat treatment, also known as heat-shocking conditions. Heat shock triggers a physiological response that leads to the synthesis of a specific set of proteins known as heat-shock proteins (HSPs) (92,93). HSPs may enhance the survival of pathogens in foods during exposure to high temperatures. Microbial cells in which HSPs are synthesized acquire enhanced thermal tolerance to a second heat challenge that would normally be lethal to them. Generally speaking, heat-shocked cells have to be heated twice as long as non–heat-shocked cells to achieve the same extent of lethality (94). Heat-shock response and induced thermotolerance has been reported in a wide range of microorganisms including *Escherichia coli* (95), *Salmonella* Typhimurium (45,96), *Salmonella* Thompson (46), *Salmonella* Enteritidis phage type 4 (97), *L. monocytogenes* Scott A (98,99), and *E. coli* O157:H7 (11,100,101). An increase in heat resistance of spores following heat shock has also been reported in spore-forming organisms such as *B. stearothermophilus* (102,103), *C. botulinum* (104), and *Clostridium sporogenes* (105). Although the scientific literature provides some evidence regarding the cause-and-effect relationship between the synthesis of HSPs and the induced thermotolerance response, this evidence is largely indirect and insufficient. Researchers have suggested that HSPs are not necessarily the major contributory agents in the development of thermotolerance but are required for recovery from heat stress (106–108). Their apparent role is to protect the cells against heat damage and to help the cells to return to their normal physiological state following the stressful event. Schlesinger (109) and Sanchez and Lindquist (110) suggested that the role of HSPs in thermotolerance may be to act as chaperones to remove denatured proteins. The primary function of classical chaperones, such as the *E. coli* DnaK (HSP 70) and its co-chaperones, DnaJ and GrpE, and GroEL (HSP60) and its chaperone, GroES, is to bind to and stabilize polypeptides already present in cells, modulate protein-folding pathways to prevent misfolding and aggregation of proteins, and promote refolding and proper assembly (111). Some *E. coli* HSPs are able to proteolyze irreversibly damaged polypeptides and assist in nucleic acid synthesis, cell division, and motility by promoting synthesis of a flagellum (112). In *E. coli*, regulation of stress response has been studied in greater detail through the transcriptional control of alternate sigma factors encoded by *rpoS* and *rpoH* in response to general stress and heat, respectively (113,114).

VI. FACTORS AFFECTING HEAT RESISTANCE

An appropriate heat treatment designed to achieve a specified lethality of microorganisms is influenced by many factors, some of which can be attributed to the inherent resistance of microorganisms, while others are due to environmental influences. Examples of inherent resistance include the differences among species and the different strains or isolates of bacteria (assessed individually or as a mixture) and the differences between spores and vegetative cells. Environmental factors include those affecting the microorganisms during growth and formation of cells or spores (e.g., stage of growth, growth temperature, growth medium, previous exposure to stress) and those active during the heating of bacterial suspension, such as the composition of the heating menstruum (amount of carbohydrate, proteins, lipids, solutes, etc.), water activity (a_w), pH, added preservatives, method of heating, and methodology used for recovery of survivors. This chapter deals with the most significant research on the factors affecting the heat resistance of foodborne pathogens.

VII. HEAT RESISTANCE OF VEGETATIVE FOODBORNE PATHOGENS

The heat resistance of foodborne pathogens has been studied in different substrates. Comparing the heat resistance of some pathogens, such as *L. monocytogenes*, *Salmonella* spp., and *E. coli* O157:H7, it appears that *L. monocytogenes* is relatively more heat resistant (Table 2). However, it is practically feasible to inactivate this pathogen by the type of mild heat treatment given to minimally processed foods without negatively impacting the product quality.

The pH of the heating menstruum is recognized as one of the most im-

Table 2 Heat Resistance of Three Foodborne Pathogens in Meat Expressed as D-Values in Minutes

Menstruum/temperature	*E. coli* O157:H7	*Salmonella* spp.	*L. monocytogenes*
Beef/60°C	3.17[a]	5.48[c]	8.32[d]
Beef/57.2°C	5.3[g]; 4.5[f]	5.4[e]	5.8[h]
Beef/62.8°C	0.5[g]; 0,4[f]	0.7[c]	1.2[h]
Chicken/60°C	1.63[a]	5.20[c]	5.29[d]
Turkey/60°C	1.89[b]	4.82[c]	—
Pork/60°C	2.01[b]	6.65[c]	—

[a] (214); [b] (215); [c] (118); [d] (216); [e] (217); [f] (119); [g] (218); [h] (219).

portant factors influencing the heat resistance of bacteria. Microorganisms usually have their maximum heat resistance at pH values close to neutrality; a decrease in the pH of the heating medium usually results in decreased D-value. Reichart (115) provided a theoretical interpretation of the effect of pH on microbial heat destruction and described a linear relationship between pH and the logarithm of the D-values for *E. coli*. The author stated that the logarithm of the heat-destruction rate increases linearly in the acid and alkaline range and has a minimum at the optimum pH for growth. High pH interacts synergistically with high temperature to destroy gram-negative foodborne pathogens (116). Growth temperature history and pH, including the type of acidulant used to adjust the pH, influence the heat resistance of *L. monocytogenes* strain Scott A (58). D-values significantly decreased ($p < 0.05$) with increased growth temperature when the pH of the growth medium was 5.4; the values significantly increased ($p < 0.05$) with increased temperature at pH 7, regardless of acid identity. At pH 5.4 adjusted with lactic acid, D-values ranged from 1.30 minute for 10°C-grown cells to 1.14 minute for 37°C-grown cells. At pH 5.4 adjusted with acetic acid, *L. monocytogenes* failed to grow at 10°C; the D-values were 1.32 minute and 1.22 minute when the organism was grown at 19 and 37°C, respectively. At pH 7 adjusted with lactic acid, D-values were 0.95, 1.12, and 1.28 minute for cells grown at 10, 19, and 37°C, respectively; the values ranged from 0.83 minute for 10°C-grown cells to 1.11 minute for cells grown at 37°C and pH adjusted with acetic acid. Alternatively, if conditions can be found that produce a more susceptible cell, they can be exploited to enhance inactivation (58).

The protective effect of fatty materials in the heating medium on the heat resistance of microorganisms is well documented (117). Theories behind increased heat resistance in foods with higher fat contents relate to reduced water activity or poor heat penetration (lower heat conductivities) in the fat portion (118). Doyle and Schoeni (119) reported a D-value at 60°C of 0.75 minute for *E. coli* O157:H7 strain 932 in ground beef containing 17–20% fat. Ahmed et al. (120) reported that D-values of *E. coli* O157:H7 in ground beef heated at 60°C ranged from 0.45 (beef, 7% fat) to 0.47 (beef, 20% fat) minute; the values ranged from 0.38 (chicken, 3% fat) to 0.55 (chicken, 11% fat) minute in chicken. These authors reported that the D-value of *E. coli* O157:H7 in ground turkey and pork sausage heated at 55°C in thermal death time tubes ranged from 6.37 (turkey, 3% fat) to 9.69 (turkey, 11% fat) minutes and 6.37 (pork sausage, 7% fat) to 11.28 (pork sausage, 30% fat) minutes; the values at 60°C ranged from 0.55 (turkey, 3% fat) to 0.58 (turkey, 11% fat) minute and 0.37 (pork sausage, 7% fat) to 0.55 (pork sausage, 30% fat) minute. In another study, Ahmed and Conner (121) reported D-values at 55°C for *E. coli* O157:H7 ranging from 12.5 (turkey, 3% fat) to 11.0 (turkey, 11% fat) minutes; the value at 60°C was 0.9 minute regardless of the percentage of fat in turkey. Ground beef contaminated with *S.* Typhimurium DT 104 heated to an internal temperature of 58°C for 53.5 (7% fat) or 208.1 minutes (24% fat) results in 7-D process for the pathogen; the heating time at 65°C to achieve the same level of reduction is 7.1 and 20.1 minutes, respectively (118). The authors reported that the pathogen does not possess any unique characteristics that would predispose it to survival during thermal processing. Vacuum-packaged pasteurized salmon fillets (10.56–17.2%, w/w, fat) had one to four times higher D-values for *L. monocytogenes* than the lower fat (0.6–0.8%, w/w, fat) cod fillets (122). However, Donnelly and Briggs (123) reported little difference in D-values for *L. monocytogenes* in skim milk, 11% milk solids, and whole milk when fat was studied as a single factor affecting heat resistance. In another study, differences were observed in D-values for *L. monocytogenes* in sheep, cow, and goat milks (124). Sheep milk fat added to cow and sheep skim milks resulted in higher D-values compared to the values obtained from the addition of cow milkfat. Obviously, it is not solely fat content but also the fatty acid composition that affects heat resistance.

Various solutes in the heating medium exert different effects on the heat resistance of microorganisms, depending upon the nature of the solutes and their concentration. The effects of solutes on thermal resistance have mainly been examined by determining the relationships between thermal resistance and either solute concentration or water activity of the heating menstruum. *S. aureus* exhibited increased heat resistance when heated in different substrates with reduced water activity (125). In a study by Reichart and Mohacsi-Farkas (126), when heat destruction of seven foodborne microorganisms as a function of temperature, pH, redox potential, and water activity was assessed in synthetic heating media, the heat destruction increased with increasing water activity and decreasing pH. While an increased heat resistance is observed when the a_w of the heating menstruum is lowered, there appears to be no direct correlation between a given a_w level and heat resistance. For example, sucrose protected salmonellae from heat destruction far more efficiently than glycerol at given a_w levels down to 0.87 (127). Tuncan and Martin (128) suggested that the effect of salts on thermal inactivation of microorganisms is mainly related to reduced water activity and increased osmotic pressure of the heating menstruum. Sodium chloride at different concentrations (3, 5, and 9% w/v) protected *S. aureus* from heat injury, with the highest concentration affording the maximum protection (129). The authors explained that sodium chloride may be involved in stabilizing membrane protein structures such as nucleic acids and nucleotides as well as increasing the melting temperatures of membrane phospholipids. Thus, the damage to the cell membrane and leakage of cell components from the cytoplasm is prevented. In another study, 4 and 8% sodium chloride protected *S. aureus* cells from heat injury at pH 7.0, while at pH 6.5 a concentration of 8% gave protection (130). The D-values at 65.6°C for *S.* Typhimurium ranged from 0.29 to 40.2 in sucrose solutions with a_w ranging from 0.98 to 0.83 (131). For *L. monocytogenes*, the D-values at 65.6°C shifted from 0.36 to 3.8 minutes in sucrose solutions with a_w ranging from 0.98 to 0.90. The values ranged from 1.2 to 3.2 minutes, while the a_w of the chocolate syrups ranged from 0.75 to 0.84 (131). The addition of small amounts of water to low-a_w foods has been noted to greatly lower the heat resistance of microorganisms. Researchers have reported that the survival of salmonellae in dry egg white over the range 50–70°C decreased rapidly when the moisture level of the product was increased from 3 to 12%, but the a_w levels of these products were not reported (132). Similarly, Van Cauwenberge et al. (133) found that heating at 47°C was more effective for a corn flour product with a moisture level of 15% than for one with a 10% moisture level in reducing the bacterial population from 10^5 to 10^3 cells per g. Archer et al. (134) reported that heat resistance of *S.* Weltevreden increased as the initial a_w of the flour prior to heating decreased and that the relative humidity of the atmosphere during preconditioning did not have a significant effect.

Both an increase in the number of viable cells capable of producing colonies and an increase in the estimated D-value are observed under optimum recovery conditions. Temperatures below the optimum for growth may enhance repair of heat damage (135). Higher recoveries of sublethally heat-damaged cells in anaerobic conditions or when the recovery media is supplemented with oxygen scavengers or reducing agents have been observed in studies with *E. coli* O157:H7 (100,117,136–139), *L. monocytogenes* (140–146), *S.* Enteritidis (147), and *S. aureus* (148). Xavier and Ingham (147) measured heat resistance of *S.* Enteritidis and reported D-values at 52°C of 11.3 and 5.4 minutes and at 58°C of 1.5 and 1.0 minutes when heated cells were recovered under anaerobic and aerobic conditions, respectively. Similarly, for *L. monocytogenes*, the D-value at 55°C was increased from 8.9 to 12.0 minutes when prereduced TSAYE plates were incubated anaerobically rather than aerobically (142). The increased resistance of *L. monocytogenes* under anaerobic conditions may explain why the pathogen appeared to survive pasteurization at 62.8°C for 15 minutes when recovered anaerobically (141) and why minimal heat treatment had only a marginal effect on the pathogen inoculated into chicken gravy (149) and *sous-vide* chicken breast (150). The D-values of *E. coli* O157:H7 at 55°C were reported to be 8.0, 11.1, and 18.3 minutes following recovery on aerobic TSA plates, on anaerobic TSA plates, and in anaerobic roll tubes, respectively (100,137). Interestingly, George et al. (139) observed that the measured heat resistance of cells was greatly influenced by oxygen concentration. The heat resistance of *E. coli* O157:H7, *S.* Enteritidis, and *L. monocytogenes* was up to eightfold greater when they were grown, heated, and recovered anaerobically rather than aerobically. The authors reported that the time at 59°C for a 6-D reduction of *E. coli* O157:H7 was 19–24 minutes; this was reduced to 5–17 minutes when 0.5–1% oxygen was included and to 3 minutes when 2–40% oxygen was used. Further, Bromberg et al. (138) reported that oxygen-sensitive cells were only sublethally heat damaged and regained their ability to grow in the presence of oxygen when allowed a recovery period in anaerobic conditions. George and Peck (151) differentiated between the effects of oxygen and that of high redox potential and reported that sublethally heat-damaged cells regained their ability to grow in media of high redox potential at a similar rate whether the redox potential was increased by the addition of potassium ferricyanide, 2,6-dichloroindophenol, or oxygen.

Addition of various concentrations of the bacteriocin nisin renders the bacteria sensitive to the lethal effect of heat, thereby enhancing the effectiveness of the thermal processes during mild heat treatments. Boziaris et al. (152) reported a reduction of required pasteurization time of up to 35% for *S.* Enteritidis when the heating menstrua (nutrient media, liquid whole egg, and egg white) was supplemented with 500–2500 IU/mL of nisin. The pathogen was most sensitive in egg white that had an alkaline pH and contained no fat. In a study by Budu-Amoako et al. (153), when nisin was added at a level of 25 mg/kg of can contents to the brine surrounding the lobster, in combination with a heat process giving internal temperatures of 60°C for 5 minutes and 65°C for 2 minutes, a 3–5 log reduction of *L. monocytogenes* was observed compared to heat or nisin alone, which resulted in 1–3 log reductions of the pathogen.

It has been well documented that bacterial cells/population in stationary phase or those that have undergone some sublethal stress undergo physiological changes that make them more resistant to subsequent heat treatment or any other potentially stressful condition (154). For example, sublethal heat stress renders an organism more resistant to subsequent heat treatment that would otherwise be lethal. Cross protections in which exposure to one stress alters resistance to another can also occur. For example, acid pH exposure or osmotic shock can induce stress responses including resistance to thermal stress (155,156). Adaptation of *L. monocytogenes* to starvation, ethanol, hydrogen peroxide, and acid significantly increased the resistance of the pathogen to heat (157).

VIII. HEAT RESISTANCE OF SPORES OF FOODBORNE PATHOGENS

The heat resistance of spores of *C. botulinum* has been studied more thoroughly than those of most other spore-formers. This is because the pathogen produces a potent neurotoxin that can cause various symptoms of paralysis and is thus the most hazardous spore-forming foodborne pathogen. *C. botulinum* type A and B (proteolytic) strains produce highly heat-resistant spores and are of primary importance to food safety. These spores are targeted for destruction to assure microbiological safety of low-acid foods. The canning industry historically adopted a D-value at 121°C of 0.2 minute (12 log reduction) as a standard for designing a required thermal process for an adequate degree of protection against

C. botulinum. In contrast, the heat resistance of *C. botulinum* type E and nonproteolytic strains of type B is comparatively low and has been thoroughly investigated in aqueous media as well as in a variety of foods (Table 3). The greatest food safety concerns in foods preserved by pasteurization and refrigeration is the survival of nonproteolytic *C. botulinum*; spores of those strains that survive the thermal process would pose a botulism hazard even under proper refrigeration temperatures if a secondary barrier, e.g., low pH, low a_w, is not present. However, the thermal process for cook/chill foods is not designed for the destruction of proteolytic strains of *C. botulinum* because these strains do not grow at or below 10°C.

Gaze and Brown (158) assessed the heat resistance over the temperature range 70–92°C of nonproteolytic *C. botulinum* type B and E spores heated in homogenates of cod (pH 6.8) and carrot (pH 5.7). It was calculated that a heat treatment at 90°C for 7 minutes or equivalent time-temperature combination would be sufficient to achieve a 6-D process for the most heat-resistant nonproteolytic *C. botulinum* spores. Juneja et al. (159) reported that turkey should be heated to an internal temperature of 80°C for at least 27.1 minutes to achieve a 6-D process for nonproteolytic *C. botulinum* spores.

Composition of the sporulation medium and the temperature used for spore preparation influence the spore heat resistance. The highest observed heat resistance corresponds to the *C. perfringens* spores obtained in the pH range 7.0–9.0 (160). Spores of *B. cereus* produced in the pH range 6.5–8.3 showed a decline in D-values at 100°C by about 65% per pH unit (161). While sporulation temperature can influence heat resistance, there is no general trend. According to some authors, spores produced at higher temperatures have greater heat resistance (26,38,162). However, some investigators have found the opposite effect

Table 3 Heat Resistance of *C. botulinum* Type E and Nonproteolytic Type B Spores in Buffer and a Variety of Foods

Strains	Menstruum	Temperature (°C)	D-value (min)	Ref.
Type E	Phosphate buffer	80	1.03–4.35	159
Type E	Phosphate buffer	82.2	0.37–0.52	220
Type E	Oyster homogenate	80	0.78	221
Type E	Mehnaden surimi	82.2	1.22	222
Type E	Turkey	70	51.89	159
Type E	Turkey	85	1.18	159
Type B	Phosphate buffer	82.2	1.49–73.6	220
Type B	Phosphate buffer	80	3.22–4.31	159
Type B	Turkey	75	32.53	159
Type B	Turkey	90	0.80	159

(163,164). It has also been shown that heat resistance increases with sporulation temperature up to a maximum, decreasing thereafter (165). Sala et al. (162) reported that the sporulation temperature induces important changes in the influence of the pH of the heating menstruum on the heat resistance of *Bacillus subtilis*. Spores sporulated at 32°C were less heat resistant at pH 4 than at pH 7 regardless of the temperature of treatment. However, the effect of acidic pH on heat resistance decreased as heating temperature increased for spores sporulated at 52°C. While the z-value of a spore suspension sporulated at 32°C did not change irrespective of the pH of heating menstruum, the z-value of suspension sporulated at 52°C increased with acidification. This potential for increase in z-value is a hazard that should be taken into account in hot climates, especially when designing sterilization processes for acidified foods.

Composition and pH of the heating menstruum and the presence of any antibacterial agents during the heating process will substantially affect the heat resistance of bacterial spores. Juneja et al. (159) reported a concomitant increase in heat resistance of nonproteolytic *C. botulinum* spores in turkey slurry as compared to phosphate buffer. While D-value at 80°C for type E spores in buffer was 4.35 minutes, turkey slurry offered protection, and the D-value was increased to 13.37 minutes. Similarly, D-values for the nonproteolytic type B strain in buffer and turkey were 4.31 and 15.21 minutes, respectively. Generally, thermal resistance of spores decreases as the pH of the heating medium is reduced and increases with decreasing a_w. Using *C. botulinum* type A and B strains, Odlaug and Pflug (166) reported D-values in tomato juice (pH 4.2) that were three times lower than in phosphate buffer (pH 7). Xezones and Hutchings (167) reported D-values at 115.5°C of 0.128, 2.6, 4.91, and 5.15 minutes for *C. botulinum* 62A spores suspended in spaghetti sauce at pH 4, 5, 6, and 7, respectively. For *C. botulinum* type E spores, D-values at 110°C of less than 0.1 second at a high a_w were increased by more than 100,000 times as the a_w was lowered to 0.2–0.3 (168). It has also been reported that the influence of a_w on the heat resistance is more pronounced for species with more heat sensitivity (169). In addition, researchers have reported that the effect of a_w of the heating menstruum depended on the treatment temperature (170). Mazas et al. (171) reported that the effects of a_w on the D-values of *B. cereus* depended on the nature of the compound used to control a_w, the strain tested, and the heating temperature. In their study, the protective effect of NaCl against heat on the *B. cereus* spores was higher than that of LiCl at the same a_w values, and that of sucrose was higher than that of glycerol.

Fat and proteins protect spores against heat lethality (172,173). Spores are more resistant to heat when suspended in oils than in buffer. This resistance is ascribed to the low water activity of oils and to their content of free fatty acids. Molin and Snygg (174) reported a D-value at 95°C of 13 minutes for *B. cereus* spores suspended in phosphate buffer and D-values at 121°C of 17.5–30 minutes for the same spores heated in olive and soybean oils. Survivor curves of the *B. subtilis* spores suspended in olive oil were concave upward with a characteristic tailing (175).

Nisin has been shown to reduce the thermal processing requirements in several food products by enhancing the thermal inactivation of bacterial spores. In a study by Wandling et al. (176), when heat resistance of *B. cereus* was assessed in skim milk supplemented with various concentrations of nisin, the D-values at 97°C were 7.0, 4.8, and 4.7 minutes for the control and 2000 and 4000 IU/mL nisin treatments, respectively. In the same study, the apparent D-values of *B. stearothermophilus* at 130°C were reduced by 13 and 21% because of the presence of 2000 and 4000 IU of nisin/mL, respectively.

Determination of spore survival after exposure to heat is characterized by the ability of the injured spores (which are more nutritionally demanding than unheated spores) to germinate, outgrow, and form colonies on the recovery media. In fact, any factor influencing repair mechanisms might also influence the measured D-value. In well-documented reviews on the recovery of injured spores, Roberts (177) and Adams (178) outlined the importance of the recovery media for injured spores and classified the additives that can increase recovery according to their mechanisms of action. Recovery conditions including the composition and pH of the medium, nature of acidulants, the presence of inhibitors, and temperature and time of incubation can significantly affect survival counts and, consequently, the values obtained for the parameters used to characterize heat resistance (179,180). Feeherry et al. (181) observed a decrease in D-values at 121°C when the recovery medium was acidified with HCl. Fernandez et al. (182) described a higher inhibition of citric acid compared with glucono-δ-lactone on spore recovery. Potassium sorbate at concentrations as low as 0.025% and sodium benzoate at 0.1% were very effective inhibitory agents for heat-injured spores of *B. stearothermophilus* (183). It is usually considered that the most adequate temperature for the recovery of heat-treated spores is lower than the optimal for unheated spores (180). Lower incubation temperatures lead to higher D-values and longer shoulders (184).

There is sufficient evidence to document that the addition of lysozyme to the recovery medium resuscitates and increases the recovery of the heat-injured spores, thereby increasing the measured or apparent heat resistance (159,185, 189). In a study by Sebald et al. (189), when *C. botulinum* type E spores were heated in phosphate buffer for 10 minutes at 80°C, surviving spores were able to form colonies on a medium containing lysozyme, but not in its absence. Alderton et al. (185) reported an estimated D-value of 1.30 minutes when type E spores were heated in phosphate buffer at 79.5°C and recovered on a medium without lysozyme. In the same study, when the recovery medium contained lysozyme, the heating temperature had to be raised to obtain measurable spore destruction; the D-values were 13.50 and 3.80 minutes at 90.5 and 93.3°C, respectively.

In a study by Peck et al. (187), when spores of nonproteolytic *C. botulinum* type B strains were heated at 85°C and survivors were enumerated on a highly nutritive medium, a 5 decimal kill in less than 2 minutes was observed. However, enumeration of survivors on a medium supplemented with lysozyme showed that heating at 85°C for 5 minutes resulted in only an estimated 2.6 decimal kill of spores; also, biphasic survivor curves indicating heat-sensitive and heat-resistant population subfractions were observed. Treatment with alkaline thioglycollate resulted in heated spores being permeable to lysozyme, and a biphasic heat-inactivation curve was converted to a logarithmic curve, the slopes of which were similar to the second part of the biphasic curves (190). This treatment is known to rupture disulfide bonds in the spore coats, thereby making the spore coat permeable to lysozyme. These findings have implications for assessing heat treatments necessary to reduce risk of nonproteolytic *C. botulinum* survival and growth in cook/chill foods.

Duncan et al. (191) suggested that the heat alteration of the spore results in inactivation of the cortex lytic enzyme system, i.e., the system responsible for cortical degradation during germination. Lysozyme in the plating medium can replace the thermally inactivated spore germination enzymes (188). Lysozyme permeates the spore coat and degrades the cortex, leading to core hydration and, consequently, spore germination (13). The lysozyme levels in foods of plant and animal origin are 1.8–27.6 and 20–160 μg/g, respectively. Since lysozyme is heat stable and is present in a variety of foods, the influence of this enzyme in the recovery of heat-damaged spores warrants further investigation to determine its effect on the efficacy of recommended heat processes.

Simulating the conditions in minimally processed refrigerated foods, Juneja and Eblen (192) recovered heated nonproteolytic *C. botulinum* spores on both reinforced clostridial medium (RCM) with lysozyme and on RCM with lysozyme and the same salt levels as the heating menstruum. This approach of heating in a food substrate and recovery under the same conditions is the closest to the real situation in food. When the recovery medium contained no salt, D-values in turkey slurry containing 1% salt were 42.1, 17.1, 7.8, and 1.1 minutes at 75, 80, 85, and 90°C, respectively. The D-values were 27.4, 13.2, 5.0, and 0.8 minutes at 75, 80, 85, and 90°C, respectively, when both the turkey slurry and the recovery medium contained 1% salt. Increasing levels (2–3%, w/v) of salt in turkey slurry resulted in parallel decrease in the D-values obtained from the recovery of spores on the media containing the same levels of salt as the heating menstruum. Hutton et al. (193) reported the presence of salt at 2% levels in the modified PA3679 agar decreased the *C. botulinum* spores D-value by 20–40% irrespective of the pH of the heating menstruum. However, the protective effect of salt leading to increased D-values is also well known. Inclusion of salt in heating menstruum results in increased heat resistance due to reduced water activity leading to spore dehydration. Juneja and Eblen (192) indicated that the decrease in D-values obtained from the recovery of heat-damaged spores on the media with added salt was a consequence of the inability of heat-injured spores to recover in the presence of salt. The heat-injured spores are sensitive to salt in the recovery medium. These data should assist food processors in designing reduced thermal processes that ensure safety against nonproteolytic *C. botulinum* type B spores in cook/chill foods while maintaining desirable organoleptic attributes of foods.

IX. HEAT-INACTIVATION KINETICS PREDICTIVE MODELS

The effectiveness of the individual effects of heat treatment, pH, salt, etc., with regard to pathogen inactivation is maximized by conducting multiple factorial experiments in which the effects and interactions of these parameters in foods are assessed in lowering the heat resistance of foodborne pathogens. Subsequently, inactivation kinetics or thermal death models are developed which predict the target pathogen's survival within a specific range of food formulation variables. These models can help either to establish an appropriate heat treatment or to understand and determine the extent to which existing/traditional thermal processes could be modified for a variety of cooked foods. The models can contribute to more effective evaluation and assessment of the impact of changes in food formulations that could affect their microbiological safety or the heat lethality of pathogens. These predictive models enable food processors and regulatory agencies to ensure critical food safety margins by predicting the combined effects of multiple food formulation variables. The food processors are able to design appropriate processing times and temperatures for the production of safe food with extended shelf life without substantially adversely affecting the sensory quality of the product. However, it is of critical importance that the D-values predicted by the models first be validated with heat resistance data obtained by actual experiments in specific foods before the predicted values can be used to design thermal processes for the production of a safe food.

Juneja et al. (194,195) and Juneja and Eblen (196) employed a fractional factorial design to assess and quantify the effects and interactions of temperature, pH, salt, and phosphate levels and found that the thermal inactivation of nonproteolytic *C. botulinum* spores, *E. coli* O157:H7, and *L. monocytogenes* was dependent on all four factors. Thermal resistance of spores or vegetative cells can be lowered by combining these intrinsic factors. The following multiple regression equations, developed in these studies, predict D-values of nonproteolytic *C. botulinum* spores for any combinations of temperature (70–90°C), salt (NaCl; 0.0–3.0%), sodium pyrophosphate (0.0–0.3%), and pH (5.0–6.5); *E. coli* O157:H7, for combinations of heating temperature (55–62.5°C), salt (0.0–6.0%, w/v), sodium pyrophosphate (0.0–0.3%, w/v), and pH (4.0–6.5); and *L. monocytogenes*, for combinations of heating temperature (55–6.5°C), salt (0.0–6.0%, w/v), sodium pyrophosphate (0.0–0.3%, w/v), and pH (4.0–8.0). The predicted D-values are for changes in the parameter values in the range tested from any combination of four environmental factors.

1. Nonproteolytic *C. botulinum* spores:

$$\begin{aligned}\text{Log}_e\ \text{D-value} = {} & -9.9161 + 0.6159(\text{temp}) - 2.8600(\text{pH}) \\ & - 0.2190(\text{salt}) + 2.7424(\text{phos}) \\ & + 0.0240(\text{temp})(\text{pH}) - 0.0041(\text{temp})(\text{salt}) \\ & - 0.0611(\text{temp})(\text{phos}) + 0.0443(\text{pH})(\text{salt}) \\ & + 0.2937(\text{pH})(\text{phos}) - 0.2705(\text{salt})(\text{phos}) \\ & - 0.0053(\text{temp})^2 + 0.1074(\text{pH})^2 \\ & + 0.0564(\text{salt})^2 - 2.7678(\text{phos})^2\end{aligned}$$

2. *E. coli* O157:H7:

$$\begin{aligned}\text{Log}_e\ \text{D-value} = {} & -43.0646 + 1.4868(\text{temp}) + 3.5737(\text{pH}) \\ & - 0.1341(\text{salt}) - 8.6391(\text{phos}) - 0.0419(\text{temp})(\text{pH}) \\ & + 0.0103(\text{temp})(\text{salt}) + 0.1512(\text{temp})(\text{phos}) \\ & - 0.0544(\text{pH})(\text{salt}) + 0.2253(\text{pH})(\text{phos}) \\ & - 0.2682(\text{salt})(\text{phos}) - 0.0137(\text{temp})^2 \\ & - 0.0799(\text{pH})^2 - 0.0101(\text{salt})^2 - 6.4356(\text{phos})^2\end{aligned}$$

3. *L. monocytogenes*:

$$\begin{aligned}\text{Log}_e\ \text{D-value} = {} & -61.4964 + 2.3019(\text{temp}) + 1.2236(\text{pH}) \\ & + 0.7728(\text{salt}) + 1.0477(\text{phos}) \\ & - 0.0102(\text{temp})(\text{pH}) - 0.0085(\text{temp})(\text{salt}) \\ & - 0.0566(\text{temp})(\text{phos}) - 0.0210(\text{pH})(\text{salt}) \\ & - 0.4160(\text{pH})(\text{phos}) + 0.1861(\text{salt})(\text{phos}) \\ & - 0.0217(\text{temp})^2 - 0.0273(\text{pH})^2 \\ & - 0.0213(\text{salt})^2 - 13.1605(\text{phos})^2\end{aligned}$$

The authors developed confidence intervals (95%) to allow microbiologists to predict the variation in the heat resistance of the pathogens. Representative observed and predicted D-values of nonproteolytic *C. botulinum* in ground turkey,

Table 4 Observed and Predicted D-Values at 70–90°C of Nonproteolytic *Clostridium botulinum* in ground turkey, *E. coli* O157:H7, and *L. monocytogenes* in beef gravy

Temperature (°C)	pH	% NaCl	% Phosphate	D-value observed (min)	D-value predicted[a] (min)
Nonproteolytic *Clostridium botulinum*					
70	6.50	0.0	0.00	57.7	66.0
70	6.50	1.5	0.15	40.1	46.5
75	6.25	1.0	0.10	39.1	42.3
75	6.25	1.0	0.20	32.9	38.6
90	5.00	0.0	0.00	5.0	6.3
90	5.00	1.5	0.15	3.1	4.8
E. coli O157:H7					
55	4	0.0	0.0	2.8	4.1
55	4	0.0	0.30	1.9	2.7
55	4	6.0	0.30	3.5	4.3
60	4	3.0	0.15	2.1	2.2
60	6	3.0	0.30	1.8	2.1
L. monocytogenes					
55	4	0.0	0.0	5.35	9.03
55	4	6.0	0.0	12.49	15.74
57.5	5	4.5	0.10	6.92	8.05
57.5	5	4.5	0.20	10.61	8.45

[a] Predicted D-values are the 95% upper confidence limits.
Source: Refs. 194–196.

E. coli O157:H7, and *L. monocytogenes* in beef gravy are provided in Table 4. Predicted D-values from the model compared well with the observed thermal death values. Thus, the model provides a valid description of the data used to generate it. Examples of secondary models used to mathematically describe the inactivation rates of microorganisms are provided in Table 5.

Periago et al. (197) used a factorial experimental design to develop a model that describes the combined effect of NaCl (S; 0.5–3%) and pH 5.75–6.7 (in the heating menstruum and recovery medium) on the apparent heat resistance of *B. stearothermophilus* spores suspended in mushroom extract, over the temperature (T) range 115–125°C. Coefficients obtained for the second order polynomial were:

$$\begin{aligned}\ln(y) = &-93.136 + 0.7413 \times T + 20.0194 \times pH + 0.6067 \times S \\ &+ 0.022153 \times T \times pH - 0.011486 \times T \times S + 0.1033 \\ &\times pH \times S - 0.0048545 \times T^2 - 1.7085 \times pH^2 - 0.09131 \times S^2\end{aligned}$$

Table 5 Examples of the Secondary Models Used to Describe Inactivation Rates of Microorganisms

Models (Ref.)	Mathematical description (equation)
Arrhenius model[a] (223)	$k = A \exp[(-E_a/R)/T]$
Model[b] z-value (223)	$D = D_r 10^{-(T-T_r)/Z}$
	$Z = 2 \cdot 303\ RTT_r/Ea$
Linear-Arrhenius model[c] (224)	$\ln k = C_0 + C_1/T + C_2\ pH \div C_3\ pH^2$
A quadratic response surface model represented a polynomial[d] (194–196)	$\ln(y) = c_1 + c_2T + c_3\ pH + c_4S \ldots c_8T^2 + c_9\ pH^2 + c_{10}S^2$
Extension of the Eyrings model[e] (115)	$k = k_0 + \kappa \frac{K_bT}{h} \cdot \left(\exp\left[\frac{\Delta G^*_H}{RT}\right] \cdot [H^+]^{n_H} + \exp\left[\frac{\Delta G^*_{OH}}{RT}\right] \cdot [OH^-]^{n_{OH}} \right)$

[a] A is the rate constant at infinite temperature, Ea is the activation energy of inactivation, R is the gas-low constant; T in degrees Kelvin.
[b] D_r is the D-value at the reference temperature Tr.
[c] $C_0 - C_3$ are fit parameters/coefficients for the particular food-bacterial system.
[d] ln(y) is the natural logarithm of the dependent variable of the model, the D-value, T is the heating temperature, pH is the pH value of the substrate, S is the NaCl concentration and $C_1 - C_{10}$ are the coefficients to be estimated.
[e] k_o, κ, k_b, h, n_H, n_{OH}, ΔG^*_H and ΔG^*_{OH} as defined by Reichart (115).

The authors observed that both pH and NaCl had a considerable influence on *B. stearothermophilus* D-values. For example, at 115°C, when spores were heated and recovered at pH 6.7 with 0.5% NaCl, the D-value obtained was 9.6 minutes, whereas at the same heating temperature, spores heated and recovered at pH 6 with 3% NaCl had a D-value of 3.3 minutes. This represents almost a 90% reduction in thermal resistance.

A log-logistic model has been applied successfully to quantify: the effects of temperature on the thermal inactivation of *L. monocytogenes* (8), the effect of heating rate on the inactivation of *L. monocytogenes* (198), the effect of pH and temperature on the survival of *Yersinia enterocolitica* (199), the effects of temperatures below 121°C on the inactivation of *C. botulinum* 213B (200), and the effect of temperature on the inactivation of *S.* Typhimurium (201).

The modified Gompertz equation was effectively used to model *L. monocytogenes* inactivation in a formulated infant product with varying temperature (50, 55, 65°C), pH (5.0, 6.0, 7.0), and NaCl concentration (0, 2, 4%) (202). More recently, Blackburn et al. (203) used a log-logistic function to develop a three-factor thermal inactivation models for *S.* Enteritidis and *E. coli* O157H7 as affected by temperature (54.5–64.5°C), pH (4.2–9.6 adjusted using HCl or NaOH), and NaCl concentration (0.5–8.5% w/w). In this study, 84% of *S.* Enteritidis and

83% of *E. coli* O157H7 survival curves represented a linear logarithmic death, with the remaining curves demonstrating shoulder and tailing regions. Chhabra et al. (204) developed a predictive model using a modified Gompertz equation to estimate *L. monocytogenes* death in a formulated and homogenized milk system as affected by milkfat (0–5%), pH (5.0–7.0), and processing temperature (55–65°C). The shoulder region of the survival curve was affected by pH; however, the maximum slope was affected by temperature, milkfat, and the interaction of temperature and milkfat, thereby suggesting that both temperature and milkfat play a role in the rate of *L. monocytogenes* inactivation during thermal processing. Regardless of which equation is used to represent data, predictive model is a useful initial tool to estimate pathogen inactivation and gives an increased understanding of how various conditions affect the death of pathogens.

Mafart and Leguerinel (205) emphasized that the use of D- and z-values is not sufficient and that the ratio of spore recovery after incubation should be considered in calculations used in thermal processing of foods. The authors derived a model describing the recovery of injured spores as a function of both the heat treatment intensity and the recovery conditions, i.e., incubation temperature, pH, and sodium chloride content. When heated spores are recovered under unfavorable conditions, the ratio of cell recovery and apparent D-value are reduced (205).

X. COMBINATION TREATMENTS

Severe heat treatments can impair the organoleptic properties and nutritional values of foods. To avoid the undesirable effects of heat, one approach is to use heat in combination with other already known preservation techniques. The use of combination preservation treatments incorporating mild heat can result in enhanced preservative action by having an additive or synergistic effect on microbial inactivation—particularly in foods with a high water content—and/or reduce the severity of one or all the treatments. A few examples include:

1. The lethal effect of heat is enhanced if bacterial cells have undergone ultrasound treatment (206). The combination treatment of heat and ultrasound is termed ''thermosonication'' (207).
2. For spores, a combination of high pressure and high temperature is necessary for inactivation. Under high pressure, bacterial spores germinate to vegetative cells and are then inactivated by the heat.
3. An irradiation dose of 5 kGy is sufficient to sensitize clostridial spores to subsequent heating (208). Prior exposure to ionizing radiation stimulates spore germination, rendering the spores sensitive to subsequent heat treatment. Irradiation can also sensitize vegetative cells to subse-

quent heating. Thayer et al. (209) reported that irradiation at a dose of 0.9 kGy caused heat sensitization of *S.* Typhimurium in mechanically deboned chicken. A dose of 0.8 kGy was sufficient to increase the heat sensitivity of *L. monocytogenes* in roast beef and gravy (210). Since the principal target of ionizing radiation is DNA, vegetative cells treated first by ionizing radiation experience damage to their DNA, and then subsequent heat treatment damages enzymes necessary for DNA repair.

4. Efficacy of the lethal effect of heat on microorganisms is increased if subsequently exposed to organic acids. This is a consequence of prior heating causing damage to the cell membrane, making it easier for weak acids to penetrate into the cytoplasm.

XI. CONCLUSIONS AND FUTURE OUTLOOK

The use of heat for the inactivation of microorganisms is the most common process in use in food preservation today. Heat treatment designed to achieve a specific lethality for foodborne pathogens is one of the fundamentally important strategies used to assure the microbiological safety of thermally processed foods. Heat resistance of microorganisms can vary depending on the species and strain of bacteria, food composition, physiological stage of microbial cells or spores, and recovery conditions (type of media, temperature, atmosphere, and time of incubation) for the detection of survivors. Food characteristics leading to increased heat resistance of an organism include water activity and the presence of carbohydrates, lipids, proteins, salt, etc. Heat resistance of spores is attributed primarily to thermal adaptation, mineralization, and dehydration. Alterations in membrane fatty acid profile results in an altered response to subsequent heat treatment. Potential targets of heat damage include nucleic acid, proteins and enzymes, and cellular membranes.

Quantitative knowledge of the factors in food systems that interact and influence the inactivation kinetics are required to accurately estimate how a particular pathogen is likely to behave in a specific food. There is a need for a better understanding of how the interaction among preservation variables can be used for predicting safety of minimally processed, ready-to-eat foods. The effects and interactions of temperature, pH, sodium chloride content, and sodium pyrophosphate concentration are among the variables that researchers have considered when attempting to assess the heat-inactivation kinetics of foodborne pathogens. Incorporation of these multiple barriers increased the sensitivity of cells/spores to heat, thereby reducing heat requirements and ensuring the safety of ready-to-eat food products.

The future of thermal death determination of bacteria will likely rely on predictive thermal inactivation kinetics modeling. Complex multifactorial experiments and analysis to quantify the effects and interactions of additional intrinsic and extrinsic factors and development of ''enhanced'' predictive models are warranted to ensure the microbiological safety of thermally processed foods. In view of the continued interest in minimally processed foods, it would be logical to define a specific lethality at low temperatures. It would be useful to determine the possible effects of injury to vegetative cells and spores that may result from mild heat treatments and the factors in foods that influence the recovery of cells/ spores heated at these low temperatures. In conclusion, future research should focus on conducting dynamic pasteurization (low-temperature, long-time cooking) studies to assess the integrated lethality of cooking and develop integrated predictive models for pathogens for the thermal inactivation, injury, repair, and behavior in ready-to-eat meats including those packaged in modified atmospheres.

REFERENCES

1. D Roberts. Sources of infection: food. In: WM Waites, JP Arbuthnott, eds. Foodborne Illness. London: Edward Arnold, 1991, pp. 31–37.
2. J Farkas. Physical methods of food preservation. In: MP Doyle, LR Beuchat, TJ Montville, eds. Food Microbiology: Fundamentals and Frontiers. Washington, DC: ASM Press, 1997, pp. 495–519.
3. MJ Lewis. Ultra-high temperature (UHT) treatments. In: RK Robinson, CA Batt, PD Patel, eds. Encyclopedia of Food Microbiology, Vol. 2. London: Academic Press, 2000, pp. 1023–1030.
4. CR Stumbo. Thermal resistance of bacteria. In: Thermobacteriology in Food Processing. New York: Academic Press, 1973, pp. 79–104.
5. RI Tomlins, ZJ Ordal. Thermal injury and inactivation in vegetative bacteria. In: FA Skinner, WB Hugo, ed. Inhibition and Inactivation of Vegetative Microbes. New York: Academic Press, 1976, pp. 153–190.
6. S McKee, GW Gould. A simple mathematical model of the thermal death of microorganisms. Bull Math Biol 50:493–501, 1988.
7. IJ Pflug, RG Holcomb. Principles of thermal destruction of microorganisms. In: SS Block, ed. Disinfection, Sterilization, and Preservation, 3rd ed. Philadelphia: Lea and Febiger, 1983, pp. 751–810.
8. MB Cole, KW Davies, G Munro, CD Holyoak, DC Kilsby. A vitalistic model to describe the thermal inactivation of *Listeria monocytogenes*. J Indust Microbiol 12: 232–239, 1993.
9. NH Hansen, H Riemann. Factors affecting the heat resistance of non-sporing organisms. J Appl Bacteriol 26:314–333, 1963.
10. O Cerf. Tailing of survival curves of bacterial spores—a review. J Appl Bacteriol 42:1, 1977.
11. VK Juneja, PG Klein, BS Marmer. Heat shock and thermotolerance of *Escherichia coli* O157:H7 in a model beef gravy system and ground beef. J Appl Microbiol 84:677–684, 1997.

12. A Casolari. Microbial death. In: MJ Bazin, JI Prosser, eds. Physiological Models in Microbiology, Vol. 2. Boca Raton, FL: CRC Press, Inc., 1988, pp. 1–44.
13. GW Gould. Heat-induced injury and inactivation. In: GW Gould, ed. Mechanisms of Action of Food Preservation Procedures. New York: Elsevier Science Publishers, Ltd., 1989, pp. 11–42.
14. IJ Pflug. Microbiology and Engineering of Sterilization Processes, 7th ed. Minneapolis: Environmental Sterilization Laboratory, 1990.
15. RC Whiting. Microbial modeling in foods. Crit Rev Food Sci Nutr 35(6):467–494, 1995.
16. G Abrahm, E Debray, Y Candau, G Piar. Mathematical model of thermal destruction of *Bacillus stearothermophilus* spores. Appl Environ Microbiol 56:3073, 1990.
17. AC Rodriguez, GH Smerage, AA Teixeira, JA Lindsay, FF Busta. Population model of bacterial spores for validation of dynamic thermal processes. J Food Proc Eng 15:1, 1992.
18. AC Rodriguez, GH Smerage, AA Teixeira, FF Busta. Kinetic effects of lethal temperatures on population dynamics of bacterial spores. Trans ASAE 31:1594, 1988.
19. AA Teixeira, AC Rodriguez. Microbial population dynamics in bioprocess sterilization. Enzyme Microb Technol 12:469–473, 1990.
20. GH Smerage, AA Teixeira. A new view on the dynamics of heat destruction of microbial spores. J Ind Microbiol 12:211, 1993.
21. V Sapru, AA Teixeira, GH Smerage, JA Lindsay. Predicting thermophilic spore population dynamics for UHT sterilization processes. J Food Sci 57:1248, 1992.
22. RE Beams, KF Girard. The effect of pasteurization on *Listeria monocytogenes*. Can J Microbiol 4:55–61, 1958.
23. JG Bradshaw, JT Peeler, JJ Corwin, JM Hunt, JT Teiney, EP Larkin, RM Twedt. Thermal resistance of *Listeria monocytogenes* in milk. J Food Prot 48:743–745, 1985.
24. CW Donnelly, EH Briggs, LS Donnelly. Comparison of the heat resistance of *Listeria monocytogenes* in milk as determined by two methods. J Food Prot 50:14–17, 1987.
25. P Gerhard. The refractory homeostasis of bacterial spores. In: R Whittenbury, JG Banks, GW Gould, RB Board, eds. Homeostatic Mechanisms in Microorganisms. Bath, UK: Bath University Press, 1988, p. 41.
26. P Gerhardt, RE Marquis. Spore thermoresistance mechanisms. In: I Smite, RS Slepecky, P Setlow, eds. Regulation of Procaryotic Development. Washington, DC: American Society Microbiology, 1989, p. 43.
27. DE Gombas. Bacterial spore resistance to heat. Food Technol 37(11):105, 1983.
28. GW Gould. Mechanism of resistance and dormancy. In: A Hurst, GW Gould, eds. The Bacterial Spore, Vol. 2. London: Academic Press, 1983, p. 173.
29. GW Gould. Water and survival of bacterial spore. In: AC Leopold, ed. Membranes, Metabolism and Dry Organisms. Ithaca, NY: Cornell University Press, 1986, p. 143.
30. JA Lindsay, WG Murrell, AD Warth. Spore resistance and the basic mechanism of heat resistance. In: LE Harris, AJ Skopek, eds. Sterilization of Medical Products. Botany, New South Wales, Australia: Johnson & Johnson, 1985, p. 162.

31. WG Murrell. Biophysical studies on the molecular mechanisms of spore heat resistance and dormancy. In: HS Levinson, AL Soneshein, DJ Tipper, eds. Sporulation and Germination. Washington, DC: American Society Microbiology, 1981, p. 64.
32. WG Murrell. Bacterial spores—nature's ultimate survival package. In: WG Murrell, IR Kennedy, eds. Microbiology in Action. New York: John Wiley, 1988, p. 311.
33. AD Warth. Mechanisms of heat resistance. In: GJ Dring, DJ Ellar, GW Gould, eds. Fundamental and Applied Aspects of Bacterial Spores. London: Academic Press, 1985, p. 209.
34. ME Doyle, AS Mazzotta. Review of studies on the thermal resistance of salmonellae. J Food Prot 63:779–795, 2000.
35. BH Belliveau, TC Beaman, P Gerhardt. Heat resistance correlated with DNA content in *Bacillus megaterium* spores. Appl Environ Microbiol 56:2919–2921, 1990.
36. JA Lindsay, WG Murrell. Changes in density of DNA after interaction with dipicolinic acid and its possible role in spore heat resistance. Curr Microbiol 12:329–334, 1985.
37. GR Bender, RE Marquis. Spore heat resistance and specific mineralization. Appl Environ Microbiol 50:1414–1421, 1985.
38. TC Beaman, P Gerhardt. Heat resistance of bacterial spores correlated with protoplast dehydration, mineralization, and thermal adaptation. Appl Environ Microbiol 52:1242–1246, 1986.
39. A Palop, FJ Sala, S Condon. Heat resistance of native and demineralized spores of *Bacillus subtilis* sporulated at different temperatures. Appl Environ Microbiol 65:1316–1319, 1999.
40. AD Warth. Relationship between the heat resistance of spores and the optimum and maximum growth temperatures of *Bacillus* species. J Bacteriol 134:699, 1978.
41. OB Williams, WJ Robertson. Studies on heat resistance. VI. Effect of temperature of incubation at which formed on heat resistance of aerobic thermophilic spores. J Bacteriol 67:377, 1954.
42. W Heinen, AM Lauwers. Growth of bacteria at 100C and beyond. Arch Microbiol 129:127, 1981.
43. JS Garcia-Alvarado, RG Labbe, MA Rodriguez. Sporulation and enterotoxin production by *Clostridium perfringens* type A at 37 and 43°C. Appl Environ Microbiol 58:1411–1414, 1992.
44. JS Novak, VK Juneja. Heat treatment adaptations in *Clostridium perfringens* vegetative cells. J Food Protect 2000 (Accepted).
45. BM Mackey, CM Derrick. Elevation of the heat resistance of *Salmonella* Typhimurium by sublethal heat shock. J Appl Bacteriol 61:389–393, 1986.
46. BM Mackey, CM Derrick. The effect of prior heat shock on the thermoresistance of *Salmonella* Thompson in foods. Lett Appl Microbiol 5:115–118, 1987.
47. RE Marquis, GR Bender. Mineralization and heat resistance of bacterial spores. J Bacteriol 161:789–791, 1985.
48. P Setlow. Mechanisms which contribute to the long term survival of spores of *Bacillus* species. J Appl Bacteriol 76(suppl):49S–60S, 1994.
49. DJ Kihm, MT Hutton, JH Hanlin, EA Johnson. Influence of transition metals added

during sporulation on heat resistance of *Clostridium botulinum* 113B spores. Appl Environ Microbiol 56:681, 1990.
50. JM Mason, P Setlow. Essential role of small, acid-soluble spore proteins in resistance of *Bacillus subtilis* spores to UV light. J Bacteriol 167:174–178, 1986.
51. B Setlow, P Setlow. Binding of small, acid-soluble spore proteins to DNA plays a significant role in the resistance of *Bacillus subtilis* spores to hydrogen peroxide. Appl Environ Microbiol 59:3418–3423, 1993.
52. P Setlow. I will survive: protecting and repairing spore DNA. J Bacteriol 174: 2737–2741, 1992.
53. H Fairhead, B Setlow, P Setlow. Prevention of DNA damage in spores and in vitro by small, acid-soluble proteins from *Bacillus* species. J Bacteriol 175:1367–1374, 1993.
54. B Setlow, P Setlow. Heat inactivation of *Bacillus subtilis* spores lacking small, acid-soluble spore proteins is accompanied by generation of abasic sites in spore DNA. J Bacteriol 176:2111–2113, 1994.
55. LR Beuchat, RE Worthington. Relationships between heat resistance and phospholipid fatty acid composition of *Vibrio parahaemolyticus*. Appl Environ Microbiol 31:389–394, 1976.
56. EW Hansen. Correlation of fatty acid composition with thermal resistance of *Escherichia coli*. Dan Tidsskr Farm 45:339–344, 1971.
57. N Katsui, T Tsuchido, N Takano, I Shibasaki. Effect of preincubation temperature on the heat resistance of *Escherichia coli* having different fatty acid compositions. J Gen Microbiol 122:357–361, 1981.
58. VK Juneja, TA Foglia, BS Marmer. Heat resistance and fatty acid composition of *Listeria monocytogenes*: effect of pH, acidulant, and growth temperature. J Food Prot 61:683–687, 1998.
59. EW Hansen, K Skadhange. The influence of growth temperature on the thermal resistance of *Escherichia coli*. Dan Tidsskr Farm 45:24–28, 1971.
60. RR Ernest. Sterilization by heat. In: SS Block, ed. Disinfection, Sterilization, and Preservation. Philadelphia: Lea & Febiger, 1977, p. 481.
61. T Hashimoto, WR Frieben, SF Conti. Kinetics of germination of heat-injured *Bacillus cereus* spores. In: HO Halvorson, R Hanson, LL Campbell, eds. Spores V. Washington, DC: American Society Microbiology, 1972, p. 409.
62. WA Moats. Kinetics of thermal death of bacteria. J Bacteriol 105:165–171, 1971.
63. RE Marquis, J Sim, SY Shin. Molecular mechanisms of resistance to heat and oxidative damage. J Appl Bacteriol Symp Suppl 76:40S–48S, 1994.
64. JH Hanlin, SJ Lombardi, RA Slepecky. Heat and UV light resistance of vegetative cells and spores of *Bacillus subtilis* Rec mutants. J Bacteriol 163:774–777, 1985.
65. BH Belliveau, TC Beaman, S Pankratz, P Gerhardt. Heat killing of bacterial spores analyzed by differential scanning calorimetry. J Bacteriol 174:4463–4474, 1992.
66. P Teixeira, H Castro, C Mohacsi-Farkas, R Kirby. Identification of sites of injury in *Lactobacillus bulgaricus* during heat stress. J Appl Microbiol 83:219–226, 1997.
67. P Setlow. Mechanisms which contribute to the long-term survival of spores of *Bacillus subtilis*. J Appl Bacteriol Symp Suppl 76:495–605, 1994.
68. AC Lee, JM Goepfert. Influence of selected solutes on thermally induced death and injury of *Salmonella* Typhimurium. J Milk Food Technol 38:175–200, 1975.

69. DR McCoy, ZJ Ordal. Thermal stress of *Pseudomonas fluorescens* in complex media. Appl Environ Microbiol 37:443–448, 1979.
70. WA Anderson, ND Hedges, MV Jones, MB Cole. Thermal inactivation of *Listeria monocytogenes* studied by differential scanning calorimetry. J Gen Microbiol 137: 1419–1424, 1991.
71. JM Kijowski, MG Mast. Effect of sodium chloride and phosphates on the thermal properties of chicken meat proteins. J Food Sci 53:367–370, 1988.
72. BM Mackey, CA Miles, SE Parsons, DA Seymour. Thermal denaturation of whole cells and cell components of *Escherichia coli* examined by differential scanning calorimetry. J Gen Microbiol 137:2361–2374, 1991.
73. MC Allwood, AD Russell. Mechanism of thermal injury in *Staphylococcus aureus*. Appl Microbiol 15:1266–1269, 1967.
74. HB Hoa, GE Begard, P Beaudry, P Maurel, M Grunberg-Manago, P Douzou. Analysis of cosolvent and divalent cation effects on association equilibrium and activity of ribosomes. Biochemistry 19:3080–3087, 1980.
75. A Hurst. Reversible heat damage. In: A Hurst, A Nasim, eds. Repairable Lesions in Microorganisms. London: Academic Press, 1984, pp. 303–318.
76. A Hurst, A Hughes. Stability of ribosomes of *Staphylococcus aureus* S6 sublethally heated in different buffers. J Bacteriol 133:564–568, 1978.
77. A Hurst, A Hughes. Repair of salt tolerance and recovery of lost D-alanine and magnesium following sublethal heating of staph aureus are independent events. Can J Microbiol 27:627–632, 1981.
78. LJ Rosenthal, JJ Iandolo. Thermally induced intracellular alteration of ribosomal ribonucleic acid. J Bacteriol 103:833–835, 1970.
79. LL Miller, ZL Ordal. Thermal injury and recovery of *Bacillus subtilis*. Appl Microbiol 24:878–884, 1972.
80. PJ Stephens, MV Jones. Reduced ribosomal thermal danaturation in *Listeria monocytogenes* following osmotic and heat shocks. FEMS Microbiol Lett 106:177–182, 1993.
81. T Tolker-Nielsen, S Molin. Role of ribosomal degradation in the death of heat stressed *Salmonella* Typhimurium. FEMS Microbiol Lett 142:155–160, 1996.
82. RG Earnshaw, J Appleyard, RM Hurst. Understanding physical inactivation processes: combined preservation opportunities using heat, ultrasound and pressure. Int J Food Microbiol 28:197–219, 1995.
83. EW Kellogg, I Fridovich. Superoxide, hydrogen peroxide, and singlet oxygen in lipid peroxidation by xanthine oxidase system. J Biol Chem 250:8812–8816, 1975.
84. V Amin, N Olson. Influence of catalase activity on resistance of a coagulase-positive staphylococci to hydrogen peroxide. Appl Microbiol 16:267–270, 1968.
85. AW Dallmier, SE Martin. Catalase and superoxide dismutase activities after heat injury of *Listeria monocytogenes*. Appl Environ Microbiol 54:581–582, 1988.
86. AD Warth. Heat stability of *Bacillus cereus* enzymes within spores and in extracts. J Bacteriol 143:27–34, 1980.
87. PJ Coote, MB Cole, MV Jones. Induction of increased thermotolerance in *Saccharomyces cerevisiae* may be triggered by a mechanism involving intracellular pH. J Gen Microbiol 137:1701–1708, 1991.

88. RS Flowers, DM Adams. Spore membrane(s) as the site of damage within heated *Clostridium perfringens* spores. J Bacteriol 125:429, 1976.
89. MR Brown, WJ Melling. Release of dipicolinic acid and calcium and activation of *Bacillus stearothermophilus* as a function of time, temperature and pH. J Pharm Pharmocol 25:478, 1973.
90. JW Hunnell, ZJ Ordal. Cytological and chemical changes in heat killed and germinated bacterial spores. In: HO Halvorson, ed. Spore II. Minneapolis, MN: Burgess Pub. Co., 1961, p. 101.
91. LJ Rode, JW Foster. Induced release of dipicolinic acid from spores of *Bacillus megaterium*. J Bacteriol 79:650, 1960.
92. MJ Schlesinger. Heat shock proteins: the search for functions. J Cell Biol 103:321–325, 1986.
93. S Lindquist. The heat-shock response. Ann Rev Biochem 55:1151–1191, 1986.
94. JM Farber, BE Brown. Effect of prior heat shock on heat resistance of *Listeria monocytogenes* in meat. Appl Environ Microbiol 56:1584–1587, 1990.
95. T Tsuchido, M Takano, I Skibasaki. Effect of temperature elevating process on the subsequent isothermal death of *E. coli* K12. J Ferment Technol 52(10):788–792, 1984.
96. BM Mackey, CM Derrick. Heat shock synthesis and thermotolerance of *Salmonella* Typhimurium. J Appl Bacteriol 69:373–383, 1990.
97. TJ Humphrey, NP Richardson, KM Statton, RJ Rowbury. Effects of temperature shift on acid and heat tolerance in *Salmonella* Enteritidis phage type 4. Appl Environ Microbiol 59(9):3120–3122, 1993.
98. WM Fedio, H Jackson. Effect of tempering on the heat resistance of *L. monocytogenes*. Lett Appl Microbiol 9:157–160, 1989.
99. RH Linton, MD Pierson, JR Bishop. Increase in heat resistance of *Listeria monocytogenes* Scott A by sublethal heat shock. J Food Prot 53:924–927, 1990.
100. EA Murano, MD Pierson. Effect of heat shock and growth atmosphere on the heat resistance of *Escherichia coli* 0157:H7. J Food Prot 55:171–175, 1992.
101. F Jorgensen, B Panaretou, PJ Stephens, S Knochel. Effect of pre- and post-heat shock temperature on the persistence of thermotolerance and heat shock-induced proteins in *Listeria monocytogenes*. J Appl Bacteriol 80:216–224, 1996.
102. FX Etoa, L Michiels. Heat induced resistance of *Bacillus stearothermophilus* spores. Lett Appl Microbiol 6:43–45, 1988.
103. TC Beaman, HS Pankratz, P Gerhardt. Heat shock affects permeability and resistance of *Bacillus stearothermophilus* spores. Appl Environ Microbiol 54(10):515–520, 1988.
104. J Appleyard, J Gaze. The effect of exposure to sublethal temperatures on the final heat resistance of *Listeria monocytogenes* and *Clostridium botulinum*. Technical Memorandum No. 683, CCFRA, Chipping Campden, Glos. UK, 1993.
105. S Alcock. Elevation of heat resistance of *Clostridium sporogenes* following heat shock. 1:39–47, 1994.
106. RE Susek, S Lindquist. HSP26 of *Saccharomyces cerevisiae* is related to the superfamily of small heat shock proteins but is without demonstrable function. Mol Cell Biol 9:5265–5271, 1989.
107. CA Barnes, GC Johnston, RA Singer. Thermotolerance is independent of induction

of the full spectrum of heat shock proteins and of cell cycle blockage in the yeast *Saccharomyces cerevisiae*. J Bacteriol 174:4352–4358, 1990.
108. BJ Smith, MP Yaffe. Uncoupling thermotolerance from the induction of heat shock proteins. Proc Natl Acad Sci USA 88:11091–11094, 1991.
109. MJ Schlesinger. Heat shock proteins. J Biol Chem 265:12111–12114, 1990.
110. Y Sanchez, SL Lindquist. HSPIO4 required for induced thermotolerance. Science 248:1112–1115, 1990.
111. C Georgopoulos, WJ Welch. Role of the major heat shock proteins as molecular chaperones. Annu Rev Cell Biol 9:601–634, 1993.
112. JG Morris. Bacterial shock response. Endeavor 17(1):2–6, 1993.
113. R Hengge-Aronis. Survival of hunger and stress: the role of *rpoS* in early stationary phase gene regulation in *E. coli*. Cell 72:165–168, 1993.
114. T Yura, T Tobe, K Ito, T Osawa. Heat shock regulatory gene (*htpR*) of *Escherichia coli* is required for growth at high temperature but is dispensable at low temperature. Proc Natl Acad Sci USA 81:6803–6807, 1984.
115. O Reichart. Modeling the destruction of *Escherichia coli* on the base of reaction kinetics. Int J Food Microbiol 23:449–465, 1994.
116. Y Teo, TJ Raynor, KR Ellajosyula, SJ Knabel. Synergistic effect of high temperature and high pH on the destruction of *Salmonella* Enteritidis and *Escherichia coli* O157:H7. J Food Prot 59:1023–1030, 1996.
117. NM Ahmed, DE Conner. Evaluation of various media for recovery of thermally injured *Escherichia coli* O157:H7. J Food Prot 58:357–360, 1995.
118. VK Juneja, BS Eblen. Heat inactivation of *Salmonella* Typhimurium DT104 in beef as affected by fat content. Lett Appl Microbiol 30:461–467, 2000.
119. MP Doyle, JL Schoeni. Survival and growth characteristics of *Escherichia coli* associated with hemorrhagic colitis. Appl Environ Microbiol 48:855–856, 1984.
120. MN Ahmed, DE Conner, DL Huffman. Heat-resistance of *Escherichia coli* O157:H7 in meat and poultry as affected by product composition. J Food Sci 60: 606–610, 1995.
121. MN Ahmed, DE Conner. Heat inactivation of *Escherichia coli* O157:H7 in turkey meat as affected by sodium chloride, sodium lactate, polyphosphate, and fat content. J Food Prot 60:898–902, 1997.
122. PK Ben Emarek, HH Huss. Heat resistance of *Listeria monocytogenes* in vacuum packaged pasteurized fish fillets. Int J Food Microbiol 20:85–95, 1993.
123. CW Donnelly, EH Briggs. Psychrotrophic growth and thermal inactivation of *Listeria monocytogenes* as a function of milk composition. J Food Prot 49:994–998, 1986.
124. F MacDonald, AD Sutherland. Effect of heat treatment on *Listeria monocytogenes* and gram negative bacteria in sheep, cow and goat milks. Lett Appl Microbiol 9: 89–94, 1993.
125. JA Troller. The water relations of foodborne bacterial pathogens: a review. J Milk Food Technol 36:276–288, 1973.
126. O Reichart, C Mohacsi-Farkas. Mathematical modelling of the combined effect of water activity, pH and redox potential on the heat destruction. Int J Food Microbiol 24:103–112, 1994.
127. JM Goepfert, IK Iskandar, CH Amundson. Relation of the dry heat resistance of

salmonellae to the water activity of the environment. Appl Microbiol 19:429–433, 1970.

128. EU Tuncan, SE Martin. Combined effects of salts and temperature on the thermal destruction of *Staphylococcus aureus* MF-31. J Food Sci 55:833–836, 1990.
129. JL Smith, RC Benedict, SA Palumbo. Protection against heat-injury in *Staphylococcus aureus* by solutes. J Food Prot 45:54–58, 1982.
130. PG Bean, TA Roberts. Effect of sodium chloride and sodium nitrite on the heat resistance of *Staphylococcus aureus* NCTC 10652 in buffer and meat macerate. J Food Technol 10:327–332, 1975.
131. SS Sumner, TM Sandros, MC Harmon, VN Scott, DT Bernard. Heat resistance of *Salmonella* Typhimurium and *Listeria monocytogenes* in sucrose solutions of various water activities. J Food Sci 56(6):1741–1743, 1991.
132. GJ Banwart, JC Ayres. The effect of high temperature storage on the content of *Salmonella* and on the functional properties of dried egg white. Food Technol 10(2): 68–73, 1956.
133. JE Van Cauwenberge, RJ Bothast, WF Kwolek. Thermal inactivation of eight *Salmonella* serotypes on dry corn flour. Appl Environ Microbiol 42:688–691, 1981.
134. J Archer, ET Jervis, J Bird, JY Gaze. Heat resistance of *Salmonella* Weltevreden in low moisture environments. J Food Prot 61:969–973, 1998.
135. N Katsui, T Tsuchido, N Takano, I Shibasaki. Viability of heat stressed cells of microorganisms as influenced by pre-incubation and post-incubation temperatures. J Appl Bacteriol 53:103–108, 1982.
136. SM Czechowicz, O Santos, EA Zottola. Recovery of thermally-stressed *Escherichia coli* O157:H7 by media supplemented by pyruvate. Int J Food Microbiol 33:275–284, 1996.
137. EA Murano, MD Pierson. Effect of heat shock and incubation atmosphere on injury and recovery of *Escherichia coli* O157:H7. J Food Prot 56:568–572, 1993.
138. R Bromberg, SM George, MW Peck. Oxygen sensitivity of heated cells of *Escherichia coli* O157:H7. J Appl Microbiol 85:231–237, 1998.
139. SM George, LCC Richardson, IE Pol, MW Peck. Effect of oxygen concentration and redox potential on recovery of sublethally heat-damaged cells of *Escherichia coli* O157:H7, *Salmonella* Enteritidis and *Listeria monocytogenes*. J Appl Microbiol 84:903–909, 1998.
140. SV Busch, CW Donnelly. Development of a repair-enrichment broth for resuscitation of heat-injured *Listeria monocytogenes* and *Listeria innocua*. Appl Environ Microbiol 58:14–20, 1992.
141. SJ Knabel, HW Walker, PA Hartman, AF Mendonca. Effects of growth temperature and strictly anaerobic recovery on the survival of *Listeria monocytogenes* during pasteurization. Appl Environ Microbiol 56:370–376, 1990.
142. RH Linton, JB Webster, MD Pierson, JR Bishop, CR Hackney. The effect of sublethal heat shock and growth atmosphere on the heat resistance of *Listeria monocytogenes* Scott A. J Food Prot 55:84–87, 1992.
143. AF Mendonca, SJ Knabel. A novel strictly anaerobic recovery and enrichment system incorporating lithium for detection of heat-injured *Listeria monocytogenes* in pasteurized milk containing background microflora. Appl Environ Microbiol 60: 4001–4008, 1994.

144. SJ Knabel, SA Thielen. Enhanced recovery of severely heat-injured, thermo-tolerant *Listeria monocytogenes* from USDA and FDA primary enrichment media using a novel, simple, strictly anaerobic method. J Food Prot 58:29–34, 1995.
145. JR Patel, CA Hwang, LR Beuchat, MP Doyle, RE Brackett. Comparison of oxygen scavengers for their ability to enhance resuscitation of heat-injured *Listeria monocytogenes*. J Food Prot 58:244–250, 1995.
146. L Yu, DYC Fung. Effect of oxyrase enzyme on *Listeria monocytogenes* and other facultative anaerobes. J Food Safety 11:163–175, 1991.
147. IJ Xavier, S Ingham. Increased D-values for *Salmonella* Enteritidis resulting from the use of anaerobic enumeration methods. Food Microbiol 10:223–228, 1993.
148. TO Ugborogho, SC Ingham. Increased D-values of *Staphylococcus aureus* resulting from anaerobic heating and enumeration of survivors. Food Microbiol 11:275–280, 1994.
149. IPD Huang, AE Yousef, EH Marth, ME Matthews. Thermal inactivation of *Listeria monocytogenes* in chicken gravy. J Food Prot 55:492–496, 1980.
150. K Shamsuzzaman, N Chauqui-Offermans, L Lucht, T McDougal, J Borsa. Microbiological and other characteristics of chicken breast meat following electron-beam and *sous-vide* treatment. J Food Protect 55:528–533, 1992.
151. SM George, MW Peck. Redox potential affects the measured heat resistance of *Escherichia coli* O157:H7 independently of oxygen concentration. Lett Appl Microbiol 27:313–317, 1998.
152. IS Boziaris, L Humpheson, MR Adams. Effect of nisin on heat injury and inactivation of *Salmonella* Enteritidis PT4. Int J Food Microbiol 43:7–13, 1998.
153. E Budu-Amoako, RF Ablett, J Harris, J Delves-Broughton. Combined effect of nisin and moderate heat on destruction of *Listeria monocytogenes* in cold-pack lobster meat. J Food Prot 62(1):46–50, 1999.
154. AW Smith. Stationary phase induction in *Escherichia coli* a new target for antimicrobial therapy. J Antimicrobial Chemother 35:359–361, 1995.
155. GJ Leyer, EA Johnson. Acid adaptation induces cross-protection against environmental stresses in *Salmonella* Typhimurium. Appl Environ Microbiol 59:1842–1847, 1993.
156. F Jorgensen, PJ Stephens, S Knochel. The effect of osmotic shock and subsequent adaptation on the thermotolerance and cell morphology of *Listeria monocytogenes*. J Appl Bacteriol 79:274–281, 1995.
157. Y Lou, AE Yousef. Resistance of *Listeria monocytogenes* to heat after adaptation to environmental stresses. J Food Prot 59:465–471, 1996.
158. JE Gaze, GD Brown. Determination of the heat resistance of a strain of non-proteolytic *Clostridium botulinum* type B and a strain of type E, heated in cod and carrot homogenate over a temperature range 70–90C. Technical Memorandum No. 592. Campden Food and Drink Research Association, Chipping Campden, Glos., UK, 1990.
159. VK Juneja, BS Eblen, BS Marmer, AC Williams, SA Palumbo, AJ Miller. Thermal resistance of non-proteolytic type B and type E *Clostridium botulinum* spores in phosphate buffer and turkey slurry. J Food Prot 58:758–763, 1995.
160. SE Craven. The effect of the pH of the sporulation environment on the heat resistance of *Clostridium perfringens* spores. Curr Microbiol 22:233–237, 1990.

161. M Mazas, M Lopez, I Gonzalez, A Bernardo, R Martin. Effects of sporulation pH on the heat resistance and sporulation of *Bacillus cereus*. Lett Appl Microbiol 25: 331–334, 1997.
162. F Sala, P Ibarz, A Palop, J Raso, S Condon. Sporulation temperature and heat resistance of *Bacillus subtilis* at different pH values. J Food Prot 58:239–243, 1995.
163. LA De Pieri, IK Ludlow. Relationship between *Bacillus sphaericus* spore heat resistance and sporulation temperature. Lett Appl Microbiol 14:121–124, 1992.
164. S Feig, AK Stersky. Characterization of a heat resistant strain of *Bacillus coagulans* isolated from cream style canned corn. J Food Sci 46:135–137, 1981.
165. JA Lindsay, LE Barton, AS Leinart, HS Pankratz. The effect of sporulation temperature on sporal characteristics of *Bacillus subtilis* A. Curr Microbiol 21:75–79, 1990.
166. TE Odlaug, IJ Pflug. Thermal destruction of *Clostridium botulinum* spores suspended in tomato juice in aluminum thermal death time tubes. Appl Environ Microbiol 34:23, 1977.
167. H Xezones, IJ Hutchings. Thermal resistance of *Clostridium botulinum* (62A) spores as affected by fundamental food constituents. Food Technol 19:1003, 1965.
168. WG Murrell, WJ Scott. The heat resistance of bacterial spores at various water activities. J Gen Microbiol 43:411, 1966.
169. AL Reyes, RG Crawford, AJ Wehby, JT Peeler, JC Wimsatt, JE Campbell, RM Twedt. Heat resistance of *Bacillus* spores at various relative humidities. Appl Environ Microbiol 42:692–697, 1981.
170. J Pfeiffer, HG Kessler. Effect of relative humidity of hot air on the heat resistance of *Bacillus* cereus spores. J Appl Bacteriol 77:121–128, 1994.
171. M Mazas, S Martinez, M Lopez, AB Alvarez, R Martin. Thermal inactivation of *Bacillus cereus* spores affected by the solutes used to control water activity of the heating medium. Int J Food Microbiol 53:61–67, 1999.
172. L Ababouch, FF Busta. Effect of thermal treatments in oils on bacterial spore survival. J Appl Bacteriol 62:491–502, 1987.
173. L Ababouch, A Dikra, FF Busta. Tailing of survivor curves of clostridial spores heated in edible oils. J Appl Bacteriol 62:503–511, 1987.
174. N Molin, BG Snygg. Effect of lipid materials on heat resistance of bacterial spores. Appl Microbiol 15:1422–1426, 1967.
175. LH Ababouch, L Grimit, R Eddafry, FF Busta. Thermal inactivation kinetics of *Bacillus subtilis* spores suspended in buffer and in oils. J Appl Bacteriol 78:669–676, 1995.
176. LR Wandling, BW Sheldon, PM Foegeding. Nisin in milk sensitizes *Bacillus* spores to heat and prevents recovery of survivors. J Food Prot 62:492–498, 1999.
177. TA Roberts. Recovering spores damaged by heat, ionizing radiations or ethylene oxide. J Appl Bacteriol 33:74–94, 1970.
178. DM Adams. Heat injury of bacterial spores. Adv Appl Microbiol 23:245–261, 1979.
179. PM Foegeding, FF Busta. Bacterial spores injury—an update. J Food Prot 45:776–786, 1981.
180. AD Russell. The Destruction of Bacterial Spores. London: Academic Press, 1982.

181. FE Feeherry, DT Munsey, DB Rowley. Thermal inactivation and injury of *Bacillus stearothermophilus* spores. Appl Environ Microbiol 43:365–370, 1987.
182. PS Fernandez, FJ Gomez, MJ Ocio, MJ Sanchez, T Rodrigo, MA Martinez. Influence of acidification and type of acidulent on the recovery medium on *Bacillus stearothermophilus* spore counts. Lett Appl Microbiol 19:146–148, 1994.
183. M Lopez, S Martinez, J Gonzalez, R Martin, A Bernardo. Sensitization of thermally injured spores of *Bacillus stearothermophilus* to sodium benzoate and potassium sorbate. Lett Appl Microbiol 27:331–335, 1998.
184. S Condon, A Palop, J Raso, FJ Sala. Influence of the temperature after heat treatment upon the estimated heat resistance values of spores of *Bacillus subtilis*. Lett Appl Microbiol 22:149–152, 1996.
185. G Alderton, JK Chen, KA Ito. Effect of lysozyme on the recovery of heated *Clostridium botulinum* spores. Appl Microbiol 27:613–615, 1974.
186. JG Bradshaw, JT Peeler, RM Twedt. Thermal inactivation of ileal loop-reactive *Clostridium perfringens* type A strains in phosphate buffer and beef gravy. Appl Environ Microbiol 34:280–284, 1977.
187. MW Peck, DA Fairbairn, BM Lund. The effect of recovery medium on the estimated heat resistance of spores of non-proteolytic *Clostridium botulinum*. Lett Appl Microbiol 15:146–151, 1992.
188. VN Scott, DT Bernard. The effect of lysozyme on the apparent heat resistance of non-proteolytic type B *Clostridium botulinum*. J Food Safety 7:145–154, 1985.
189. M Sebald, H Ionesco, AR Prevot. Germination IzP-dependante des spores de *Clostridium botulinum* type E. E.C.R. Acad Sci Paris (Serie D) 275:2175–2182, 1972.
190. MW Peck, DA Fairbairn, BM Lund. Heat resistance of spores of non-proteolytic *Clostridium botulinum* estimated on medium containing lysozyme. Lett Appl Microbiol 16:126–131, 1993.
191. CL Duncan, RG Labbe, RR Reich. Germination of heat- and alkali-altered spores of *Clostridium perfringens* type A by lysozyme and an initiation protein. J Bacteriol 109:550–559, 1972.
192. VK Juneja, BS Eblen. Influence of sodium chloride on thermal inactivation and recovery of non-proteolytic *Clostridium botulinum* type B spores. J Food Prot 58: 813–816, 1995.
193. MT Hutton, PA Dhehak, JH Hanlin. Inhibition of botulinum toxin production by *Pedicoccus acidilacti* in temperature abused refrigerated foods. J Food Safety 11: 255–267, 1991.
194. VK Juneja, BS Marmer, JG Phillips, AJ Miller. Influence of the intrinsic properties of food on thermal inactivation of spores of non-proteolytic *Clostridium botulinum*: development of a predictive model. J Food Safety 15:349–364, 1995.
195. VK Juneja, BS Marmer, BS Eblen. Predictive model for the combined effect of temperature, PH, sodium chloride, and sodium pyrophosphate on the heat resistance of *Escherichia coli* O157:H7. J Food Safety 19:147–160, 1999.
196. VK Juneja, BS Eblen. Predictive thermal inactivation model for *Listeria monocytogenes* with temperature, pH, NaCl and sodium pyrophosphate as controlling factors. J Food Prot 62:986–993, 1999.
197. PM Periago, PS Fernandez, MC Salmeron, A Martinez. Predictive model to de-

scribe the combined effect of pH and NaCl on apparent heat resistance of *Bacillus stearothermophilus*. Int J Food Microbiol 44:21–30, 1998.

198. PT Stephens, MB Cole, MV Jones. Effect of heating rate on the thermal inactivation of *Listeria monocytogenes*. J Appl Bacteriol 77:702–708, 1994.
199. CL Little, MR Adams, WA Anderson, MB Cole. Application of a log-logistic model to describe the survival of *Yersinia enterocolitica* at sub-optimal pH and temperature. Int J Food Microbiol 22:63–71.
200. WA Anderson, PJ McClure, AC Baird-Parker, MB Cole. The application of a log-logistic model to describe the thermal inactivation of *Clostridium botulinum* 213B at temperatures below 121.1C. J Appl Bacteriol 80:282–290, 1996.
201. A Ellison, WA Anderson, MB Cole, GSB Stewart. Modeling the thermal inactivation of *Salmonella* Typhimurium using bioluminescencedata. J Food Microbiol 23: 467–477, 1994.
202. RH Linton, WH Carter, MD Pierson, CR Hackney, JD Eifert. Use of a modified Gompertz equation to predict the effects of temperature, pH, and NaCl on the inactivation of *Listeria monocytogenes* Scott A heated in infant formula. J Food Prot 59:16–23, 1996.
203. CW Blackburn, LM Curtis, L Humpheson, C Billon, PJ Mcclure. Development of thermal inactivation models for *Salmonella* Enteritidis and *Escherichia coli* O157:H7 with temperature, pH and NaCl as controlling factors. Int J Food Microbiol 38:31–44, 1997.
204. AT Chhabra, WH Carter, RH Linton, MA Cousin. A predictive model to determine the effects of pH, milkfat, and temperature on thermal inactivation of *Listeria monocytogenes*. J Food Prot 62(10):1143–1149, 1999.
205. O Mafart, I Leguerinel. Modeling the heat stress and the recovery of bacterial spores. Int J Food Microbiol 37:131–135, 1997.
206. DM Wrigley, NG Llorca. Decrease of *Salmonella* Typhimurium in skim milk and egg by heat and ultrasonic wave treatment. J Food Prot 55:678–680, 1992.
207. RM Hurst, GD Betts, RG Earnshaw. The Antimicrobial Effect of Power Ultrasound. R&D Report No. 4, Chipping Campden, Glos., UK, 1995.
208. I Kiss, GY Zachariev, J Farkas. The use of irradiation in food processing. In: Food Preservation by Irradiation, Vol 1. Proceedings of a symposium on food preservation by irradiation, Vienna, 1977. Vienna: IAEA, 1978, pp. 263–272.
209. DW Thayer, S Songprasertchai, G Boyd. Effect of heat and ionizing radiation on *Salmonella* Typhimurium in mechanically deboned chicken meat. J Food Prot 54: 718–724, 1991.
210. IR Grant, MF Patterson. Combined effect of gamma radiation and heating on the destruction of *Listeria monocytogenes* and *Salmonella* Typhimurium in cook-chill roast beef and gravy. Int J Food Microbiol 27:117–128, 1995.
211. RC Whiting. Modeling bacterial survival in unfavorable environments. J Indust Microbiol 12:240–246, 1993.
212. DN Kamau, S Doores, KM Pruitt. Enhanced thermal destruction of *Listeria monocytogenes* and *Staphylococcus aureus* by the lactoperoxidase system. Appl Environ Microbiol 56:2711–2716, 1990.
213. AC Rodriguez, GH Smerage. System analysis of the dynamics of bacterial spore populations during lethal heat treatment. Trans ASAE 39:595–603, 1996.

214. VK Juneja, OP Snyder, Jr., BS Marmer. Thermal destruction of *Escherichia coli* O157:H7 in beef and chicken: determination of D- and z-values. Int J Food Microbiol 35:231–237, 1997.
215. VK Juneja, BS Marmer. Lethality of heat to *Escherichia coli* O157:H7: D- and z-values determinations in turkey, lamb and pork. Food Res Int 32:23–28, 1999.
216. JE Gaze, GD Brown, DE Gaskell, JG Banks. Heat resistance of *Listeria monocytogenes* in homogenates of chicken, beef steaks and carrots. Food Microbiol 6:251–259, 1989.
217. SJ Goodfellow, WL Brown. Fate of *Salmonella* inoculated into beef for cooking. J Food Prot 41:598–605, 1978.
218. JE Line, AR Fain, AB Mogan, LM Martin, RV Lechowich, JM Carosella, WL Brown. Lethality of heat to *Escherichia coli* O157:H7: D-value and z-value determination in ground beef. J Food Prot 54:762–766, 1991.
219. AR Fain, Jr., JE Line, AB Moran, LM Martin, RV Lechowich, JM Carosella, WL Brown. Lethality of heat to *Listeria monocytogenes* scott A: D-value and Z-value determinations in ground beef and turkey. J Food Prot 54(10):756–761, 1991.
220. VN Scott, DT Bernard. Heat resistance of spores of non-proteolytic type B *Clostridium botulinum*. J Food Prot 45:909–912, 1982.
221. MW Bucknavage, MD Pierson, CR Hackney, JR Bishop. Thermal inactivation of *Clostridium botulinum* type E spores in oyster homogenates at minimal processing temperatures. J Food Sci 55:372–373, 1990.
222. EJ Rhodehamel, HM Solomon, T Lilly, Jr., DA Kautter, JT Peeler. Incidence and heat resistance of *Clostridium botulinum* type E spores in menhaden surimi. J Food Sci 56:1562–1563, 1592, 1991.
223. H Fugikawa, T Itoh. Thermal inactivation analysis of mesophiles using the arrhenius and z-value models. J Food Prot 61(7):910–912, 1998.
224. KR Davey. Linear-arrhenius models for bacterial growth and death and vitamin denaturations. J Ind Microbiol 12:172–179, 1993.

3
Microbial Control with Cold Temperatures

Colin O. Gill
Agriculture and Agri-Food Canada
Lacombe, Alberta, Canada

I. INTRODUCTION

Any microorganism will grow over a limited range of temperatures, usually no greater than 40°C. Microorganism are usually assigned to one of five groups according to the optimum and maximum temperatures for growth (Table 1). Of those groups, hypothermophiles are bacteria associated with geothermally heated environments in which they grow at temperatures around or above the normal boiling point of water (1), while psychrophilic bacteria and yeasts are rare outside natural environments where the temperatures are consistently close to 0°C, such as the southern oceans or permafrost regions on land (2). Thus, the microorganisms associated with foods are classified as thermophiles, mesophiles, or psychrotrophs.

Because those organisms grow over different temperatures ranges, the term ''cold'' can be used with respect to their behavior only in a relative sense, with reference to the minimum temperature for growth of each organism. Although a defining minimum temperature is often given for each group—40, 5, and <0°C for thermophiles, mesophiles, and psychrotrophs, respectively (3)—the minimum temperatures for growth of individual organisms within each group vary widely (Table 2). The range of minimum temperatures for organisms regarded as mesophiles is particularly large, with the inclusion of some organisms that grow over only narrow temperature ranges, such as the pathogen *Campylobacter jejuni*, which grows between 32 and 45°C (4), and others that grow over wide tempera-

Table 1 Grouping of Microorganisms According to Their Optimum and Maximum Temperatures for Growth

	Temperature (°C)	
Group	Optimum	Maximum
Hyperthermophiles	>80	>90
Thermophiles	>50	<90
Mesophiles	>30	<50
Psychrotrophs	<30	<35
Psychrophiles	<15	<20

ture ranges, such as the pathogen *Yersinia enterocolitica*, which grows between >40 and <0°C (5).

Moreover, the minimum temperature for growth of a species under particular sets of conditions may be difficult to define. Variations between strains in the minimum temperatures for growth are commonly reported: for example, minimum growth temperatures for strains of *Salmonella* in laboratory media of from 4 to >7°C (6). Whatever the minimum temperature for growth under conditions that are otherwise optimal, the minimum temperature at which growth occurs will tend to increase as other factors, such as nutrient availability, water activity

Table 2 Minimum Temperatures for Growth of Some Food-Associated Microorganisms

Organism	Importance for food	Group	Minimum temperature (°C)
Bacillus stearothermophilus	Spoilage	Thermophile	40
Campylobacter jejuni	Pathogen	Mesophile	32
Clostridium thermosaccharolyticum	Spoilage	Thermophile	30
Streptococcus thermophilus	Fermentation	Thermophile	19
Clostridium perfringens	Pathogen	Mesophile	15
Clostridium botulinum type A	Pathogen	Mesophile	10
Escherichia coli	Indicator	Mesophile	7
Aeromonas hydrophila	Pathogen	Mesophile	2
Listeria monocytogenes	Pathogen	Mesophile	0
Pseudomonas fragii	Spoilage	Psychrotroph	−3
Cryptococcus laurentii	Spoilage	Psychrotroph	−5
Thamnidium elegans	Spoilage	Psychrotroph	−7

Table 3 Minimum Temperatures (Tmin) for Growth of *Escherichia coli* K12 in Minimal or Rich Media With or Without the Addition of Glycerol or NaCl

Medium	Additive	Tmin (°C)
Minimal	None	5
	Glycerol (20%)	10
	NaCl (2%)	13
Rich	None	5
	Glycerol (20%)	6
	NaCl (5%)	6

Source: Ref. 7

(a_w), salt concentration, etc., deviate from the optimum (Table 3). Thus, minimum temperatures for growth in foods can be higher or lower than those observed for the same organism in a laboratory medium (8,9). There is also the possibility that mistakenly low minimum temperatures for persisting growth are sometimes reported because with broth cultures at temperatures a degree or two below the minimum, increases in optical densities at progressively declining rates may continue for some time after the medium has attained such temperatures (Fig. 1).

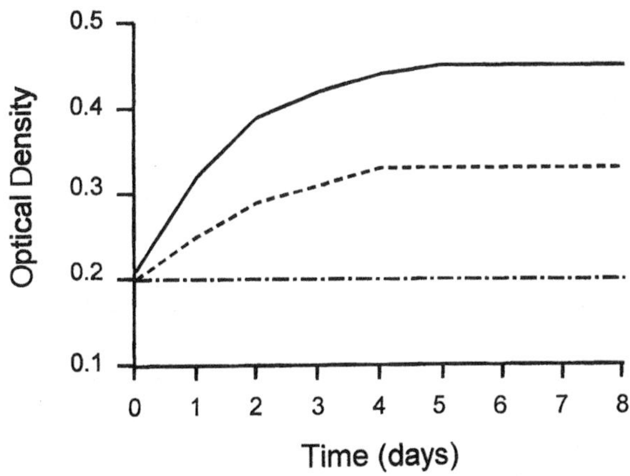

Figure 1 Optical densities of log phase cultures of wild-type strain of *Escherichia coli*, minimum growth temperature 7°C, after being moved from 12°C to temperatures of 5 (—), 3 (--), or 2 (·-·-·)°C.

Establishment of the minimum temperature for growth is further complicated when the growth temperature range of an organism extends below 0°C, because of changes in a_w and solute concentrations that occur as consequences of freezing. The minimum temperatures for growth of psychrotrophic organisms should ideally be determined in supercooled media. However, that has rarely been done; so the minimum temperatures reported for psychrotrophs often reflect the xerotolerances of the organisms as well as their tolerances of freezing temperatures per se. For example, the mold *Thamnidium elegans* will not grow on frozen meat or on unfrozen medium of equivalent a_w at temperatures below −5°C; but growth at −7°C on a supercooled medium has been observed, with indication that the minimum temperature for growth in the absence of any other stress may be as low as −10°C (10).

Although the minimum temperature for growth of any organism under particular conditions may be more or less uncertain, cold temperatures must be defined, and subdivided if necessary, by reference to minimum temperatures for growth if their effects on microorganisms are to be discussed. Therefore, for that purpose, cold temperatures can be considered as encompassing cool, low, and freezing temperatures when those temperatures are defined, respectively, as (a) temperatures that approach the minimum for growth of a specified organism on a particular, unfrozen substrate, (b) temperatures below the minimum for growth of the organism on a particular, unfrozen substrate, and (c) temperatures below that at which freezing of a substrate commences irrespective of whether or not an organism can grow on the substrate below that temperature.

II. PHYSIOLOGICAL RESPONSES OF MICROORGANISMS TO ENVIRONMENTAL STRESSES

It has long been known that microorganisms can adapt to adverse environmental conditions by inducing enzymes and/or transport systems that act to counter particular stresses, for example, the induction in *E. coli* of amino acid deaminases or decarboxylases in response to high or low pH, respectively (11), or the accumulation of polyols by yeasts and fungi in response to osmotic stress (12). In recent years it has become apparent that environmental stresses each induce specific suites of proteins in adapting microorganisms, although the functions of most stress-induced proteins are poorly or not understood (13). Two suites of proteins are induced by cold temperatures. Abrupt shifts to cool temperatures of cells growing at near optimum temperatures induce cold-shock proteins, while cold acclimation proteins are synthesized at increased levels during balanced growth at cold temperatures as compared with their levels during growth at higher temperatures.

The induction of cold-shock proteins in bacteria and yeasts has been extensively investigated. When *E. coli* growing in a rich medium is shifted abruptly from 37 to 10°C, there is a lag of several hours before growth resumes at a slow rate (14). During the lag, 13 proteins are produced at rates that are increased as compared with their rates of production at 37°C, although the synthesis of most proteins is greatly reduced. The protein produced in the greatest amounts as a result of cold shock to *E. coli* has been termed CspA. CspA is only one of nine similar proteins produced by *E. coli*, some of which are also induced by cold shock, but others of which are or may be induced by other environmental stresses (15). Similar families of CspA-like proteins have been identified in other bacteria (16) as well as other cold-shock proteins in various bacteria and eukaryotes (17,18).

In *E. coli*, the messenger RNA for CspA is produced constitutively, but the m-RNA is so unstable at 37°C that no detectable CspA is synthesized (19). When *E. coli* is cold shocked, the stabilization of the m-RNA at cool temperatures allows its efficient transcription with consequently rapid accumulation of CspA. The control mechanisms for the production of CspA-like or other cold-shock proteins in other organisms have not been identified.

During balanced growth of bacteria at cool temperatures, some cold-shock proteins continue to be synthesized at high levels (20). That is not so for other cold-shock proteins, while yet other proteins not induced by cold shock exhibit increased synthesis at low temperatures (21,22). The functions of those latter, cold acclimation proteins are largely unknown. However, fatty acid desaturases may be one class of enzymes that is common among cold acclimation proteins. Increased fractions of unsaturated fatty acids in lipids have been observed to result from growth at cool temperatures in many organisms (23). It is postulated that such changes are necessary to maintain membrane function and integrity at low temperatures (24). The recent identification of a desaturase induced in *Bacillus subtilis* by growth at temperatures in the lower half of its growth temperature range would seem to support such a view (25). Despite that, the absolute need to increase the unsaturated fatty acid fractions of lipids in response to low temperatures is not apparent in various microorganisms that do not significantly modify their lipids in response to temperature change (26).

Some proteins are induced in response to several different types of stress. Such general responses to stress may be the basis for stress cross protection, i.e., the induction by one environmental stress of resistance to one or more other stresses (27). In particular, starvation for carbon, nitrogen, or phosphorus enhances resistance to a number of other environmental stresses (28). However, the induction of some common proteins in response to two different types of stress does not necessarily imply cross protection. For example, *E. coli* and *Salmonella* produce proteins associated with heat shock in response to carbon starvation, ultraviolet light, oxidizing agents, acid or alkaline conditions, etc. (29), but apparently only starvation and acid shock produce cross protection to heat (13). Conversely, the induction of tolerance to some stresses may induce sensitivity to others, such as the induction in *E. coli* of sensitivity to acid conditions by adaptation to alkaline pH or of sensitivity to alkaline conditions by adaptation to acid pH (30,31), or enhanced susceptibility to sublethal heat injury in *E. coli* adapted to growth at cool temperatures (32).

III. EFFECTS OF COLD TEMPERATURES

It is apparent that the effects of cold temperatures on any microorganism are likely to vary with its physiological state at the time that a particular cold temperature is experienced. The physiological state will be affected by the medium composition and the temperature before cooling commenced. The medium composition may determine whether the final temperature is experienced as a cool or low temperature, if freezing does not occur, while the rate of cooling will determine whether or not the organism can adapt in any way to cold temperatures before those are experienced. In addition, the cooling process may involve the organism being exposed to stresses other than just cold temperatures.

As an organism can by definition grow at cool temperatures, its viability can be affected by such temperatures only if the exposure to cool temperatures results in the organism being cold shocked. If cold shock is induced by diluting a bacterial culture into a relatively large volume of cold water or a minimal medium, cell death and lysis can occur (33,34). Despite that, less sudden reductions of temperatures, as are obtained, for example, by transferring a flask or tube of a culture from a warm to a cool water bath, or rapid cooling in a rich medium do not cause cell death. Instead, if the temperatures remain within the range for growth, a lag phase of several hours may be induced, as with *E. coli* and *Salmonella* (16). With other organisms there may be no lag. For example, Panoff et al. (35) observed with *Lactococcus lactis* that following cold shock the organism grew for several hours at a rate intermediate between the sustained growth rates at the higher and lower temperatures (Fig. 2), although generally in the absence of cold shock the growth rate of a microorganism rapidly adjusts to the sustainable rate for any temperature within the growth temperature range (36).

Thus, induction of cold shock by rapid cooling to cool temperatures of a food in which an organism is growing may induce a lag in some organisms but is unlikely to cause the death of any. Some greater effects might occur if cooling is accompanied by alterations to the medium composition that could additionally stress an organism, although cold-shocked organisms may be more resistant to inhibitory substances than those that have not been shocked (Table 4). The rapid establishment of cool temperatures during the production of foods is then likely to have little effect on microorganisms. That speculative conclusion is seemingly supported by the apparent lack of reports on substantial reductions in the numbers of any microorganisms in a food subjected to a procedure for rapid cooling to a cool temperature.

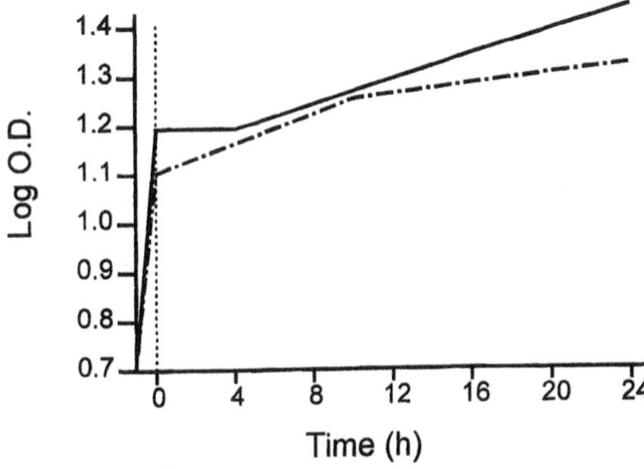

Figure 2 The effect of a shift from 37 to 10°C on the growth of *Escherichia coli* (—) and of a shift from 30 to 8°C on the growth of *Lactococcus lactis* subsp. *lactis* (---). (From Refs. 14, 35.)

IV. EFFECTS OF LOW TEMPERATURES

Microorganisms exposed to low temperatures will enter a stationary phase. The cessation of growth does not necessarily signify the cessation of metabolic activity. Certainly, physiological adjustments that facilitate survival during the stationary phase occur in bacteria after growth has ceased as a result of starvation (38).

Table 4 Effects of Cold Shock on Survival of *Escherichia coli* Treated with Acrylic Acid or Cu^{2+} Ions

	Survival (%)			
	Acrylic acid treated		Cu^{2+} ion treated	
Strain	Nonshocked	Shocked	Nonshocked	Shocked
1	2	17	32	59
2	30	65	25	75
3	50	74	24	62
4	24	40	21	79
5	31	51	11	73
6	17	44	3	69

Source: Ref. 37.

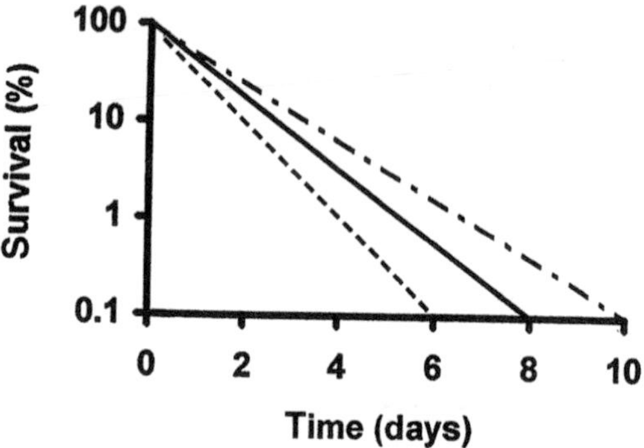

Figure 3 Survival at 4°C of a wild-type *Escherichia coli* (—), a mutant that overproduces the cold-shock induced trigger factor (---), and a mutant that overproduces heat shock proteins (·--·). (From Ref. 39.)

However, the physiological changes associated with the inhibition of growth by low temperatures do not appear to have been investigated.

Whether or not physiological adjustments occur at low temperatures, the physiological state of an organism when it is exposed to low temperatures can affect its survival at persisting low temperatures. In broth cultures at low temperatures numbers of bacteria have been observed to decline exponentially. In a study with mutants of *E. coli* it was shown that a mutant that constitutively overproduced heat-shock proteins declined in numbers more rapidly than the wild type when held at a low temperature (Fig. 3) Conversely, a mutant that constitutively overproduced the trigger factor, which in the wild types is produced in increased amounts in the latter part of a cold shock–induced lag, showed a slowed rate of decline when held at the same low temperature (39). With *L. monocytogenes* it was shown that growth at cool temperatures slowed the rate at which number decreased at low temperatures in a medium of a pH that inhibited growth (40), while growth under acid conditions enhanced the survival at low temperatures of *Salmonella typhimurium* in cheese, which provides an acid environment (41).

Studies with *Campylobacter* growing in either broth or on chicken meat indicated that, at low temperatures, the rates at which numbers decreased increased with increasing temperature (4,42). Similar findings have been reported for *Salmonella* grown on trypticase soy agar (43) and for *Vibrio parahaemolyticus* grown in an oyster meat homegenate (44). However, reductions in numbers at low temperatures may occur only slowly or not at all. Thus, in an appropriate medium, the numbers of *Campylobacter* remained unchanged during storage for 14 days at 4°C (45), and during storage at chiller temperatures for the same time the numbers of various mesophilic bacteria in ground beef apparently did not

Table 5 Log Mean Numbers of Total Aerobic Counts and *Escherichia coli* on Sheep Carcasses Entering or Leaving an Air-Chilling Process

	Log mean numbers	
Process stage	Aerobes (cfu/cm^2)	*E. coli* (cfu/100 cm^2)
Entry	3.33	3.57
Exit	2.86	1.49

Source: Ref. 51

alter (46). The latter finding is supported by the observation that numbers of *E. coli* in chilled beef trimmings remained unaltered during commercial distribution and storage for periods of up to 18 days (47).

In some processes for chilling foods, the microorganisms that contaminate the product may be exposed to environmental stresses additional to low temperatures. Such a situation arises, for example, in the cooling of red meat carcasses when the carcasses are not sprayed with water. The surfaces of carcasses are contaminated with bacteria during the dressing process, but the deep tissues remain sterile (48). Water evaporates from the warm carcass, with desiccation of the surface tissues (49). Subsequently, as the deep tissues cool, the muscle tissue surfaces will rehydrate as the rate of water movement from the underlying tissue begins to exceed the rate of evaporation. It has long been known that the transient dehydration associated with carcass cooling can result in some small decreases in the total aerobic counts due to the loss of viability of the gram-negative rather than gram-positive fractions of flora (50). However, it has recently been shown that the decrease in *E. coli* counts can be very much larger than the decrease in total counts might suggest (Table 5). The combination of cooling to low temperatures with one or more other stresses can then in practice produce large decreases in the numbers of some microorganisms.

V. EFFECTS OF FREEZING TEMPERATURES

Unlike cooling to low temperatures, freezing will always result in the imposition on microorganisms of stresses additional to that of temperature alone. The mechanisms by which microorganisms may be stressed or damaged by freezing have been extensively reviewed (52–55).

The water in foods is mainly in the form of more or less complex solutions. The temperature at which water in a food starts to freeze will then be determined by the concentration of solutes in the water, with the ice fraction increasing over a range of temperatures. Usually, as the temperature falls below the point at which freezing commences, the ice fraction will at first increase rapidly, but, subsequently, relatively large decreases in temperatures may be required to produce small increases in the ice fraction. For example, in muscle tissue freezing commences at about −1°C and about half the water in the tissue is frozen at −2°C. However, even at −40°C some 10% of the water, which is associated with structural proteins, remains unfrozen (56). The formation of ice in any food affects the water activity (a_w), which becomes identical to the a_w of pure water at the same temperature (Table 6). Only if water sublimates from a frozen food with desiccation of some areas will the a_w in those areas fall below the value determined by the ice vapor pressure (59).

Because of the behavior of water in a freezing food, microorganisms in a food could in principle be affected by extracellular ice formation, intracellular ice formation, concentration of extracellular solutes in the unfrozen medium, and dehydration of the microbial cell. In practice it appears that mechanical damage to microorganisms by the extracellular formation of ice is limited (60). Damage from intracellular freezing is also limited because the cytoplasm of microbial cells will invariably supercool at freezing temperatures down to about −10°C (52). The difference in osmotic pressure between the freezing medium and the supercooled cytoplasm will result in the movement of water from the cell. The rate of water loss from a cell will depend upon its size, small cells with relatively large surface area–to–volume ratios being able to lose a given fraction of their water faster than larger cells (61). If the water loss is sufficiently rapid, osmotic

Table 6 Effects of Temperatures on Ice Fraction of Total Water in Tissue and Water Activity of Frozen Muscle

Temperature (°C)	Ice fraction (%)	Water activity, a_w
0	0	0.993
−1	2	0.990
−2	48	0.981
−3	64	0.971
−4	71	0.962
−5	74	0.953
−10	83	0.907
−20	88	0.823
−30	89	0.746

Source: Refs. 57, 58.

equilibrium will be maintained between the cytoplasm and the environment, the intracellular solute concentration will increase proportionally with the decreasing temperature, and intracellular freezing of water will never occur. Thus, cooling rates of 1°C/min or less are too slow for the intercellular formation of ice in microorganisms. Cooling rates in excess of 10°C/min are apparently required for the formation of intracellular ice in bacteria (62). The former rates of cooling may be obtained at surfaces treated with cryogenic liquids, but the latter rates of cooling an unlikely in any food-production process.

When rates of cooling are slow, as would be usual in a mass of food undergoing freezing, some microorganisms may be trapped within forming crystals of pure ice, but most will apparently partition into the unfrozen water to be exposed to increasing concentrations of solutes in the residual liquid phase (63). When equilibrium is maintained, the concentration of solutes in the residual water will be such as to give a solution of water vapor pressure equal to that of the ice fraction. However, at surfaces or in thin layers of food, the rate of freezing can be so rapid as to preclude the maintenance of equilibrium between the solution and pure ice. Then, the water will tend to freeze as a solution, with little concentration of solutes in residual water. Thus, slow freezing, when some microorganism can be exposed to increasing solute concentrations as well as dehydration, is likely to have more severe effects on microorganisms than rapid freezing, during which they will be exposed only to dehydration.

In general, the survival of microorganisms tends to be enhanced by increased rates of cooling in the range likely to be encountered in foods (Fig. 4). At faster rates of cooling, which would be unusual for foods but at which intracellular ice may form, survival is likely to decline with increasing rates of cooling.

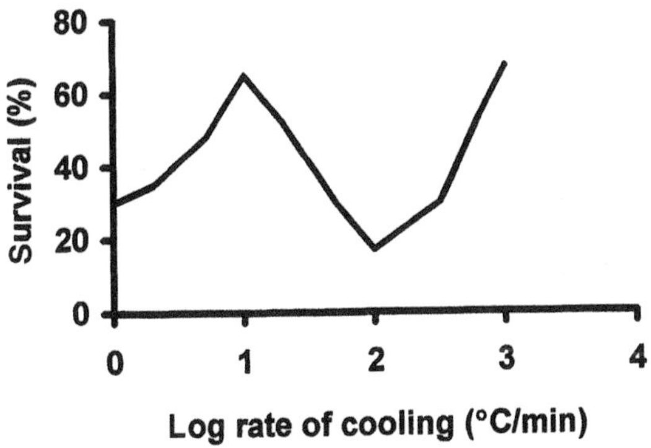

Figure 4 The effect of the rate of cooling on the survival of *Escherichia coli* in freezing water. (From Ref. 66.)

However, with ultra-rapid cooling only a few small ice crystals form within cells, and survival tends to increase with increased rates of cooling (64). Survival of ultra-rapidly cooled cells may be reduced by slow thawing, because intracellular ice crystals can grown during the thawing process (65). The rate of thawing will generally have little effect upon the survival of microorganisms that were frozen with slower rates of cooling, although exponential but not log phase cells of *Salmonella* have been reported to be more affected by slow than rapid thawing after freezing (62).

The available information therefore indicates that during freezing microorganisms are usually damaged by osmotic stress and elevated solute concentrations rather than being damaged mechanically by ice crystals. After freezing, the numbers of viable cells will tend to decline further as frozen storage is prolonged. The rate of decline in numbers during frozen storage may be far slower than the rate of decline associated with freezing, and the rate of decline may decrease with time until the numbers of viable cells are essentially stable (Fig. 5). However, decreases in numbers at an approximately constant exponential rate during freezing and frozen storage have been reported for *Salmonella* (68), while continuing declines in the numbers of strains of *E. coli* O157:H7 during frozen storage for several months have been reported (32).

Loss of viability is likely to be more rapid at higher than at lower freezing temperatures. For example, *L. monocytogenes* in phosphate buffer was little affected by freezing and storage in liquid nitrogen at −198°C, but substantial decreases in numbers were observed with freezing and storage at −18°C or freezing with liquid nitrogen followed by storage at −18°C (69). However, *V. parahaemolyticus* has been reported to survive similarly at −18 or −34°C (70).

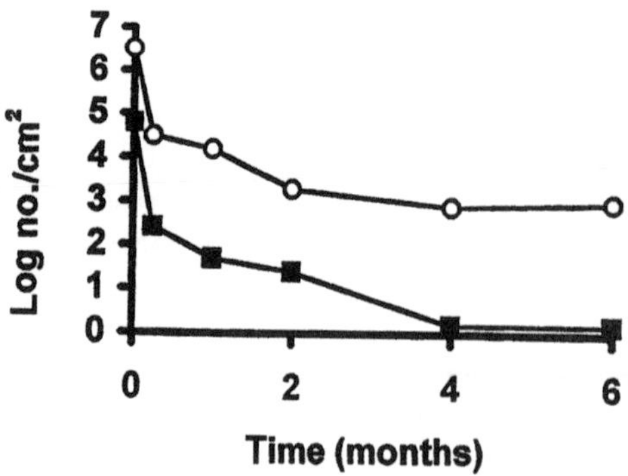

Figure 5 The effects of freezing and frozen storage at −29°C on the numbers of total aerobes (○) and coliform (■) on turkeys. (From Ref. 67.)

Repeated freeze-thaw cycles could be expected to produce larger decreases in the numbers of microorganisms than storage at a constant, frozen temperature. Such behavior has indeed been reported for *L. monocytogenes* and for *S. aureus* in an acid medium (70,71). However, *S. aureus* and other organisms in other media have been observed to be no more affected by fluctuating temperatures than by storage at a single, relatively low, freezing temperature (72,73).

The medium in which freezing occurs can greatly affect the survival of microorganisms. Generally, the more complex the medium, the less likely microorganisms are to be affected by freezing. Survival of freezing has been observed to be greater in saline solutions than in water, greater in milk or broth than in phosphate buffer, and greater in foods than in laboratory media (74–76). The enhanced cryotolerance of microorganisms in many foods is likely due in part to the present of cryoprotective agents such as glycerol, glycine, betaine, etc. (77,78). However, the solutes in some foods, such as lactic acid in raw muscle tissue, may exacerbate freezing injury if they are concentrated in the liquid phase during a slow freezing process (79).

The physiological state of a microorganism will modify the effects that freezing has on it. It is well recognized that cells in the exponential phase of growth are more sensitive to freezing than are cells in the stationary phase (63). Exponential phase cells of *Salmonella* grown in a minimal medium were found to be less sensitive to freezing than exponential phase cells from the faster-growing culture in a rich medium (62), while *Enterococcus faecalis* was observed to be less sensitive to freezing after cultivation at cool than at warm temperatures (80). All those observations indicate that the starvation and cold adaptation responses that enhance the viability of organisms exposed to low temperatures also tend to protect the organisms against damage from freezing. Moreover, organisms adapted to osmotic stress by the accumulation of compatible solutes are likely to be resistant to freezing damage also (81).

In addition to the factors of rate of freezing, medium composition, and physiological adaptation, there are wide, intrinsic differences between microorganisms in their susceptibility to damage from freezing. In general, gram-negative bacteria are more susceptible to freezing injury than are gram-positive species. Thus, freezing tends to enrich the flora of raw meat for gram-positive species (Table 7). However, the relative insensitivity to freezing of gram-positive species is by no means uniform, as some lactic acid bacteria used for food fermentation and vegetative cells of clostridial species are highly susceptible to the lethal effects of freezing (83,84). In contrast, the spores of clostridia as well as those of other spore-forming bacteria are essentially unaffected by freezing (85).

The responses to freezing of yeasts and molds appear to be similarly variable. Fungal spores are little affected by freezing (52), while extensive loss of viability in vegetative fungi during freezing and frozen storage has been reported

Table 7 Numbers and Compositions of Flora on Beef Cuts or in Ground Beef Before and After Freezing and Thawing

Beef product	Processing stage	Total counts (No./g)	Gram-negative species (%)	Gram-positive species (%)
Cuts	Before freezing	3.9×10^5	85	15
	After thawing	7.7×10^4	30	70
Ground	Before freezing	4.0×10^5	78	22
	After thawing	4.7×10^4	23	77

Source: Ref. 82.

(86). However, others have found that vegetative, psychrotrophic fungi are little affected by the freezing and frozen storage of food (87,88), while large differences in responses to freezing between conventional and cryotolerant strains of bakers' yeast are observed (89).

VI. EXPLOITATION OF THE INACTIVATION OF MICROORGANISMS BY COLD TEMPERATURES

Inactivation by cold temperatures of microorganisms used for fermenting or raising foods can restrict commercial options in the preparation and processing of some foods or the maintenance of starter cultures (84,90). Conversely, inactivation by cold might in principle be exploited to enhance the microbiological safety and/or storage stability of foods. In practice, however, factors of processing convenience and maintenance of desirable organoleptic properties dictate the cooling processes used with foods. Consequently, any destruction of microorganisms as a result of cooling and/or cold storage is fortuitous rather than planned, and food processors are probably little aware of any inactiviting effects on microorganisms of their cooling and cold storage practices.

Given the extreme variability of the effects of cold temperatures on the viability of microorganisms, attempts to adjust cooling and cold storage processes to obtain reliable inactivations of microorganisms of concern would likely be unrewarding with many foods. However, with others, useful reductions in the numbers of some pathogenic or potent spoilage organism associated with the food might be possible with only minor modification of existing practices. For example, spray-chilling processes for red meat carcasses can apparently be carried out to achieve substantial reductions in the numbers of *E. coli* on the carcass surfaces, although most such processes have little effect on the numbers of those or other bacteria (91). Appropriate examination of commercial processes might then reveal situations in which inactivation of specific microorganisms by cold could be exploited in practice to enhance the safety and/or storage stability of some foods.

REFERENCES

1. MWW Adams. Enzymes and proteins from organisms that grow near and above 100°C. Ann Rev Microbiol 47:627–658, 1993.
2. DD Wynn-Williams. Ecological aspects of Antartic microbiology. Adv Microbial Ecol 11:71–146, 1990.
3. JC Olson, Jr., PM Nottingham. Temperature. In: International Commission on Microbiological Specifications for Foods. Microbial Ecology of Foods, Vol. 1. London: Academic Press, 1980, pp. 1–37.
4. MP Doyle, DJ Roman. Growth and survival of *Campylobacter fetus* subsp. *jejuni* as a function of temperature and pH. J Food Prot 44:596–601, 1981.
5. NJ Stern, MD Pierson. *Yersinia enterocolitica*: a review of the psychrotrophic water and foodborne pathogen. J Food Sci 44:1736–1742, 1979.
6. J-Y D'Aoust. Psychrotrophy and foodborne *Salmonella*. Int J Food Microbiol 13: 207–216, 1991.
7. CO Gill, DM Phillips. The effects of media composition on the relationship between temperature and growth rate of *Escherichia coli*. Food Microbiol 2:285–290, 1985.
8. JY D' Aoust. *Salmonella*. In: MP Doyle, ed. Foodborne Bacterial Pathogens. New York: Marcel Dekker, 1989, pp. 327–445.
9. CO Gill, GG Greer, BD Dilts. The aerobic growth of *Aeromonas hydrophila* and *Listeria monocytogenes* in broths and on pork. Int J Food Microbiol 35:67–74, 1997.
10. PD Lowry, CO Gill. Temperature and water activity minima for growth of spoilage moulds from meat. J Appl Bacteriol 56:193–199, 1984.
11. EF Gale, HMR Epps. The effect of the pH of the medium during growth on the enzymic activities of bacteria (*Escherichia coli* and *Micrococcus lysodiekticus*) and the biological significance of the changes produced. Biochem J 36:600–618, 1942.
12. AD Brown. Microbial water stress. Bacteriol Revs 40:803–846, 1976.
13. JW Foster, MP Spector. How *Salmonella* survive against the odds. Ann Rev Microbiol 49:145–174, 1995.
14. PG Jones, RA Van Bogelen, FC Neidhardt. Induction of proteins in response to low temperature in *Escherichia coli*. J Bacteriol 169:2092–2095, 1987.
15. K Yamanaka, L Fang, M Inouye. The CspA family in *Escherichia coli*: multiple gene duplication for stress adaptation. Mol Microbiol 27:247–255, 1998.
16. AG Jeffreys, KM Hak, RJ Steffan, JW Foster, AK Bej. Growth, survival and characterization of CspA in *Salmonella enteritidis* following cold shock. Curr Microbiol 36:29–35, 1998.
17. J Muñoz-Dorado, K. Kondo, M Inouye, H Sope. Identification of *cis* and *trans*-acting elements involved in the expression of cold shock inducible TIP 1 gene of yeast *Saccharomyces cerevisiae*. Nucleic Acids Res 2:560–568, 1994.
18. MP Chapot-Chartier, C Schouler, A-S LePeuple, J-C Gripon, MC Chopin. Charac-

terization of Csp B, a cold-shock inducible gene from *Lactococcus lactis*, and evidence for a family of genes homologous to the *Escherichia coli csp*A major cold shock gene. J Bacteriol 179:5589–5593, 1997.

19. L Fang, W Jiang, W Bae, M Inouye. Promoter-independent cold-shock induction of *csp*A and its derepression at 37°C by mRNA stabilization. Mol Microbiol 23: 355–364, 1997.
20. DO Bayles, BA Annous, BJ Wilkinson. Cold stress proteins induced in *Listeria monocytogenes* in response to temperatures down shock and growth at low temperature. Appl Environ Microbiol 62:1116–1119, 1996.
21. M Hebraud, E Dubois, P Potier, J Labadie. Effect of growth temperatures on the protein levels in a psychrotrophic bacterium, *Pseudomonas fragi*. J Bacteriol 176: 4017–4024, 1994.
22. ME Roberts, WE Inniss. The synthesis of cold shock proteins and cold acclimation proteins in the psychrophilic bacterium *Aquaspirillum arcticum*. Curr Microbiol 25: 275–278, 1992.
23. J Farrell, AH Rose. Temperature effects on microorganisms. In: AH Rose, ed. Thermobiology. London: Academic Press, 1967, pp. 147–219.
24. AR Cossins, M Sinensky. Adaptation of membranes to temperature, pressure and exogenous lipids. In: M Shinitzky, ed. Physiology of Membrane Fluidity. Boca Raton, FL: CRC Press, 1984, pp. 1–20.
25. PS Aguilar, JE Cronan Jr, D DeMendoza. A *Bacillus subtilis* gene induced by cold shock encodes a membrane phosphlopid desaturase. J Bacteriol 180:219–2200, 1998.
26. RA Herbert. Microbial growth at low temperatures. In: GW Gould, ed. Mechanisms of Action of Food Preservation Procedures. Barking, U.K.: Elsevier, 1989, pp. 71–96.
27. HK Hall, KL Karem, JW Foster. Molecular responses of microbes to environmental pH stress. Adv Microbial Physiol 37:229–272, 1995.
28. MP McCann, JP Kidwell, A Martin. The putative sigma factor Kat F has a central role in development of starvation-mediated general resistance in *Escherichia coli*. J Bacteriol 173:4188–4194, 1991.
29. J Yura, H Nagai, H Mori. Regulation of the heat shock response in bacteria. Ann Rev Microbiol 47:321–350, 1993.
30. RJ Rowburn, NH Hassian. Exposure of *Escherichia coli* to acid habituation conditions sensitizes it to alkaline stress. Lett Appl Microbiol 22:57–61, 1996.
31. RJ Rowbury, M Goodson, JT Humphrey. Novel acid sensitivity induced in *Escherichia coli* at alkaline pH. Lett Appl Microbiol 16:223–227, 1993.
32. JJ Semenchek, DA Golden. Influence of growth temperature on inactivation and injury of *Escherichia coli* O157:H7 by heat, acid and freezing. J Food Prot 61:395–401, 1998.
33. BM McKay. Lethal and sublethal effects of refrigeration, freezing and freeze drying on microorganisms. In:MHE Andrews, AD Russell, eds. The Revival of Injured Microbes. London: Academic Press, 1984, pp. 45–75.
34. T Tsuchido, T Nishino, Y Kato, M Takano. Involvement of membrane lipids in cold shock-induced autolysis of *Bacillus subtilis* cells. Biosci Biotech Biochem 59:1636–1640, 1995.

35. J-M Panoff, S Legrand, R Thammavongs, P Boutibonnes. The cold shock response in *Lactococcus lactis* subsp. *lactis*. Curr Microbiol 29:213–216, 1994.
36. CO Gill. Cold storage temperature fluctuations and predicting microbial growth. J Food Prot 59 (suppl):43–47, 1996.
37. GC Whiting, RJ Rowbury. Increased resistance of *Escherichia coli* to acrylic acid and to copper ions after cold-shock. Lett Appl Microbiol 20:240–242, 1995.
38. R Kolter, DA Siegele, A Tormo. The stationary phase of the bacterial life cycle. Ann Rev Microbiol 47:855–874, 1993.
39. O Kandror, AL Goldberg. Trigger factor is induced upon cold shock and enhances viability of *Escherichia coli* at low temperature. Proc Natl Acad Sci USA 94:4978–4981, 1997.
40. A Gray, O Cerf. Significance of temperature and pre-incubation temperature on survival of *Listeria monocytogenes* at pH 4.8. Lett Appl Microbiol 25:257–260, 1997.
41. GJ Leyer, EA Johnson. Acid adaptation promotes survival of *Salmonella* spp. in cheese. Appl Envion Microbiol 58:2075–2080, 1992.
42. LC Blankenship, SE Craven. *Campylobacter jejuni* survival in chicken meat as a function of temperature. Appl Environ Microbiol 44:88–92, 1982.
43. JR Matches, J Liston. Low temperature growth of *Salmonella*. J Food Sci 33:641–645, 1968.
44. JM Muntada-Garriga, JJ Rodriguez-Jerez, EI Lopez-Sabater, M Teresa Mara-Ventara. Effect of chill and freezing temperatures on survival of *Vibrio parahaemolyticus* inoculated in homogenates of oyster meat. Lett Appl Microbiol 20:225–227, 1995.
45. NJ Stern, AW Kotula. Survival of *Campylobacter jejuni* inoculated into ground beef. Appl Environ Microbiol 44:1150–1153, 1982.
46. JM Goepfert, HU Kim. Behaviour of selected food-borne pathogens in raw ground beef. J Milk Food Technol 38:449–452, 1975.
47. CO Gill, JC McGinnis. Charges in the microflora on commercial beef trimmings during their collection, distribution and preparation for retail sale as ground beef. Int J Food Microbiol 18:321–332, 1993.
48. CO Gill. Current and emerging approaches to assuring the hygienic condition of red meat. Can J Anim Sci 75:1–13, 1995.
49. SJ James, C Bailey. Process design data for beef chilling. Int J Refrig 12:42–49, 1989.
50. PM Nottingham. Microbiology of carcass meats In: MH Brown, ed. Meat Microbiology. London: Applied Science Publishers, 1982, pp. 13–66.
51. CO Gill, T Jones. Assessment of the hygienic performances of an air-cooling process for lamb carcasses and a spray-cooling process for pig carcasses. Int J Food Microbiol 38:85–93, 1997.
52. P Mazur. Physical and chemical basis of injury in single-celled microorganism subjected to freezing and thawing. In: HT Meryman, ed. Cryobiology. London: Academic Press, 1966, pp. 213–315.
53. PO Hagen. The effect of low temperatures on microorganisms: conditions under which cold becomes lethal. In: WB Hugo, ed. Inhibition and Destruction of the Microbial Cell. London: Academic Press, 1971, pp. 39–76.

54. ML Speck, B Ray. Effects of freezing and storage of microorganisms in frozen foods: a review. J Food Prot 40:333–336, 1977.
55. SE EL-Kest, EH Marth. Freezing of *Listeria monocytogenes* and other microorganism: a review. J Food Prot 55:639–648, 1992.
56. R Rosset. Chilling, freezing and thawing. In: MH Brown, ed. Meat Microbiology. London: Academic Press, 1982, pp. 265–318.
57. L Riedel. Kalorimetrische Untersuchungen über das Gefrieren von Fleisch. Kaltetechnik 9:38–40, 1957.
58. L Leistner, W Rodel, K Krispien. Microbiology of meat and meat products in high and intermediate moisture ranges. In: LB Rockland, GF Stewart, eds. Water Activity: Influences on Food Quality. London: Academic Press, 1981, pp. 885–916.
59. L Van den Berg. Physiochemical changes in foods during freezing and subsequent storage. In: J Hawthorne, EJ Rolfe, eds. Oxford: Pergemon Press, 1968, pp. 205–219.
60. AP Harrison, Jr., RE Cerroni. Fallacy of crushing death in frozen bacterial suspension. Proc Soc Exp Biol NY 91:577–579, 1956.
61. P Mazur. Cryobiology: the freezing of biological systems. Science 168:939–969, 1970.
62. R Davies, A Obafemi. Responses of micro-organisms to freeze-thaw stress. In: RK Robinson, ed. Microbiology of Frozen Foods. London: Elsevier Applied Science, 1985, pp. 83–107.
63. PH Calcott. Manifestations of freeze-thaw injury. In: JG Cook, ed. Freezing and Thawing of Microbes. Shildon, UK: Meadowfield Press, 1978, pp. 26–40.
64. RA MacLeod, PH Calcott. Cold shock and freezing damage to microbes. In: TRG Gray, JR Postgate, eds. Survival of Vegitative Microbes. Cambridge: Cambridge University Press, 1976, pp. 81–109.
65. H Bank, P Mazur. Visualization of freezing damage. J Cell Biol 57:729–742, 1973.
66. PH Calcott, RA MacLeod. Survival of *Escherichia coli* from freeze-thaw damage: a theoretical and practical study. Can J Microbiol 20:671–681, 1974.
67. AA Kraft, JC Ayres, JC Weiss, WW Marion, SL Balloun, RH Forsythe. Effect of method of freezing on survival of microorganisms on turkey. Poultry Sci 42:128–137, 1963.
68. GM Farrell, ME Upton. The effect of low temperature on the growth and survival of *Staphylococcus aureus* and *Salmonella typhimurium* when inoculated onto bacon. J Food Technol 13:15–23, 1978.
69. SE El-Kest, AE Yousef, EH Marth. Fate of *Listeria monocytogenes* during freezing and frozen storage. J Food Sci 56:1068–1071, 1991.
70. JR Matches, J Liston, LP Daneault. Survival of *Vibrio parahaemolyticus* in fish homogenate during storage at low temperatures. Appl Microbiol 21:951–952, 1971.
71. SE El-Kest, EH Marth. Strains and suspending menstrua as factors affecting death and injury of *Listeria monocytogenes* during freezing and frozen storage. J Dairy Sci 74:1209–1213, 1990.
72. PH Demchick, SA Palumbo, JL Smith. Influence of pH on freeze-thaw lethality in *Staphylococcus aureus*. J Food Safety 4:185–189, 1982.
73. DI Georgala, A Hurst. The survival of food poisoning bacteria in frozen foods. J Appl Bacteriol 26:346–358, 1963.

74. D Covert, M Woodburn. Relationship of temperature and sodium chloride concentration to the survival of *Vibrio parahaemolvticus* in broth and fish homogenate. Appl Microbiol 23:321–325, 1972.
75. SE El-Kest, EH Marth. Injury and death of frozen *Listeria monocytogenes* as affected by glycerol and milk components. J Dairy Sci 74:1201–1208, 1991.
76. M Gianfrancheschi, P Aureli. Effects of freezing and frozen storage on the survival of *Listeria monocytogenes* in different foods. Ital J Food Sci 8:303–309, 1996.
77. JE Graham, BL Wilkinson. *Staphylococcus aureus* osmoregulation: roles for chlorine, glycine betaine, proline and taurine. J Bacteriol 174:2711–2716, 1992.
78. R Ho, LT Smith, GM Smith. Glycine betaine confers enhanced osmotolerance and crytolerance on *Listeria monocytogenes*. J Bacteriol 176:426–431, 1994.
79. PD Lowry, CO Gill. Microbiology of frozen meat and meat products. In: RK Robinson, ed. Microbiology of Frozen Foods. London: Elsevier Applied Science, 1985, pp. 109–168.
80. B Thammavongs, D Corroler, J-M Panoff, Y Auffray, P Boutibonnes. Physiological response of *Enterococcus faecalis* JH 2-2 to cold shock: growth at low temperatures and freezing/thawing challenge. Lett Appl Microbiol 23:398–402, 1996.
81. AD Brown. Compatible solutes. In: Microbial Water Stress Physiology: Principles and Perspectives. Chichester, UK: Wiley, 1990, pp. 241–279.
82. W Partmann. The effects of freezing and thawing on food quality. In: RB Duckworth, ed. Water Relations of Foods. London, Academic Press, 1975, pp. 505–537.
83. EM Barnes, JE Despaul, M Ingram. The behaviour of a food poisoning strain of *Clostridium welchii* in beef. J Appl Bacteriol 26:415–427, 1963.
84. WS Kim, NW Dunn. Identification of a cold shock gene in lactic acid bacteria and the effect of cold shock on cryotolerance. Curr Microbiol 35:59–63, 1977.
85. MF Gunderson, AC Peterson. Microbiology of frozen foods In: N Derosier, D Tressler, eds. Fundamentals of Food Freezing. Westport, CT: AVI Publishing Co., 1977, pp. 476–505.
86. J Christophersen. Effect of freezing and thawing on the microbial population of foodstuffs. In: J Hawthorn, EJ Rolfe, eds. Low Temperature Biology of Foodstuffs. New York: Pergamon Press, 1968, pp. 251–269.
87. CO Gill, PD Lowry. Growth at sub-zero temperatures of black spot fungi from meat. J Appl Bacteriol 52:245–250, 1982.
88. JL Oblinger, JE Kennedy Jr. Microflora isolated from imported frozen lean beef. J Food Prot 41:251–253, 1978.
89. S Hatano, M Udou, N Koga, K Honjgh, J Miyumoto. Improvement of the glycolytic system and lactic in baker's yeast during frozen storage. Biosci Biotech Biochem 60:61–64, 1996.
90. MJ Wolt, BL D'Appolonia. Factors involved in the stability of frozen dough. I. The influence of yeast reducing compounds on frozen-dough stability. Cereal Chem 61: 209–212, 1984.
91. CO Gill, J Bryant. Assessment of the hygienic performances of two beef carcass cooling processes from product temperature history data or enumeration of bacteria on carcass surfaces. Food Microbiol 14:593–602, 1997.

4
Inactivation by Irradiation

Aubrey F. Mendonca
Iowa State University
Ames, Iowa

I. IRRADIATION

The electromagnetic spectrum is composed of at least six separate forms of radiation that differ in wavelength, frequency, and penetrating power. Of these forms, gamma radiation, ultraviolet (UV) radiation, and microwaves are of interest to the food industry. Microwave radiation is utilized primarily for its heating properties; therefore, the antimicrobial effects of microwaves are not covered in this chapter. A comprehensive discussion of the inactivation of microorganisms by this form of radiation can be found in Chapter 5. Irradiation refers mainly to any process involving the application of ionizing radiation, including alpha particles, beta rays or electrons, and x-rays generated by machines or gamma rays from radioisotopes. Ionizing radiation that is of interest for food preservation includes x-rays, beta rays, and gamma rays that have wavelengths of 2000. Å or less and are very energetic (1,2). They possess enough energy to ionize molecules in their paths and can inactivate foodborne microorganisms without increasing the temperature of the irradiated food.

X-rays, beta rays, and gamma rays differ in their capacity to penetrate foods. Beta rays have poor penetration power and are inadequate for food preservation. X-rays have stronger penetration power than beta rays; however, the difficulty in focusing them on foods limits their application in food preservation. The very high penetration power of gamma rays makes them attractive for use in food preservation (2). Gamma rays have approximately 1–2 million electron volts (MeV) of energy and can penetrate materials with a thickness of about 40 cm. These rays are emitted by radioisotopes such as cobalt-60 and cesium-137. Cobalt-60 is prepared for use in nuclear medicine and is used for food irradiation when it decays to an energy level that is inadequate for medical needs. This radioisotope is more available than cesium-137. Electrons, which are relatively low in energy, can be accelerated with a linear accelerator or a Van de Graaf generator to produce energy levels of 10 MeV or higher. High-energy electrons can be used to produce X-rays via their bombardment of heavy metals such as tungsten. Doses of ionizing radiation that are adequate to produce positive and negative charges in food can be used to inactivate foodborne microorganisms.

The irradiation dose applied to a food is the most important factor of the irradiation process. The unit of absorbed dose used in the past is called a rad. A rad is equivalent to 100 ergs of energy absorbed per gram of irradiated material. The currently used unit of absorbed dose is the Gray (Gy), which is equivalent to 100 rads or the absorption of 1 joule of energy per kilogram of irradiated material. One kilogray (kGy) is equivalent to 100,000 rads (2).

II. APPLICATION OF RADIATION

The application of radiation to food may be categorized as low dose, medium dose, or high dose. These categories of doses are applied to foods for different purposes. Low doses (<1 kGy) may be used for destroying insects and pests in grains and fruits, inactivating parasites in fresh meat, and delaying ripening in fruits or sprouting in vegetables. Medium doses (1–10 kGy) impart a pasteurizing effect in food by inactivating most foodborne pathogens and spoilage microorganisms to improve the safety and shelf life of refrigerated foods. High doses (10–50 kGy) can be applied to achieve commercial sterilization of food as in commercial canning operations and for sterilizing spices and vegetable seasonings that are used in very small amounts in foods. Also, high irradiation doses sterilize foods for various purposes, such as for use by astronauts during space missions and for immunocompromised patients who are highly susceptible to microbial infection (3).

Various levels of radiation that might be applied to foods can be defined as radurization, radicidation, and radappertization. Radurization is called radiation pasteurization and involves use of doses that range from 0.75 to 2.5 kGy to reduce populations of viable spoilage microorganisms in foods such as fresh meats, poultry, seafood, vegetables, fruits, and cereal grains. However, the effectiveness of this method for improving the safety and shelf life of foods can be limited by the survival of psychrotrophic pathogens and psychrotrophic gram-positive spoilage bacteria. Radiation pasteruized foods should be stored at ≤4°C to prevent microbial growth. Radicidation involves irradiating foods to destroy vegetative foodborne pathogens. This method is equivalent to milk pasteurization in that it reduces the number of vegetative pathogenic bacteria so that none can be detected by standard methods. Typical doses range from 2.5 to 10 kGy. Viruses and spores of pathogens are not destroyed, and some radiation-resistant strains of bacteria can survive. Foods irradiated by this method should be stored ≤4°C to prevent germination and outgrowth of *Clostridium botulinum* spores. Radappertization involves the application of high radiation doses (30–40 kGy) for destruction of *C. botulinum* spores and is equivalent to a 12-D heat treatment. This method is not recommended for use in foods.

III. FACTORS AFFECTING MICROBIAL INACTIVATION BY IRRADIATION

Inactivation of foodborne microorganisms by ionizing radiation is influenced by several factors. These factors include irradiation dose, numbers and types of microorganisms, food composition and preservation method, temperature, and atmospheric gas composition.

A. Irradiation Dose

In general, higher doses of ionizing radiation cause greater destruction of microorganisms. However, microbial destruction at a given irradiation dose is decreased under anaerobic or dry conditions due to the lower rate of oxidizing reactions that produce free radicals and toxic oxygen derivatives.

B. Numbers and Types of Microorganisms

As with heating, use of chemical preservatives, and certain other food preservation methods, microbial numbers have the same impact on the effectiveness of irradiation: large numbers of microorganisms reduce the effectiveness of a given irradiation dose. With regard to microbial types, viruses are more resistant than bacterial spores, which are more resistant than bacterial vegetative cells. Vegetative cells of bacteria are more resistant to irradiation than yeast and molds. A general rule is that more complex life forms are more sensitive to irradiation than simpler life forms. For example, humans are more sensitive to irradiation than foodborne animal parasites, such as roundworms and tapeworms, which are more sensitive than microorganisms. In fact, an irradiation dose as high as 40 kGy is required to inactivate viruses, whereas a dose as low as 0.01 kGy could kill a human being (4).

Among bacteria, gram-negative bacteria are more sensitive to irradiation than gram-positives. Several researchers have demonstrated that irradiation doses of at least 1.0 kGy, which could virtually eliminate gram-negative bacteria in food, have a much smaller effect on gram-positive, lactic acid–producing bacteria (5–7). Non–spore-formers are more sensitive than spore-formers. With respect to physiological state of bacteria, exponential phase cells tend to be more sensitive to irradiation than lag phase cells or stationary phase cells.

C. Food Composition and Preservation Methods

The composition of food, including liquid or solids content, protein content, and thickness, affects the inactivation of microorganisms by irradiation. Solid foods offer greater protection to microorganisms against irradiation than phosphate buffer or other liquid media. Similarly, increasing amounts of protein in foods tend to provide a protective effect. Diehl et al. (1) reported that proteins exert a neutralizing effect on free radicals. This neutralizing effect of protein may explain the relatively high radiation resistance of microorganisms in meats and dairy products. Proteins and other food components, including natural antioxidants such as vitamin C and vitamin E, compete for free radicals formed from activated molecules and the radiolysis of water. This competition for free radicals miimizes the antimicrobial effect of irradiation. Interestingly, the sensitivity of *Listeria monocytogens*, *Bacillus cereus*, *Salmonella* Typhimurium, and *Yersinia enterocolitica* F5692 to gamma radiation was unaffected by increasing levels of fat (0.39–32.5%) in various varieties of fish (8). The ineffectiveness of fat content of foods to alter the sensitivity of foodborne pathogens to irradiation has been reported by other researchers (9–12).

The major effect of ionizing radiation on food lipids is the formation of free fatty acids, carbonyl compounds, hydrogen peroxide, and hydroperoxides. The formation of hydrogen peroxide and other toxic oxygen derivatives would be expected to increase the antimicrobial effect of irradiation in lipid-containing foods. However, the ineffectiveness of increasing fat levels to alter the radiation sensitivity of foodborne pathogens may be attributed to other food components. These food components, mainly proteins, may protect microorganisms from the harmful products of radiation-induced chemical changes in fat (1).

Food preservation methods such as heating, acidification, high hydrostatic pressure, and addition of chemical food preservatives increase the sensitivity of microorganisms to ionizing radiation by decreasing the number of survivors. In contrast, the drying of food, a traditional food preservation method, increases the resistance of microorganisms to ionizing radiation. This increased resistance is attributed to lowered water activity, which reduces the production of free radicals in food during irradiation (13).

D. Temperature

Temperature is a major extrinsic factor that influences the survival of microorganisms during irradiation. Significant effects of irradiation temperature on survival of foodborne microorganisms have been reported for *Salmonella* (14,15), *Campy-*

lobacter jejuni (10), *Escherichia coli* O157:H7 (10,16,17), *C. botulinum* spores (18), *Staphylococcus aureus* (19), and *L. monocytogenes* (20). The vast majority of published research regarding the effect of temperature on the antimicrobial efficacy of ionizing radiation indicate that microbial sensitivity to irradiation is higher at ambient temperatures than at subfreezing temperatures. Matsuyama et al. (21) reported that an 8.5-fold higher radiation dose was required for a 90% reduction in *Pseudomonas* spp. at subfreezing temperature than an ambient temperature. Irradiation doses of 20 and 50 kGy were required to destroy similar numbers of irradiation-resistant *Moraxella-Acinetobacter* in ground beef at 35 and 30°C, respectively (22). The D-values (decimal reduction, or dose required to destroy 90% of the microbial population) for *E. coli* O157:H7 in mechanically deboned chicken meat were 0.28 kGy at 5°C and 0.44 kGy at −5°C (16). In a more recent study (17), significantly higher D-values were reported for *E. coli* O157:H7 in ground beef patties irradiated at −15°C as compared with 5°C.

Subfreezing temperatures in food causes a reduction in water activity, which is associated with increased irradiation resistance of microorganisms. This can be explained by the fact that the production of free radicals from the radiolysis of water is decreased at subfreezing temperature due to a reduction of reaction rates (1). Also, the frozen state of food inhibits the migration of free radicals to other areas beyond sites of free radical production (23). Free radicals such as the hydroxyl radical (OH·) and the hydrogen radical (·H), were linked to approximately 85% of the potential damage in irradiated *E. coli* (24).

E. Atmospheric Gas Composition

The gaseous composition of the atmosphere in contact with microorganisms influences their inactivation by irradiation under specific conditions. Earlier published studies (25–28) indicated that most microorganisms exhibit increased sensitivity to irradiation in the presence of oxygen. In a recent study, gamma radiation treatments were shown to be significantly more lethal to *L. monocytogenes* in turkey meat packaged in air than in either vacuum or modified atmospheres (29). The type(s) of gas in modified-atmosphere packaging (MAP) may also affect microbial sensitivity to irradiation. Hastings et al. (30) observed that *Lactobaccillus sake*, *Lactobacillus alimentarius*, and *Lactobacillus curvatus* were more sensitive to gamma radiation in ground meat packaged under 100% carbon dioxide (CO_2) than under nitrogen (N_2). Patterson (31) reported greater sensitivities of *S.* Typhimurium and *E. coli* to ionizing radiation in poultry meat packaged under vacuum or CO_2 compared to aerobic packaging. These findings vary from those of other researchers who demonstrated increased sensitivity of *Salmonella* in the presence of oxygen (29). Some studies have reported no significant differences in total numbers of bacteria or *E. coli* O157:H7 that survived electron beam irradiation of ground beef packaged under air versus vacuum (17,32). More recently, Thayer and Boyd (29) observed a small but significant increase in the radiation sensitivity of *L. monocytogenes* on turkey meat packaged in 100% CO_2 compared to 100% N_2.

The inconsistency in some of the published research may be attributed to other factors that were not considered in most of the studies. For example, variations in techniques used by researchers for recovering microorganisms that survived irradiation treatments might have contributed to variations in published D_{10} values. Results of two independent studies (33,34) indicated that the type of plating media used for enumerating *L. monocytogenes* following irradiation significantly affected D_{10} values for this pathogen. Various food matrices can give different amounts of protection to microorganisms during irradiation (1). Also, differences in irradiation temperatures used in various studies may account for variations in microbial sensitivity to irradiation under anaerobic conditions. Hollaender et al. (25) demonstrated that low temperatures increased microbial sensitivity to irradiation under anoxic conditions. In a review of the effects of irradiation and packaging of fresh meat and poultry, Lee et al. (35) concluded that more complete information is needed to ensure the appropriate use of vacuum packaging or MAP combined with irradiation for the microbial safety of fresh meat and poultry.

IV. INACTIVATION OF MICROORGANISMS

Contamination of food with pathogenic microorganisms, including bacteria of public health significance, parasitic worms, certain protozoa, and enteric viruses, is a major food safety problem and an important cause of illness worldwide. In the United States, foodborne pathogenic bacteria, including *Salmonella, Campylobacter jejuni, E. coli* O157:H7, *L. monocytogens, and Y. enterocolitica*, and certain parasites such as *toxoplasma gondii* account for approximately 6.5–81 million cases of diarrheal diseases and 9000 deaths annually (36,37). Apart from causing much human suffering, foodborne pathogens cause substantial economic loss that result in lower productivity, increased medical costs, litigation, and loss of business (38,39).

The growing awareness of foodborne diseases caused by microorganisms has greatly increased our interest in the use of food irradiation for destroying foodborne pathogens. Destruction of foodborne pathogens in foods is most important to immunocompromised persons, including the elderly, pregnant women, the very young, AIDS patients, and organ transplant patients. Despite past efforts to reduce microbial contamination of foods via thermal, chemical, or other treatment, a relatively large percentage of foods of animal origin is contaminated with pathogenic bacteria and cause increasing foodborne disease in many countries (40). Over the past two decades human diseases caused by an increased number of foodborne pathogens have been well documented. This increase in foodborne disease has been linked to various factors, including large-scale production of animals used for human consumption, increased environmental pollution that causes contamination of food and feed, large-scale processing of foods, increased movement of people among countries, and increased international trade of food and feeds.

Although thermal pasteurization has a long history of use for destroying pathogens and spoilage microorganisms, this process is best suited for liquid foods. The use of chemical treatments for sanitizing foods poses problems related to levels of chemical residues in food, waste disposal, and environmental pollution. These disadvantages warrant using alternative intervention strategies such as irradiation for destroying pathogenic microorganisms in foods (41).

A. *Salmonella*

Compared to other gram-negative food-borne pathogenic bacteria, *Salmonella* is most resistant to irradiation and has a D-value of 0.6 kGy. Therefore, any irradiation treatment that destroys *Salmonella* would also destroy other gram-negative pathogens (Table 1). Recommended doses of radiation for treatment of chilled (1.5–2.5 kGy) and frozen poultry (3.0–5.0 kGy) have proven effective in reducing the most resistant serotype of *Salmonella* in these products by approximately 3 log-cycles (42). Gamma radiation doses of 1.5–3.0 kGy reduced the numbers of *S.* Typhimurium in mechanically deboned chicken meat by 2.8–5.1 log-cycles (15). Thayer et al. (43) did not detect any viable *S.* Typhimurium cells on chicken wings inoculated with this pathogen at 10^3 or 10^4 CFU/g and irradiated at 2.7 kGy. The same authors reported that a dose of 1.8 kGy resulted in an estimated 19% survival rate; however, *S.* Typhimurium failed to recover from radiation injury during 3 days of refrigeration. Beuchat et al. (44) were able to reduce the

Table 1 D-Values for Radiation Resistance of Some Foodborne Pathogenic Bacteria

Pathogen	D-value	Product	Temperature	Ref.
Aeromonas hydrophilia	0.14–0.19	Ground fish	2	133
Campylobacter jejuni	0.186	Ground turkey	0–5	134
Clostridium perfringens (vegetative cells)	0.826	Ground pork	10	98
Escherichia coli O157:H7	0.24	Beef	2–4	10
Listeria monocytogenes	0.42–0.44	Ground pork	4	78
Salmonella spp.	0.61–0.66	Ground beef	4	10
Shigella dysenteria	0.40	Oysters	5	135
Staphylococcus aureus	0.40–0.46	Chicken	0	12
Vibrio parahaemolyticus	0.053–0.357	Crab meat	24	136
Yersinia enterocolitica	0.164–0.204	Ground pork	10	70

numbers of *Salmonella* in chilled ground beef by 2.0–3.0 log-cycles by applying radiation doses as low as 1.0–3.0 kGy. In a recent study, Chung et al. (45) compared the radurization effects of low-dose gamma irradiation and electron beam irradiation of fresh beef cuts (8°C) inoculated with *S.* Typhimurium to give 10^6 CFU/g. *S.* Typhimurium was undetected in irradiated samples stored aerobically at 5°C for 8 days. Electron beam irradiation at 1.5 kGy reduced *S.* Enteritidis cells on the surface of whole shell eggs and in liquid whole eggs and in liquid whole eggs by 4 log-cycles without significant alteration to egg quality (46). Gamma irradiation at 2.0 kGy was adequate for eliminating *Salmonella* in powdered egg at 20°C (47).

Rajkowski and Thayer (48) investigated the effects of low dose gamma irradiation at 19 ± 1°C for eliminating selected gram-negative pathogens on sprouts. A radiation dose as low as 0.5 kGy eliminated *Salmonella* from naturally contaminated radish sprouts, and the radiation D-value (0.46 kGy) was similar to previously reported values for this pathogen on moist food products.

B. *Campyobacter jejuni*

Campylobacter jejuni is frequently isolated from foods of animal origin and is believed to be the leading cause of foodborne gastroenteritis in the United States (49,50). This organism is among the most radiation-sensitive foodborne enteric pathogen. Collins et al. (51) reported a D-value of 0.19 for *C. jejuni* treated with electron beam radiation in vacuum-packaged ground pork. Clavero et al. (10) determined the sensitivity of *C. jejuni* to gamma irradiation in refrigerated (4 ± 1°C) or frozen (−16 ± 1°C) ground beef. D-values for *C. jejuni* in low-fat ground beef were 0.235 kGy at −16 ± 1°C and 0.175 at 4 ± 1°C. D-values in high-fat ground beef were 0.207 kGy at −16 ± 1°C and 0.178–0.199 kGy at 4 ± 1°C. Tarkowski et al. (52) reported D-values ranging from 0.08 to 0.11 kGy and 0.14 to 0.16 kGy for *C. jejuni* in filet americain (raw, ground beef with mayonnaise) or corresponding ground beef, respectively. The authors concluded that irradiation at 1 kGy was adequate for eliminating *C. jejuni* in foods because of the relatively low D-values (0.08–0.16 kGy) for this organism.

Lambert and Maxcy (53) investigated the effects of temperature, type of menstruum, and physiological state of cells on resistance of *C. jejuni* to gamma irradiation. Irradiation of *C. jejuni* in frozen meat was less effective in eliminating this pathogen than irradiation at refrigeration or an ambient temperature. The D-values for *C. jejuni* in brain heart infusion broth (BHI) ranged from 0.24 kGy at −30 ± 10°C to 0.30 kGy at 30 ± 10°C, with an intermediate value of 0.26 kGy at 0–5°C. D-values in ground beef at 30 ± 10°C, 0–5°C, and 30 ± 10°C were 0.315 kGy, 1.16 kGy, and 0.17 kGy, respectively. Early log phase *C. jejuni* cells were more sensitive to irradiation in BHI than log- or stationary phase cells. In contrast, no sinificant effect of physiological state of the culture was observed when cells were irradiated in ground beef or turkey meat.

C. *Escherichia coli*

Escherichia coli is part of the normal intestinal microflora of warm-blooded animals. The presence of this organism in foods is an indication of fecal contamination and the possible presence of enteric pathogens. Enterohemorrhagic *E. coli* (O157:H7) is a pathogen of major public health concern. It has a low infectious dose (~15 CFU/g) and causes hemorrhagic colitis, hemolytic uremic syndrome (HUS), and thrombotic thrombocytopenic pupura (TTP) (54). Outbreaks of foodborne illness caused by *E. coli* O157:H7 have been associated mainly with the consumption of undercooked beef (55). Outbreaks have also been linked to certain acidic foods including fermented sausages, apple cider, and yogurt (56–58). Recently this pathogen was incriminated in foodborne outbreaks involving seed sprouts (59).

The irrdiation sensitivity of *E. coli* O157:H7 does not differ greatly from that of other gram-negative foodborne pathogens. Radiation D-values of 0.241 kGy and 0.307 kGy were reported for this organism in ground beef at 4 ± 1°C and −16 ± 1°C, respectively (10). Lopez-Gonzalez et al. (17) observed higher D-values for *E. coli* O157:H7 in ground beef patties irradiated by gamma rays compared to electron beam irradiation. The authors suggested that the differences in D-values could be attributed to dose rates of 1.0 kGy/h for gamma rays and 17 kGy/min for electron beam. The relatively low gamma irradiation dose rates may allow microbial enzymes more time to repair radiation-induced damage in cells, thus resulting in higher D-values. Thayer and Boyd (16) investigated the effects of gamma irradiation on *E. coli* O157:H7 in chicken meat. D-values for this organism in mechanically deboned chicken meat were 0.27 kGy at 5°C and 0.42 kGy at −5°C. The authors concluded that the minimum irradiation dose for poultry (1.5 kGy), set by the U.S. Food and Drug Administration (FDA), would give a 4–5 log reduction of *E. coli* O157:H7. Since this organism is found in relatively low numbers (100 CFU/g), an irradiation dose of 1.5 kGy would be adequate to eliminate it from ground meat. In clarified apple juice at 2°C, D-values for acid-adapted and non–acid-adapted *E. coli* O157:H7 ranged from 0.22 to 0.31 kGy and 0.12 to 0.21 kGy, respectively (60). In radish, alfalfa, and broccoli sprouts, D-values for vegetable and meat isolates of *E. coli* O157:H7 were 0.30 and 0.34 kGy, respectively (48).

D. *Yersinia enterocolitica*

Yersinia enterocolitica has been isolated from a variety of food products, including vacuum-packaged meats, seafood, milk, and vegetables. This enteric pathogen is facultatively anaerobic and can grow at refrigeration temperature. Several foodborne disease outbreaks caused by *Y. enterocolitica* have been reported (61).

Several researchers have demonstrated the sensitivity of *Y. enterocolitica* to irradiation in different types of meat. D-values for three strains of this organism in filet americain and ground beef, respectively, ranged from 0.043 to 0.080 kGy and 0.10 to 0.21 kGy (52). Kamat and Thomas (8) reported a D-value of 0.09 kGy for *Y. enterocolitica* F5692 in 10% fish homogenate at 0–1°C. A gamma irradiation dose of 1.0 kGy was sufficient for eliminating low numbers ($<10^3$/g) of *Y. enterocolitica* on naturally contaminated fresh pork without any recovery of the pathogen during refrigeration storage (62). No survivors of *Y. enterocolitica* were detected in beefsteaks and ground beef irradiated at 1.5–2.0 kGy and then stored at 7°C for 1 week (32). Abu-Tarboush et al. (63) reported that a gamma irradiation dose of 2.5 kGy eliminated *Yersinia* spp. on naturally contaminated raw chicken. No *Yersinia* cells were detected on the irradiated chicken stored at 4°C for up to 21 days.

Shenoy et al. (64) investigated the survival of heat-shocked *Y. enterocolitica* following electron beam irradiation in ground pork. Irradiation of inoculated pork with 1.0 kGy was adequate to eliminate the organism. The D-value for both heat-shocked and non–heat-shocked *Y. enterocolitica* in fresh ground pork was 0.15 kGy. Neither heat shocking the organism nor packaging the inoculated pork under vacuum increased the radiation resistance of this pathogen.

Lepebe et al. (65) irradiated vacuum-packaged pork loins at 3 kGy and then stored them at 2–4°C. After 7 weeks of storage, 2.5 cm^2 thick pork chops were aseptically excised, wrapped in oxygen-permeable film (Saran Wrap™), then held at 5–7°C in a display case to simulate retail storage conditions. Microbiological analysis of the pork chops were conducted at 2-day intervals for 10 days. No Enterobacteriaceae were detected by direct planting; however, *Yersinia* spp. were recovered from some pork chop samples via enrichment. The detection of the pathogen via enrichment indicated that low numbers (<2 cells/cm^2) of *Yersinia* spp. survived in pork loins following irradiation at 3 kGy. The survivors most likely failed to initiate growth in pork during refrigeration storage. The use of enrichment to detect *Yersinia* survivors following irradiation indicates the need for improved detection methods for low numbers of pathogens that survive irradiation.

Initial populations of *Yersinia* (log 3.03) on chicken carcasses individually packaged in air [according to FDA regulation for irradiated poultry [66]) and irradiated at 2.5 kGy were reduced to undetectable levels. During 18 days of storage at 4°C, some *Yersinia* cells recovered from irradiation injury and grew under aerobic conditions (67). The packaging of irradiation under air as stipulated by FDA to prevent the growth of *C. botulinium* allowed growth of *Yersinia* cells that survived irradiation on chicken carcasses.

E. *Listeria monocytogenes*

Listeria monocytogenes is an enteric pathogen of major food safety concern. The high fatality rate associated with this organism and its ability to grow at refrigeration temperatures (68,69) have prompted FDA and the Food Safety and Inspection Services (FSIS) of the U.S. Department of Agriculture (USDA) to establish zero tolerances for this pathogen in ready-to-eat foods.

Elimination of *L. monocytogenes* from foods by use of gamma irradiation has been proposed (44,70,71), and several studies have demonstrated the efficacy of irradiation for inactivating this pathogen. Patterson (72) irradiated phosphate buffered saline (PBS) and ground poultry meat each containing four strains of *L. monocytogenes*. The D-values for *L. monocytogenes* in PBS and ground poultry meat, respectively, ranged from 0.32 to 0.49 kGy and 0.42 to 0.55 kGy. Cursel and Gurakau (73) reported that D-values for *L. monocytogenes* in trypticase soy broth supplemented with 0.6% yeast extract (TSBYE), in chicken meat slurry, and in raw ground beef, respectively, were 0.364, 0.599, and 0.699 kGy. The authors suggested that irradiation of meat at 2.5 kGy prior to refrigeration is an efficient way for preserving meat products contaminated with *L. monocytogenes* at 10^3–10^4 CFU/g. Huhtanen et al. (74) investigated the resistance of seven strains of *L. monocytogenes* to gamma irradiation in mechanically deboned chicken meat. D-values ranged from 0.27 to 0.77 kGy, with an average D-value of 0.45 kGy. Those researchers concluded that an irradiation dose of 2.0 kGy would be adequate to reduce the numbers of *L. monocytogenes* by 4 log-cycles.

Mead et al. (75) studied the effects of gamma irradiation on survival of *L. monocytogenes* on poultry carcasses and on the behavior of this pathogen during refrigeration at 5 or 10°C. Freshly processed, air-chilled broiler carcasses were inoculated with a four-strain cocktail of *L. monocytogenes* to give approximately, 10^2 or 10^4 CFU/cm^2 then irradiated at 2.5 kGy. About 62% of the unirradiated, freshly processed carcasses were naturally contamined with *L. monocytogenes*; counts of the pathogen ranged from 0.36 to 24 CFU/cm^2. After irradiation treatment, inoculated carcasses were tested for *L. monocytogenes* at 0, 7, 14, and 21 days of storage at 5°C, and at 0, 3, and 7 days for carcasses stored at 10°C. Immediately after irradiation, *L. monocytogenes* were detected in only 1 of 12 carcasses. No further detection of this pathogen occurred until day 14 at 5°C and day 5 at 10°C. At the end of each storage period, *L. monocytogenes* were detected in 8 of 36 carcasses that received the higher inoculum level. Also, *L. monocytogenes* was detected in only 1 of 18 carcasses that received the lower inoculum level, which is more typical of the numbers of the pathogen found on naturally contaminated meat. The study demonstrated that *L. monocytogenes* could be inactivated to a large extent by 2.5 kGy of gamma irradiation. However, it did not confirm the results of Huhtanen et al. (74) that 2.0 kGy is adequate to destroy *L. monocytogenes* at a level of 10^4 CFU/cm^2 of poultry meat.

Thayer and Boyd (20) studied the effect of gamma irradiation on the survival of a 4-strain cocktail of *L. monocytogenes* on beef at −60 to +15°C. D-values were determined at 5°C intervals from −20 to 5°C. The data were then used for developing an equation to predict the response of the pathogen to gamma irradiation within that temperature range. There was a significant response of *L. monocytogenes* on beef to the temperature during irradiation. In fact, $10^{2.9}$ more cells would survive an irradiation dose of 2.4 kGy at −20°C than at 0°C. A linear increase in the $\log_{10}$ D-values was observed from −5 to 20°C. This linear trend supported the hypothesis that destruction of microbial cells at temperatures even below freezing are directly related to reaction temperature as evident in chemical reactions.

Thayer and Boyd (76) demonstrated the combined effect of gamma irradiation and MAP for controlling or eliminating *L. monocytogenes* on ground turkey meat. Radiation-sterilized ground turkey meat was inoculated with *L. monocytogenes*, packaged under mixtures of N_2 and CO_2 and irradiated at 0–3 kGy. The authors reported a statistically significant, but probably not biologically significant, lower (0.39 log) predicted survival of *L. monocytogenes* in 100% CO_2 than in 100% N_2. Irradiation treatments were significantly more destructive to *L. monocytogenes* in meat packaged in air than in either MAP or vacuum. A concentration-dependent CO_2 inhibition of *L. monocytogenes* multiplication and/or recovery was observed in meat samples treated at doses greater than 1.0 kGy.

More recent studies involving the use of electron beam irradiation for improving the microbial safety of foods have demonstrated the efficacy of this type of irradiation for inactivating *L. monocytogenes*. Medium doses (1.8–2.0 kGy) of electron beam irradiation redued the population of *L. monocytogenes* in cooked pork chops (pumped with salt/polyphosphate brine or untreated) and cured hams to undetectable levels (77). Tarte et al. (78) investigated the sensitivity of three strains of *L. monocytogenes* along with *Listeria innocua* and *Listeria ivanovii* in ground pork treated with electron beam irradiation. Ground pork was inoculated with one of five strains of *Listeria* and irradiated from 0 to 1.25 kGy at 0.25 kGy intervals. D-values for the three *L. monocytogenes* strains ranged from 0.424 to 0.447 kGy. D-values for *L. innocua* and *L. ivanovii* were 0.638 kGy and 0.372 kGy, respectively. The authors concluded that the dose range (1.5–4.5 kGy) being considered by the FDA for irradiation of fresh beef and pork was sufficient for eliminating *L. monocytogenes* in pork.

F. *Staphylococcus aureus*

Enterotoxigenic *S. aureus* is a common cause of foodborne intoxication. Generally *S. aureus* may be present, at least in low numbers, in food products of animal origin or in foods handled by humans. A large number of food products, handled directly by humans and improperly refrigerated after preparation, have been incriminated in outbreaks of *S. aureus* foodborne intoxication (79).

Treatment of foods with irradiation can eliminate *S. aureus*. Thayer and Boyd (19) investigated the effectiveness of gamma irradiation for eliminating *S. aureus* in mechanically deboned chicken. The authors reported that doses of 1.5

kGy and 3.0 kGy could reduce population of *S.* aureus by 3.2 and 6.3 log-cycles, respectively. Populations of *S. aureus* in roast beef and gravy were reduced by 3–4 log-cycles following irradiation at 2 kGy (80). Inoculated pack studies performed on pedha, a fermented dairy product, confirmed that 10^5 CFU of *S. aureus* per gram were completely eliminated by gamma irradiation at 3 kGy (81). Thayer et al. (82) irradiated meat (5°C) from bison, ostrich, alligator, or caiman that was inoculated with three-strain mixture of *S. aureus*. The type of meat did not significantly change the resistance of *S. aureus* to irridation; therefore, all of the results were combined to obtain a radiation D-value of 0.37 ± 0.01 kGy. The authors concluded that *S. aureus* could be greatly reduced or eliminated from meats by doses of gamma irradiation between 1.5 and 3.0 kGy at 5°C. Doses of 1.5–3.0 kGy have been approved by FDA and USDA for irradiating poultry.

Monk et al. (83) reported D-values ranging from 0.435 to 0.453 kGy for *S. aureus* in frozen (−17 to −15°C) or refrigerated (3–5°C) ground beef patties treated with gamma irradiation. Neither temperature of meat during irradiation nor fat content of the meat [11.1–13.9% (low fat) and 27.1–27.9% (high fat)] affected the inactivation rate of *S. aureus*. The authors concluded that irradiation at 2.5 kGy would theoretically destroy 5.2 log of *S. aureus* per gram of ground beef. The significance of those results is more meaningful to the meat industry because the investigation was conducted under commercial processing and irradiation treatment conditions.

The efficacy of gamma irradiation (1.0 kGy), with or without high hydrostatic pressure (200 MPa for 30 min), for improving the microbiological quality of ground lamb meat during storage at 0–3°C was assessed by Paul et al. (84). Initial counts of total aerobic bacteria, coliforms, and *Staphylococcus* spp. on treated and control samples were 10^5, 10^2, and 10^4 CFU/g, respectively. Individual or combination treatments eliminated coliforms; however, irradiation or high hydrostatic pressure alone reduced numbers of staphylococci by only 1 log-cycle. Only combination treatments eliminated *Staphylococcus* spp.

G. *Bacillus cereus*

Bacillus cereus is an aerobic, spore-forming enteric pathogen that is normally present in soil, dust, and water. This pathogen may cause food-poisoning symptoms such as diarrhea and vomiting (85). Low numbers of *B. cereus* can be isolated from a variety of food products, including raw milk, dairy products, and rice products (86,87).

Grant and Patterson (70) investigated the sensitivity of *B. cereus* vegetative cells to irradiation in the components of a roast beef meal (beef, gravy, cauliflower, roast potato, and mashed potato). D-values for this pathogen ranged from 0.126 to 0.288 kGy. The pathogen exhibited the greatest sensitivity to irradiation in gravy compared to the other components of the meal. In another study, Grant et al. (80) reported that *B. cereus* vegetative cells were reduced by 3–4 log-cycles when roast beef and gravy were irradiated with 2.0 kGy.

Kamat and Thomas (8) reported D-values ranging from 0.15 to 0.25 kGy for *B. cereus* vegetative cells in 10% fish homogenate at 0–1°C. Inoculated pack studies, involving fatty (Indian sardine, 7.1% fat) and lean (Golden anchovy, 0.39% fat) fish indicated no difference in survival of *B. cereus* vegetative cells following irradiation at 1.0 kGy and 3.0 kGy. In a more recent study (88), high- or low-fat fish inoculated with vegetative cells of *B. cereus* was irradiated at 1.0 and 3.0 kGy. Numbers of *B.* cereus survivors were monitored during storage of the fish at 2–4°C for 2 weeks following irradiaiton. No increase in growth of *B. cereus* was detected in irradiated fish compared to controls.

Thayer and Boyd (11) evaluated the resistance of *B. cereus* vegetative cells and endospores to gamma irradiation in mechanically deboned chicken, ground pork loin, and beef gravy. The D-values for *B. cereus* ATCC 33018 logarithmic-phase cells, stationary-phase cells, and endospores in mechanically deboned chicken meat at 5°C were 0.184, 0.431, and 2.56 kGy, respectively. At 5°C, a dose of 7.5 kGy was required to eliminate 4.0×10^3 *B. cereus* cells in mechanically deboned chicken meat that was temperature-abused (30°C for 24 h) following irradiation treatment. The D-value for a mixture of endospores from six strains of *B. cereus* (five enterotoxic and one emetic) was 2.78 kGy in ground beef round, ground pork loin, or beef gravy, whereas the D-value for these strains in mechanically deboned chicken meat was 1.91 kGy. The authors concluded that irradiation of meat or poultry with approved doses could offer significanty protection against *B. cereus* vegetative cells but not against spores of this organism.

The irradiation resistance of *B. cereus* spores with or without dipicolinic acid was unaffected by gamma irradiation; D-values for both types of spores was 4.0 kGy. Although the presence of dipicolinic acid is linked to heat resistance in bacterial spores, this spore component was not shown to be associated with irradiation resistance (89). Ma and Maxcy (90) studied the factors that influence the radiation resistance of vegetation bacteria and spores during radappertization of meat. The authors reported that temperature, dessication, or suspending medium did not affect the irradiation resistance of *B. cereus* spores to the same extent as vegetative cells. Unlike vegetative cells, the spores were more resistant to irradiation at ambient temperatures than at subfreezing temperature.

H. Clostridia

The canning industry has employed the 12-D concept for eliminating *C. botulinum* spores from canned foods. This concept involves the use of an appropriate heating protocol to reduce numbers of *C. botulinum* spores from 10^2 to 10^0. The 12-D concept has been applied to radiation sterilization or radappertization. Radappertization involves the application of an appropriate dose of irradiation under the proper conditions to destroy all viable cells, including spores of food spoilage organisms as well as pathogens. Typical doses used for radappertization are within the 30–40 kGy range and are not allowed for irradiation of foods except for sterilizing spices and vegetable seasonings. Irradiation of foods with doses up to 10 kGy, the maximum dose allowed for food irradiation, is inadequate for destruction of all bacterial spores, including those of clostridia, unless spores are present in low numbers (91).

Firstenberg-Eden et al. (92) investigated the extent of growth of natural microflora and *C. botulinum* type E on chicken following irradiation at 3.0 kGy. Populations of natural microflora were reduced from 10^4 to 10^6 to approximately 110–500 cells per 7 cm^2, whereas *C. botulinium* type E (Beluga) spores were reduced by only 1 log-cycle. Within 8 days of storage at 10°C, microbial survivors grew and produced off-odors on the irradiated chicken skin under aerobic or anaerobic conditions; however, no toxin production by *C. botulinum* survivors was detected until 14 days. Under temperature-abuse conditions (30°C), the natural microflora rapidly increased in numbers and produced off-odors before the skin became toxic.

Thayer et al. (93) assessed the effects of irradiation and anaerobic refrigerated storage on the natural microflora, *Salmonella*, and *C. botulinum* types A and B in vacuum-canned, mechanically deboned chicken meat. The meat product was inoculated with *C. botulinum* spores (20 strains of type A and B; final spore concentration of about 400/g) and then irradiated with 0, 1.5, or 3.0 kGy. The irradiated samples were stored at 5°C for 0, 2, and 4 weeks. *C. botulinum* toxin was not detected in any of the samples stored at 5°C; however, samples became toxic within 18 hours of temperature abuse at 28°C and exhibited signs of spoilage such as swelling of the cans and off-odors. It should be noted that some strains of *C. botulinum* types B, E, and F could grow and produce toxin at temperatures as low as 3.5–5°C (94); therefore, it is advisable to store irradiated foods at temperatures below 3°C.

Lambert et al. (95–97) studied the effect of MAP in combination with irradiation at 0.5 and 1.0 kGy on toxin production in fresh pork inoculated with *C. botulinum* type A and B spores and stored at 5, 15, and 25°C following irradiation treatment. After 14 days at 15°C, *C. botulinum* toxin was detected in both irradiated and unirradiated pork packaged at 10 or 20% headspace oxygen. Pork packaged in 0% oxygen and oxygen absorbent and irradiated with 1.0 klGy became toxic after 43 days at 15°C, whereas corresponding unirradiated samples became toxic in 21 days. The earlier onset of toxin production in pork initially packaged under oxygen could be attributed to respiratory activities of aerobic microflora and meat tissue resulting in the production of CO_2, which enhances spore germination. Toxin production was delayed in irradiated samples that were packaged under five levels (15, 30, 45, 60, and 75%) of CO_2 or 20% O_2 and stored at

15°C. *C. botulinum* toxin was not detected in any of the samples of pork stored at 15°C.

Vegetative cells of *Clostridium perfringens* exhibited greater resistance to irradiation than *L. monocytogenes*, *S.* Typhimurium, *E. coli*, and *Y. enterocolitica* when irradiated with 2.5 kGy in ground pork with 1.75 kGy did not significantly reduce the numbers of *C. perfringens*. Monk et al. (99) suggested that the presence of spores could account for the observed irradiation resistance of *C. perfringens*, although it was assumed that the inocula used by Grant and Patterson (98) consisted mainly of *C. perfringens* vegetative cells. In a later study Grant and Patterson (70) reported D-values ranging from 0.342 to 0.586 kGy for *C. perfringens* cells in a roast beef meal, which consisted of beef, gravy, cauliflower, roast potato, and mashed potato. Harewood et al. (100) reported a D-value of 2.71 kGy for *C. perfringens* in hard-shelled clams.

Naik et al. (101) treated prepacked, fresh buffalo meat with irradiation at 2.5 kGy and then exposed the irradiated meat to ambient temperatures (28–30°C). *Clostridium* spp. were not detected in the irradiated samples throughout the storage period, whereas unirradiated samples tested positive for *Clostridium* spp. after 12 hours.

Clostridia spores are more easily inactivated by irradiation after they have been exposed to certain physical treatments, such as heating and high hydrostatic pressure. Crawford et al. (102) reported a D-value of 4.1 kGy for *Clostridium sporogenes* PA 3679 spores in chicken breast treated with electron beam irradiation alone. When the inoculated chicken breast was subjected to high hydrostatic pressure (6800 atm, 80°C, 20 min) followed by irradiation with 3.0 kGy, the D-value for *C. sporogenes* spores was reduced to approximately 2.0 kGy. Garcia-Zepeda et al. (103) investigated the combined effects of extrusion cooking and electron beam irradiation on *C. sporogenes* PA 3679 spores in vacuum-packaged beef-based steaks, which contained beef cardiac muscle. During formulation of the meat product, *C. sporogenes* spores were added to give a final concentration of 10^4 CFU/g. After extrusion cooking (72°C), surviving *C. sporogenes* spores ranged from 3.17 to 3.50 log CFU/g. Irradiation of the extruded inoculated product with 3.5 kGy reduced the surviving spores to undetectable levels.

I. Molds and Yeast

Microbial contamination of food products from ingredients such as spices and vegetable seasonings is of major concern to food processors. Spices are usually highly contaminated and invariably have microbial populations ranging from 10^4 to 10^8 CFU/g. Although spices, grains, and dried vegetables may not be conducive to long-term survival of bacterial vegetative cells, molds are often the dominant microorganisms. Decontamination of spices, grains, and other dry agricultural products by irradiation is a suitable, cost-effective alternative to the traditionally use fumigation process involving ethylene oxide (104).

Irradiation has proven to be effective for inactivating molds and yeast in a wide variety of raw agricultural products. Treatment of cacao beans with an irradiation dose of 5.0 kGy reduced *Penicillium citrinum* spores by approximately 5 log-cycles, whereas 4.0 kGy reduced *Aspergillus flavus* spores by approximately 7 log-cycles (105). Irradiation at 4.0 kGy destroyed most *Alternaria, Fusarium*, and *Epicoccum* spp. in barley seeds (106). Byun et al. (107) reported that irradiation of animal feed at 3.0–7.0 kGy eliminated *Aspergillus*, *Penicillium*, *Cladosporium*, *Aureobasidium*, and *Rhizopus* spp. The authors concluded that, unlike thermal treatment, irradiation was less detrimental to the nutritional content of animal feeds. No growth of *A. flavus* or production of aflatoxin occurred in ground beef that was irradiated with 1.50 kGy and stored at 5°C for 2 weeks (108). Irradiation of dry red kidney beans at 1.5 kGy and 3.0 kGy was necessary for complete inactivation of *Aspergillus* spp. and *Penicillium* spp., respectively (109).

Hassan and Aziz (110) investigated the influence of moisture content and storage temperature on aflatoxin production by *Aspergillus flavus* EA-81 in maize following gamma irradiation. At 8°C, the viability of *A. flavus* on irradiated maize decreased over time with increasing moisture content. At 28°C, the numbers of viable conidiospores of the organism increased from 4.5×10^7 to about 3.0×10^8/g of maize with 40% moisture. Irradiation of maize at 1.0 or 2.0 kGy greatly reduced mold growth relative to unirradiated controls, whereas 4.0 kGy eliminated all viable fungi. Increased doses of irradiation resulted in decreased production of alfatoxin B_1; negligible amounts of this mycotoxin were detected in maize irradiated with 4.0 kGy. Maize inoculated with *A. flavus* following irradiation and storage had higher spore counts and aflatoxin levels than unirradiated and inoculated controls after 30 days. The authors suggested that the natural competitive microflora in control samples prevented the growth of *A. flavus* and limited aflatoxin production.

Narvaiz et al. (111) tested the effectiveness of irradiation for inactivating molds and yeast on almonds and cashews nuts. Doses of 1.0–1.5 kGy and 2.0 kGy reduced the fungal population on almonds by 10-fold and 100-fold, respectively. The population of fungi on cashew nuts was reduced by about 10-fold with 1.0–2.0 kGy and by 100-fold with 3.0 kGy. Differences in genera of molds and yeast on the nuts could have contributed to differences in the sensitivity of these organisms to irradiation.

A comparison of the resistance of fungal spores to gamma and electron beam irradiation in distilled water was conducted by Blank and Corrigan (112). D-values of spores treated with gamma irradiation ranged from 0.236 to 0.416 kGy and from 0.209 to 0.319 kGy for *Penicillium* spp. and *Aspergillus* spp., respectively. D-values of spores treated with electron beam irradiation ranged from 0.192 to 0.341 kGy and from 0.198 to 0.243 kGy for *Penicillium* spp. and *Aspergillus* spp., respectively. About 50% of the *Aspergillus* spp. and 66% of *Penicillium* spp. demonstrated significantly greater sensitivity to electron beam irradiation compared to gamma irradiation.

The use of electron beam irradiation for eliminating molds on cheddar cheese was studied by Blank et al. (113). The lowest irradiation dose required to inactivate 50–60 spores of *Aspergillus ochraceus* or *Penicillium cyclopium* per cm^2 of cheese was about 0.42 kGy and 0.95 kGy, respectively. D-values for *A. ochraceus* and *P. cyclopium* spores were 0.21 and 0.42 kGy, respectively.

McCarthy and Damoglou (114) evaluated the effectiveness of low-dose gamma irradiation for inactivating yeast in fresh sausage with or without sulfite. Populations of *Debaromyces* spp. were greatly reduced by irradiation at 1.5 kGy and were only isolated from unirradiated samples after storage at 4°C for 7 and 14 days. *Trichosporon* spp. exhibited greater resistance to irradiation and were isolated from unsulfited sausages that were irradiated at 3.0 kGy and stored at 4°C for 7 and 14 days; none was isolated from irradiated (3.0 kGy) sausage that contained sulfite. *Candida* spp. represented the major microflora of both types of sausage initially and persisted after irradiation at 3.0 kGy. Of all the yeast strains tested, *Candida* spp. were the most resistant to irradiation, whereas *Sporobolomyces roseus* was the least resistant. Since there were no significant differences between the number of yeast in unirradiated sulfited samples and unsulfited samples irradiated at 1.5 kGy, the authors concluded that low-dose irradiation could be an alternative treatment to sulfite preservation of sausages.

Some researchers have evaluated the effectiveness of low-dose irradiation for inactivating yeast on fruits and vegetables. Toraskar and Modi (115) demonstrated that irradiation of mangoes with 0.5 kGy resulted in approximately 1 log-cycle reduction of *Saccharomyces cerevisiae*. Irradiation of fresh-cut lettuce with 0.19 kGy reduced populations of yeast from 1400 to 60 CFU/g (116).

J. Viruses

Several studies have focused on the use of irradiation for inactivating foodborne enteric viruses. Sullivan et al. (117) studied the radiation resistance of 30 viruses in Eagle's minimal essential medium supplemented with 2% serum. Coxsackievirus, echovirus, and poliovirus were among the 30 viruses tested. D-values for the viruses ranged from 3.9 to 5.3 kGy. D-values were lower for five selected viruses irradiated in distilled water and ranged from 1.0 to 1.4 kGy. In another study, Sullivan et al. (118) assessed the radiation resistance of coxsackievirus in raw and cooked ground beef. The D-values for the virus in raw ground beef at −90 to 16°C ranged from 6.8 to 7.5 kGy; D-values in cooked ground beef ranged from 6.8 to 8.1 kGy. Generally, D-values increased as temperature during irradiation decreased from −10 to −90°C.

Mallett et al. (119) investigated the effectiveness of irradiation for inactivating viruses and improving the sanitary quality of live hard-shell clams, *Mercenaria mercenaria*, and oysters, *Crassostrea virginica*. Gamma irradiation of the shellfish significantly reduced the numbers of viruses without affecting the shellfish survival rates or desirable sensory qualities. D-values for hepatitis A virus and rotavirus were 2.0 and 2.4 kGy, respectively. Results of the investigation indicated that irradiation, in combination with depuration treatment, could greatly improve the sanitary quality of shellfish.

Some researchers explored the use of irradiation and heating for inactivating viruses. Lasta et al. (120) reported that a combination of heating (78°C for 20 min) and gamma irradiation at 0.5 and 0.24 kGy was effective in destroying A, O, and C serotypes of foot-and-mouth disease virus. Pirtle et al. (121) evaluated the extent of heat sensitivity of four viruses following irradiation treatment. The viruses were irradiated in ground pork at doses of 4.4 and 5.2 kGy. Irradiated and nonirradiated pork samples were heated at four temperatures for four time intervals and tested for surviving viruses. Irradiation did not increase the heat sensitivity of the viruses to a level that has practical application for destruction of viruses in meat.

K. Parasites

Animal parasites that cause foodborne illness in humans belong to three distinct groups, namely, roundworms, flatworms, and protozoa. Irradiation of meatborne trichina larvae, cysticerci, and parasitic protozoa could result in loss of infectivity, loss of pathogenicity, sexual sterility, interruption of the life cycle, and death of the parasite. Generally, the application of relatively high doses (4–10 kGy) of irradiation to destroy foodborne parasites could cause undesirable organoleptic changes in raw foods (122). Alternatively, the application of much lower irradiation doses could eliminate infectivity of the parasite while minimizing undesirable changes in the irradiated foods.

Wilkinson and Gould (2) reported that lower doses (0.1–2.0 kGy) of irradiation were sufficient for eliminating infectivity of a wide variety of foodborne parasites by preventing their maturation and ability to reproduce. Gamma irradiation at doses of 7.0–9.3 kGy destroyed *Trichinella spiralis* in situ, whereas 0.18 kGy stopped the maturation of the larvae to the adult stage (123). Sivinski et al. (124,125) reported that gamma irradiation of *T. spiralis*–infected pork with 0.15–0.30 kGy caused the parasite to become sexually sterile and prevented maturation of the larvae in the host's gut. The irradiation sensitivity of encysted trichina in pork was not significantly affected by age of the muscle or oxygen tension in the meat. Results of these studies indicated that irradiation of trichina-infested pork with 0.3 kGy could greatly improve the safety of this meat for human consumption. Irradiation at 0.3–1.0 kGy was approved by FDA for control of *T. spiralis* in pork (126).

Humans could become infected by adult tapeworms from beef (*Taenia saginata*) and from pork (*Taenia solium*). The larval forms of *T. saginata* and *T. solium* are known as *Cysticercus bovis* and *Cystericercus cellulosae*, respectively. Cysticerci could infect humans who ingest undercooked, infested beef or pork. *C. bovis* and *C. cellulosae* could be inactivated by irradiating infested carcasses with 0.2–0.6 kGy (127). Irradiation beef infested with *C. bovis* at 0.4 kGy prevented the development of the parasite in humans (128).

Toxoplasma gondii is a coccidian protozoan, which causes toxoplasmosis in humans. Irradiation has proven effective for inactivating *T. gondii* as well as its infective oocysts. Dubey and Thayer (129) reported that *T. gondii* protozoa could be inactivated by irradiation at 0.3–0.7 kGy. Song et al. (130) irradiated tissues of mice and pigs that contained *T. gondii* oocysts. Irradiated tissue, unlike unirradiated controls, did not produce infection when fed to animals. The minimal effective irradiation dose for eliminating the infectivity of the Chinese NT strain and the American ME-49 and TS-2 strains of *T. gondii* oocysts in mouse and pig tissues was approximately 0.6 kGy.

Dubey et al. (131) assessed the effect of gamma irradiation on unsporulated and sporulated oocysts of *T. gondii*. Unsporulated oocysts irradiated with doses greater than or equal to 0.4–0.8 kGy were able to sporulate but lost their ability to infect mice. Sporulated oocysts irradiated with those same doses were able to excyst; however, the sporozoites were incapable of producing infection in mice. *T. gondii* was detected in tissues of mice up to 5 days but not 7 days after feeding the mice oocysts irradiated at 0.5 kGy. Results of the study indicated that irradiation at 0.5 kGy is adequate for inactivating oocysts of *T. gondii*.

V. MECHANISM OF INACTIVATION

When microorganisms are exposed to ionizing radiation, high-energy rays and particles collide with components of the microbial cells and produce changes at both molecular and atomic levels. These collisions result in very rapid absorption of energy by thousands of atoms and molecules in a fraction of a second. Changes at the molecular level occur when collisions produce enough energy to break chemical bonds between atoms resulting in the production of free radicals. Free radicals have unpaired electrons, which make them highly unstable. They tend to react with each other and with other molecules to gain stability via pairing of their odd electrons. Changes at the atomic level occur when the energy from collisions between ionizing radiation and cellular components are enough to expel an electron from an atomic orbit to produce ion pairs. The production of free radicals and ion pairs, reaction of free radicals with components of the microbial cell, and recombination of free radicals involved in the mechanism by which irradiation inactivates microorganisms.

The antimicrobial mechanism of action of ionizing radiation occurs via both direct and indirect effects of irradiation on cell components, including DNA and the cytoplasmic membrane (132). The direct effect involves removal of electrons from DNA resulting in damage to this vital genetic material. High-energy rays and particles make direct contact with DNA and other cell components much as a bullet hits a target. Direct damage to vital cell components does not need to occur for irradiation to inactivate microorganisms. Microbial inactivation could result from the indirect action of irradiation involving the radiolysis of water in the cell as well as in the suspending mentstuum. During the radiolysis of water, water molecules are altered to produce highly reactive hydrogen and hydroxyl radicals. These radicals can alter bases such as thymine to form dihydroxydihydrothymine. In addition, they can cause oxidation, reduction, and breakage of carbon-to-carbon bonds of cell components, including DNA. Radiation damage to the DNA of microorganisms includes both single-strand and double-strand breaks. Free radicals can also react with each other, with dissolved oxygen, and with other molecules and ions that may be present in water to form toxic oxygen derivatives and other reaction species that are lethal to microbial cells. For example, the combination of hydrogen radicals produces hydrogen gas.

$$\cdot H + \cdot H \rightarrow H_2$$

two hhydroxyl radicals combine to form hydrogen peroxide,

$$\cdot OH + \cdot OH \rightarrow H_2O_2$$

a hydrogen radical combines with dissolved oxygen to produce a peroxide radical,

$$\cdot H + O_2 \rightarrow \cdot HO_2$$

and two peroxide radicals interact to form hydrogen peroxide and oxygen,

$$\cdot HO_2 + \cdot HO_2 \rightarrow H_2O_2 + O_2$$

Hydrogen peroxide and the hydroxyl radical are powerful oxidizing agents, whereas hydrogen radicals are reducing agents. These reactive species are deleterious to microorganisms and cause damage to DNA, cytoplasmic membrane, and protein. The inability of microorganisms to repair lesions caused by free radicals and other reactive species results in the loss of their ability to replicate DNA and reproduce, which leads to cell death.

REFERENCES

1. JF Diehl. Safety of Irradiated Food. 2nd ed. New York: Marcel Dekker, 1995.
2. VM Wilkinson, GW Gould. Food Irradiation. A Reference Guide. Oxford: Butterworth-Heinemann, 1996.
3. DG Olson. Irradiation of food. Food Technol 52:56–62, 1998.
4. M Satin. Food Irradiation: A Guidebook. Lancaster, PA: Technomic Publishing Co., 1993.
5. RM Ehioba, AA Kraft, RA Molins, HW Walker, DG Olson, G Subbaraman, RP Skowronski. Identification of microbial isolates from vacuum-packaged ground pork irradiated at 1 kGy. J Food Sci 53:278–279, 281, 1988.
6. AD Lambert, JP Smith, KL Dodds. Physical, chemical, and sensory changes in irradiated fresh pork packaged in modified atmosphere. J Food Sci 57:1294–1299, 1992.
7. DW Thayer, G Boyd, RK Jenkins. Low-dose gamma irradiation and refrigerated storage in vacuo affect microbial flora of fresh pork. J Food Sci 58:717–719, 733, 1993.
8. A Kamat, P Thomas. Radiation inactivation of some foodborne pathogens in fish as influenced by fat levels. J Appl Microbiol 84:478–484, 1998.
9. JD Monk, RS Clavero, LR Beuchat, MP Doyle, RE Brackett. Irradiation inactivation of *Listeria monocytogenes* and *Staphylococcus aureus* in low and high fat frozen and refrigerated ground beef. J Food Prot 51:969–974, 1994.
10. MR Clavero, JD Monk, LR Beauchat, MP Doyle, RE Brackett. Inactivation of *Escherichia coli* O157:H7, salmonellae, and *Campylobacter jejuni* in raw ground beef by gamma irradiation. Appl Environ Microbiol 60:2069–2075, 1994.
11. DW Thayer, G Boyd. Control of enterotoxic *Bacillus cereus* on poultry or red meats and in beef gravy by gamma irradiation. J Food Prot 57:758–764, 1994.
12. DW Thayer, G Boyd, JB Fox Jr., L Lakritz, JW Hampson. Variations in radiation sensitivity of foodborne pathogens associated with the suspending meat. J Food Sci 60:63–67, 1995.
13. DW Thayer, G Boyd, JB Fox Jr., L Lakritz. Effects of NaCl, sucrose, and water content on the survival of *Salmonella trphimurium* on irradiated pork and chicken. J Food Prot 58:490–496, 1995.
14. DW Thayer, G Boyd. Effect of ionizing radiation dose, temperature, and atmosphere on the survival of *Salmonella typhimurium* in sterile mechanically deboned chicken meat. Poult Sci 70:381–388, 1991.
15. DW Thayer, G Boyd. Survival of *Salmonella typhimurium* ATCC 14028 on the surface of chicken legs or in mechanically deboned chicken meat gamma irradiated in air or vacuum at temperatures of -20 to +20°C. Poult Sci 70:1026–1033, 1991.
16. DW Thayer, G Boyd. Elimination of *Escherichia coli* O157 in meats by gamma irradiation. Appl Environ Microbiol 59:1030–1034, 1993.
17. V Lopez-Gonzalez, PS Murano, RE Brennan, EA Murano. Influence of various commercial packaging conditions on survival of *Escherichia coli* O157:H7 to irradiation by electron beam versus gamma rays. J Food Prot 62:10–15, 1999.
18. HM EL-Bisi, OP Snyder, RE Levin. Radiation death kinetics of *C. botulinum* spores

at cryogenic temperatures. In: M Ingram, TA Roberts, eds. Botulism 1966, Proc 5th Intl Symp Food Microbiol. London: Chapman and Hall, Ltd., 1966, pp. 89–94.
19. DW Thayer, G Boyd. Gamma ray processing to destroy *Staphylococcus aureus* in mechanically deboned chicken meat. J Food Sci 57:848–851, 1992.
20. DW Thayer, G Boyd. Radiation sensitivity of *Listeria monocytogenes* on beef as affected by temperature. J Food Sci 60:237–240, 1995.
21. AT Matsuyama, MJ Thornley, M Ingram. The effect of freezing on the radiation sensitivity of vegetative bacteria. J Appl Bacteriol 27:110–124, 1964.
22. MA Bruns, RB Maxcy. Effect of irradiation temperature and drying on survival of highly radiation resistant bacteria in complex menstrual. J Food Sci 44:1743–1746, 1979.
23. IA Taub, RA Kaprielan, JW Halliday, JE Walker, P Angelini, C Meritt Jr. Factors affecting radiolytic effects in food. Radiat Physics Chem 14:639–653, 1979.
24. D Billen. Free radical scavenging and the expression of potentially lethal damage in x-irradiated repair deficient *Escherichia coli*. Radiat Res 11:354–360, 1987.
25. H Hollaender, GE Stapleton, FL Martin. X-ray sensitivity of *E. coli* as modified by oxygen tension. Nature 167:103–104, 1951.
26. M Thornley. Microbiological aspects of the use of radiation for elimination of salmonellae from foods and feeding stuffs. In: Radiation Control of Salmonellae in Food and Feed Products. Technical Report Series No. 22. IAEA, Vienna, 1963, pp. 81–106.
27. ER Epp, H Weiss, A Santomasso. The oxygen effect in bacterial cells irradiated with high-intensity pulsed electrons. Radiat Res 84:320–325, 1968.
28. A Samuni, C Czapski. Radiation-induced damage in *Escherichia coli* B: the effect of superoxide radicals and molecular oxygen. Radiat Res 76:624–632, 1978.
29. DW Thayer, G Boyd. Irradiation and modified atmosphere packaging for the control of *Listeria monocytogenes* on turkey meat. J Food Prot 62:1136–1142, 1999.
30. JW Hastings, WH Holzapfel, JG Niemand. Radiation resistance of lactobacilli isolated from radurized meat relative to growth and environment. Appl Environ Microbiol 52:898–901, 1986.
31. M Patterson. Sensitivity of bacteria to irradiation on poultry meat under various atmospheres. Lett Appl Microbiol 7:55–58, 1988.
32. AH Fu, JG Sebranek, EA Murano. Survival of *Listeria monocytogenes*, *Yersinia enterocolitica*, and *Escherichia coli* O157:H7 and quality changes after irradiation of beef steak and ground beef. J Food Sci 60:972–977, 1995.
33. MA El-Shenawy, AE Yousef, EH Marth. Inactivation and injury of *Listeria monocytogenes* in tryptic soy broth or ground beef treated with gamma irradiation. Lebensm Wiss Technol 22:387–390, 1989.
34. MF Patterson. Sensitivity of *Listeria monocytogenes* to irradiation on poultry meat and in phosphate-buffered saline. Lett Appl Microbiol 8:181–184, 1989.
35. M Lee, JG Sebranek, JG Olson, JS Dickson. Irradiation and packaging of fresh meat and poultry. J Food Prot 59:62–72, 1996.
36. DL Archer, JE Kvenberg. Incidence and cost of foodborne diarrheal disease in the United States. J Food Prot 48:887–894, 1985.
37. PR Lee. Irradiation to prevent foodborne illness. From the Assistant Secretary for Health, U.S. Public Health Service. J Am Med Assoc 272:261, 1994.

38. P Loaharanu, D Murrell. A Role for irradiation in control of foodborne parasites. Trends Food Sci Technol 5:190–195, 1994.
39. JC Buzby, T Roberts, CT Jordan Lin, JM MacDonald. Bacterial foodborne disease: medical costs and productivity losses. USDA Agric Econ Rep No. 741, 1996.
40. FK Kaferstein. The contribution of food irradiation to food safety and food security. In: DAE Ehlermann, WEL Spiess, W Wolf, eds. Lebensmittelbestrahlung. BFE-R-92-01. Karlsruhe: Bundesforschungsanstalt für Ernahrung, 1992.
41. JE Corry, C James, SJ James, M Hinton. *Salmonella*, *Campylobacter* and *Escherichia coli* O157:H7 decontamination techniques for the future. Int J Food Microbiol 28:187–196, 1995.
42. EH Kampelmacher. Irradiation of food: a new technology for preserving and ensuring the hygiene of foods. Fleischwirtschaft 64:322–327, 1984.
43. DW Thayer, CY Dickerson, DR Rao, G Boyd, CB Chawan. Destruction of *Salmonella typhimurium* on chicken wings by gamma radiation. J Food Sci 57:586–589, 1992.
44. LB Beuchat, MP Doyle, RE Brackett. Irradiation inactivation of bacterial pathogens in ground beef. Report to the American Meat Institute, University of Georgia Center for Food Safety and Quality Enhancement, September 1993.
45. MS Chung, YT Ko, WS Kim. Survival of *Pseodomonas fluorescens* and *Salmonella* Typhimurium after electron beam and gamma irradiation of refrigerated beef. J Food Prot 63:162–166, 2000.
46. LE Serrano, EA Murano, K Shenoy, DG Olson. D values of *Salmonella enteritidis* isolates and quality attributes of shell eggs and liquid whole eggs treated with irradiation. Poult Sci 76:202–206, 1997.
47. P Narvaiz, G Lescano, E Kairiyama. Physiocochemical and sensory analysis of egg powder irradiated to inactivate *Salmonella* and reduce microbial load. J Food Safety 12:263–282, 1992.
48. KT Rajkowski, DW Thayer. Reduction of *Salmonella* spp. and strains of *Escherichia coli* O157:H7 by gamma radiation of inoculated sprouts. J Food Prot 7:871–875, 2000.
49. I Nachamkin, MJ Blaser, LS Tompkins, eds. *Campylobacter jejuni*. Current status and future trends. Washington, DC. American Society of Microbiologists, 1992.
50. Centers for Disease Control and Prevention. Incidence of foodborne illnesses—FoodNet. MMWR 47:782–786, 1998.
51. CI Collins, EA Murano, IV Wesley. Survival of *Arcobacter butzeri* and *Campylobacter jenuni* after irradiation treatment in vacuum-packaged ground pork. J Food Prot 11:1164–1166, 1996.
52. JA Tarkowski, CC Stofer, RR Beumer, EH Kampelmacher. Low dose gamma irradiation of raw meat. I. Bacteriological and sensory quality effects in artificially contaminated samples. Int J Food Microbiol 1:13–23, 1984.
53. AD Lambert, RB Maxcy. Effect of gamma radiation on *Campylobacter jejuni*. J Food Sci 49:665–667, 1984.
54. MP Doyle. *Escherichia coli* O157:H7 and its significance in foods. Int J Food Microbiol 12:289–302, 1991.
55. MP Doyle, JL Schoeni. Isolation of *Escherichia coli* O157:H7 from retail fresh meats and poultry. Appl Environ Microbiol 53:2394–2396, 1987.

56. Centers for Disease Control and Prevention. *Escherichia coli* O157:H7 outbreak linked to commercially distributed dry-cured salami—Washington and California. MMWR 44:157–160, 1995.
57. RE Besser, SM Lett, JT Weber, MP Doyle TJ Barrett, JG Wells, PM Griffin. An outbreak of diarrhea and hemorrhagic uremic syndrome from *Escherichia coli* O157:H7 in fresh pressed apple cider. JAMA 269:2217–2220, 1993.
58. D Morgan, CP Newman, DN Hutchinson, AM Walker, B Rowe, F Majid. Verotoxin producing *Escherichia coli* O157:H7 infections associated with the consumption of yoghurt. Epidemiol Infect 111:181–187, 1993.
59. National Advisory Committee on Microbiological Criteria for Foods. Microbiological safety evaluation and recommendations on fresh produce. Food Control 10: 117–134, 1999.
60. RL Buchanan, SG Edelson, K Snipes, G Boyd. Inactivation of *Escherichia coli* O157:H7 in apple juice by irradiation. Appl Environ Microbiol 64:4533–4535, 1998.
61. DA Schiemann. *Yersinia enterocolitica* and *Yersinia pseudotuberculosis*. In: MP Doyle, ed. Foodborne Bacterial Pathogens. New York: Marcel Dekker, Inc., 1989, pp. 601–672.
62. AS Kamat, S Khare, T Doctor, PM Nair. Control of *Yersinia enterocolitica* in raw pork and pork products by gamma irradiation. Int J Food Microbiol 36:69–76, 1997.
63. HM Abu-Tarboush, HA Al-Khatani, M Atia, A Abu-Arab, AS Bajaber. Sensory and microbial quality of chicken as affected by irradiation and postirradiation storage at 4°C. J Food Prot 60:761–770, 1997.
64. K Shenoy, EA Murano, DG Olson. Survival of heat-shocked *Yersinia enterocolitica* after irradiation in ground pork. Int J Food Microbiol 39:133–137, 1998.
65. S Lebepe, RA Molins, SP Charoen, H Farrar IV, RP Skowronski. Changes in microflora and other characteristics of vacuum-packaged pork loins irradiated at 3.0 kGy. J Food Sci 55:918–924, 1990.
66. FDA. Irradiation in the production, processing and handling of food. Fed Reg 55: 18538, 19701, 1990.
67. PO Lamuka, GR Sunki, DR Chawan, DR Rao, LA Shackelford. Bacteriological quality of freshly processed broiler chickens as affected by carcass pre-treatment and gamma irradiation. J Food Sci 57:330–332, 1992.
68. SA Palumbo. Is refrigeration enough to restrain foodborne pathogens? J Food Prot 49:1003–1009, 1986.
69. JR Juntilla, SI Niemla, J Hirn. Minimum growth temperatures of *Listeria monocytogenes* and nonhaemolytic *Listeria*. J Appl Bacteriol 65:321–327, 1988.
70. IR Grant, MF Patterson. Sensitivity of foodborne pathogens to irradiation in the components of a chilled ready meal. Food Microbiol 9:95–103, 1992.
71. MF Patterson, AP Damoglou, RR Buick. Effects of irradiation dose and storage temperature on the growth of *Listeria monocytogenes* on poultry meat. Food Microbiol 10:197–203, 1993.
72. MF Patterson. Sensitivity of *Listeria monocytogenes* to irradiation on poultrymeat and in phosphate-buffered saline. Lett Appl Microbiol 8:181–184, 1989.
73. B Gursel, GC Gurakan. Effects of gamma irradiation on the survival of *Listeria*

monocytogenes and on its growth at refrigeration temperature in poultry and red meat. Poult Sci 76:1661–1664, 1997.
74. CN Huhtanen, RK Jenkins, DW Thayer. Gamma radiation sensitivity of *Listeria monocytogenes*. J Food Prot 52:610–613, 1989.
75. GC Mead, WR Hudson, R Arafin. Survival and growth of *Listeria monocytogenes* on irradiated poultry carcasses. Lancet I:1036, 1990.
76. DW Thayer, G Boyd. Irradiation and modified atmosphere packaging for the control of *Listeria monocytogenes* on turkey meat. J Food Prot 62:1136–1142, 1999.
77. AH Fu, JG Sebranek, EA Murano. Survival of *Listeria monocytogenes* and *Salmonella typhimurium* and quality attributes of cooked pork chops and cured ham after irradiation. J Food Sci 60:1001–1005; 1008, 1995.
78. RR Tarte, EA Murano, DH Olson. Survival and injury of *Listeria monocytogenes, Listeria innocua*, and *Listeria ivanovii* in ground pork following electron beam irradiation. J Food Prot 59:596–600, 1996.
79. JM Jay. Modern Food Microbiology. 6th ed. Gaithersburg, MD: Aspen Publishers Inc., 2000, pp. 443–444.
80. IR Grant, CR Nixon, MF Patterson. Effect of low dose irradiation on growth and toxin production by *Staphylococcus* and *Bacillus cereus* in roast beef and gravy. Int J Food Microbiol 18:25–36, 1993.
81. J Bandekar, A Kamat, P Thomas. Microbiological quality of the dairy product pedha and its improvement using gamma radiation. J Food Safety 18:221–230, 1998.
82. DW Thayer, G Boyd, JB Fox Jr., L. Lakritz. Elimination by gamma irradiation of *Salmonella* spp. and strains of *Staphylococcus aureus* inoculated in bision, ostrich, alligator, and caiman meat. J Food Prot 60:756–760, 1997.
83. JD Monk, MR Clavero, LR Beuchat, MP Doyle. Irradiation inactivation of *Listeria monocytogenes* and *Staphylococcus aureus* in low- and high-fat frozen and refrigerated ground beef. J Food Prot 57:969–974, 1994.
84. P Paul, SP Chawla, P Thomas, PC Kesavan, R Fotedar, RN Arya. Effect of high hydrostatic pressure, gamma irradiation and combination treatments on the microbiological quality of lamb meat during chilled storage. J Food Safety 16:263–271, 1997.
85. FA Drobniewski. *Bacillus cereus* and related species. Clin Microbiol Rev 6:324–338, 1993.
86. HD Larsen, K Jorgensen. The occurrence of *Bacillus cereus* in Danish pasteurized milk. Int J Food Microbiol 34:179–186, 1997.
87. JM Kramer and RJ Gilbert. *Bacillus cereus* and other *Bacillus* species. In: MP Doyle, ed. Foodborne Bacterial Pathogens. New York: Marcel Dekker, Inc., 1989, pp. 21–77.
88. A Kamat, P Thomas. Recovery of some foodborne pathogens in irradiated low and high fat fish during storage at 2–4°C. J Food Safety 19:35–44, 1999.
89. AS Kamat, NF Lewis. Influence of heat and radiation on the germinability and viability of B. cereus BIS-59 spores. Indian J Microbiol 23:32–36, 1983.
90. K Ma, RB Maxcy. Factors influencing radiation resistance of vegetative bacteria and spores associated with radappertization of meat. J Food Sci 46:612–616, 1981.

91. J Farkas. Microbiological safety of irradiated foods. Int J Food Microbiol 9:1–15, 1989.
92. R Firstenberg-eden, DB Rowley, GE Shatluck. Factors affecting growth and toxin production by *Clostridium botulinum* type E on irradiated (0.3 Mrad) chicken skin. J Food Sci 47:867–870, 1982.
93. DW Thayer, G Boyd, CN Huhtanen. Effects of ionizing radiation and anaerobic refrigerated storage on indigenous microflora, *salmonella*, and *Clostridium botulinum* types A and B in vacuum-canned, mechanically beboned chicken meat. J Food Prot 58:752–757, 1995.
94. ICMSF. Temperature. In: Microbiological Ecology of Foods. Vol. 1. New York: Academic Press, 1980, pp. 1–37.
95. JD Lambert, JP Smith, K1 Dodds. Combined effect of modified atmosphere packaging and low dose irradiation on toxin production by *Clostridium botulinum* in fresh pork. J Food Prot 54:94–101, 1991.
96. JD Lambert, JP Smith, KL Dodds. Effect of headspace CO_2 concentration on toxin production by *Clostridium botulinum* in MAP, irradiated fresh pork. J Food Prot 54:588–592, 1991.
97. JD Lambert, JP Smith, KL Dodds. Effect of initial O_2 and CO_2 and low dose irradiation on toxin production by *Clostridium botulinum* in MAP fresh pork. J Food Prot 54:939–944, 1991.
98. IR Grant, MF Patterson. Effect of irradiation and modified atmosphere packaging on the microbiological safety of minced pork under temperature abuse conditions. Int J Food Sci Technol 26:521–533, 1991.
99. JD Monk, LR Beuchat, MP Doyle. Irradiation inactivation of foodborne microorganisms. J Food Prot 58:197–208, 1995.
100. P Harewood, S Rippey, M Montesalvo. Effect of gamma irradiation on shelf life and bacterial and viral loads in hard-shelled clams *Mercenaria mercenaria*). Appl Environ Microbiol 60:2666–2670, 1994.
101. GN Naik, P Paul, SP Chawla, AT Sherikar, PM Nair. Improvement in microbiological quality and shelf life of buffalo meat at ambient temperature by gamma irradiation. J Food Saf 13:177–183, 1993.
102. Y Crawford, EA Murano, DG Olson, K Shenoy. Use of high hydrostatic pressure and irradiation to eliminate *Clostridium sporogenes* spores in chicken breast. J Food Prot 59:711–715, 1996.
103. CM Garcia-Zeoeda, CL Kastner, JR Wolf, JE Boyer, DH Kropf, MC Hunt, CS Setser. Extrusion and low-dose irradiation effects of destruction of *Clostridium sporogenes* spores in a beef-based product. J Food Prot 60:777–785, 1997.
104. J Farkas. Irradiation of Dry Food Ingredients. Boca Raton, FL: CRC Press, 1988.
105. L Restaino, JJ Myron, LM Lenovich, S Bills, K Tscherneff. Antimicrobial effects of ionizing radiation on artificially and naturally contaminated cacao beans. Appl Environ Microbiol 47:886–887, 1984.
106. N Ramakrishna, J Lacey, JE Smith. Effect of surface sterilization, fumigation and gamma irradiation on the microflora and germination of barley seeds. Int J Food Microbiol 13:47–54, 1991.
107. MW Byun, JN Kwon, BS Cha, YB Kim. Effect of gamma irradiation on the decon-

tamination of animal feeds: sterilization of protein sources. Korean J Food Sci Technol 20:112–118, 1988.
108. BM Youssef, SR Mahrous, NH Aziz. Effect of gamma irradiation on aflatoxin B production by *Aspergillus flavus* in ground beef stored at 5°C. J Food Safety 19:231–239, 1999.
109. MK Dogbevi, C Vachon, M Lacroix. Physiochemical properties of red dry kidney beans and natural microflora as affected by gamma irradiation. J Food Sci 64:540–542, 1999.
110. AA Hassan, NH Aziz. Influence of moisture content and storage temperature on the production of aflatoxin by *Aspergillus flavus* EA-81 in maize after exposure to gamma radiation. J Food Safety 18:159–171, 1998.
111. P Narvaiz, G Lescano, E Kairiyama. Irradiation of almonds and cashew nuts. Lebensm Wiss Technol 25:232–235, 1992.
112. G Blank, D Corrigan. Comparison of resistance of fungal spores to gamma and electron beam radiation. Int J Food Microbiol 26:269–277, 1995.
113. G Blank, K Shamsuzzaman, S Sohal. Use of electron beam irradiation for mold decontamination of cheddar cheese. J Dairy Sci 75:13–17, 1992.
114. JA McCarthy, AP Damoglou. The effect of low dose irradiation on the yeasts of British fresh sausage. Food Microbiol 10:439–446, 1993.
115. MV Toraskar, VV Modi. Studies on the spoilage of Alphonso mangoes by *Saccharomyces cerevisiae*. Acta Hortic Wageningen 231:697–708, 1989.
116. RD Hagenmaier, RA Baker. Low-dose irradiation of cut iceberg lettuce in modified atmosphere packaging. J Agric Food Chem 45:2864–2868, 1997.
117. R Sullivan, AC Fassolitis, EP Larkin, RB Read Jr, JT Peeler. Inactivation of thirty viruses by gamma radiation. Appl Microbiol 26:61–65, 1971.
118. R Sullivan, PV Scarpino, AC Fassolitis, EP Larkin, JT Peeler. Gamma radiation inactivation of coxsackievirus B-2. Appl Microbiol 26:14–17, 1973.
119. JC Mallet, LE Beghian, TG Metcalf, JD Kaylor. Potential of irradiation technology for improved shellfish sanitation. J Food Safety 11:231–245, 1991.
120. J Lasta, JH Blackwell, A Sadir, M Gallinger, F Marcoveccio, M Zamorano, B Ludden, R Rodriguez. Combined treatments of heat, irradiation, and pH effects on infectivity of foot-and-mouth disease virus in bovine tissues. J Food Sci 57:36–39, 1992.
121. EC Pirtle, TA Proescholt, GW Beram. Trial of heat inactivation of selected viruses following irradiation. J Food Prot 60:426–429, 1997.
122. WM Urbain. Food irradiation. In: CO Chichester, EM Mrak, GF Stewart, eds. Advances in Food Research. New York: Academic Press, 1978, pp. 115–227.
123. JF Dempster. Radiation preservation of meat and meat products: a review. Meat Sci 12:61–89, 1985.
124. JS Sivinsky. Control of trichinosis by low-dose irradiation of pork. Food Irradiat Newsl 9(2):8, 1985.
125. JS Sivinsky. Efficacy testing and market research for the pork industry. Radiat Phys Chem 25:263–269, 1985.
126. FDA. Irradiation in the production, processing and handling of food. Fed Reg 50(14), (21 CFR Part 179)-29658–29659, 1985.
127. A Verster, AJ DuPlessis, LW Van den Heever. Sterilization of cystcerci with

gamma radiation. In: Proceedings of the National Symposium on Food Irradiation, Pretoria, South Africa. Atomic Energy Board, 1997, pp. 10p1–10p4.

128. BL King, ES Josephson. Action of radiations on protozoa and helminths. In: ES Josephson, MS Peterson, eds. Preservation of Foods by Ionizing Radiation. Vol. 2. Boca Raton, FL: CRC Press, 1983, pp. 245–267.
129. JP Dubey, DW Thayer. Killing different strains of *Toxoplasma gondii* tissue cysts by irradiation under defined conditions. J Parasitol 80:764–767, 1994.
130. CC Song, XZ Yuan, LY Shen, XX Gan, JZ Ding. The effect of cobalt-60 irradiation on the infectivity of *Toxoplasma gondii*. Int J Parasitol 23:89–93, 1993.
131. JP Dubey, DW Thayer, CA Speer, SK Shen. Effect of gamma irradiation on unsporulated *Toxoplasma gondii* oocysts. Int J Parasitol 28:369–375, 1998.
132. M Ingram, TA Roberts. Ionizing irradiation. In: International Commission for Microbiological Safety of Foods, ed. Microbial Ecology of Foods, Vol. 1. New York: Academic Press, 1980, pp. 46–69.
133. SA Palumbo, RK Jenkins, RL Buchanan, DW Thayer. Determination of irradiation D-values for *Aeromonas hydrophilia*. J Food Prot 49:189–191, 1986.
134. JD Lambert, RB Maxcy. Effect of gamma radiation on *Campylobacter jejuni*. J Food Sci 49:665–667, 674, 1984.
135. DJ Quinn, AW Anderson, JF Dyer. The inactivation of infection and intoxication microorganisms by irradiation in seafood. In: Microbiological Problems in Food Preservation by Irradiation. Vienna: IAEA, 1967, pp. 1–13.
136. JR Matches, J Liston. Radiation destruction of *Vibrio parahaemolyticus*. J Food Sci 36:339, 1971.

5

Microwave Inactivation of Pathogens

Stephanie Doores
The Pennsylvania State University
University Park, Pennsylvania

I. INTRODUCTION

The application of microwave energy to heat foods was patented in 1945 by Percy Spencer of Raytheon Corporation as an offshoot of radar technology developed during World War II (1). The first Radarange™ became available for food-service use in 1947 and commercial ovens were introduced in 1955. Approximately 93% of U.S. households own a microwave oven, primarily for use in rewarming previously cooked, chilled, or frozen foods (2). Microwave energy is more efficient and rapid than conventional heating, allowing food to be heated in as little as one quarter of the normal time. Because of the unusual way in which microwaves heat, some foods require reformulation to accommodate these changes (3).

There are additional commercial applications of microwave heating other than cooking and reheating operations typically in use by consumers. Microwaves can be used to inactive enzymes (4,5), blanch foods (6), defrost, temper, or thaw products (7), bake, pasteurize, or sterilize (6,8–11), and evaporate, dry, and freeze-dry (12).

Like ultraviolet, visible light, infrared, and radio waves, microwave energy is a form of nonionizing electromagnetic radiation in the frequency range from 300 MHz (0.3 GHz) to 300 GHz as opposed to ionizing radiation, such as gamma and x-rays, which operates in the frequencies of 50,000 THz or above (13). Microwaves differ from other energy forms, in addition to frequency of oscillation, in wavelength, and wave velocity (14–17). Frequency is defined as the number of cycles per second where one cycle per second equals 1 Hz and a million of cycles per second equals 1 MHz (18). Wavelength is defined as the length of one cycle with wavelengths of 1 to 0.1 m generally associated with microwave ovens (Table 1). The intensity of a microwave radiation field is expressed in watts (W) or kilowatts (kW), which are transmitted through a unit cross section of space.

In the United States, frequencies used for microwave heating in an industrial, scientific, and medical context are controlled by the Federal Communications Commission (FCC). Of the commonly used frequencies (915,2450,5800,

Table 1 Microwave and Radiofrequency Band Designations

Designation	Wavelengths	Frequencies	Typical uses
Radiofrequency bands			
Low frequency (LF)	10^4–10^3 m	30–300 KHz	Radionavigation, radio beacon
Medium frequency (MF)	10^3–10^2 m	0.3–3 MHz	Marine radiotelephone, loran, AM broadcast
High frequency (HF)	10^2–10 m	3–30 MHz	Amateur radio, worldwide broadcasting, medical diathermy, radio astronomy
Microwave bands			
Very high frequency (VHF)	10–1 m	30–300 MHz	FM broadcast, television, air traffic control, radionavigation
Ultra-high frequency (UHF)	1–0.1 m	0.3–3 GHz	Television, citizens band, microwave point-to-point, microwave ovens, telemetry, tropo scatter, meteorological radar
Super high frequency (SHF)	10–1 cm	3–30 GHz	Satellite communication, airborne weather radar, altimeters, shipborne navigational radar, microwave point-to-point
Extra high frequency (EHF)	10–1 mm	30–300 GHz	Radio astronomy, cloud detection radar, space research, HCN (hydrogen cyanide) emission

Source: Ref. 15.

and 22125 MHz), the U.S. Food and Drug Administration (FDA) currently allows only 915 and 2450 MHz frequencies for commercial and home ovens (14).

A. Principles of Microwave Energy

The magnetron or generator produces microwaves through interaction of strong electric and magnetic fields. The magnetron is a vacuum tube that uses a magnetic field to affect the flow of electrons from the cathode to the anode. When power is supplied, an electron-emitting material at the cathode becomes excited and emits electrons into the vacuum space between the cathode and anode. The anode is composed of resonant cavities that act as oscillators, which generate electric fields. Microwave energy from the magnetron flows down the waveguide, a hollow metallic tube, into the oven. As the waves enter the cavity, they are dispersed by a mode stirrer, causing multiple reflections of the energy to minimize hot and cold spots in the oven cavity (18).

Once inside the microwave oven, waves can be reflected off the oven sides and floor, can be transmitted through containers and lids made of ceramic, glass, paper, and plastic material, and can be absorbed by a medium such as food (18). Cookware remains cool unless heat from the cooking product is absorbed. In the case of plastics, sufficient heat can soften or melt the container; absorption of moisture by paperboard may result in sogginess. Metals such as aluminum foil and steel reflect microwaves and can be used to shield parts of foods, but arcing can occur if metal touches metal or the oven wall (17). Excessive reflection can harm the magnetron.

B. Production of Heat

The electrical field of normal alternating current changes direction about 100 times a second, whereas the magnetron produces an electromagnetic field with positive and negative charges that change direction about 5 billion times each second (23). Production of heat occurs primarily through dipolar rotation and ionic polarization (14). Polar molecules such as water have negatively and positive charged ends. In the presence of a microwave electric field, the water molecule attempts to line up with the field. Since the microwave field is reversing polarity millions of times per second, the water molecule moves first in one direction and then reverses to move in the opposite direction. The friction caused by constantly oscillating charged particles produces heat. Most foods are composed of 50–90% water (19), and it is the unbound water that contributes to the heating effect. Ions, because of their electrical charge, flow in one direction, then in the opposite direction in an electromagnetic field. This effect leads to a higher temperature at the surface of water or foods containing salt. It is only when microwave energy is absorbed by the food that the energy is converted to heat.

The ability of microwaves to heat depends upon the dielectric properties of the food as described by the dielectric constant and the dielectric loss factor (3). The dielectric constant (ε') describes the ability of a material to store electrical energy and varies significantly with temperature and frequency. The dielectric loss factor (ε'') describes the ability of the material to dissipate electrical energy as heat. The dielectric loss tangent ($\tan \delta$) determines the ability of the material to absorb electrical energy as well as its ability to dissipate the energy as heat. This is described by the following equation (20):

$$\frac{\varepsilon''}{\varepsilon'} = \tan \delta$$

Polar materials such as food and water that exhibit a high degree of intermolecular motion are characterized as "lossy" (derived from the dielectric loss tangent). The amount of lossiness varies depending upon the radiation frequency, temperature, and nature of the material. Increased lossiness of a material can be directly related to increased absorption of microwave energy and, hence, heat production. Water serves as a good example of a lossy material that heats easily in a microwave field. Foods with large loss factors are heated more rapidly by microwaves, whereas foods with low loss factors heat poorly.

C. Heating Characteristics—Microwave Versus Conventional Heating

Heat transfer by microwaves is markedly different from conventional heating using an electric or gas oven. In conventional heating, hot walls and heating elements radiate infrared waves, which are absorbed and heat a very thin layer at the surface of the food. This heat is transferred from the exterior of the food to the interior by convection or conduction currents depending upon the structure and composition of the food. The thermal conductivity of foods is not high, thus it takes quite a long time for the interior of the food to reach cooking temperatures. Therefore, the food heats from the outside in, and, because of surface evaporation, the food can develop a browned appearance or crust (21).

In contrast, microwaves do not generally heat from the inside out, nor do they deposit all of their power at the penetration depth. The air inside the microwave oven cavity is cool, which causes surface cooling of the microwave-cooked foods. Consequently, the maximum temperature is reached somewhat below the surface of the cooked food, leaving the surface cooler and moister than those foods cooked conventionally. The speed of microwave cooking results from the ability of microwaves to deposit more energy at greater depths in the food than can be achieved by conventional techniques (22). Therefore, it is often recommended that final cooking of microwave-heated foods should include a "standing time'' during which foods continue to transfer heat from the internal portion to the surface.

Because of the rapidity of heating by microwaves, some foods will not have the proper time to develop flavors, textures, or other physiochemical attributes typically associated with foods heated by conventional means. Therefore, reactions that must occur in a certain sequence or for a specific time within that sequence may not happen, and product quality suffers. This can be seen with the tenderization process that occurs during heating of muscle foods. In conventional heating, meats are cooked slowly, resulting in a softened texture and enhanced flavor development; in microwave cooking, although the same temperature may be achieved, the shortened heating time leads to a rubbery texture and lack of flavor. In baking operations, expansion of gases occurs much faster than formation of a stable structure in some baked goods, resulting in a collapse of the product (21). Heating can cause staling or toughness in baked goods because interior temperatures are higher than normal boiling temperatures, which leads to greater movement of water to the surface.

D. Factors Affecting Microwave Heating

1. Penetration Depth of Microwaves

The penetration depth of microwaves into a material is governed by its chemical composition, physical properties, temperature, and oven frequency (23). Penetration depth is infinite in substances that are transparent to microwaves and zero in substances that are reflective to microwaves. In the frequency range found in commercial and domestic microwave ovens, all of the microwaves contacting the surface of the product will be absorbed (24). The penetration depth is arbitrarily defined as the depth in a material at which microwave power level is 37% of the surface value. The penetration depth for water at 2450 MHz is approximately 1.4 cm, while at 915 MHz the depth increases to 3.7 cm.

Materials possessing a high dielectric loss factor, high moisture content, or increased concentration of dissociated salts demonstrate a decrease in penetration depth (25–28).

2. Microwave Frequency

There is an inverse relationship between microwave frequency and penetration depth of microwaves into material. This is demonstrated as a peripheral heating effect with 2450 MHz ovens compared to a core heating effect with 915 MHz ovens (3,24,29). Although the frequency of microwave ovens is fixed at 915 or 2450 MHz, its wavelength changes as it passes through air or materials. This, in turn, affects the penetration depth. In addition, as the microwave frequency changes, so do the dielectric constant, temperature, and moisture content.

3. Product Mass and Volume

As with conventional heating, the larger the total mass of food to be heated, the longer the time needed to heat that product, although efficiency improves as the mass becomes larger. If the thickness of the food exceeds the penetration depth of the microwaves, then more energy is absorbed by surface layers and less energy by deeper layers, leading to nonuniform heating profiles and longer processing times. However, small, thin, less dense portions of food may overcook before other areas reach the desired temperature (21). Thus, uniform heating profiles will be optimized when product thickness does not exceed the penetration depth of the microwaves.

Placement of multiple foods in a microwave field determines the overall heating process. Because thick foods like solid pieces of meat heat slowly, they should be placed near the perimeter of the plate, while foods having large surface area such as cubed potatoes are placed at the center. However, if potato is reformed into a mashed product, it now possesses greater mass and it should be placed towards the outside.

Microwave heating also demonstrates a ''load factor,'' which determines the amount of microwave energy absorbed by the product and is defined as a relationship between the mass of the product and the size of the oven (30). Therefore, the greater the load, the greater the microwave output of the oven (31). If the oven is operated without a load then a great deal of the power is reflected. To avoid this situation, oven manufacturers place a glass or ceramic tray in the cavity that will absorb some power (30). Uniform cooking results when food is positioned above the oven floor. If the oven walls or the space beneath the shelf becomes contaminated with grease or filled with water, then the power absorbed by the oven cavity will increase, having a corresponding effect on the power delivered to the load. Therefore, the oven should be kept as clean as possible in order to achieve its maximum efficiency, particularly with light loads.

If two foods of different microwave absorption are heated together in a microwave oven, there will be competition for the microwave power between the two foods. This competitive heating situation is quite different from a single food that is heated, and all microwave power will be absorbed by the food no matter how well it absorbs microwave energy (19).

4. Moisture Content

Unbound water is the major contributor to dipolar rotation, one of the primary mechanisms that translates microwave energy into heat. Therefore, heating rate is highly dependent on moisture content. Products with higher moisture content have a higher dielectric loss factor, which causes more of the absorbed microwave energy to be dissipated as heat, thereby causing the product to heat faster (3). There are instances where low-moisture food products may have similar heating rates as high-moisture foods, depending upon the specific heat capacity of the food.

5. Product Density

The density of a material also affects the heating rate of the product. Air reduces the dielectric constant of a product and acts as an insulator. Therefore, as density of a material increases during heating, the dielectric constant also increases. The denser the food, the longer it takes to cook by microwaves.

The more homogeneous the food, the greater and more even the absorption of microwaves and the less time required for cooling. The presence of bone affects the uniformity of microwave heating in that internal doneness is accelerated by the presence of bone and heating is greater at the bone surface than in other areas of the meat.

6. Product Temperature

Because the dielectric loss factor is a function of viscosity, which, in turn, is a function of temperature, then the temperature of a product affects the dielectric loss factor (24). The higher the initial temperature of the food, the faster it will be heated by microwaves (32). The temperature change during heating also affects the moisture content of a material, which affects the dielectric constant and dielectric loss factor of the material.

Because the generation of heat is continuous, the rise in temperature is also continuous with no automatic limit to the temperature reached. Therefore, the longer the microwaves are applied, the higher the temperature that is reached. In foods having a high percentage of water, microwaves raise the temperature to 100°C, after which the continued generation of heat boils off free water. With the reduction of free water, temperature of the nearly dry food can rise rapidly with the risk of burning (33). In contrast, conventional heating is limited by the temperature of the heat-transfer medium.

Postprocessing temperature rise (PPTR) is the increase in temperature after microwave heating as measured at the coolest site in a product. The duration and extent of PPTR vary with the size, shape, and internal temperature of the food, food product, and power level (18).

7. Physical Geometry

The shape of the food often determines the heating pattern. Spherical objects with diameters of 25–60 mm, such as beets, onions, and potatoes, demonstrate a phenomenon termed "focusing," resulting in greater heating in the interior of the product, while the exterior remains cool (1). The practical consequence of this center heating effect is that a meatball may be cold on the outside, but steaming hot in the center. If conduction heating does not dissipate the temperature gradient, then the mass could erupt. Such is the case with an explosion of a microwave-cooked egg. The rapid temperature increase builds up steam in the center of the egg; when the steam is released, the egg explodes (19). Rapid steam formation makes venting essential for products with skins and necessitates removal of shells.

Materials with a large surface area heat more quickly because microwaves penetrate over a larger area, but these same materials experience a greater degree of surface cooling due to increased surface evaporation. If microwave heating proceeds unchecked, gross overcooking of the surface will result, but cooking to a lower internal temperature allows conduction to complete the heating cycle.

Next to spherical shapes, cylindrical shapes such as rolled meat are an optimal form in terms of heating performance.

Foods with slab geometry are difficult to heat, particularly at the corners and edges because they are exposed to microwaves coming from numerous directions. Edges tend to heat less than corners, and centers remain quite cold. These heating effects become more pronounced if the penetration depth is shortened as in products containing more salt. Corners can be shielded with aluminum foil to reduce the heating rate in these areas (19). Food products heat more evenly if contained in packages with rounded corners.

Shielding, or the use of metal to reflect microwave energy, is effective in preventing foods from being heated to too high a temperature. This is used practically to even out the heating process in foods such as frozen dinners where sauces may heat differently than protein foods. Foods touching each other can "shadow" or shield, thus reducing the heating rate in the adjacent product.

An optimum geometry would be a configuration in the form of an annulus or donut (34) since heating to the center of the mass could easily be achieved.

8. Specific Heat and Thermal Conductivity

The specific heat and thermal conductivity of a product also affect microwave heating. These two variables influence the magnitude of local temperature rise at a particular microwave power absorption level. In nonhomogeneous foods, these variables change from point to point within the food, which translates into different local heating rates. Low-moisture products that have a low dielectric loss factor may still heat well in the microwave oven due to a low specific heat value.

Microwave absorption by fat or cooking oil is very low compared to water and other foods because of low dielectric losses. Therefore, it should be difficult to heat fat in a microwave oven, e.g., fat on a piece of meat. However, fat requires only small amounts of energy to heat well, and it will even overheat in a microwave oven because of the difference in the specific heat between fat and other foods (19,29).

Table 2 Microwave Oven Characteristics

Power output	400–750 W
Cavity size	0.4–1.8 cubic feet
Microwave feed system	Mode stirrer, rotating antenna, rotating waveguide
Location of microwave input into the oven	At top of oven cavity or both top and bottom of oven or sides of the oven
Cavity wall construction	Stainless steel or painted cold-rolled steel
Turntable	Present or absent
Extra items	Glass shelf or metal rack in some ovens
Additional types of ovens	Microwave only, microwave-convection, or microwave with browning element
Power settings	1–99

Source: Ref. 3.

Thermal conductivity describes the ability of a substance to conduct heat throughout its volume. Thermal conductivity will differ in thick items as opposed to thin items, which translates into a longer time for heating or transfer of heat to the interior. Thermal conductivity allows conduction and convection currents to aid in the elimination of temperature gradients during long heating times but does not contribute as much during short heating periods (3).

E. Problems Related to Microwave Ovens

Each oven has its own power output, wattage, frequency, and electromagnetic field pattern that leads to hot and cold spots (Table 2) (3). These variables can be easily determined using the methods of Buffler (35). Furthermore, there are no manufacturer's standards for the description of "high" or "medium" setting or the measurement of power output, further complicating the ability to compare one model versus another. To circumvent this difficulty, ovens can be produced with field stirrers, turntables, or other devices that help to minimize these variations. The consumer can also overcome these problems by rotating foods once or twice during cooking, changing power levels, covering or stirring food, and including the standing time after cooking.

F. Problems Related to the Interaction of Microwaves with Food Materials

1. Frozen Products

Heating of frozen foods presents an interesting challenge. Ice is transparent to microwaves and, as such, does not absorb microwave energy, whereas water is highly absorptive and heats well (19,36). The difference in loss factor between ice and water is very great, and the differences between various components of a frozen food are sufficiently great to cause uneven heating. Approximately 5–10% of water in frozen food is liquid, forming a concentrated solution of solutes such as salt, sugar, etc. that absorbs microwaves rather well. The microwaves will heat the water preferentially, leading to ''runaway'' heating. This can result in a product having one area that is boiling with surrounding areas still frozen. Therefore, stirring of products, if possible, is advisable during cooking. The use of a defrost cycle, usually set at 300 W, allows time for heat to diffuse inward.

Tempering, on the other hand, is a process whereby temperatures are raised to about −3°C without forming water. Runaway heating does not occur, but irregularities within the food can give rise to uneven heating.

2. Multicomponent Foods

For composite or heterogeneous foods, microwave processing becomes more difficult because of the different loss characteristics, specific heats, and thermal conductivities. For example, sausage sandwiches or fillings of frozen pizzas have been observed to heat at different rates, with overheating of the filling and underheating of the dough (28). Similar considerations apply to other food products having suspended phases with essentially different electrical properties, e.g., soups and jelly doughnuts (28). Thus, these differences can lead to overcooking in one area and undercooking in another.

G. Problems Related to Measurement of Microwave Temperatures

In early studies comparing microwave heating with conventional heating, a tremendous effort was focused on obtaining an accurate temperature analysis of heated samples. Although the time of heating could be easily determined, temperature measurement was available only at the end of the heating cycle. The most common procedure was to use a mercury thermometer to gauge the endpoint temperature immediately after heating. The sample was removed from the oven, mixed or stirred, and measured. This temperature comprised a representative reading of the entire product with no delineation of any variations in heating pattern throughout the product. Thermocouples, while commonplace in other temperature-measurement settings, were observed to arc. Even with proper shielding, thermocouples did not provide sophisticated temperature measurements and often altered the heating parameters. Infrared imaging or thermography provided a detailed map of surface temperature distributions but could not be used within the oven to map heating patterns as they were developing or changing.

With the advent of fluorescent fiberoptic or ''fluoroptic'' thermometry, foods can be measured for temperature during heating, and a somewhat detailed temperature profile can be characterized (36,37). The temperature pattern in foods, however, is dependent upon the number of probes used and their position in the product, as these probes only provide information on discrete points within the sample. The probes are constructed of a phosphor sensor attached to the tip of an optical fiber, neither of which adversely affects temperature measurement. Probes are generally inserted into the microwave through a small (1/8 inch) hole drilled in the oven wall, which does not affect microwave heating. Depending upon the placement of the probes in foods, localized hot or cold spots can be described, spatial field intensity patterns can be mapped out, and heating uniformity can be compared between ovens. Therefore, the use of fluoroptic thermometry allows for the monitoring of heating rates as well as final temperatures. This is of practical use for multicomponent foods, such as frozen entrees.

It is common for food to continue to cook after removal from a heating source. This may be termed ''dwell time'' (38), ''standing time'' (39–42) or ''postprocessing temperature rise'' (PPTR) (43). If these temperatures are consistent, the PPTR could be considered in future heating recommendations.

In a study by Sawyer (43), microwave heating was compared to heating in a hot air oven set at 160°C. Temperatures were monitored in the geometric center of the food using fluoroptic thermometry in microwave heating and using potentiometer and iron-constantan thermocouples in conventional heating. Foods were chosen based on homogeneous composition, consumer preference, and microwaveability. Beef loaf patties (~150 g initial weight) were microwave heated for 90 seconds in a donut-shaped Pyrex dish, and boneless tied raw pork loin roasts, boneless ready-to-serve turkey roll, and turkey casserole were placed in a Pyrex baking dish and rotated every 20 minutes to decrease the effect of hot spots developing in the foods. The PPTR varied depending upon the product heated, but no PPTR occurred in foods heated by hot-air ovens. The temperature rise of fabricated foods varied with the food, with turkey roast exhibiting the largest PPTR of 6.1°C, followed by pork roast (2.9°C); turkey casserole did not demonstrate any additional rise in temperature.

Other temperature-sensing systems were developed by Goedeken et al. (44), using an infrared system to measure surface temperatures of products, and by Tong et al. (45) using a feedback controller. Ramaswamy et al. (46) developed an aluminum-shielded thermocouple to monitor product temperature continuously within ovens. A preset temperature could be maintained by providing feedback to turn the magnetron on or off.

Packaging can also be used to modify power distribution within a food (36). Electric field susceptors constructed of conducting metal films or foil can be used to absorb additional power locally.

II. MICROBIAL DESTRUCTION

A. Foodborne Illness Attributed to Consumption of Microwave-Heated Foods

Three outbreaks of foodborne illness attributed to consumption of microwave-heated foods contaminated with *Salmonella* spp. raise concern about this process as a sole means of eliminating microbial pathogens from foods. Considering that *Salmonella* spp. are not particularly heat-resistant organisms, these outbreaks point to the necessity of using excellent preparation and storage techniques to reduce the likelihood of contaminating foods and the proper temperature measurement during reheating to ensure an adequate heat process.

Pork served at a picnic and taken home as leftovers was implicated as the vehicle for *Salmonella* Typhimurium (47). Of 30 persons who brought home pork, 20 used a conventional oven or skillet and 10 used a microwave oven to reheat the leftovers. All 10 individuals consuming microwave-reheated pork became ill.

Poached eggs were thought to be a source of salmonellosis for four members of a single household. To confirm this observation, *Salmonella* Enteritidis, Typhimurium, Virchow, and Mbandaka were inoculated at various levels into the egg yolk. The eggs were heated in the microwave oven either in a greased dish or in boiling water adjusted to low- or high-power settings until the white of the egg was solid and the yolk still soft. Temperatures reached a maximal level of 47–52°C in the yolk after heating. Salmonellae were more likely to be recovered from eggs heated in a dish than in boiling water, even with an inoculum level as low as 10^1 cfu/mL. Bates and Spencer (48) found that survival of salmonellae might be anticipated if eggs are microwave heated and the yolk remains soft or "runny."

Salmonella Enteritidis phage type 4 (PT4) was isolated at a level of 6×10^3 cfu/g from rice salad containing boiled rice, raw carrots, eggs, cheese, and curry powder (49). The rice dish had been heated for 5 minutes in a 500 W microwave oven equipped with a rotating turntable. Leftover cheese and onion quiche also contained PT4, but at substantially reduced levels, suggesting cross-contamination from the rice salad. Improper reheating contributed to this outbreak.

B. Athermal Effects

Microwave heating is assumed to inactivate microorganisms by a thermal effect, but some researchers have implied that athermal effects could also play a role in inactivation. This hypothesis is championed by many of the papers that use a sublethal temperature to determine athermal effects. Early research on the ability of microwaves to inactivate microorganisms by a mechanism other than thermal effects were conducted with frequencies outside of the microwave range (50,51). Chief among the problems concerning the identification of an athermal effect in earlier studies was the lack of specific experimental details, particularly concerning measurement of temperature. Furthermore, interpretation of killing due to athermal effects was compounded by the inability to distinguish an athermal effect in the presence of a thermal effect.

Webb et al. (57,58) cited several difficulties in determining the existence of athermal effects. The inability to measure and control on-line temperatures in a microwave field is probably the most difficult variable to control in experiments. Several studies compared organisms heated at a single temperature using traditional methods with microwave heating over a range of temperatures. Early attempts to maintain temperatures included using ice to cool solutions (54) but were more recently replaced by using solvent that is cooled and circulated around the container to be heated. Temperature measurements after heating have been taken using a mercury thermometer on mixed solutions rather than assessing the thermal profile during heating. Uneven heating due to microwave field distribution and sample characteristics can also be difficult to control. In addition, concentration of solute due to evaporative losses does occur, but its impact on destruction of microorganisms has been overlooked. Finally, sublethal injury and repair of injured cells has not been identified as a possible explanation for the supposed inactivation of cells.

One of the major arguments against athermal effects of microwaves centers on the ability of microwaves to break chemical bonds. Gamma rays and x-rays possess sufficient quantum energy to break most chemical bonds, whereas UV and visible light break only weak hydrogen bonds. Microwaves and radio waves fall short by several orders of magnitude of breaking chemical bonds (18,55,56). Despite this information, some studies have specifically focused on interference with essential enzymes or metabolic activities or alteration of the cell structure.

The work of Webb and coworkers (57,58) using *Escherichia coli* determined that microwaves in the 65,000–75,000 MHz and 136,000 MHz range slowed down cell division by retarding uptake of essential nutrients into the cell, but some frequencies stimulated uptake. Cell-replication rate was also unaffected by exposure of *E. coli* and *Pseudomonas aeruginosa* to low-level microwave radiation at constant temperatures of 37°C for 12 hours (60 mW/cm^2) as compared with similarly inoculated flasks incubated in a conventional incubator (59).

Mayers and Habeshaw (60) proposed that microwave radiation of 2450 MHz and a power level of 50 mW/cm led to a decrease in macrophage phagocytic activity in cell cultures. When the radiation was discontinued, the phagocytic activity was restored.

Barnes and Hu (61) contended that exposure to radiowave or microwave frequencies could cause biological changes in membranes. The size of the currents generated from the electric field was large enough to produce increased ion transport across membranes and reorientation of long-chain molecules thus potentially changing membrane porosity.

Dreyfuss and Chipley (62) examined the metabolic activities of specific enzymes as a function of exposure to conventionally heated, microwave-heated and nonheated cells of *Staphylococcus aureus*. Cultures were exposed for 10, 20, 30, and 40 seconds at 2450 MHz or for comparable times equivalent to temperatures attained by microwave heating. Cell lysates from all three treatments were assayed for specific enzymes. Differences occurred in several essential enzymes after cells were exposed to microwave radiation that cannot be explained by thermal effects alone. Details of temperature measurements taken during the course of these experiments were not given.

Jahngen et al. (63) examined the hydrolysis of adenosine triphosphate by conventional and microwave heating. Reaction sample temperatures were more difficult to account for mathematically in microwave-heated samples. They concluded that no special microwave heating effects were observed for the hydrolysis of ATP.

Numerous studies were undertaken to determine specifically the presence of an athermal effect, but data indicate that no athermal force was measured. Inactivation of *E. coli* and *Bacillus subtilis* var. *niger* spores was identical between conventional heating and microwave heating. Furthermore, Goldblith and Wang (54) did not support the theory that microwaves per se exhibit an inherent or athermal effect. Lechowich et al. (64) also concluded that no lethal effects were noted other than those produced during heating of *Streptococcus faecalis* and *Saccharomyces cerevisiae* at variable temperatures from 25 to 55°C in modified 2450 MHz oven constructed with a heat exchanger in which kerosene circulated around the samples to maintain constant temperature.

Roberts (65) compared conventional heating with that of microwave heating using $\sim 2 \times 10^5$ cells or ascospores/ml of *S. cerevisiae*. Hexane was used as a heat-transfer medium. The thermal death point (temperature used to kill all organisms in 4 minutes) was 56 and 60°C for vegetative cells and ascospores, respectively, for conventional heating. A microwave power level of 40 W with no coolant added resulted in a total loss of viability after 30 seconds for vegetative cells and 40 seconds for ascospores. With cooling, the time to complete destruction was 8 and 12 minutes, respectively, but temperatures did not exceed 33°C. The differences in thermal destruction were attributed to superheating at the microthermal level. Cope (66) similarly theorized that athermal effects might occur by a superconductive medium. This phenomenon is supported by the discovery of high-temperature superconductivity in lipids and protein.

Staphylococcus aureus was sublethally heated in phosphate buffer to 50°C for 30 minutes in a water bath or a microwave oven modified to include continuously circulating kerosene to remove heat from the sample so that a constant temperature is maintained (67). Detection of injured cells of *S. aureus* was conducted by plating cells on trypticase soy agar (TSA) or TSA containing 7% salt and by measuring the release of 260 nm absorbing material, an indicator of heat damage to the cytoplasmic membrane. Results by both analyses indicate that staphylococci were more injured by microwave heating than by conventional heating. Furthermore, during the recovery period, toxin production was regained more slowly for microwave-heated cells than the conventionally heated cells. Cells were heated for additional temperatures ranging from 5 to 50°C to measure nonthermal effects of microwaves. There was no injurious effect from temperatures of 5–35°C by either heating method; however, with temperatures of 40, 45, and 50°C, injury rates were higher for microwave-heated cells. As a follow-up to these experiments, Khalil and Villota (68) examined the stability of ribosomal RNA, a cellular component affected by sublethal heating. Cells were again heated at 50°C for 30 minutes in magnesium-chelating buffer, and the stability of 16S and 23S ribosomal RNA was measured. Although both heating methods destroyed 16S RNA, microwave heating affected 23S RNA more than conventional heating. Recovery was complete within 180 minutes for both 16S and 23S RNA from conventionally heated cells, but only 16S RNA was recovered from microwave-heated cells; 23S RNA was recovered within 270 minutes. The authors suggest that thermal gradients may play a role within the cell.

Saeed and Gilbert (69) attempted to correlate changes in the cell envelope that would affect growth environment after a low-level exposure to microwaves over a 16-hour period. *P. aeruginosa* and *S. aureus* were exposed to microwave energy for varying times and were assessed for their sensitivity against inactivating agents such as 2-phenoxyethanol, chlorhexidine diacetate, and benzalkonium chloride. Although some structural changes were noted in the cell envelopes, these changes did not translate into differences in cell morphology, growth rate, or motility, suggesting that low-level microwave exposure did not cause any negative effects.

In order to separate any athermal from thermal effects, Jeng et al. (70) examined the effect of microwave sterilization on $\sim 10^5$ spores of *B. subtilis* subsp. *niger* coated onto the inside wall of a borosilicate glass vial. After heat exposure by convection oven or microwave energy up to 6 kW, the contents of the vials were subcultured in broth. Temperatures were recorded during heating and automatically controlled in both heating systems using fluorometric thermometry probes. Temperature profiles of the two heating systems were nearly identical for heating profiles conducted at 107, 117, 130, and 137°C. There was no significant difference observed in inactivation kinetics between the convection and microwave heat systems at all temperatures tested. Microwave temperatures below 117°C were not effective for sterilization of dry material. The results of this study strongly supported that destruction is a thermal process.

Tomato, vegetable, and beef soups served as the foods for inoculation of *S.* Typhimurium and *E. coli* at levels of $\sim 10^7$ cells/mL (71). Soups were reconstituted according to manufacturer's directions and microwave heated at 915 MHz in beakers. Temperature strips were positioned at various levels within the beaker or in a graduated cylinder as an attempt to measure the location of maximum killing. Samples were removed sequentially from the bottom, middle, or top of the container and assessed for pathogen numbers. The temperature determined for the middle portion of the cylinder containing tomato or vegetable soups was comparable and was the warmest of the three regions, with the top of the container being the coolest. In beef broth, the middle portion was the warmest region, with the coolest temperature registered in the bottom portion and intermediate temperature in the top region. Survival patterns of the two pathogens were similar between beef broth and the other two soups. Inactivation occurred in the central region when temperatures reached 62–67°C. Because of the disparity between temperatures achieved in the three regions and the number of organisms recovered, the authors suggested that a factor other than heat was responsible for inactivation.

Kozempel et al. (72) proposed several theories to explain athermal effects of sublethally heated pediococci. They postulated that microorganisms could selectively absorb electromagnetic energy such that their inactivation could occur during microwave heating systems at below inactivation temperatures. Electroporation, or the production of pores across the cell membrane, could lead to leakage of materials; voltage across the membrane would be sufficient to disrupt membranes and critical elements within the cell would be inactivated, thus leading to death.

C. Microbial Destruction—Applied Studies

Destruction of microorganisms and their toxins has been described in terms of exposure to a lethal agent or process for a specified time. The length of time dictates the type of effect, with shorter times resulting in injury and longer times, death. Furthermore, resistance of microorganisms varies as to metabolic state, gram type, and spore formation. In general, logarithmic phase cells are more susceptible to adverse situations than stationary cells, gram-negative bacteria are more sensitive than gram-positive cells, and vegetative cells are more vulnerable than sporulated cells. With exposure to microwaves, the overwhelming body of evidence points to thermal effects as the underlying means of destruction, but numerous reports of "athermal" effects may also provide a killing effect. Since microwave energy, by its very definition, generates heat from frictional movement of molecules, the question becomes how much heat is produced for how long an exposure time. Furthermore, separation of athermal from thermal effects is problematic in that most experiments are not carefully controlled to examine such differences. Since traditional thermal destruction or reduction of microbial load is a time-temperature relationship, and the temperatures achieved during heating can attain the same level as in conventional heating, the most important variable becomes time. Because exposure time is less than that used in conventional heating, organisms survive.

Since microwave ovens have become commonplace in the home, numerous studies have been conducted that compare conventional heating methods to microwave heating and focus on nutrient retention, yield, palatability, and microbial safety. Depending on the focus of each study, a number of variables can be considered related to microorganisms, heating regimens, and foodstuffs. Some studies examine indigenous populations, whereas others use inoculated organisms, usually foodborne pathogens, and follow the kinetics of their destruction. Heating variables include temperature-distribution measurements and the necessity to achieve comparable heating profiles, adjustment of power levels, thawing prior to heating, standing times, and load distribution. Food variables are almost too numerous to mention but do include composition differences, dielectric properties, structure, and specific heat.

The studies designed to compare microwave heating with conventional heating are presented by categories of food as these groupings take into consideration the food or ingredient effects on microwave heating. The majority of studies use 2450 MHz for heating, and this frequency will not be routinely stated; however, those studies using other microwave frequencies, e.g., 915 MHz, will be so indicated. Unfortunately, many of the studies do not provide specifics as to the power level or means of temperature measurement.

1. Plant Foods

Plant Products. In one of the few studies focusing on plant products, Causey and Fenton (73) reheated previously cooked and frozen broccoli, green beans, Swiss chard, diced carrots, shredded beets, and baked potatoes by conventional oven, boiling water, or microwave oven to internal temperatures of 180–185°C. Reheating times were shortest in the microwave oven (3.5–5 min). It was difficult to demonstrate any significant difference among the reheating methods given that bacterial load was low before the reheating process.

The use of microwave heating as a pasteurization process for juices was investigated by Nikdel et al. (74) and Tajchakavit et al. (75). Nikdel et al. (74) modified a 600 W microwave oven for continuous flow to pasteurize orange juice. Microwave heating for 15 seconds at 70°C was sufficient to reduce $\sim 10^8$ cfu/mL *Lactobacillus plantarum* by 6 logs, but an increase in temperature to 80°C for the same time was necessary to reduce the population to nondetectable levels. No taste difference was detected between microwave-processed juice and nonheated samples or juice treated by conventional hot water pasteurization. Tajchakavit et al. (75) compared a batch heat treatment with a continuous-flow 700 W microwave process for the inactivation of *S. cerevisiae* and *L. plantarum* in apple juice. Comparison of thermal resistance values indicated a faster microbial destruction rate with microwave heating compared to batch heating. The authors concluded that some type of enhanced effect accounted for the increased destruction by microwave heating.

Grain Products. Microwave heating has been examined as a means to extend the shelf life of finished bakery products. Because loaves of bread leave the oven containing very low levels of microorganisms, fungal contamination usually reoccurs on the surface of the bread during cooling, slicing, and wrapping operations. Cathcart (76) and Cathcart et al. (77) used a 50-second heat treatment to raise the temperature of a 20-ounce loaf of packaged bread to 60°C. Although mold growth was eliminated for 3 weeks, the moisture released from the bread condensed on the crust of the packaged loaf, leaving a soggy surface.

Olsen (8) inoculated the surface of freshly baked and sliced, enriched white bread with a spore suspension of *Aspergillus niger*, *Penicillium* sp., or *Rhizopus nigricans*. Bread slices treated with microwaves at 500 W power achieved temperatures of 62.6–65.5°C for approximately 2 minutes. Counts for *Penicillium* and *A. niger* were reduced by approximately 3 logs, while *R. nigricans* was recovered from 21.4% of the bread slices. Using noninoculated bread, microwave heating retained freshness for 21 days, whereas nonheated slices were covered with fungal growth. Even bread dough formulated with 0.1% sodium propionate, a commonly used antimycotic agent, was completely covered by mold within 12 days. Because of the thermal resistance of mold spores used in these studies (68–71°C for 20 minutes) and the relatively large decrease in fungal numbers seen after microwave heating, Olsen (8) suggested that an additional athermal effect accounted for the enhanced reduction of mold count.

Packaged cakes have also been treated with microwaves to prevent mold growth (78).

2. Muscle Foods

Several studies have examined the use of microwave heating to preserve or cook muscle foods, including red meat, pork, chicken, and seafood. Solid muscle foods carry not only an indigenous microflora residing primarily on the surface, but also a unique microbial population transferred by environmental contamination. The microbial population becomes homogeneously redistributed throughout the product during a grinding process. Further processed products, i.e., cured meats such as wieners, are formulated with salt, nitrites, and other antimicrobial compounds that not only select for specific groups of spoilage organisms, e.g., streptococci and lactobacilli, but also enhance the destructive effect of conventional cooking.

Beef Products. The use of microwave heating to extend the shelf life of raw and vacuum-packaged beef muscle was examined by Paterson et al. (79) and Hague et al. (80). Paterson et al. (79) vacuumed-sealed raw steaks in polyvinylidene chloride (PVDC, Saran™) nylon bags and heated them by water bath or microwave oven for various times at 700 W power to 50, 52.5, 55, and 60°C. Changes in the appearance of the steak as well as increases in the destruction of aerobes and anaerobes were a function of temperature. A 50°C heat treatment showed a minimal difference in appearance compared with the fresh meat sample and ~1-log reduction in microbial numbers. Although microbial counts continued to decrease (up to 4 logs) with increased temperatures of heating, a definite cooked appearance was noted. Hague et al. (80) investigated the use of microwave treatment and/or the addition of nisin to preserve precooked beef semitendinosus muscle. The surface of the muscle was inoculated with 7.5×10^4 *Clostridium sporogenes* spores/mL followed by microwave heating for 20 seconds in a 700 W oven. Meat was held at 4 and 10°C and examined after 21 and 70 days of storage. Although postmicrowave surface temperatures reached as high as 55°C, a 3-log reduction was achieved by all treatments. Heat treatment did not provide any additional preservative effect that was not afforded by nisin alone.

Dessel et al. (81) may have been the first to study the use of microwave heating to reduce microbial populations in hamburger. Hamburger patties were inoculated with $\sim 5 \times 10^5$ cfu/g *Salmonella* Typhosa and cooked to doneness by broiling and microwave heating. *Salmonella* was effectively killed by both cooking methods for hamburger patties.

Because the resident microflora of raw hamburger meat is so varied, Crespo and coworkers (82) examined the dynamics of thermal destruction of naturally occurring bacteria. Microwave heating was compared to conventional oven heating at low (149°C) and high (232°C) temperatures to achieve internal temperatures of 34, 61, and 75°C. Microwave heating to 34°C afforded very little destruction of overall microbial load in hamburger (~0.18 $\log_{10}$ cfu/g) compared to oven heating (0.39 $\log_{10}$ cfu/g, low setting; 0.54 $\log_{10}$ cfu/g, high setting). When the temperature was increased to 61°C, a product considered "rare," significantly greater reductions of 2.39 $\log_{10}$ cfu/g were achieved at a low temperature setting and 2.79 $\log_{10}$ cfu/g at a high temperature compared to 1.48 $\log_{10}$ cfu/g for microwave heating. When the temperature was increased still further to 75°C, the same trend was produced with significant reductions of 4.32 and 5.24 for the oven set at low and high temperatures, respectively, compared to 3.61 $\log_{10}$ cfu/g for microwave heating. These differences could be explained by ineffective temperature distribution in microwave-heated hamburger and lack of time at killing temperatures. In subsequent work (83), hamburgers were inoculated with pure cultures of *Pseudomonas putrefaciens*, *Lactobacillus plantarum*, or *Streptococcus faecalis*. To achieve a 2-log reduction in counts in hamburgers when using a conventional oven, it was necessary to heat *P. putrefaciens* to internal temperatures of 51°C, *L. plantarum* to 57°C, and *S. faecalis* to 62°C. Higher temperatures of 63, 78, and 74°C, respectively, were necessary to achieve comparable reductions by microwave heating. Crespo et al. (83) concluded that because of the variability in temperature distribution within microwave ovens, heating by this method was less effective for microbial destruction than conventional heating. The differences in survival among genera were attributed to heat sensitivity of the individual organisms and the location of the load within the oven.

Ground beef patties containing ~10^7 vegetative cells of *Clostridium perfringens*/g were heated to internal temperatures of 65–71°C for rare and 77–93°C for well done (84). Microwave heating reduced aerobic bacteria counts by ~3 logs as compared to a 5-log reduction for conventional heating. Log reductions of *C. perfringens* ranged from 0.75 to 1.48 by microwave heating and 3.06 to 3.51 by conventional heating. Wright Rudolph et al. (84) concluded that conventional cooking was a far more reliable method than microwave heating to reduce microbial populations when foods were heated to the same internal temperature. In similar research, the effect of microwave and conventional cooking using minced beef samples cooked to varying doneness levels was investigated by Hollywood et al. (39). Beef samples were formed in cylindrical shapes and heated to rare, medium, or well-done states by microwave and conventional oven. After microwave heating, samples were wrapped in aluminum foil and allowed to stand for 30 minutes; standing time for conventionally cooked beef varied with the cooking time. Destruction by microwave heating of bacterial groups including mesophiles, psychrotrophs, gram-negative bacteria, enterococci, coagulase-positive staphylococci, coliforms, yeasts and molds, anaerobic spore formers, *Pseudomonas*, *Lactobacillus*, *Brochothrix*, and *Bacillus cereus* was a function of doneness level, with all groups being reduced from 1 to 5 logs with postheating standing time. *Bacillus cereus* was not recovered from any samples. *Listeria monocytogenes* was recovered after microwave heating in all of the rare and one of the medium-cooked samples prior to the standing period and in one of the samples after standing time; no listeriae were recovered from conventionally cooked beef. It was concluded that standing time plays a vital role in further reducing microbial loads in microwave-heated foods.

Ground beef experimentally inoculated with ~10^5–10^6 cells/g of *S.* Mbandaka was microwave heated at 780 W for 3.5 or 4 minutes, covered with metal foil upon removal from the oven, and held for 2 minutes (85). No salmonellae were recovered after the 4-minute exposure time, indicating a 5-log reduction, although 43.3% of the samples heated for 3.5 minutes were positive for salmonellae. Internal temperatures were always >70°C after 4 minutes, but variability in temperature was noted for the briefer heat treatment.

Ground beef mixed with dehydrated onion, oatmeal, salt, and spices was inoculated with *S. aureus* and *E. coli* and portioned as 150, 600, and 1200 g loaves (86). Beef was placed into doughnut-shaped Pyrex dishes designed to eliminate cold spots and covered or not covered with a wrap of PVDC film. The loaves were heated in a microwave oven rated at 713 and 356 W at 100 and 50% power, respectively. Temperature was measured using fluoroptic probes along the geometric center of the loaf, and postprocessing temperature rise was monitored.

Survival of *S. aureus*, *E. coli*, and aerobic bacteria was lower in beef loaves wrapped with PVDC film, suggesting that the film retains heat and acts as a barrier to evaporation from the surface of products. Furthermore, the presence of film increased the endpoint temperature and the postprocessing temperature rise. Load size did not affect bacterial survival when the exposed microwave dose method (EMD; watts × min/g of food) was used. It was also noted that as the size of the load decreased, the uniformity of temperature in the loaf increased. The microwave power level (100% vs. 50%) did not affect the survival of bacteria when exposed to the same microwave dose. Lin and Sawyer (86) recommended that EMD and endpoint temperature were more accurate methods to calculate thermal processes than measurement of heating time as is currently practiced.

The use of microwaves to thaw frozen ground beef patties inoculated with $\sim 10^6$ cfu/g of *E. coli* O157:H7 was investigated by Sage and Ingham (87). Inoculated beef patties were frozen at −20°C for 24 hours and then thawed by storage at 4°C for 12 hours, 23°C for 3 hours, or microwave heating for 120 seconds at 700 W. Destruction was variable depending upon the strain, the recovery method, and the thawing regimen.

Beef wieners inoculated with natural microflora from spoiled wieners were heated for 34–48 seconds to temperatures of 45–56°C (88). Counts of yeasts, coliforms, and lactobacilli decreased with heating time, with very little reduction in the numbers of streptococci. Isolates representative of each of the bacterial or yeast groups were reinoculated into additional wieners and heated; destruction followed a similar trend. A two-stage process involving exposure to microwave energy for 39 seconds, followed by a 2-minute hold period, then a second exposure of 18 seconds, eliminated most of the organisms except for the streptococci.

Pork Products. Pork loins were surface inoculated and injected with *Bacillus subtilis*, *P. putrefaciens*, or *Leuconostoc mesenteroides* and heated to internal temperatures of 60, 68, 77, and 85°C (89,90). Microwave heating showed less destruction of all microorganisms at each final temperature compared to conventional heating, and this effect was attributed to the reduced exposure time afforded by microwave heating. *Bacillus subtilis* was the most heat-resistant organism, followed by *L. mesenteroides* and *P. putrefaciens* when subjected to either heating regimen.

Meat loaf consisting of ground pork, bread, onions, celery, eggs, bell pepper, and salt was inoculated with $\sim 3 \times 10^7$ cfu/g of *C. perfringens*, *E. coli*, *Streptococcus faecalis*, and *S. aureus* (91). The mixture was fashioned into 500 g loaves and heated by microwave, conventional electrical oven, or slow cooker for 10, 60, and 90 minutes, respectively, regardless of endpoint temperature. Aerobic plate counts were reduced by 1.29, 2.27, and 2.75 logs, respectively, by the three heating methods. Fruin and Guthertz (91) noted that microwave heating of nonfluid products is highly irregular, leading to increased survival of microorganisms as compared with conventional oven or slow cooker methods, which were considered more efficient in ensuring destruction of bacterial loads.

Cooked, sliced sausage composed of beef and pork were surface inoculated with *Candida albicans*, *C. perfringens*, *Enterobacter sakazakii*, *Enterococcus faecium*, *E. coli*, *Lactobacillus alimentarius*, *Lactobacillus viridescens*, *L. monocytogenes*, *Micrococcus luteus*, *P. fluorescens*, *S.* Enteritidis, *S.* Senftenberg, *S.* Typhimurium, *Serratia liquefaciens*, or *S. aureus* and then vacuum-packaged (92). Sausage slices were heated at 750 W in 10- or 20-second episodes with 3-minute breaks between heating, after which samples were held for 7 days at 7°C. Only *C. perfringens*, *L. alimentarius*, *L. monocytogenes*, *L. viridescens*, and *E. faecium* were recovered after heating, with the least destruction of the latter two organisms. Their survival is especially important in that they are noted for their storage defects of cured meats.

Poultry Products. Microwave energy has been used to cook whole chicken and turkey carcasses and parts as well as comminuted products, such as patties or loaves.

Consistent reduction of microbial numbers on poultry carcasses has been difficult because of differences in specific heat due to physical structure and whether the carcass contains stuffing. Whole chickens rinsed with 5×10^5 *Salmonella* Typhimurium/mL were halved and heated in either an electric oven at 163°C or a microwave oven to an internal temperature of 85–91°C (93). None of the 9 oven-heated chicken halves was positive for *Salmonella*, but 5 of 9 samples from the microwave-heated samples were positive by a direct plating method. Furthermore, the most probable number method to assess actual numbers of salmonellae showed that 1.1×10^2 to 2.4×10^2 MPN/mL were recovered from microwave-heated samples compared to <0.3–2.3 MPN/mL for oven-heated samples. Lindsay et al. (93) concluded that measurement of internal temperature was an unreliable means of ensuring destruction of salmonellae.

In another study (94), whole chickens were rinsed with 2×10^7 cfu/mL of a *S.* Typhimurium culture and heated to an internal temperature of 74, 77, 79, or 85°C in a microwave, convection microwave, or conventional electric oven. Survival of salmonellae was assessed by ELISA methods after heating. Of 11 chickens tested for each oven, one bird was positive for salmonellae cooked by conventional heating, 2 by convection microwave oven, and 9 by a microwave oven. Survival of salmonellae was linked to internal temperatures with 50% of chickens testing positive at 74°C and 33% at 85°C.

Lund et al. (40) inoculated *L. monocytogenes* at a level of 10^6–10^7 cells/mL on the surface of the skin of whole chickens and in stuffing. The chickens were enclosed within cooking bags and microwave heated at 650 W in an oven equipped with a turntable, followed by a standing time of 20 minutes. Temperatures taken with a thermocouple inserted between the breast and the leg varied from 60–87°C after 15 minutes of heating to 80–99°C after 38 minutes. *Listeria monocytogenes* was recovered by enrichment from one of the samples, but not after a 5-minute standing time. *Listeria monocytogenes* was reduced by <1 log by exposure to temperatures of 52–78°C measured in a central location in stuffing; after 10 minutes of additional heating, temperatures ranged from 63 to 90°C and the listeriae count was <10/g. After the full cooking time and a 20-minute standing time, temperatures reached 72–85°C and *L. monocytogenes* was recovered by enrichment from one sample. Uneven temperatures generated throughout the food led to survival of this organism. In follow-up work (95), *L. monocytogenes* was inoculated either on excised chicken skin or skin immersed in broth and exposed to a programmed temperature profile comparable to that achieved during the microwave cooking of whole chickens (40). There was a large variation in the number of survivors for the *Listeria*-inoculated skin particularly when the temperatures rose above 65°C. The temperatures programmed during the simulated heating trials were more consistent than would normally occur during heating of a solid surface like chicken. In contrast, the immersed skin samples showed a more even temperature profile resulting in a 6–8 log reduction after 2 minutes of heating. Coote et al. (95) concluded that temperature measurement of microwave-heated foods, particularly solid or complex ones, should be based on the temperatures at which the microorganisms will be exposed. Temperatures should be taken at several different points in the product and cooking times should be based on the minimum measured temperatures taken.

Further work using chickens naturally contaminated with *L. monocytogenes* was conducted by Farber et al. (96) using a variety of microwave models. Broilers (≤1.8 kg) and roasters (≥1.8 kg) were microwave heated according to manufacturers' directions with a standing time of 10–15 minutes. One of 81 broilers and 9 of 93 roasters tested positive for *L. monocytogenes* after microwave heating. When compared with conventional cooking, one of the broilers and none of the roasters were positive for *Listeria*. Ovens equipped with turntables, different cavity-size ovens, or various wattage levels did not appear to affect destruction of *L. monocytogenes*. Farber et al. (96) recommended that chickens be heated in a microwave oven to an internal temperature of ≥74°C, frozen foods should be thawed before cooking, stuffing should be cooked separately from the bird, foods should be rotated manually during cooking, and thicker portions of the food should be placed closer to the exterior of the microwave dish.

Previously frozen turkeys were defrosted and were dipped for 5 minutes in a suspension of *S.* Typhimurium, *S. aureus*, or *C. perfringens* (97). In some cases, turkeys were stuffed with a breadcrumb, margarine, and water mixture containing a suspension of these organisms. Turkeys were heated for 30–45 minutes depending upon weight to final temperatures of 76.7°C for an uncovered bird and 68.3°C for a bird contained within a roast-in bag. A 2-log reduction was noted with *C. perfringens* cells, which corresponds to the data of Blanco and Dawson (98). A 4- to 7-log reduction was seen with *S.* Typhimurium and *S. aureus*. Increased reduction of *C. perfringens* and *S.* Typhimurium was demonstrated when turkeys were contained within a brown-in bag, but increased survival of *S. aureus* occurred. The inclusion of margarine in the bread crumb stuffing mixture led to a 4-log reduction of *S. aureus* after 75 minutes of heating, while stuffing made without margarine led to only a 2-log reduction. Temperatures of these mixtures were not taken. Holding turkeys for an additional 30 minutes at a warm setting (10% power) to an endpoint of 76.6°C to allow for temperature equilibrium after cooking further reduced levels of salmonellae and staphylococci.

Microwave radiation was examined as a means of reducing salmonella load on raw poultry without cooking the meat. Turkey hen carcasses or drumsticks were inoculated with $\sim 10^5$ cfu/mL *S.* Senftenberg or *S.* Typhimurium (99). Salmonellae were not recovered from drumsticks after 2 minutes of heating or from carcasses after 10 minutes. Partial cooking of the product occurred, and therefore this treatment would not be considered as an alternative method to reduce microbial loads on raw turkey parts.

Chicken breasts were precooked by steam-jacketed kettle for 20 minutes at 87.8°C or 6 minutes by microwave energy (1600 W) (100). Mesophilic and psychrophilic bacteria belonging to the staphylococci and micrococci groups were recovered at significantly higher levels from the skin area of microwave-heated samples compared to water-heated samples. There was no indication that the two heating processes provided equivalent lethality.

Blanco and Dawson (98) examined the survival of vegetative cells and spores of *C. perfringens* inoculated on chicken thigh and leg meat and heated by a variety of methods including microwave energy at 915 and 2450 MHz. Microwave heating at 915 MHz resulted in a 2-log reduction of vegetative cells; subsequent treatments of freezing and thawing did not affect clostridial numbers, but browning in oil did reduce numbers to nondetectable levels. Freezing and thawing of chicken pieces before microwave heating, however, led to increased reduction of clostridia. Blanco and Dawson (98) suggested that microwave heating at 2450 MHz accomplished activation and germination of sporulated cells such that subsequent heat shocking destroyed the sensitive cells, but this effect was not noted with heating at 915 MHz. They suggested that this difference might be due to slower heating in the commercial oven (915 MHz; 4 min) versus the household oven (2450 MHz; 1–1.25 min). Similar work by Craven and Lillard (101) studied microwave reheating of precooked chicken thigh meat inoculated with spores or vegetative cells of *C. perfringens.* Exposure times of 45–90 seconds achieved internal temperatures of 49–84°C. Although the weight of chicken pieces was similar, variation in internal temperatures arose from the differences in conformation and composition of the pieces and the presence of ''cold spots'' in the oven. Microwave heating of the chicken followed by blending and heat shocking to recover spores increased destruction of spores as compared with spores that were only heat shocked. Craven and Lillard (101) suggested that microwave heating stimulated germination and outgrowth of spores that were further reduced by exposure to a second heat treatment. Furthermore, the authors concluded that the reduction in clostridial numbers was due to a thermal effect only.

Harrison and Carpenter (102) examined the survival of *L. monocytogenes* on chicken skin heated by microwave energy. Chicken breast portions were surface inoculated with *L. monocytogenes* and heated at 700 W power to internal endpoint temperatures of 65.6, 71.7, 73.9, 76.7, and 82.2°C. After heating, portions of chicken were sampled immediately or vacuum-packed or wrapped in 0.7 mL thick oxygen permeable PVDC film and stored for up to 4 weeks at 4°C or up to 10 days at 10°C. *Listeria monocytogenes* was reduced by 2.5–3.8 logs at temperatures of 73.9–82.2°C, while at the lower heating temperatures *L. monocytogenes* was reduced by <2 logs. Furthermore, samples heated at the lower temperatures of cooking were recovered throughout the 4°C storage period; however, their numbers did not increase. At the abuse storage temperature of 10°C, numbers of listeriae multiplied to reach the population levels similar to or higher than the original inoculum levels. With either storage temperature, vacuum packaging led to decreased numbers of listeriae compared with oxygen permeable samples. Harrison and Carpenter (102) noted that chicken appeared dry and cracked in some areas and moist and slightly undercooked in other areas, suggesting that chicken breasts cooked unevenly. Furthermore, temperature rise was so quick that there may not have been sufficient time to eliminate microbial loads.

Comminuted poultry products present different challenges to microwave heating than whole-carcass heating in that microbial load may or may not be equally distributed throughout the food. Naturally contaminated turkey rolls were assessed for standard plate count and coliform count after precooking by microwave energy or water bath (103). Standard plate counts were reduced by greater than 3–4 logs, and coliforms were reduced to nondetectable levels from initial counts of $\sim 10^2$ cfu/g. Both methods effectively reduced microbial populations, but the microwave-heated product was judged as more tender, flavorful, and juicier with greater yield.

Cunningham (104) used microwave heating (915 MHz) to decrease total bacterial and coliform numbers associated with chicken patties. Reduction of total counts was seen with as little as 10 seconds of heating with a 2-log reduction after 40 seconds. Coliform counts showed a similar trend, but only a 1-log reduction was seen after 20 seconds before a leveling off occurred. The presence of an athermal effect might explain the decrease in microbial numbers with such a brief exposure time at an internal temperature of only 40°C. Cunningham (104) concluded that subcooking processes using a microwave might be possible without altering the flavor or appearance. Further work by Cunningham (105) examined the use of microwave heating to decrease the resident psychrotrophic microflora associated with spoiled chicken products. After 10 seconds of heating, pure cultures of psychrotrophs, e.g., *Pseudomonas*, *Flavobacterium*, *Achromobacter*, *Alcaligenes*, were reduced by 4 logs from an initial level of 10^7 to 10^8 cfu/mL. Total and coliform counts were reduced on cut-up chicken pieces by almost 2 logs after 40 seconds of treatment. Numbers also decreased when heating chicken skin samples in a microwave oven. Treatment of skin with microwave radiation for 30 seconds not only reduced the population initially after heating, but numbers remained lower after refrigeration of samples for 7 days. Cunningham (105) suggested that psychrotrophic bacteria were particularly susceptible to microwave heating.

Sterilized chicken, chicken and broth, and chicken and white sauce were packaged in cook-in pouches composed of polyester-polyethylene laminate (106). Samples were inoculated with *S.* Senftenberg, *S.* Typhimurium, and *S. aureus* at levels of 1.3–2.3 $\times$ 10^6 cells/g. Pouches were heated in a microwave oven for 40, 60, 90, 120, or 150 seconds or boiling water bath for 4, 6, 8, or 10 minutes. No salmonellae or staphylococci were recovered from chicken or chicken plus broth after microwave heating for 120 seconds; *S.* Senftenberg showed a 5-log reduction and *S. aureus* exhibited a 4-log reduction in the chicken plus white sauce combination. No salmonellae were recovered after 8 minutes of heating by the boiling water method, but *S. aureus* was recovered in chicken alone after 10 minutes of heating.

Salmonella Typhimurium and *S. aureus* were the test organisms to demonstrate the effectiveness of microwave heating on chicken burgers, chicken loaves, or roasted chicken (107). The burgers and loaves were mixed with 10^8 cfu/g of cultures, while the roasts were injected intramuscularly. In general, *S.* Typhimurium was detected only in chicken burgers heated by microwave energy, but counts were still reduced by a level of 4 logs. This may have been due to differences in endpoint temperatures of 64°C for chicken burgers compared with 82–83°C for chicken loaves and 76–77°C for roasted chicken. *Staphylococcus aureus* was generally recovered from all microwave-heated products.

Egg Products. Baked custard and scrambled eggs were chosen for study by Dessel et al. (81). *Serratia marcescens*, *S. aureus*, and *Bacillus cereus* were added to custards and heated in a conventional gas range in 70°C water at an oven temperature of 176°C or at a high setting in a microwave oven. Microwave cooking of custard for 4–5 minutes was comparable to conventional baking for 30–40 minutes in destroying *S. marcescens* and *S. aureus*, with only a 3-log reduction for *B. cereus*. There was no indication whether vegetative cells or spores of *B. cereus* were used. Scrambled eggs inoculated with *S. aureus* and *B. cereus* were cooked for 1–3 minutes in a skillet on a thermostatically controlled burner set to 71.7°C and on low microwave frequency for 35–40 seconds until doneness was observed. *Staphylococcus aureus* was eliminated by microwave heating but not by conventional heating; neither method was effective for reducing *B. cereus*. The authors suggested that microwave heating was superior to conventional heating and was more efficient in causing destruction of bacterial numbers.

Baker et al. (107) heated egg custard containing *Salmonella* Typhimurium and *Staphylococcus aureus* in an oven at 163°C for 47 minutes to an endpoint temperature of 82–84°C or in a microwave oven for 7 minutes. *Staphylococcus aureus* was reduced by >4–6 logs for conventionally heated product compared with 3 logs for microwave-heated product. *Salmonella* Typhimurium was not recovered in any product.

Baldwin et al. (108) found that ingredients or pH level can influence the destruction of salmonellae in lemon and chocolate pies. *Salmonella* Typhimurium was inoculated at a level of 10^5 cells/g into foam products and heated for 100 seconds, a time chosen to yield a desirable product. The final temperature of product measured 15 seconds after heating was 85°C. The solids content of the lemon foam pie ranged from 47.1 to 47.8% and pH level of 3.5 to 3.8 compared to chocolate, which was 54.9–57.1% and pH 7.5–8.0, respectively. *Salmonella* Typhimurium was not detected in chocolate foam pies by agar plate counts after microwave heating but was recovered by qualitative methods for up to 72 hours postheating. *Salmonella* Typhimurium was not recovered by either procedure from lemon foam pies. It was suggested that the high solids content coupled with the low pH did not provide a conducive environment for survival of salmonellae. To test this theory, Baldwin et al. (108) formulated model broth systems containing 0, 35, or 37% sugar at pH levels of 7.3 and 3.7 adjusted with citric acid. Sugar concentration did not affect growth of salmonellae at neutral pH but did significantly retard its growth at pH 3.7.

A process to pasteurize intact shell eggs using microwave technology resulted in a 7-log reduction of *S.* Enteritidis (109). Hot water, hot air, and microwave energy were compared using various inoculation levels and holding times. The combination of an initial rapid microwave-heating step to 55°C followed by holding at that temperature in hot air or water was desirable. Furthermore, microwave heating preferentially heated the egg yolks without adversely affecting functionality of the egg proteins.

Seafood Products. A study by Baldwin et al. (110) focused on microwave destruction of *S.* Typhimurium in carp. Carp was surface inoculated with two strains of *S.* Typhimurium at a level of ~10^6 cfu/g and heated by microwaves for 195 seconds to an internal temperature of 73°C, a time chosen for degree of doneness and palatability. Salmonellae were recovered in 8 of 10 trials, although numbers were reduced by over 5 logs. In a second study, temperature measurements taken by thermocouple at 15-second intervals indicated that although temperature increased as exposure to microwaves increased, there was considerable variation within individual tuna pies, tuna casseroles, fish fillets, and fish sticks. This variation was attributed to ingredient variation and location of foods within the oven. The temperature was coolest near the center of the sample, but temperature was not necessarily related to the size of the sample.

Huang et al. (111) examined the fate of *Aeromonas hydrophila* and *L. monocytogenes* in channel catfish heated by microwave energy. The two pathogens were added at a level of 10^8 cells/mL to water in which the catfish were dipped for 1 minute and allowed to dry for 5 minutes. Fillets were placed in polystyrene trays and either overwrapped with PVDC film or left unwrapped. Samples were incubated at 37°C for 5 hours to allow the bacteria to increase to 10^6 cells/cm^2. Fillets were heated at 650 W for 3–5 min/0.45 kg of fish depending upon thickness to internal temperatures of 55, 60, or 70°C as measured by fluoroptic probes. Neither *L. monocytogenes* nor *A. hydrophila* were recovered from overwrapped samples following cooking to 70°C, but *L. monocytogenes* was reduced by 3 logs in uncovered samples. At 60°C, *L. monocytogenes* was reduced by 4 logs for wrapped fillets and 2 logs for uncovered fillets. At 60°C, *A. hydrophila* was reduced by 5 logs in unwrapped fillets; cooking at 55°C reduced populations by 2.5 logs.

Shrimp inoculated with approximately 5×10^5 cfu/g of a five-strain cocktail of *L. monocytogenes* was placed in a covered microwaveable plastic container (112). Samples were heated at four power levels (200, 400, 560, and 800 W) for 140, 70, 52, and 40 seconds and held for 2 minutes after heating to equilibrate. At least one replication in four tested positive for *L. monocytogenes* after heating for each cooking time. After increasing the cooking times by 20%, no *L. monocytogenes* were recovered. Survival of *L. monocytogenes* may have occurred because of the centralized location of the inoculum and the use of a mathematical model to predict cooking times.

3. Dairy Products

Microwave heating has been considered as an alternate means of milk pasteurization. Microwave energy has the potential to be more energy efficient than conventional pasteurization because of shorter heating times. In addition, flavor characteristics are retained and heat-labile vitamins are preserved. Because of the ability of microwaves to penetrate paperboard, in-package heating of milk may provide an added benefit. Excellent review articles have been prepared by Young and Jolly (113), Sieber et al. (114), and Heddleson and Doores (115).

Merin and Rosenthal (116) compared low-temperature, long-time (LTLT, 65°C, 30 min) batch pasteurization processes using microwave energy and the conventional method. A 6-log reduction in standard plate, coliform, and psychrotrophic counts was achieved by both systems. The chemical composition of the milks was not substantially different, although the whey proteins were slightly more denatured by the microwave method. The phosphatase reaction, a chemical marker of an effective pasteurization process, was negative by both heat treatments.

Microwave heating for 50 seconds at 700 W was compared to conventional batch pasteurization at 62.5°C for 30 minutes to reduce microbial load in breast milk (117). Both pasteurization methods were effective in reducing microbial populations, but destruction of immunoglobulin A, an important immune component in breast milk, was highly variable in microwave-treated milk, most likely because of the 24°C range of endpoint temperatures. The use of microwave heating to warm bottles resulted in highly variable temperatures depending upon the formula used (118). Heating of multiple bottles simultaneously produced temperatures of 72°C for the center bottle and 48°C for surrounding bottles. Furthermore, some materials used in the manufacture of the bottles did not lend themselves to be heated by microwaves and ruptured during normal heating regimens. In another study (119), enterotoxigenic *B. cereus* inoculated into reconstituted infant formula survived normal washing procedures but were reduced by up to 5 logs when steamed in a microwave oven at 100°C for 9 minutes.

Jaynes (120) used a variable-power microwave generator designed to deliver a continuous heat treatment of 72°C for 15 seconds comparable to that provided by the conventional high-temperature, short-time pasteurization (HTST) process. Flow rates of 200, 300, and 400 mL/min were used. At each flow rate, standard plate and coliform counts were reduced to comparable levels by microwave and HTST treatment. Furthermore, the phosphatase test was negative by both treatments.

Knutson et al. (121) examined microwave energy as a means of delivering an LTLT and HTST process to pasteurize raw milk. Raw milk was steamed to reduce initial microbial flora, then inoculated with *S.* Typhimurium, *E. coli*, or *P. fluorescens* at a level of 10^3–10^4 cfu/mL prior to a simulated microwave HTST treatment of 59 seconds. Even with a microwave treatment of 65 seconds to a temperature of 78.6°C, there was recovery of all three organisms. To simulate LTLT treatment of 62.8°C for 30 minutes, raw milk was inoculated with 10^6 cfu/mL of *S. faecalis*. Because of the inability to maintain the temperature consistently for this time, milk was heated to above 62.8° but less than 71.7°C and slowly cooled. Although *S. faecalis* was reduced by 3–4 logs after heating, survivors remained. Knutson et al. (121) suggested that survival of microorganisms was due to nonuniform distribution of heat through the milk and the formation of cold spots.

Milk has frequently been implicated in the transmission of foodborne pathogens, especially psychrotrophic organisms. Choi et al. (122,123) examined the ability of microwave energy to inactivate *Y. enterocolitica*, *C. jejuni*, and *L. monocytogenes* in milk and their recovery postpasteurization under refrigerated conditions. Milk samples inoculated with each pathogen at levels of 10^6–10^7

cells/mL were dispensed as 20 mL volumes into glass vials, sealed, immersed into a jar containing water, and heated at 71.1°C for various times. Although some heating times appeared to destroy the organisms immediately after heating, storage at 4°C for various times allowed for recovery of stressed cells. Inactivation of *Y. enterocolitica* occurred after 5 minutes of heating for 50% of the trials, and no viable yersiniae were recovered after 8 minutes. *Campylobacter jejuni* was more heat sensitive than *Yersinia* with a 5-log reduction in the population after only 1 minute of heating, but some campylobacter were recovered during storage; a 3-minute exposure resulted in complete inactivation. *Listeria monocytogenes* was destroyed after 10 minutes of heating.

Chiu et al. (124) used microwave heating to extend the shelf life of pasteurized fluid milk. Milk was microwave heated for 85, 97, 108, and 120 seconds, stored for 8 and 11 days at 6.7°C, and assessed for standard plate and psychrotrophic bacteria counts. Heat treatment for 120 seconds to achieve 60°C was effective in reducing psychrotrophic bacteria by 6 logs with standard plate counts reduced from 4 to 6 logs depending upon the initial contamination level. Considering that milk flavor defects arise from proteolytic and lipolytic enzymatic activity of psychrotrophic bacteria, a postpasteurization heat treatment can not only reduce microbial load, but also extend shelf life and reduce stale, bitter, and rancid off-flavors.

A similar study was undertaken by Stearns and Vasavada (125) in which quality and shelf life of milk were examined. Raw milk was exposed to microwave treatment for 0–90 seconds and assessed for standard plate and coliform counts. With a heat treatment of 65 seconds, >99.9% of the organisms were eliminated. Storage of raw milk at 4°C was extended from 1 week for unheated milk to 2 weeks for milk heated for ≥35 seconds.

Injury of *E. coli* by microwave heating was explored in the work of Aktas and Özilgen (126). *Escherichia coli* was added to nonfat dry milk medium at a level of 10^5 cfu/mL and pasteurized through a tubular flow reactor in a microwave oven with flow velocities of 159, 211, 233, and 247 cm/min. Approximately 15–25% of microorganisms pasteurized in this system was considered injured. Cells injured during pasteurization have the potential to repair and potentially cause spoilage problems or foodborne illness depending upon the organism.

Thompson and Thompson (127) attained a 5-log reduction in standard plate counts taken from goat's milk pasteurized by using microwave energy on the "medium" setting set to rise to 65°C with a temperature hold feature to maintain milk for 30 minutes. In other studies using goat milk, Villamiel et al. (128) achieved a pasteurization process of 72.5 and 80.1°C for 15 seconds employing a continuous flow microwave heating system and a laboratory-scale plate heat exchanger designed to simulate an HTST process. Microbial counts of finished product did not substantially differ between the two methods. Proteins and enzymes important in cheese manufacture were denatured less by microwave heating.

Microwave energy can also be used to vacuum dry yogurt while maintaining the viability of lactic acid bacteria (129). Decimal reduction times were dependent upon water activity and temperature of heating with the lower temperature range of 35–45°C providing a greater degree of survival of *Streptococcus salivarius* subsp. *thermophilus* and *Lactobacillus delbrueckii* subsp. *bulgaricus*.

In-package microwave heating of cottage cheese was effective in extending the shelf life of this perishable dairy food (130). Cottage cheese contained in polystyrene tubs and flexible packages (ethylene vinyl acetate/PVDC/ethylene vinyl acetate) was heated to various temperatures using microwave heating at 500 and 2860 W, then stored at 4°C. A heating temperature of 48.8°C at 500 W was optimal in reducing coliform, psychrotroph, yeast and mold, and total counts while maintaining acceptable sensory properties and pH levels. Furthermore, shelf life was extended from one week to 42 days as compared with nonheated samples.

4. Multicomponent Foods

Heat-and-serve products traditionally encompass a variety of products that are precooked, chilled or frozen, and then reheated in the home. These products enjoy a wide audience because they are convenient, varied, and time-saving. Typically, microwave ovens are used for reheating, although conventional heating could also be used. Pathogenic psychrotrophic microorganisms, particularly *L. monocytogenes*, may be of paramount concern.

Creamed chicken, rice ring, chicken paprika, chicken breast and leg with gravy, spaghetti and meatballs, and ham patties were heated from a frozen state by conventional oven, double boiler, boiling water, and microwave oven (131). Weight losses were greatest when foods were microwave heated. Bacterial counts were reduced to acceptable levels by all reheating methods, but spore-formers and staphylococci were isolated.

Postprocessing temperature rise was measured in water, chicken frankfurters, beef loaf patty, pork roast, turkey roast, turkey casserole, and cake cones after heating by conventional hot air oven (160°C) and microwave (50 and 100% power, 645 W). Temperature rise was evident in products heated by microwave energy and was product dependent. Temperature rise occurred depending upon the power level, but endpoint temperatures were achieved sooner when products were heated at 100% power (43). Overwrapping of chicken drumsticks, ham slices, or pork slices with PVDC wrap did not significantly influence mean internal temperatures after microwaving, but the temperature range was lengthened for the drumsticks. *Salmonella* Senftenberg was not recovered from any of pork samples or the wrapped chicken samples but was isolated from one unwrapped sample. *Staphylococcus aureus* inoculated at an initial level of 6.6 $\log_{10}$ cfu/mL was not isolated from any chicken but was recovered from one unwrapped ham sample (2.36 $\log_{10}$ cfu/g).

In a survey study, Hollywood et al. (132) purchased a range of prepared frozen meals including steak pies, roast lamb or roast beef, Asian dishes, and Italian meals. Meals were heated according to package directions at full power (650 W). Total bacteria and psychrotroph counts, coliforms, yeast and mold, lactobacilli, enterococci, *Pseudomonas*, *Brochothrix thermosphacta* as well as the pathogens *A. hydrophila*, *B. cereus*, *Salmonella*, *C. perfringens*, *L. monocytogenes* and coagulase-positive *Staphylococcus* were analyzed from samples of the frozen product and the heated product. In general, levels of spore-forming organisms were unaffected by the heat treatment and appeared to be the dominant microflora reflected in the total count. Although these meals were precooked and frozen, some pathogens, e.g., *C. perfringens*, *B. cereus*, staphylococci, were recovered at low levels unlikely to cause illness. However, these recoveries point to the necessity of preparing safe food for the consumer.

Kerr et al. (133) suggested that the high numbers of cases of listeriosis reported in the United Kingdom might be due to the consumption of cook-chill foods that are intended for take-out use and reheated in the home. Of the 21 foods examined, 24% contained detectable levels of *L. monocytogenes*. Sheeran et al. (134) inoculated 10^6 *L. monocytogenes*/g into these foods and microwave-heated the products according to package directions using two different domestic ovens. Temperatures after heating ranged from 48 to 100°C. Of the 27 foods sampled, 81% were positive for listeriae after heating. Although undetectable levels (below the 10^2/g limit of detection) of *L. monocytogenes* were noted after heating in 19% the foods, subsequent refrigeration at 6°C for 5–8 days resulted in detectable levels, indicating that *L. monocytogenes* could recover from sublethal microwave heat injury.

Walker et al. (135) examined the survival of *L. monocytogenes* in eight foods (Chinese chicken breasts, enchiladas, chicken and sweet corn soup, turkey breast joints with stuffing, roast boneless breast of chicken, cannelloni di carne, chicken Korma, and chicken cacciatora) that had been purchased from retail outlets. A three-strain cocktail of *L. monocytogenes* was inoculated into the foods at a level of $\sim 10^5$–10^6 cells/g, mL, or piece of food and heated at a power setting of "high" (660 W) or "medium" (320 W). Temperature measurements were taken by using four fluoroptic thermometry probes. After heating, *L. monocytogenes* was recovered from one enchilada sample and from all four samples of chicken and sweet corn soup and roast chicken breasts. Enclosure in foil packaging led to greater survival of listeriae most likely due to the inability of microwaves to penetrate the foil. Temperature gradients differed by as much as 50°C.

Again, temperature variation within products could have led to differences in survival of *L. monocytogenes*.

In a follow-up study, Dealler et al. (41) purchased 30 preprepared, microwaveable foods belonging to the cook/chill, sterilized, and cook/freeze classes of food products from major supermarkets. Foods were inoculated with 10^5–10^6 cells/g of *S.* Typhimurium and *L. monocytogenes* and heated at the highest microwave power setting (650 W) for the time recommended on the package. The postheating temperatures of the foods ranged from 43 to 102°C as measured by a temperature probe with surface temperatures being uniformly hot and core temperatures often much cooler. *Salmonella* Typhimurium was isolated from 70% of the 30 samples by direct plating with 77% isolated by the enrichment method. *Listeria monocytogenes* was isolated from 80% of the samples by direct plating with 97% isolated by the enrichment method. Nonuniform heating temperatures and cold pockets most likely gave rise to the recovery of the two pathogens. Dealler et al. (41) suggested that longer cooking times be stipulated with stirring instructions and standing times included in package directions. In addition, added salt or monosodium glutamate could decrease the core temperatures such that pathogenic organisms are not destroyed upon reheating.

Dos Reis Tassinari and Landgraf (136) compared two microwave ovens: one equipped with a turntable and a second oven with a preset program based on final temperature of the food. Thirty formulations each of baby food, mashed potatoes, and home-made beef stroganoff were inoculated with $\sim 10^4$ cfu/mL final concentration of *S.* Typhimurium. Samples were heated in the control microwave oven for 50 seconds for baby food and 75 seconds for mashed potatoes and stroganoff; times were based on sensory analysis of the palatability of the products. The second oven used heating times determined by the preset temperature sensor. Temperatures of mixed samples were taken 10 seconds after heating. Of the 90 samples tested, *S.* Typhimurium was recovered from 83.3% of the baby food, 20% of the mashed potato, and 40% of the beef stroganoff samples for the control microwave compared to 80, 100, and 100%, respectively, for the microwave with the preset sensor. The oven equipped with the temperature sensors automatically chose lower temperatures, which in turn led to survival of salmonellae.

5. Foodservice Products

Centralized foodservice operations, such as those found in hospitals or cafeterias, typically cook foods by conventional means, chill or freeze foods, and then reheat prior to serving. These operations require rapid reheating of foods, and microwave power offers just such an option. Concerns center on uneven heat distribution of products, product formulations, portion sizes, and geometry of a food. Of particular importance with these products as opposed to similar multicomponent foods is that target consumers may be more at risk for foodborne illness than the general population.

Lacey et al. (137) compared microwave heating to infrared and conventional electrical heating for use in hospitals. Preliminary experiments showed that *B. cereus* from soil samples survived 3 minutes of microwave treatment but was destroyed after a 2-minute exposure to an infrared-microwave oven preheated to 150°C and 0.5 minute when the oven was preheated to 160°C. Spore count survival in mashed potatoes heated by the three methods for 2 minutes resulted in 10.3–21.4%, 9.2–38.8%, and 10.7–20.0% survival, respectively, for the three heating methods.

Work by Creamer and Chipley (138–141) demonstrated that microwave heating could be used in food service systems as a means of reheating "ready prepared" foods. As a model food, frozen, pasteurized whole eggs were scrambled on a grill preset to a temperature of 149°C. After cooking, eggs were transferred to pans and refrigerated for ~26 hours. Eggs were reheated in microwave ovens to internal temperatures ranging from 35 to 92°C with an average of 67.1°C. Low numbers of *Bacillus* spp., *Clostridium sporogenes*, *Staphylococcus epidermidis*, *E. coli*, and *Enterobacter aerogenes* were present, indicating inadequate heat processing or recontamination of cooked eggs from contaminated serving utensils. In a second, more controlled study, Cremer (140) reheated eggs by microwave power to average temperatures of 85°C. Time for reheating eggs was only 30 seconds, but voltage levels varied significantly during the trials. Although considerably more energy was used in heating eggs and beef patties by convection ovens than institutional microwave ovens (1300 W), sensory analysis demonstrated that overall quality was significantly better for microwave-heated eggs, but convection heating was preferable for beef patties (141).

In a similar study, Cremer and Chipley (139) investigated the product flow of raw roast beef from initial cooking by convection heating to 135°C, chilling for 45 hours, then onto the final reheating by microwave energy, ranging from 43 to 93°C. Although portion size and placement of meat slices on dishes were uniform, the composition of each piece, e.g., ratio of fat to lean portion, the weight of meat, and the amount of gravy added, was not controlled. Microwave heating reduced total microbial load by about 10- to 25-fold. *Bacillus* spp., *Clostridium sporogenes*, *C. perfringens*, *S. epidermidis*, and *S. aureus* were recovered after microwave heating, suggesting that temperatures were insufficient to kill these organisms.

The work by Dahl and coworkers (142) examined cook/chill foodservice systems prevalent in hospitals. Fresh ground beef was blended with raw eggs, formed into beef loaves, heated to varying internal temperatures (45, 60, 75, or 90°C) within a forced-air convection oven, and then stored for 24 hours at 6°C.

Although loaves were sliced into 100 g pieces and reheated in a microwave oven to ≥74°C, microwave heating did not reduce the aerobic mesophile plate count. Initial cooking temperature of ≤60°C did increase yield, moisture content, and possibly nutrient content.

Beef loaf, mashed potatoes, and frozen and canned beans were chosen as typical menu items served to hospital patients (143). Foods were heated for 20, 50, 80, or 110 seconds, and temperature was measured in the center of the food 30 seconds after the completion of heating using a mercury thermometer. Even after 85 seconds of microwave heating, a 40°C range in endpoint temperature occurred, suggesting that variations in temperatures achieved through microwave heating were inconsistent with producing a microbiologically sound product. In a second study, *S. aureus* was inoculated onto the surfaces of these foods to achieve an initial concentration of $\geq 10^7$ cfu/g of sample (144). Microwave heating of beef loaf for 20 seconds had little effect, but heating for 50 seconds resulted in reduction to $\leq 30–5.4 \times 10^7$ cfu/g, and no organisms were recovered after 80 seconds of heating. Counts of *S. aureus* after microwave heating mashed potatoes for 85 seconds were reduced to $\leq 30–7.9 \times 10^3$ cfu/g. In frozen green beans, *S. aureus* decreased to $\leq 30–1.4 \times 10^4$ cfu/g after 80 seconds of microwave heating but decreased to ≤30 cfu/g in 4 of 6 samples in canned green beans. The discrepancy of heating effects and destruction of staphylococci was thought to be related to the dielectric properties of the individual foods coupled with the unevenness of heating. A third study conducted by Dahl et al. (145) focused on survival of *Streptococcus faecium*, considered a more ''radiation-resistant'' bacterium, in beef loaf and mashed potatoes. *Streptococcus faecium* was surface inoculated at a level of 1×10^8 cfu/g onto beef loaf and 1×10^6 cfu/g stirred into mashed potatoes. Foods were microwave-heated for 80–100 seconds for beef and 85 seconds for potatoes to achieve in internal endpoint temperature of 74–77°C. *Streptococcus faecium* survived heating at 50 seconds but was undetectable after 80 seconds of heating in beef loaf and was undetectable after 85 seconds of heating in mashed potatoes, whereas *S. aureus* was recoverable after this time in a previous study (144).

Further work by Sawyer et al. (146) examined microbiological quality as measured by the aerobic plate count method and endpoint temperature of beef loaves, peas, and mashed potatoes by the conduction, convection, and microwave heating. Reheating of beef loaves by microwave heating led to inconsistent endpoint temperatures that varied as much as 18°C and were lower than the 74°C standard suggested by FDA standards for reheating of foods. The authors were unable to predict the endpoint temperatures dependably despite preparing beef loaf portions having the same characteristics of weight and shape. Furthermore, 75% of the endpoint temperatures were lower than the targeted level of 74°C when foods were combined on the same plate. Microbial counts were not significantly different for any of the reheating methods, but variation in counts did occur during the preparation, initial heating, and chilling stages.

Microwave reheating frequently leads to variable endpoint temperatures. Turkey rolls were cooked by forced-air convection oven to an internal temperature of 77°C, chilled for 24 hours at 1°C, sliced in 1 cm portions, and placed on plates. Slices were microwave heated for 30 or 40 seconds to end temperatures of 48–63°C and 59–71°C, respectively. Aerobic plate counts averaged a 2- to 5-log reduction during the initial cooking of turkey rolls. Coliform counts ranging from 3.2 to 4.58 $\log_{10}$ cfu/g were reduced to negligible numbers after the completed process. It was concluded that microwave reheating of these products was an acceptable means of warming products and would not pose any unusual food safety risks (147).

Beef-soy loaves inoculated with *Staphylococcus aureus* at a level of 4.5–5.5 $\times 10^3$ cells/g were baked to 60°C in a convection oven at 121°C, then stored at 5°C for up to 3 days (148). Slices were then reheated in a microwave oven for 55 seconds to achieve an internal temperature of 80°C. After the initial cooking, levels of *S. aureus* decreased to 3–430 cfu/g depending upon the location of sampling. Numbers continued to decrease during chilled storage to <20 cfu/g after 72 hours. No *S. aureus* was recovered after microwave heating. Although during the heating and chilling periods for the loaves, there was opportunity for slight growth of *S. aureus*, but preformed enterotoxin was not tested for, nor was it expected to be destroyed by the final microwave heating step.

In the cook/chill foodservice system, microbial lethality is arrived at cumulatively through two cooking stages and one chilling stage. With this process, the possibility exists for bacteria to become injured, thus underrepresented by microbiological methods unless recovery procedures are instituted. Since it has been shown that bacteria survive microwave heating to a greater degree than conventional cooking because of decreased times of heating, the possibility exists that these bacteria could be in an injured state. Therefore, Dahl et al. (149) initially cooked beef loaves by convection heating to an internal temperature of 66°C, then stored them for 24 hours at 6°C, followed by microwave heating for 20, 50, 80, or 110 seconds; frozen green beans were thawed for 24 hours at 6°C, then microwave-heated. Samples of both products were plated using plate count agar (PCA) to measure aerobic plate count and PCA containing 3.0% salt for beans or 4.5% salt for beef loaf to measure the level of injured organisms. A higher number of bacteria were injured before microwave heating in the beef loaves than in green beans. This would be expected since an initial cooking step was incorporated for the beef loaf resulting in a more susceptible population that was killed by the second heating step. Lengthening the time of microwave heating increased the number of injured cells, as expected, resulting in a more uniform microbial population after microwave heating.

6. Miscellaneous Products

Jasnow and Smith (150) examined the sanitizing effect of microwaves on reducing background populations found in dry color additives used in cosmetics. Exposure time varied with amount of material and the microbial populations found in the sample. Microbial reductions to <1 organism per 75 g for gram-negative bacteria and <10/g for fungi were achieved. Counts for spore-forming bacilli were reduced 10-fold.

Bookwalter et al. (151) examined the use of microwave energy to reduce process times for and eliminate salmonellae from corn–soy–milk (CSM) foods intended for use by infants and toddlers. *Salmonella* Senftenberg was mixed with CSM in 50-pound bags, exposed to a 60 kW continuous microwave tunnel, and palletized during the cooling period. A 2-log reduction of salmonellae was achieved after exposure of CSM to 56.7°C followed by a 9.7-hour cooling period. A combination of different process times and cooling periods led to a 3- to 5-log reduction, but nutritional damage to the blends was noted.

The manufacture of soymilk yogurt is complicated by the survival of *Bacillus* species during the preparation of soybeans (152). Soymilk was microwave-heated at full power for 60 seconds or steam-heated for 20 minutes at 110°C to reduce bacilli levels prior to fermentation. Soymilk was fermented using a 5:1 ratio of *Streptococcus thermophilus* and *Lactobacillus bulgaricus* until the soymilk reached pH 4.5. Survival of bacilli by microwave and steam-heating interfered with subsequent fermentation. It was recommended that soymilk be autoclaved if it is to be used as the basis for yogurt manufacture.

Tarhana, a soup prepared by lactic acid fermentation, containing yogurt, wheat flour, vegetables, and herbs was inoculated with 1.8×10^5 *E. coli* O157:H7/g and 9.2×10^4/g of *Yersinia enterocolitica*. After 7 days of fermentation, the soup was dried in a hot-air oven at 38°C for 72 hours or by microwave oven (1000 W) at the lowest power level for 30 minutes. *Escherichia coli* survived the fermentation process at a level of $\sim 1.5 \times 10^2$ cfu/g but was not recovered by either drying method; *Y. enterocolitica* was eliminated by the fermentation process alone (153).

Fung and Cunningham (154) investigated the use of microwave heating on microorganisms suspended in a variety of soups, hot dogs, hamburgers, and fish fillets. They observed variability in the percentage of bacterial survival recovered in different regions of the food. Although covering appeared to enhance the efficiency of microbial destruction, microwave heating may not reduce high levels of microorganisms. *Streptococcus faecalis* was more heat resistant than other bacteria, confirming the results by Watanabe and Tape (88).

Park and Cliver (155) recommended that microwave treatment of wooden and plastic cutting boards was an excellent means of sanitizing cutting surfaces, thus reducing cross-contamination in the home. *Escherichia coli* and *Staphylococcus aureus* were inoculated onto the surface of sample wood blocks and exposed to the highest power setting (800 W). Within 120 seconds of exposure, up to 10^9 cfu/25 cm^2 of *E. coli* were eliminated from the surface of the wooden blocks, but even after 180 seconds *E. coli* was reduced by $<90\%$ for plastic blocks. Other means to sample wood blocks, including perfusion and removal of a sliver of wood, did not lead to recovery of any *E. coli*. Temperatures reached 95°C on the surface and 200°C for an interior portion of the wood after 4 minutes of heating, whereas plastic ranged from 35 to 38°C. Although the moisture level of the wood did not influence destruction level of organisms, the moisture level in the wood did decrease and the wood began to smoke after 11 minutes of heating.

D. Microbial Systems—Microbial Groups

Several organisms other than bacteria present unusual resistances to microwave heating or are studied in model systems and foods. Because of the nature of these studies, they are presented in this section.

1. Parasites

Trichinella. Several papers, most notably those by Kotula, Zimmermann, and coworkers, describe conditions for inactivation of *Trichinella spiralis*, the causative agent of trichinosis in pork products. The majority of these studies focus on the destruction of trichinae larvae in pork by comparing microwave heating to conventional cooking methods, e.g., oven, frying, char broiling. Traditionally, the endpoint temperature of 85°C (185°F) is used, but products heated to such high temperatures appear dried and unappetizing. Reduction of the cooking temperature to 77°C (170°F) leads to a more tender and juicy product and is recommended by The National Live Stock and Meat Board and the National Pork Producer's Council. The U.S. Department of Agriculture, Meat Inspection Division (156), recommends that pork products achieve a temperature of 76.7°C after standing, which provides a 17°C safety margin over the endpoint temperature of 58.3°C for ready-to-eat product containing pork and heated by conventional cooking. The endpoint temperature of 55°C is deemed adequate if that temperature is arrived at by slow heating; the remaining 3.3°C are added as a safety factor. Thus, the endpoint temperature for destruction of trichinae becomes important to ensure a zero tolerance level and the elimination of the threat from trichinosis.

All of the studies suggest that complete cooking of pork products likely to be contaminated with *T. spiralis* is dependent upon ensuring that the cold spot in the pork product reaches a prescribed temperature setpoint and is held for a sufficient length of time. Furthermore, microwave power and wave distribution, cut, and weight of meat affects temperatures achieved. Visual interpretation of well-done meat is not sufficient to ensure complete destruction of trichinae.

Kotula et al. (157) examined the destruction of *T. spiralis* in pork obtained from pigs experimentally infected with the parasite. Pork chops of varying thickness were heated by six methods: roast and hold at 107°C, conventional or convection heating at 162°C, flat grill at 204°C, microwave to thaw (−23°C to 0°C, 700 W, medium power) then char broil to final temperature, and precook by microwave heating to 49, 54, or 60°C and deep-fat fry at 162°C. Internal temperatures in the geometric center of the chop were targeted to 66, 71, or 77°C except for the roast-and-hold method, where 60 and 66°C were chosen. Infectivity of recovered larvae was determined by a rat bioassay method. No chops cooked to 71 or 77°C by conventional oven, convection oven, or flat grill contained viable larvae, but lower temperatures did not guarantee killing of all trichinae. Infective trichinae were found in chops thawed by microwave energy before cooking by char broiling to 71 to 77°C. Also, chops cooked partially by microwave heating then finished by deep-fat frying to 66, 71, or 77°C contained motile, but noninfective trichinae based on visual observation. This study was further expanded by the addition of microwave heating, char broiling, or deep-fat frying alone (38). Chops thawed after frozen storage and heated by microwaves to 66 or 71°C contained infective larvae, but larvae were not infective after heating at 77°C. In addition, trichinae survived microwave heating alone to 71, 77, or 82°C. Other methods of cooking, however, adequately destroyed *T. spiralis* in pork chops.

Zimmermann and Beach (42) similarly examined the survival of trichinae in pork products, including loin, shoulder, and fresh ham roasts, and three thicknesses of rib chops. In their studies, six household ovens were used, ranging from 625 to 700 W of power, with various other features such as temperature probes, rotating trays, or variable power settings. Of 51 microwave-heated pork products, infective trichinae were found in nine products including six rib roasts, one sirloin roast, and pork chops. The survival of trichinae was ascribed to unevenness of cooking, particularly with cold spots near the rib bones. Occasionally cold spots occurred just under the surface of fatty areas, possibly due to crust formation. In a second, expanded study, Zimmermann (158) found that 50 of 189 roasts contained infective trichinae, even in some samples where temperature exceeded the 76.7°C standard. However, Zimmermann (158,159) recommended that risks could be minimized by cooking roasts at lower power levels (50%) for longer times, cooking smaller roasts (<2 kg), using boneless loin and bone-in-center loin cuts, allowing the roast to stand covered with foil for 10 minutes or more until temperatures approached the 76.7°C mark, measuring the temperature at several locations throughout the roast, and visually inspecting cooked product for pink or red meat. Moreover, well-done roasts exhibiting gray coloration throughout were still found infective on occasion.

As a result of preliminary research in the area of survival of trichinae in pork products heated by microwave energy, the USDA (160,161) recommended that in addition to following manufacturer's directions, the product should be rotated during cooking and allowed to stand after cooking to reach a final temperature of 76.7°C. Zimmermann (162) determined that trichinae in pork roasts could be inactivated by combining set power levels and time depending upon the wattage oven used. For 700 W ovens, 100% power for 19.8 min/kg, 70% for 24.2 min/kg, and 30% for 55 min/kg achieved comparable levels of safety. For 625 W ovens, 100% power for 24.2 min/kg, 70% for 26.4 min/kg, and 30% for 55 min/kg were comparable methods. A 20-minute standing time did not predictably increase the destruction of trichinae in these pork products. When 16 trichina-infected pork roasts were cooked at suboptimal temperatures of 48.4 min/kg, only seven attained the desired temperature of 76.7°C (163). Of the nine undercooked roasts, five achieved the desired endpoint temperature after one reheating at 8.8 min/kg, two roasts were recooked twice, and two roasts were recooked three times. None of the trichinae in these roasts was infective. This study points out the difficulty in achieving endpoint temperatures on a consistent basis.

Meat loaves formulated from meat and pork were made in three shapes—ring, oval, and loaf—and heated to comparable temperatures at various power settings in 650 W and 625 W microwave ovens. Of the 30 samples tested, 8 were positive by the bioassay method with 5 of the positive tests coming from the loaf shape, 2 from the oval shape, and one from the ring shape (164).

Cooking of pork in microwave ovens was deemed problematic because of unevenness of temperature distribution. Despite the measurement of temperatures using thermocouples throughout the product, no consistent direction could be given to consumers to ensure a safely prepared product. The exact thermal death point of *T. spiralis* has not been established. Changes in the geometry of the food may lead to increased assurance that destructive temperatures would be reached.

Toxoplasma. *Toxoplasma gondii* is a common parasite in warm-blooded animals and can be spread to humans through exposure to infected cats or ingestion of infected raw or undercooked meat (165). Conventional heating is effective in eliminating this organism when temperatures reach ≥65°C. Meat from experimentally infected lambs was cooked by microwave heating (600 W) for 10–15 minutes, depending on weight, then reduced to medium power (330 W) until the temperature reached 65°C (0–43 min). Despite reaching this temperature endpoint, infective *T. gondii* was isolated from two of four samples.

Anisakis. The U.S. Food and Drug Administration (166) recommended using microwave energy to cook fish provided that internal temperature of 74°C was achieved in a covered product with a rotation of the product in the microwave field midway in the cooking process and a standing time of 2 minutes. Work by Adams (167) demonstrated that this protocol did not necessarily ensure a safe product, free of anisakid parasites. Arrowtooth flounder fillets naturally contaminated with *Anisakis simplex* nematodes were ''sandwiched'' to provide a 1.75 cm thick fish portion. Temperature probes were inserted into the thickest portion of the sandwich and microwave heated to 63°C using the highest heat setting (700 W) with a 90° rotation midway through heating. Pepsin digestion of cooked fish was used to recover the nematodes with subsequent testing for viability. No viable nematodes were recovered from fillets heated to 77°C when fish was not rotated during heating or in sandwiches with rotation at 65°C. Variability in mass, thickness, and shape of fillets did affect inactivation.

2. Viruses

Very little work has been done with microwave inactivation of viruses. Exposure of Phage PL-1 of *Lactobacillus casei* to microwave radiation (500 W) followed first-order thermal kinetics, as has been shown with bacteria (168). However, comparison micrographs of phage heated by standard thermal treatment for 75 seconds to 70°C and microwave radiation for 60 seconds to 75°C showed a marked change in microwave-heated phage particles, suggesting that the DNA is broken. These data imply that DNA breakage not only occurs as a temperature-dependent process, but also depends upon treatment source.

3. Spores

In general, spores are more heat-resistant than vegetative cells, although heat resistance can vary depending upon the conditions under which spore-forming bacteria are grown and heated. In some of the literature, it is not clear whether the spore or vegetative form has been used.

Information presented in this section focuses on thermal destruction of spore-formers in model systems rather than in food systems. The heat resistance and survival of the anaerobic spore-formers *Clostridium sporogenes* in beef (80) and *Clostridium perfringens* in beef (84,91,92), poultry (97,98,101), and multicomponent foods (132) have been presented previously. Research has also been cited for the aerobic spore-formers *Bacillus subtilis* in pork (89,90) and *Bacillus cereus* in eggs (81), infant formula (119), and multicomponent foods (132,137).

Grecz et al. (169) may have been the first to examine the effects of microwave heating on spores of *Clostridium sporogenes* PA3679. Spores were suspended in phosphate buffer at pH 7.0 and heated for 65, 75, 85, 95, and 100°C in a constant-temperature microwave device. Microwave heating appeared to be more lethal than conventional heating as the temperature of heating increased. A 5% beef extract solution protected spores, leading to less destruction while an acidity of pH 3.1 sensitized the spores. Spores in the dry state were unaffected by microwaves. Pretreatment of spores with 0.4 Mrad of gamma radiation decreased heat resistance of the spores to microwave heating, while exposing spores to 85°C for 7 hours only slightly affected radiation resistance of spores.

The use of microwaves to heat-activate spores was first studied by Vela et al. (170) and Chipley et al. (171). Activation is the first step in the germination process for sporulated organisms and can be achieved by exposure to heat, chemical germinants, and nutritional factors. Typical heat-activation steps include exposure of the organism to 60–95°C for 10–60 minutes. Vela et al. (170), working with soil microorganisms, determined that microwaves could induce heat activation of bacterial and fungal spores. Chipley et al. (171) examined activation at 60°C of *Bacillus subtilis*, *licheniformis*, *cereus*, *brevis*, and *stearothermophilus* by the water-bath method for 60 minutes and microwave radiation for 30 seconds. Spore suspensions heat-activated by microwave radiation resulted in increased numbers of 2–24% higher than suspensions heated by water bath. However, there was no indication whether the temperature processes were comparable between methods.

Welt et al. (172,173) compared inactivation of *Clostridium sporogenes* spores at 90, 100, and 100°C in pH 6.5 phosphate buffer using equivalent thermal treatments afforded by conventional heating in an oil bath and a specially designed microwave apparatus used to provide steady-state temperature conditions. A comparison of D-values obtained at all three temperatures by both methods were indistinguishable, suggesting that only a thermal effect was functioning in decreasing microbial numbers.

Bacillus cereus naturally occurring in dried foods was isolated, then inoculated into infant formula, nonfat dry milk, and instant mashed potatoes at a level of $\sim 10^4$ spores/mL or g and heated to a mean final temperature of 97.5°C (174). Vegetative cells exhibited a 3-log reduction of cells, spores, a 0.2-log reduction. Infant formula and nonfat dry milk were then temperature abused for 2, 4, or 6 hours at 25°C and held for 4 days at 10°C. *Bacillus cereus* diarrheal toxin was detected in both the microwave-heated and the unheated control samples of nonfat dry milk incubated for 6 hours, then refrigerated. Montemayor (174) concluded that *B. cereus* in dried products that are reconstituted, microwave heated, then temperature abused posed a potential risk for foodborne illness.

Bacillus cereus spores were also heated in phosphate buffer of varying pH levels (3, 4, 5, 7) and lipid (1, 5, 10%) or sodium chloride (0.1, 0.5, 0.75, 1.0, 1.25%) concentrations (175). Microwave heating for 121 seconds to a final temperature of 98°C of buffer adjusted 1 N acetic, lactic, citric, or hydrochloric acids resulted in a wide range of inactivation levels, with lactic acid at pH 5.0 causing maximum destruction of 69.6%. Sodium chloride (1.25%) and corn oil (10%) showed maximum destruction levels of 59.7 and 62.1%, respectively. Spores heated in 1 mM L-alanine were not affected by microwave heating. Given the ability of *B. cereus* to survive adverse conditions during microwave heating, Montemayor and Doores (175) concluded that temperature abuse of foods with subsequent microwave heating could lead to survival of *B. cereus* and the potential for foodborne illness.

E. Microbial Studies—Model Systems

The majority of studies presented thus far have examined the survival of microorganisms in complex food products. Not all of these studies have been well controlled such that temperature measurements, power levels, formulation of foods, and microwave parameters have been adequately described. As such, it is difficult to relate particular killing effects to specific ingredients in the food or exact heating regimens. Identification of such components could potentially lead to reformulation of foods that are safer. The following studies were designed to identify such components.

1. Food Ingredients

The effect of mass and food components in frozen foods was the subject of work conducted by Spite (176). Using mashed potatoes as a base material, additional foods were manufactured by the addition of margarine, nonfat dry milk, salt, and lean ground round beef. Potato mixtures were apportioned in 120 or 360 g amounts in Pyrex dishes. Cooking was performed using an oven rated at 1500 W for 3, 3.5, and 4 minutes for a 120 g size and 9 minutes for a 360 g size. Samples were inoculated with 10^5–10^7 cells/g of *S.* Cubana, *S. aureus*, *B. cereus*, and *C. perfringens*. *Salmonella* Cubana and *S. aureus* did not survive the minimal cook time of 3 minutes, but the spore-formers *B. cereus* and *C. perfringens* survived the 4-minute time. Endpoint temperatures were 76°C after 3 minutes, 83°C after 3.5 minutes, and 88°C after 4 minutes, with browning and desiccation of the potatoes appearing after 3.5 minutes. Increasing water content of the potatoes led to decreasing survival of *S.* Cubana and *S. aureus* but had no effect on the spore-formers, while increased lipid and protein content had no effect on *S.* Cubana and *S. aureus* but did exert a protective effect for the *C. perfringens* despite the higher temperatures of heating. The longer cooking time associated with the larger portion size also led to greater survival of *C. perfringens*, suggesting that when mass increases, irregular heating patterns can occur. Inoculation of a commercially prepared macaroni and cheese product with *S.* Cubana, *S. aureus*, and *B. cereus* resulted in survival of 7.4, 13.3, and 27.5% of the initial population, respectively. These studies suggest that the quality of frozen foods must be of the highest level because there is no assurance that microwave heating will destroy all pathogens.

Kudra et al. (177) examined the effect of fat, lactose, water, and protein on their relative contribution to elevation of temperature during microwave heating. Although milk heated more rapidly than water, it was determined that the protein component exerted the major effect, with fat and lactose contributing minimally to elevation of temperature.

Heddleson et al. (178) investigated the survival of multiple strains of *Salmonella* species, *L. monocytogenes*, and *S. aureus* in five food systems chosen for their chemical and physical properties. The bacteria were inoculated at a level of $\sim 10^8$ cfu/mL into beef broth, UHT-processed whole milk, vanilla pudding, pasteurized liquid whole egg, and cream sauce and microwave heated to 60 (*Salmonella* and *L. monocytogenes*) or 65°C (*S. aureus*) at 630 W. Moisture, fat, protein, sodium chloride, water activity, pH, and ash contents of the five foods were measured. Temperatures were taken at four locations along the central axis using fluoroptic thermometry. The temperature profile of each food indicated that as the salt concentration increased, a surface heating effect was noted, resulting in as much as an 18°C difference between the surface temperature and the next recorded temperature, ~1.0 cm below the surface. The temperature heating patterns also affected the survival of the pathogens, with beef broth containing 0.29% sodium leading to a 64.76% destruction of *Salmonella* species while milk containing 0.05% sodium exhibited a 99.95% destruction. *Staphylococcus aureus* destruction ranged from 95.17 to 99.25%, and *L. monocytogenes* ranged from 97.77 to 99.58% destruction (cheese sauce and pudding). The pH and water activity did not affect the survival of these organisms.

UHT milk and beef broth served as the model systems for estimating the amount of injury that occurs during microwave heating of *Salmonella* spp. (179). Salmonellae were suspended in milk and heated to temperatures of 66–74°C and in beef broth from 64 to 72°C. After heating, samples were either stirred for 5 seconds to eliminate temperature gradients and sampled or allowed to sit for 5 and 10 minutes before sampling. Heated cells were allowed to recover for 72 hours at 25°C. Stirring increased the level of destruction of salmonellae, whereas a 5- or 10-minute holding time without stirring led to recovery of viable salmonellae regardless of heating medium. The level of injured cells did increase with increased holding time and no stirring. The importance of elimination of large temperature gradients that develop within microwave foods is an important factor from a food-safety standpoint.

Work by Fujikawa and coworkers (180,181) examined the effect of ingredients on the heating patterns and injury of microorganisms. *Escherichia coli* was suspended in 0.01 M phosphate buffer adjusted to varying pH levels, salt, and sucrose concentrations and suspended in different shaped containers. Cells were heated in a microwave oven equipped with a turntable and with powers of 100, 200, 300, and 500 W at 2450 MHz. When microwave treatment was compared with conventional heating, the two destruction profiles were almost identical. Small differences were noted in the initial and final parts of the destruction patterns, but they concluded that destruction of *E. coli* was due to thermal effects, not athermal ones. Fujikawa and Ohta (181) continued their studies by examining the factors involved in the patterns of destruction for *E. coli*, *S. aureus*, *B. cereus*, and *P. fluorescens*. Organisms were suspended in 0.01 M phosphate buffer or 0.85% sodium chloride and heated at several different wattages for various periods in a microwave equipped with a turntable. Bacteria were recovered in plate count agar (PCA) and PCA containing 0–7.5% sodium chloride to measure the incidence of injured cells. The rate of temperature rise for phosphate buffer and saline was a function of the wattage level with rates of 0.28, 0.50, and 0.86°C/sec for 200, 300, and 500 W, respectively. Destruction of the mesophiles *S. aureus* and *E. coli* suspended in phosphate buffer was nearly identical at 200 W power with the psychrotroph *P. fluorescens*, mimicking the shape of the pattern but at lower temperatures. Data verified that destruction was a first-order reaction. The destruction pattern for *B. cereus* approximated a straight line at higher wattages. Injury was demonstrated in PCA plus 6.0% salt with no recovery on PCA containing 7.5% salt. Midlogarithmic phase cells of *E. coli* were more sensitive than stationary phase cells. Cells grown to stationary phase at 22, 28, 36, and 44°C were indistinguishable except for the highest temperature, which conferred a higher heat resistance to the cells.

Woo et al. (182) studied the effects of microwave heating and bacterial cell wall structure, using *E. coli*, a gram-negative bacterium, and *B. subtilis*, a gram-positive bacterium. Pure cultures were suspended in 0.9% sodium chloride solution at 10^9–10^{10} cfu/mL, heated at full power (600 W), and examined for cell survival, injury, and damage to cell wall structure. The authors suggested that the thermal sensitivity of *B. subtilis* was greater than *E. coli* in the 40–50°C range; the reverse was true for 50–60°C. There is no indication whether *B. subtilis* had formed spores, which would naturally skew heating results if there was a mixed cell-spore population. The leakage of nucleic acid material, an indication of sublethal injury, was greater with *B. subtilis* than *E. coli*, but the opposite was true with leakage of protein. The surface structure of microwave-heated *E. coli* appeared damaged; however, structural differences were not seen in *B. subtilis* cells.

2. Dielectric Properties

The dielectric properties of liquid and semi-solid foods depend primarily on their moisture, solids, and salt contents and vary considerably with frequency and temperature (183). Water and ionic components appear to play a significant role in the development of heating patterns, suggesting that the measurement of moisture levels and conductivity could be used to model heating characteristics (183–188).

Dealler and Lacey (184) made mashed potatoes with and without added sodium chloride, potassium chloride, ammonium chloride, and monosodium glutamate. Potatoes were heated for 1 minute at 650 W in a 1250 MHz domestic oven. The core temperature of the mashed potatoes changed with increasing concentration of salts and differed by as much as 62% from the control to the highest level of ionic concentration (0.45 M). A survey of commercially processed meals ranged in ionic concentration from 0.024 to 0.15 for sodium and 0.012 to 0.12 for potassium. Dealler and Lacey (184) suggested that the differing levels of ionic concentration might explain why commercial food heated in microwave ovens boils on the surface but is cool on the inside.

Dealler et al. (25) further examined the interaction of ionic molecules and penetration of microwaves. In addition to the salts previously studied (184), glutamine, aspartic acid, D-lysine, monohydrochloride, sodium nitrite, potassium nitrite, sodium citrate, sucrose, and glucose were added at levels of 0.01–0.45 M, depending upon the solute, to mashed potatoes. Potatoes were heated at 650 W in a 2450 MHz oven. Temperatures at the core of the mashed potatoes decreased rapidly as the concentration of salts increased. This effect was not observed with sugars, which are not ionized. For eight of the additives, the temperature increased <10°C compared with 50°C rise for the controls. The results of this work emphasize the need to stir heated products and to include standing times to distribute the temperature throughout the food.

Mashed potatoes containing 0, 0.3, 1.5, and 3.0% sodium chloride were inoculated with *Vibrio parahaemolyticus*, *S. aureus*, *S.* Enteritidis, *E. coli*, and *B. cereus* and heated for 1 minute at 800 W (189). Significantly higher rates of destruction were observed in the core than on the surface when no salt was added. No microorganisms except for *B. cereus* survived a 2-minute heating process in mashed potatoes containing no salt. It was noted that core temperatures and rate of destruction of microorganisms decreased significantly when salt concentration increased. *Staphylococcus aureus* appeared to be more resistant to salt concentration after 2 minutes of heating compared to other bacteria.

Heddleson et al. (26) developed a model system to determine the effect of food components on the survival of *Salmonella* spp. in an aqueous menstruum. Potassium phosphate buffer (0.3 mM) was formulated with 1% concentrations of sucrose, sodium calcium caseinate, corn oil, and/or sodium chloride in 16 possible combinations. The solutions were inoculated with a cocktail of 10^6 cfu/mL of *Salmonella* species and heated for 47 seconds to 60°C in a 700 W oven. Fluoroptic thermometry temperature profiles demonstrated a surface heating effect in all solutions containing sodium chloride resulting in a lower temperature at subsurface areas and a greater survival of salmonellae. Upon further examination of the effect of sodium chloride on the destruction of *Salmonella* species, Heddleson et al. (27) found that changes in the temperature profile appeared more noticeable as the salt concentration increased from 0 to 1.25% with concomitant decreases in percentage destruction of salmonellae (94.36 vs. 77.27%). Sodium chloride was more effective in altering the temperature profile than potassium chloride, calcium chloride, and magnesium chloride resulting in greater survival of salmonellae.

Wayland et al. (190) may have been the first group to describe the interrelation between heat resistance of spores and moisture content of the heating material. Samples of *Bacillus subtilis* var. *niger* were exposed to heat alone and to a microwave field at 2450 MHz with power levels up to 30 kW. Inactivation rates were measurably different between the two methods, lending support to the idea of an athermal effect.

Page and Martin (191) investigated the survival of microorganisms within films. *Escherichia coli*, *S. cerevisiae*, and *B. subtilis* were collected on sterile filters and allowed to air dry. Filters were contained within a glass Petri dish and exposed to microwave energy at full power (650 W). After exposure, the filters were assayed for microbial numbers. Results indicate that *B. subtilis* was not killed even after 10 minutes of heating on full power, whereas *S. cerevisiae* was reduced by ~2 logs and *E. coli* by about 5 logs. In all cases, *S. cerevisiae* and *E. coli* suspended in phosphate buffer were reduced by 8 logs within 30 seconds and 45 seconds, respectively, and 10 minutes for spores. Any killing effect was attributed to humidity from a container placed in the oven to protect the magnetron during heating. Organisms were found on the walls or tray within the microwave oven after heating, suggesting that contaminated food could transfer organisms to the environment. Thus, microwave ovens should be cleaned properly.

Vela and Wu (192) examined the destructive effect of microwave heating on a variety of microorganisms in the presence and absence of water. Lethal effects were determined by comparing viable counts of all exposed organisms. The temperature achieved and the inactivation of organisms were dependent upon the presence of water. Lyophilized cultures were then exposed to microwave radiation in their dry and moistened state. As no microorganisms appeared to die in the lyophilized state, Vela and Wu (192) concluded that bacteria do not absorb microwave energy in the absence of water, and their findings support that destruction is a thermal effect only.

3. Physical Parameters

The effect of microwave power level, postheating holding time, volume of heating medium, and container shape and covering was studied using *Salmonella* spp. suspended in UHT processed milk (193). Temperatures were monitored by fluoroptic thermometry and mercury thermometer. Top feed and dual wave ovens were significantly more effective at reducing populations of salmonellae than a turntable model. When salmonellae were heated for a constant time of 47 seconds in a microwave oven at different power settings, the most destruction occurred at high power (level 10, 93.4%) compared with medium setting (level 5, 11.5%) or low setting (level 3, 8.3%); however, when milk was heated to a constant end temperature of 60°C, there was no significant difference in percentage destruction of salmonellae among any of the power settings. Destruction of salmonellae was significantly greater after a 2-minute postheating holding time than after no holding time, and this effect continued up through 4 minutes of holding after microwave heating. No significant differences in destruction rates of salmonellae were seen when cells were suspended in milk and heated to 60°C in a circular container compared to a rectangular container, but more time was required for the contents of the circular container to reach the endpoint temperature. In addition, no differences were seen in destruction rates due to covering samples or container volumes after heating to 60°C. In general, differences between all variables were minimized by using an endpoint temperature as the measurement of the heating process rather than time.

F. Microbial Studies—Sterilization Processes

Initial studies using microwave heating as a sterilization process were not very successful. Jackson (194) used a batch process at low frequencies in which pouches open to the atmosphere were heated and then sealed. This process was limited to temperatures below 100°C. Jeppson (195) attempted to design an aseptic process in which foods were quickly heated before sealing, but solid and liquid portions heated at differing rates.

A system for continuous microwave sterilization was described by Kenyon et al. (10). Water, chicken à la king, and frankfurters were packaged in sealed plastic pouches and passed through a continuous microwave process, cooled and overwrapped with plastic-foil laminate to ensure sterility, physical protection, and storage stability. The equipment was designed to include a pressurized system and a continuous conveyance system with an entrance and exit. After heating, pouches were cooled to 100°C to reduce internal pressure. Assurance of sterility was maintained by removing products aseptically from the cooling process and overwrapping in a "clean" area.

Ayoub et al. (9) continued the work of Kenyon et al. (10) to achieve a commercially viable microwave sterilization process. Food packaged in sealed flexible pouches was subjected to external air counter pressure ranging from 30 to 45 psi to maintain the structural integrity of the pouch. Beef slices and gravy were packaged in a laminated polypropylene-Mylar pouch overwrapped with insulating paper to reduce heat loss at ambient temperatures. Although distribution of microwave energy was nonuniform, movement of pouches through the microwave field produced a greater uniformity of energy distribution throughout the product. Seam failures in pouches occurred and attributed to inadequate overpressure.

Shin and Pyun (196) proposed using pulsed microwaves for pasteurizing foods. *Lactobacillus plantarum* inoculated into MRS broth at a level of $\sim 10^8$–10^9 cells/mL was sublethally heated at 50°C for 30 minutes using both conventional heating in a water bath, domestic microwave oven, or pulsed microwave systems. Pulsed microwaves were produced in a microwave oven equipped with a pulse generator and pulse driver adjusted to provide 500 pulses/sec. Lactobacilli were recovered on MRS agar and MRS containing 4% sodium chloride to measure injury. Minimal reduction of cell numbers occurred after exposure to conventional heating, whereas conventional microwave heating offered a 4-log reduction. An additional 2-to 3-log reduction was seen with pulsed microwaves. Injury rates for pulsed cells as measured by plating methods as well as leakage of 260 nm–absorbing materials from the cell, an indicator of sublethal injury, were higher compared to the continuous microwave-heated cells, with little effect seen with conventional heating. Normal growth rates and acid production resumed upon subculture of conventional or continuous microwave-heated cells with only a difference in lag phase. Pulsed cells took additional time to recover and achieved lower mass of cells and lass acid production.

Özilgen and Özilgen (197) examined the thermal death kinetics of *Escherichia coli* in a tubular flow reactor in a microwave field or conventional constant temperature water bath. The death rate constants for *Escherichia coli* were considerably higher with microwave heating, suggesting that microwaves may be more efficient than conventional means for processing at low temperature such as that used in fruit juices and alcoholic beverages. Further work by Özilgen and Özilgen (198) implied that death of *E. coli* might be occurring by different mechanisms in the microwave field and constant temperature water bath.

Sterilization of plastic tissue culture vessels for reuse was also accomplished using microwave heating (199). Vessels were rinsed with broth cultures of $\sim 10^6$ cells/mL of *E. coli*, *P. fluorescens*, *Klebsiella pneumoniae*, *Proteus vulgaris*, *Sarcina lutea*, *Corynebacterium equi*, *Bacillus alvei*, *Bacillus globigii*, and *Streptococcus faecium* and allowed to drain. Polio, parainfluenza, and bacteriophage T4 viruses were also tested. All cultures were eliminated after a 3-minute exposure. Microwave sterilization did not damage the attachment properties of plastics and supported the growth of tissue cultures.

Sterilization of glass vials was accomplished by Lohmann and Manique (200). Borosilicate glass vials containing 3×10^6 to 3×10^7 *Bacillus stearothermophilus* and *B. subtilis* var. *niger* spores were heated for 2.7–4.4 minutes. Spore destruction was achieved between 4.0 and 4.4 minutes but was dependent upon the location of the vials within the microwave oven. Less time was required to reduce spore loads if the oven contained a fully loaded turntable as opposed to a partially filled one.

Microwave heating was also shown to reduce or eliminate vegetative and spore-forming bacteria in glass tubes, suggesting that this method could decontaminate glassware (201). The vegetative bacteria, *Streptococcus pyogenes* Group A, *S. aureus*, *Enterococcus faecalis*, *E. coli*, and *Pseudomonas aeruginosa*, and the spore-formers, *Bacillus subtilis* and *B. stearothermophilus*, were microwave heated at 650 W on low and high power and at 1400 W. The organisms were introduced as aqueous suspensions or dried onto the inside surface of a test tube. The vegetative bacteria were mostly killed within 5 minutes of heating at low power with *E. faecalis* surviving at high power; all cultures were killed within 10 minutes at either power setting. Furthermore, *E. faecalis* was killed as an aqueous suspension but survived a 3-minute but not a 5-minute exposure at 1400 W. Interestingly, the spore-formers survived heat treatment in the aqueous suspension and in a dried form, surviving up to 40 minutes of heating at 650 W and 10–20 minutes at 1400 W. These findings suggest that water is necessary for destruction of bacteria and that domestic ovens cannot be reliably used to decontaminate highly contaminated laboratory glassware.

Latimer and Matsen (202) were able to reduce microbial populations of *B. subtilis*, *Enterobacter cloacae*, *K. pneumoniae*, *Serratia marcescens*, *Proteus mirabilis*, *P. aeruginosa*, *Enterococcus* spp., *S. aureus*, and *S. epidermidis* within 60 seconds. The spore-formers *Bacillus stearothermophilus* and *B. subtilis* required more than 11 minutes for a 9-log reduction. Similar work by Diaz-Cinco and Martinelli (203) demonstrated that microwave ovens could be used to eliminate populations of *Aspergillus nidulans*, *E. coli*, and T4 bacteriophage for *E. coli* and reduce the numbers of *B. subtilis*. They concluded that only a thermal effect was seen with microwave heating.

REFERENCES

1. CR Buffler. Microwave Cooking and Processing. New York: Van Nostrand Reinhold, 1993.
2. D Fusaro. Catching the next microwave. Prep Foods 163:33–35, 1994.
3. RF Schiffmann. Microwave foods: Basic design considerations. Tappi J 73:209–212, 1990.
4. DA Copson. Microwave irradiation of orange juice concentrate for enzyme inactivation. Food Technol 8:397–399, 1954.
5. A Pour-El, SO Nelson, EE Peck, B Tjhio, LE Stetson. Biological properties of VHF- and microwave-heated soybeans. J Food Sci 46:880–885, 895, 1981.

6. SA Goldblith. Basic principles of microwaves and recent developments. Adv Food Res 15:277–301, 1966.
7. RV Decareau. Microwaves in the Food Processing Industry. New York: Academic Press, Inc., 1985.
8. CM Olsen. Microwaves inhibit bread mold. Food Eng 37(7):51–53, 1965.
9. JA Ayoub, D Berkowitz, EM Kenyon, CK Wadsworth. Continuous microwave sterilization of meat in flexible pouches. J Food Sci 39:309–313, 1974.
10. EM Kenyon, DE Westcott, P La Casse, JW Gould. A system for continuous thermal processing of food products using microwave energy. J Food Sci 36:289–293, 1971.
11. AJH Sale. A review of microwaves for food processing. J Food Technol 11:319–329, 1976.
12. JE Sunderland. An economic study of microwave freeze-drying. Food Technol 36: 50–52, 54–56, 1982.
13. P Risman. Terminology and notation of microwave power and electromagnetic energy. J Microwave Power Electromagnetic Energy 26:243–248, 1991.
14. JR Chipley. Effects of microwave irradiation on microorganisms. Adv Appl Microbiol 26:129–145, 1980.
15. SF Cleary. Biological effects of microwave and radiofrequency radiation. CRC Crit Rev Environ Control 1:257–306, 1970.
16. SF Cleary. Biological effects of microwave and radiofrequency radiation. CRC Crit Rev Environ Control 7:121–166, 1977.
17. B Curnutte. Principles of microwave radiation. J Food Prot 43:618–624, 632, 1980.
18. KM Knutson, EH Marth, MK Wagner. Microwave heating of food. Lebensm Wiss Technol 20:101–110, 1987.
19. T Ohlsson. Fundamentals of microwave cooking. Microwave World 4:4–9, 1983.
20. A von Hippel. Dielectric Materials and Applications. Cambridge, MA: MIT Press, 1954.
21. RE Baldwin. Microwave cooking: An overview. J Food Prot 46:266–269, 1983.
22. SM Bakanowski, JM Zoller. Endpoint temperature distributions in microwave and conventionally cooked pork. Food Technol 38:45–51, 1984.
23. RE Mudgett. Microwave food processing. Food Technol 43:117–126, 1989.
24. DA Copson. Microwave Heating. Westport, CT: The AVI Publishing Company, Inc., 1975.
25. SF Dealler, NA Rotowa, RW Lacey. Ionized molecules reduce penetration of microwaves into food. Int J Food Sci Technol 27:153–157, 1992.
26. RA Heddleson, S Doores, RC Anantheswaran, GD Kuhn, MG Mast. Survival of *Salmonella* species heated by microwave energy in a liquid menstruum containing food components. J Food Prot 54:637–642, 1991.
27. RA Heddleson, S Doores, RC Anantheswaran, G Kuhn. Destruction of *Salmonella* species heated in aqueous salt solutions by microwave energy. J Food Prot 56:763–768, 1993.
28. RE Mudgett. Microwave properties and heating characteristics of foods. Food Technol 40:84–93, 98, 1986.
29. RF Schiffmann. Food product development for microwave processing. Food Technol 40:94–98, 1986.

30. JE Gerling. Microwave oven power: A technical review. J Microwave Power Electromagnetic Energy 22:199–207, 1987.
31. RF Schiffmann. Problems in standardizing microwave oven performance. Microwave World 11(3):20–24, 1990b.
32. DL Harrison. Microwave versus conventional cooking methods: Effects on food quality attributes. J Food Prot 43:633–637, 1980.
33. WR Tinga. Interactions of microwaves with materials. Proc IMPI short course for users of microwave power. Int Microwave Power Inst, Vienna, VA, 1970.
34. CR Buffler, MA Stanford. Effects of dielectric and thermal properties on the microwave heating of foods. Microwave World 12:15–23, 1991.
35. CR Buffler. A simple home test to determine microwave oven power output. Microwave World 9(2):5–8, 1988.
36. HE Berek, KA Wickersheim. Measuring temperatures in microwaveable packages. J Packaging Technol 2:164–168, 1988.
37. KA Wickersheim, MH Sun. Fiberoptic thermometry and its applications. J Microwave Power Electromagnetic Energy 22:85–94, 1987.
38. AW Kotula, KD Murrell, L Acosta-Stein, L Lamb, L Douglass. Destruction of *Trichinella spiralis* during cooking. J Food Sci 48:765–768, 1983.
39. NW Hollywood, Y Varabioff, GE Mitchell. The effect of microwave and conventional cooking on the temperature profiles and microbial flora of minced beef. Int J Food Microbiol 14:67–76, 1991.
40. BM Lund, MR Knox, MR Cole. Destruction of *Listeria monocytogenes* during microwave cooking. Lancet I:218, 1989.
41. S Dealler, N Rotowa, R Lacey. Microwave reheating of convenience meals. British Food J 92(3):19–21, 1991.
42. WJ Zimmermann, PJ Beach. Efficacy of microwave cooking for devitalizing trichinae in pork roasts and chops. J Food Prot 45:405–409, 1982.
43. CA Sawyer. Post-processing temperature rise in foods: Conventional hot air and microwave ovens. J Food Prot 48:429–434, 1985.
44. DL Goedeken, CH Tong, RR Lentz. Design and calibration of a continuous temperature measurement system in a microwave cavity by infrared imaging. J Food Proc Pres 15:331–337, 1991.
45. CH Tong, RR Lentz, DB Lund. A microwave oven with variable continuous power and a feedback temperature controller. Biotechnol Prog 9:488–496, 1993.
46. H Ramaswamy, FR van de Voort, GSV Raghavan, D Lightfoot, G Timbers. Feedback temperature control system for microwave ovens using a shielded thermocouple. J Food Sci 56:550–552, 555, 1991.
47. BD Gessner, M Beller. Protective effect of conventional cooking versus use of microwave ovens in an outbreak of salmonellosis. Am J Epidemiol 139:903–909, 1994.
48. CJ Bates, RC Spencer. Survival of *Salmonella* species in eggs poached using a microwave oven. J Hosp Inf 29:121–127, 1995.
49. MR Evans, SM Parry, CD Ribeiro. *Salmonella* outbreak from microwave cooked food. Epidemiol Infect 115:227–230, 1995.
50. H Fleming. Effect of high-frequency fields on micro-organisms. Electron Eng 63(1):18–21, 1944.

51. M Ingram, LJ Page. The survival of microbes in modulated high-frequency voltage fields. Proc Soc Appl Bacteriol 16:69–87, 1953.
52. Reference deleted.
53. Reference deleted.
54. SA Goldblith, DIC Wang. Effect of microwaves on *Escherichia coli* and *Bacillus subtilis*. Appl Microbiol 15:1371–1375, 1967.
55. C-G Rosén. Effects of microwaves on food related materials. Food Technol 26: 36–40, 55, 1972.
56. RF Schiffmann. Understanding microwaves—problem-solving key in developing foods for growing market. Food Prod Dev 13:38, 40, 1979.
57. SJ Webb, DD Dodds. Inhibition of bacterial cell growth by 136 gc microwaves. Nature 218:374–375, 1968.
58. SJ Webb, AD Booth. Absorption of microwaves by microorganisms. Nature 222: 1199–1200, 1969.
59. PE Hamrick, BT Butler. Exposure of bacteria to 2450 MHz microwave radiation. J Microwave Power 8:227–233, 1973.
60. CP Mayers, JA Habeshaw. Depression of phagocytosis: A non-thermal effect of microwave radiation as a potential hazard to health. Int J Radiat Biol 24:449–461, 1973.
61. FS Barnes, C-LJ Hu. Model for some nonthermal effects of radio and microwave fields on biological membranes. IEEE Trans Microwave Theory Tech 25:742–746, 1977.
62. MS Dreyfuss, JR Chipley. Comparison of effects of sublethal microwave radiation and conventional heating on the metabolic activity of *Staphylococcus aureus*. Appl Environ Microbiol 39:13–16, 1980.
63. EGE Jahngen, RR Lentz, PS Pescheck, PH Sackett. Hydrolysis of adenosine triphosphate by conventional or microwave heating. J Org Chem 55:3406–3409, 1990.
64. RV Lechowich, LR Beuchat, KI Fox, FH Webster. Procedure for evaluating the effects of 2,450-megahertz microwaves upon *Streptococcus faecalis* and *Saccharomyces cerevisiae*. Appl Microbiol 17:106–110, 1969.
65. PCB Roberts. Viability studies on ascospores and vegetative cells of *Saccharomyces cerevisiae* exposed to microwaves at 2450 MHz. J Sci Food Agric 23:544, 1972.
66. FW Cope. Superconductivity—a possible mechanism for non-thermal biological effects of microwaves. J Microwave Power 11(3):267–270, 1976.
67. H Khalil, R Villota. Comparative study on injury and recovery of *Staphylococcus aureus* using microwaves and conventional heating. J Food Prot 51:181–186, 1988.
68. H Khalil, R Villota. The effect of microwave sublethal heating on the ribonucleic acids of *Staphylococcus aureus*. J Food Prot 52:544–548, 1989.
69. MAJ Saeed, P Gilbert. Influence of low intensity 2,450 MHz microwave radiation upon the growth of various micro-organisms and their sensitivity towards chemical inactivation. Microbios 32:135–142, 1981.
70. DKH Jeng, KA Kaczmarek, AG Woodworth, G Balasky. Mechanism of microwave sterilization in the dry state. Appl Environ Microbiol 53:2133–2137, 1987.

71. KA Culkin, DYC Fung. Destruction of *Escherichia coli* and *Salmonella typhimurium* in microwave-cooked soups. J Milk Food Technol 38:8–15, 1975.
72. MF Kozempel, BA Annour, RD Cook, OJ Scullen, RC Whiting. Inactivation of microorganisms with microwaves at reduced temperatures. J Food Prot 61:582–585, 1998.
73. K Causey, F Fenton. Effect of reheating on palatability, nutritive value, and bacterial count of frozen cooked foods. I. Vegetables. J Am Diet Assoc 27:390–395, 1951.
74. S Nikdel, CS Chen, ME Parish, DG MacKellar, LM Friedrich. Pasteurization of citrus juice with microwave energy in a continuous-flow unit. J Agric Food Chem 41:2116–2119, 1993.
75. S Tajchakavit, HS Ramaswamy, P Fustier. Enhanced destruction of spoilage microorganisms in apple juice during continuous flow microwave heating. Food Res Int 31:713–722, 1998.
76. WH Cathcart. High frequency heating produces mold-free bread. Food Ind 18:864–865, 1946.
77. WH Cathcart, JJ Parker, HG Beattie. The treatment of packaged bread with high frequency heat. Food Technol 1:174–177, 1947.
78. KA Evans, HB Taylor. Microwaves extend shelf life of cakes. Food Manuf 42(10): 50–51, 1967.
79. JL Paterson, PM Cranston, WH Loh. Extending the storage life of chilled beef: Microwave processing. J Microwave Power Electomagnetic Energy 30(2):97–101, 1995.
80. MA Hague, CL Kastner, DYC Fung, K Kone, JR Schwenke. Use of nisin and microwave treatment reduces *Clostridium sporogenes* outgrowth in precooked vacuum-packaged beef. J Food Prot 60:1072–1074, 1997.
81. MM Dessel, EM Bowersox, WS Jeter. Bacteria in electronically cooked foods. J Am Diet Assoc 37:230–233, 1960.
82. FL Crespo, HW Ockerman. Thermal destruction of microorganisms in meat by microwave and conventional cooking. J Food Prot 40:442–444, 1977.
83. FL Crespo, HW Ockerman, KM Irvin. Effect of conventional and microwave heating on *Pseudomonas putrefaciens*, *Streptococcus faecalis* and *Lactobacillus plantarum* in meat tissue. J Food Prot 40:588–591, 1977.
84. L Wright-Rudolph, HW Walker, FC Parrish, Jr. Survival of *Clostridium perfringens* and aerobic bacteria in ground beef patties during microwave and conventional cooking. J Food Prot 49:203–206, 1986.
85. E Levre, P Valentini. Inactivation of *Salmonella* during microwave cooking. Zentralb Hyg Umweltmed 201:431–436, 1998.
86. W Lin, C Sawyer. Bacterial survival and thermal responses of beef loaf after microwave processing. Int Microwave Power Inst 23:183–194, 1988.
87. JR Sage, SC Ingham. Survival of *Escherichia coli* O157:H7 after freezing and thawing in ground beef patties. J Food Prot 61:1181–1183, 1998.
88. W Watanabe, NW Tape. Microwave processing of wieners. 2. Effect on microorganisms. Can Inst Food Technol 2(3):104–106, 1969.
89. HW Ockerman, F Leon Crespo, VR Cahill, RF Plimpton, KM Irvin. Microorganism survival in meat cooked in microwave ovens. Ohio Report 62:38–41, 1977.

90. HW Ockerman, VR Cahill, RF Plimpton, NA Parrett. Cooking inoculated pork in microwave and conventional ovens. J Milk Food Technol 39:771–773, 1976.
91. JT Fruin, LS Guthertz. Survival of bacteria in food cooked by microwave oven, conventional oven and slow cookers. J Food Prot 45:695–698, 702, 1982.
92. B Schalch, H Eisgruber, A Stolle. Reduction of numbers of bacteria in vacuum-packed sliced sausage by means of microwave heating. Dairy Food Environ Sanit 17:25–29, 1997.
93. RE Lindsay, WA Krissinger, BF Fields. Microwave vs. conventional oven cooking of chicken: Relationship of internal temperature to surface contamination by *Salmonella typhimurium*. J Am Diet Assoc 86:373–374, 1986.
94. M Schnepf, WE Barbeau. Survival of *Salmonella* Typhimurium in roasting chickens cooked in a microwave, convection microwave, and a conventional oven. J Food Safety 9:245–252, 1989.
95. PJ Coote, CD Holyoak, MB Cole. Thermal inactivation of *Listeria monocytogenes* during a process simulating temperatures achieved during microwave heating. J Appl Bacteriol 70:489–494, 1991.
96. JM Farber, J-Y D'Aoust, M Diotte, A Sewell, E Daley. Survival of *Listeria* spp. on raw whole chickens cooked in microwave ovens. J Food Prot 61:1465–1469, 1998.
97. JAG Aleixo, B Swaminathan, KS Jamesen, DE Pratt. Destruction of pathogenic bacteria in turkeys roasted in microwave ovens. J Food Sci 50:873–875, 880, 1985.
98. JF Blanco, LE Dawson. Survival of *Clostridium perfringens* on chicken cooked with microwave energy. Poult Sci 53:1823–1830, 1974.
99. JS Teotia, BF Miller. Destruction of salmonellae on poultry meat with lysozyme, EDTA, x-ray, microwave and chlorine. Poult Sci 54:1388–1394, 1975.
100. TC Chen, JT Culotta, WS Wang. Effects of water and microwave energy pre-cooking on microbiological quality of chicken parts. J Food Sci 38:155–157, 1973.
101. SE Craven, HS Lillard. Effect of microwave heating of precooked chicken on *Clostridium perfringens*. J Food Sci 39:211–212, 1974.
102. MA Harrison, SL Carpenter. Survival of *Listeria monocytogenes* on microwave cooked poultry. Food Microbiol 6:153–157, 1989.
103. DM Janky, JL Oblinger. Microwave versus water-bath precooking of turkey rolls. Poult Sci 55:1549–1553, 1976.
104. FE Cunningham. The effect of brief microwave treatment on numbers of bacteria in fresh chicken patties. Poult Sci 57:296–297, 1978.
105. FE Cunningham. Influence of microwave radiation on psychrotrophic bacteria. J Food Prot 43:651–655, 1980.
106. M Woodburn, M Bennion, GE Vail. Destruction of salmonellae and staphylococci in precooked poultry products by heat treatment before freezing. Food Technol 16: 98–100, 1962.
107. RC Baker, W Poon, DV Vadehra. Destruction of *Salmonella typhimurium* and *Staphylococcus aureus* in poultry products cooked in a conventional and microwave oven. Poult Sci 62:805–810, 1983.
108. RE Baldwin, M Cloninger, ML Fields. Growth and destruction of *Salmonella typhimurium* in egg white foam products cooked by microwaves. Appl Microbiol 16: 1929–1934, 1968.

109. WJ Stadelman, RK Singh, PM Muriana, H Hou. Pasteurization of eggs in the shell. Poult Sci 75:1122–1125, 1996.
110. RE Baldwin, ML Fields, WC Poon, B Korschgen. Destruction of salmonellae by microwave heating of fish with implications for fish products. J Milk Food Technol 34:467–470, 1971.
111. Y-W Huang, C-K Leung, MA Harrison, KW Gates. Fate of *Listeria monocytogenes* and *Aeromonas hydrophila* on catfish fillets cooked in a microwave oven. J Food Sci 58:519–521, 1993.
112. S Gundavarapu, Y-C Hung, RE Brackett, P Mallikarjunan. Evaluation of microbiological safety of shrimp cooked in a microwave oven. J Food Prot 58:742–747, 1995.
113. GS Young, PG Jolly. Microwaves: The potential for use in dairy processing. Aust J Dairy Technol 45(1):34–37, 1990.
114. R Sieber, P Eberhard, PU Gallmann. Heat treatment of milk in domestic microwave ovens. Int Dairy J 6:231–246, 1996.
115. RA Heddleson, S Doores. Factors affecting microwave heating of foods and microwave induced destruction of foodborne pathogens—a review. J Food Prot 57: 1025–1037, 1994.
116. U Merin, I Rosenthal. Pasteurization of milk by microwave irradiation. Milchwissenschaft 39:643–644, 1984.
117. M Sigman, KI Burke, OW Swarner. Effects of microwaving human milk: Changes in IgA content and bacterial count. J Am Diet Assoc 89:690–692, 1989.
118. RH George. Killing activity of microwaves in milk. J Hosp Inf 35:319–320, 1997.
119. NJ Rowan, JG Anderson. Effectiveness of cleaning and disinfection procedures on the removal of enterotoxigenic *Bacillus cereus* from infant feeding bottles. J Food Prot 61:196–200, 1998.
120. HO Jaynes. Microwave pasteurization of milk. J Milk Food Technol 38:386–387, 1975.
121. KM Knutson, EH Marth, MK Wagner. Use of microwave ovens to pasteurize milk. J Food Prot 51:715–719, 1988.
122. K Choi, EH Marth, PC Vasavada. Use of microwave energy to inactivate *Yersinia enterocolitica* and *Campylobacter jejuni* in milk. Milchwissenschaft 48:134–136, 1993.
123. K Choi, EH Marth, PC Vasavada. Use of microwave energy to inactivate *Listeria monocytogenes*. Milchwissenschaft 48:200–203, 1993.
124. P Chiu, K Tateishi, FV Kosikowski, G Armbruster. Microwave treatment of pasteurized milk. J Microwave Power 19:269–272, 1984.
125. G Stearns, PC Vasavada. Effect of microwave-processing on quality of milk. J Food Prot 49:853, 1986.
126. SN Aktas, M Özilgen. Injury of *E. coli* and degradation of riboflavin during pasteurization with microwaves in a tubular flow reactor. Lebensm Wiss Technol 25:422–425, 1992.
127. JS Thompson, A Thompson. In-home pasteurization of raw goat's milk by microwave treatment. Int J Food Microbiol 10:59–64, 1990.
128. M Villamiel, R López-Fandiño, A Olano. Microwave pasteurization of milk in a

continuous flow unit. Effects on the cheese-making properties of goat's milk. Milchwissenschaft 52(1):29–32, 1997.

129. SS Kim, KS Chang, BS Noh. Survival of lactic acid bacteria during microwave vacuum-drying of plain yoghurt. Lebensmitt Wiss Technol 30:573–577, 1997.
130. LM Tochman, CM Stine, BR Harte. Thermal treatment of cottage cheese "in-package" by microwave heating. J Food Prot 48:932–938, 1985.
131. K Causey, F Fenton. Effect of reheating on palatability, nutritive value, and bacterial count of frozen cooked foods. II. Meat dishes. J Am Diet Assoc 27:491–495, 1951.
132. NW Hollywood, RI Naidoo, GE Mitchell, TW Dommett. Effect of microwave heating on microbial flora of frozen convenience foods. Food Australia 43(4):160–163, 1991.
133. K Kerr, SF Dealler, RW Lacey. *Listeria* in cook-chill food. Lancet ii:37–38, 1988.
134. MRM Sheeran, KG Kerr, SF Dealler, PR Hayes, RW Lacey. *Listeria* survives microwave heating. J Hosp Inf 14:84–86, 1989.
135. SJ Walker, J Bows, P Richardson, JG Banks. Effect of recommended cooking on the survival of *Listeria monocytogenes* in chilled retail products. Campden Food and Drink Research Association, Chipping Campden, Glos GL55 6LD, 1989.
136. A Dos Reis Tassinari, M Landgraf. Effect of microwave heating on survival of *Salmonella typhimurium* in artificially contaminated ready-to-eat foods. J Food Safety 17:239–248, 1977.
137. BA Lacey, HI Winner, ME McLellan, KD Bagshawe. Effects of microwave cookery on the bacterial counts of food. J Appl Bacteriol 28:331–335, 1965.
138. ML Creamer, JR Chipley. Hospital ready-prepared type foodservice system: Time and temperature conditions, sensory and microbiological quality of scrambled eggs. J Food Sci 45:1422–1424, 1429, 1980.
139. ML Creamer, JR Chipley. Time and temperature, microbiological, and sensory assessment of roast beef in a hospital foodservice system. J Food Sci 45:1472–1477, 1980.
140. ML Cremer. Microwave heating of scrambled eggs in a hospital foodservice system. J Food Sci 46:1573–1576, 1581, 1981.
141. ML Cremer. Sensory quality and energy use for scrambled eggs and beef patties heated in institutional microwave and convection ovens. J Food Sci 47:871–874, 1982.
142. CA Dahl, ME Matthews, EH Marth. Cook/chill foodservice systems: Microbiological quality of beef loaf at five process stages. J Food Prot 41:788–793, 1978.
143. CA Dahl, ME Matthews. Effect of microwave heating in cook/chill foodservice systems. J Am Diet Assoc 77:289–295, 1980.
144. CA Dahl, ME Matthews, EH Marth. Fate of *Staphylococcus aureus* in beef loaf, potatoes and frozen and canned green beans after microwave-heating in a simulated cook/chill hospital foodservice system. J Food Prot 43:916–924, 1980.
145. CA Dahl, ME Matthews, EH Marth. Survival of *Streptococcus faecium* in beef loaf and potatoes after microwave-heating in a simulated cook/chill foodservice system. J Food Prot 44:128–134, 1981.
146. CA Sawyer, YM Naidu, S Thompson. Cook/chill foodservice systems: Microbiological quality and end-point temperature of beef loaf, peas and potatoes after re-

heating by conduction, convection and microwave radiation. J Food Prot 46:1036–1043, 1983.

147. PA Ollinger-Snyder, ME Matthews. Cook/chill foodservice system with a microwave oven: Coliforms and aerobic counts from turkey rolls and slices. J Food Prot 51:84–86, 1988.
148. WL Bunch, ME Matthews, EH Marth. Fate of *Staphylococcus aureus* in beef-soy loaves subjected to procedures used in hospital chill foodservice systems. J Food Sci 42:565–566, 1977.
149. CA Dahl, ME Matthews, EH Marth. Cook/chill foodservice system with a microwave oven: Injured aerobic bacteria during food product flow. Eur J Appl Microbiol Biotechnol 11:125–130, 1981.
150. SB Jasnow, JL Smith. Microwave sanitization of color additives used in cosmetics: Feasibility study. Appl Microbiol 30:205–211, 1975.
151. GN Bookwalter, TP Shulka, WF Kwolek. Microwave processing to destroy salmonellae in corn-soy-milk blends and effect on product quality. J Food Sci 47:1683–1686, 1982.
152. MA Bouno, F Niroomand, DYC Fung, LE Erickson. Destruction of indigenous *Bacillus* spores in soymilk by heat. J Food Prot 52:825–826, 1989.
153. SA Aytac. Fate of *Escherichia coli* O157:H7 and *Yersinia enterocolitica* in the preparation of tarhana. Adv Food Sci 18(1/2):28–31, 1996.
154. DYC Fung, FE Cunningham. Effect of microwaves on microorganisms in foods. J Food Prot 43:641–650, 1980.
155. PK Park, DO Cliver. Disinfection of household cutting boards with a microwave oven. J Food Prot 59:1049–1054, 1996.
156. USDA, Meat Inspection Division. Regulations Governing the Meat Inspection of the United States Department of Agriculture. Washington, DC: U.S. Department of Agriculture, 1960.
157. AW Kotula, KD Murrell, L Acosta-Stein, I Tennent. Influence of rapid cooking methods on the survival of *Trichinella spiralis* in pork chops from experimentally infected pigs. J Food Sci 47:1006–1007, 1982.
158. WJ Zimmermann. Evaluation of microwave cooking procedures and ovens for devitalizing trichinae in pork roasts. J Food Sci 48:856–890, 899, 1983.
159. WJ Zimmermann. An approach to safe microwave cooking of pork roasts containing *Trichinella spiralis*. J Food Sci 48:1715–1718, 1722, 1983.
160. USDA advises consumers to microwave pork to uniform 170 degrees Fahrenheit (News release) Washington, DC: U.S. Department of Agriculture, 1981.
161. USDA advises cooking to 170 degrees Fahrenheit throughout (News release) Washington, DC: U.S. Department of Agriculture, 1982.
162. WJ Zimmermann. Power and cooking time relationships for devitilization of trichinae in pork roasts cooked in microwave ovens. J Food Sci 49:824–826, 1984.
163. WJ Zimmermann. Microwave recooking of pork roasts to attain 76.7°C throughout. J Food Sci 49:970–971, 974, 1984.
164. F Carlin, W Zimmermann, A Sundberg. Destruction of trichinae larvae in beef-pork loaves cooked in microwave ovens. J Food Sci 47:1096–1099, 1118, 1982.
165. A Lundén, A Uggla. Infectivity of *Toxoplasma gondii* in mutton flowing curing, smoking, freezing or microwave cooking. Int J Food Microbiol 15:357–363, 1992.
166. U.S. Food and Drug Administration. Food code, sections 3-401.11 (A1) and 3-

401.12. In 1997 Recommendations of the United States Public Health Service Food and Drug Administration. Washington, DC: U.S. Food and Drug Administration, 1997.

167. AM Adams. Survival of Anisakis simplex in microwave-processed arrowtooth flounder (*Atheresthes stomias*). J Food Prot 62:403–409, 1999.
168. Y Kakita, N Kashige, K Murata, A Kuroiwa, M Funatsu, K Watanabe. Inactivation of *Lactobacillus* bacteriophage PL-1 by microwave irradiation. Microbiol Immunol 39:571–576, 1995.
169. N Grecz, AA Walker, A Anellis. Effect of radiofrequency energy (2,450 mc) on bacterial spores. Bacteriol Proc 1964:145, 1964.
170. GR Vela, JF Wu, D Smith. Effect of 2450 MHz microwave radiation on some soil microorganisms in situ. Soil Sci 121:44–51, 1976.
171. JR Chipley, LA Rohlfs, CL Ford. Heat activation of *Bacillus* spores by the use of microwave irradiation. Microbios 29:105–108, 1980.
172. BA Welt, CH Tong, JL Rossen. An apparatus for providing constant and homogeneous temperatures in low viscosity liquids during microwave heating. Microwave World 13(2):9–13, 1992.
173. BA Welt, CH Tong, JL Rossen, DB Lund. Effect of microwave radiation on inactivation of *Clostridium sporogenes* (PA 3679) spores. Appl Environ Microbiol 60: 482–488, 1994.
174. R Montemayor, S Doores. Effect of microwave heating on *Bacillus cereus* spores isolated from commercially available dry food products. Annual Meeting of the American Society for Microbiology, Washington, DC, Abstract P29, 1995.
175. RC Montemayor, S Doores. Effect of microwave heating on viability of *Bacillus cereus* spores in liquid menstrua. Annual meeting of the American Society for Microbiology, Abstract P29, Los Angeles, CA, 2000.
176. GT Spite. Microwave-inactivation of bacterial pathogens in various controlled frozen food compositions and in a commercially available frozen food product. J Food Prot 47:458–462, 1984.
177. T Kudra, FR van de Voort, GSV Raghavan, HS Ramaswamy. Heating characteristics of milk constituents in a microwave pasteurization system. J Food Sci 56:931–934, 937, 1991.
178. RA Heddleson, S Doores, RC Anantheswaran, G Kuhn. Viability loss of *Salmonella* species, *Staphylococcus aureus*, and *Listeria monocytogenes* in complex foods heated by microwave energy. J Food Prot 59:813–818, 1996.
179. RA Heddleson, S Doores. Injury of *Salmonella* species heated by microwave energy. J Food Prot 57:1068–1073, 1994.
180. H Fujikawa, H Ushioda, Y Kudo. Kinetics of *Escherichia coli* destruction by microwave irradiation. Appl Environ Microbiol 58:920–924, 1992.
181. H Fujikawa, K Ohta. Patterns of bacterial destruction in solutions by microwave irradiation. J Appl Bacteriol 76:389–394, 1994.
182. I-S Woo, I-K Rhee, H-D Park. Differential damage in bacterial cells by microwave radiation on the basis of cell wall structure. Appl Environ Microbiol 66:2243–2247, 2000.
183. S Swami, RE Mudgett. Effect of moisture and salt contents on the dielectric behavior of liquid and semi-solid foods. Proc Int Microwave Power Inst 16:48–50, 1981.
184. SF Dealler, RW Lacey. Superficial microwave heating. Nature 344:496, 1990.

185. RR Lentz. On the microwave heating of saline solutions. J Microwave Power 15(2): 107–111, 1980.
186. RE Mudgett, AC Smith, DIC Wang, SA Goldblith. Prediction on the relative dielectric loss factor in aqueous solutions of nonfat dried milk through chemical simulation. J Food Sci 36:915–918, 1971.
187. RE Mudgett, DIC Wang, SA Goldblith. Prediction of dielectric properties in oil-water and alcohol-water mixtures at 3,000 MHz, 25°C based on pure component properties. J Food Sci 39:632–635, 1974.
188. BD Roebuck, SA Goldblith. Dielectric properties of carbohydrate-water mixtures at microwave frequencies. J Food Sci 37:199–204, 1972.
189. M Hayashi, Y Shimazaki, S Kamata, N Kakiichi. Effects of sodium chloride on destruction of microorganisms by microwave heating in potatoes (English abstract). Nippon Koshu Eisei Zasshi 38:431–437, 1991.
190. R Wayland, JP Brannen, ME Morris. On the interdependence of thermal and electromagnetic effects in the response of *Bacillus subtilis* spores. Radiation Res 71: 251–258, 1977.
191. WJ Page, WG Martin. Survival of microbial films in the microwave oven. Can J Microbiol 24:1431–1433, 1978.
192. GR Vela, JF Wu. Mechanism of lethal action of 2,450-MHz radiation on microorganisms. Appl Environ Microbiol 37:550–553, 1979.
193. RA Heddleson, S Doores, RC Anantheswaran. Parameters affecting destruction of *Salmonella* spp. by microwave heating. J Food Sci 59:447–451, 1994.
194. JM Jackson. Electronic sterilization of canned foods. Food Indust 19:124–125, 224, 1947.
195. MR Jeppson. Techniques of continuous microwave food processing. Cornell Hotel Restaurant Admin Quart 5(1):60–64, 1964.
196. J-K Shin, Y-R Pyun. Inactivation of *Lactobacillus plantarum* by pulsed-microwave irradiation. J Food Sci 62:163–166, 1997.
197. S Özilgen, M Özilgen. A model for pasteurization with microwaves in a tubular flow reactor. Enzyme Microb Technol 13:419–423, 1991.
198. S Özilgen, M Özilgen. Enthalpy-entropy and frequency factor-activation energy compensation relations for death of *Escherichia coli* with microwaves in a tubular flow reactor. Acta Alimentaria 21:195–203, 1992.
199. MR Sanborn, SK Wan, R Bulard. Microwave sterilization of plastic tissue culture vessels for reuse. Appl Environ Microbiol 44:960–964, 1982.
200. S Lohmann, F Manique. Microwave sterilization of vials. J Parenteral Sci Technol 40:25–30, 1986.
201. L Najdovski, AZ Dragas, V Kotnik. The killing activity of microwaves on some non-sporogenic and sporogenic medically important bacterial strains. J Hosp Inf 19:239–247, 1991.
202. JM Latimer, JM Matsen. Microwave oven irradiation as a method for bacterial decontamination in a clinical microbiology laboratory. J Clin Microbiol 6:340–342, 1977.
203. M Diaz-Cinco, S Martinelli. The use of microwaves in sterilization. Dairy Food Environ Sanit 11:722–724, 1991.

6
Control of Microorganisms with Chemicals

P. Michael Davidson
University of Tennessee
Knoxville, Tennessee

I. INTRODUCTION

Certain chemical compounds may be added to foods to inhibit growth of, or kill, microorganisms. These chemicals are food additives that may be termed "food preservatives" or, more precisely, "food antimicrobials." The purpose of food antimicrobials is to extend shelf life by inhibiting spoilage microorganisms or improve food safety by inhibiting pathogens. While most food antimicrobials are capable of inactivation, at normal use concentrations they cause inhibition rather than inactivation. This inhibition is often reversible. Whether inactivation or inhibition is achieved is also highly dependent upon the initial number of target microorganisms and, to a lesser extent, use conditions. The inhibitory effect of food antimicrobials on microorganisms is finite, and, therefore, foods will not be preserved indefinitely. In addition, with few exceptions, chemical antimicrobials are not capable of concealing spoilage of a food product by neutralizing the effect of the detrimental end products of spoilage microorganisms. Rather, the food remains wholesome during its extended shelf life since microorganisms are not reproducing and therefore are not producing off-flavors, odors, and textures. Because chemical food antimicrobials do not sterilize a product, they are often used in combination with other food-preservation procedures such as heat or refrigeration.

The objective of this chapter is to review the characteristics, antimicrobial activity, and uses in food of U.S. Food and Drug Administration (FDA) or U.S. Department of Agriculture Food Safety and Inspection Service regulatory-approved chemical antimicrobials added directly to foods. These include organic acids, dimethyl dicarbonate, nitrites, parabens, phosphates, and sulfites. Naturally occurring antimicrobials (Chapter 10), microbially derived antimicrobials (Chapter 11), and sanitizers (also chemical antimicrobials) (Chapter 13) are covered in this volume.

II. FACTORS AFFECTING ACTIVITY

As might be expected, chemical antimicrobials have inherent differences in their activity against microorganisms. Surprisingly little information is available, however, on structure/function relationships outside gross classifications such as organic acids, oxidizing agents, or phenolic compounds. Food antimicrobial effectiveness is also dependent on other major factors including susceptibility of the target microorganisms, characteristics of the food product, the storage environment and type of process used (Gould, 1989).

Microbial factors that may affect antimicrobial activity include inherent resistance of a microorganism, initial number, growth stage, cellular composition (e.g., gram reaction), and previous exposure to stress and injury. For example, it is well known that bacterial spores are generally more resistant to chemical antimicrobials than vegetative cells and that gram negative bacteria are often more resistant than gram positive bacteria due to the presence of an outer membrane. Recently there has been much research done on the effect of environmental stress factors such as heat, cold, starvation and low pH/organic acids, on developed resistance to the same or other subsequent stressors. This developed resistance is termed tolerance, adaptation, or habituation depending upon how the microorganism is exposed to the stress and the physiological conditions that lead to enhanced survival (Foster, 1995; Buchanan and Edelson, 1999). In contrast, a lack of increased resistance to lactic acid was demonstrated by Van Netten et al. (1998) for *Escherichia coli* O157:H7, *Salmonella* Typhimurium, *Staphylococcus aureus*, or *Campylobacter jejuni* when acid-adapted cells were inoculated onto pork bellies and treated with 2% lactic acid as a sanitizer. A possible method for overcoming acid tolerance has been suggested by Jordan et al. (1999) utilizing a combination of lactate and ethanol.

Factors affecting activity of chemical antimicrobials associated with the food product include pH, product composition, oxidation reduction potential, and water activity. pH is the most important factor influencing the effectiveness of many food antimicrobials including organic acids, nitrite, and sulfites. Antimicrobials that are weak acids are most effective in their undissociated or protonated form because they are able to penetrate the cytoplasmic membrane of a microorganism more effectively. The lower the pH of a food, the greater the proportion of acid in the protonated form and the greater the antimicrobial activity. The pK_a of a weak acid indicates how much of the compound will be in the most active form in a given food application. For example, the use of organic acids is generally limited to foods with a pH less than 5.5, since most organic acids have pK_a's of pH 3.0–5.0 (Doores, 1993). Another factor affecting activity is food composition. This relates to the hydrophobicity of the antimicrobial molecule. Antimicrobials must be lipophilic to attach and pass through the cell membrane, but also should be at least partially soluble in the aqueous phase (Davidson, 1997). A food high in fat will cause the antimicrobial to be solubilized in the lipid phase which will reduce its availability to inhibit a microorganism in the water phase (Rico-Munoz and Davidson, 1983).

Storage factors affecting antimicrobial activity include time, temperature, and atmosphere. Processing of foods using heat, non-thermal methods (e.g., high pressure, pulsed electric fields) leads to reduction in microbial numbers and shifts in microflora which will influence the effectiveness of food antimicrobials. Some non-thermal processing methods have utilized food antimicrobials as an integral part of the process. Most factors that influence microbial inhibition or inactivation act in an interactive manner.

III. ACETIC ACID

Acetic acid (CH_3COOH; pK_a = 4.75; MW 60.05 Da) is the primary component of vinegar. Its sodium, potassium and calcium salts are some of the oldest known food antimicrobials. Derivatives of acetic acid including diacetate salts and dehydroacetic acid have also been used as food antimicrobials. In contrast to most organic acids, acetic acid is generally more effective against yeasts and bacteria than against molds (Ingram et al., 1956). Bacteria inhibited include *Bacillus*, *Clostridium*, *L. monocytogenes*, *S. aureus*, and *Salmonella*. Entani et al. (1998) found that 0.1% (w/v) acetic acid from vinegar was bacteriostatic to multiple strains of *E. coli* O157:H7, *Salmonella* Enteritidis, *S.* Typhimurium, and *Aeromonas hydrophila* after 4 days at 30°C on the surface of nutrient agar. The same concentration of acetic acid did not inhibit growth of *Bacillus cereus* or *S. aureus*. Enterohemorrhagic *E. coli* strains were less susceptible to bactericidal concentrations (2.5%) of acetic acid than an enteropathogenic strain. Finally, Entani et al. (1998) showed that stationary phase cells of *E. coli* O157:H7 were more resistant to acetic acid than log phase cells. Only *Acetobacter* species (microorganisms involved in vinegar production), lactic acid bacteria, and butyric acid bacteria are tolerent to acetic acid (Baird-Parker, 1980; Doores, 1993). Some yeasts and molds are sensitive to acetic acid and include species of *Aspergillus*, *Penicillium*, *Rhizopus* and some strains of *Saccharomyces* (Doores, 1993).

Acetic acid and its salts have shown variable success as antimicrobials in food applications. For example, acetic acid can increase poultry shelflife and decrease heat resistance of *Salmonella* Newport, *S.* Typhimurium, and *C. jejuni* in poultry scald tank water (Mountney and O'Malley, 1985; Okrend et al., 1986). The compound has been used as a spray sanitizer at 1.5–2.5% on meat carcasses and as an effective antimicrobial dip for beef, lamp and catfish fillets (Anderson et al., 1988; Bala and Marshall, 1998). Use of 2% acetic acid resulted in reductions in viable *E. coli* O157:H7 on beef after 7 days at 5°C (Siragusa and Dickson, 1993). Acetic acid was the most effective antimicrobial in ground roasted beef slurries against *E. coli* O157:H7 growth in comparison with citric or lactic acid (Abdul-Raouf et al., 1993). Acetic acid added at 0.1% to bread dough inhibited growth of 6 log CFU/g rope-forming *Bacillus subtilis* in wheat bread (pH 5.14) stored at 30°C for > 6 days (Rosenquist and Hansen, 1998). In brain heart infusion broth (BHI), 0.2% acetic acid at pH 5.1 or 0.1% at pH 4.8 inhibited rope-forming strains of *B. subtilis* and *Bacillus licheniformis* for > 6 days at 30°C (Rosenquist and Hansen, 1998).

Sodium acetate is also an effective inhibitor of rope-forming *Bacillus* in baked goods and of the molds, *Aspergillus flavus, A. fumigatus, A. niger, A. glaucus, Penicillium expansum*, and *Mucor pusillus* at pH 3.5–4.5 (Glabe and Maryanski, 1981). It is useful in the baking industry because it has little effect on the yeast used in baking. Al-Dagal and Bazaraa (1999) found that whole or peeled shrimp dipped in a 10% sodium acetate (w/w) solution for 2 min had extended microbiological and sensory shelflife compared to controls.

Sodium diacetate (pK_a = 4.75) is effective at 0.1–2.0% in inhibiting mold growth in cheese spread (Doores, 1993). At 32 mM (0.45%) in BHI (pH 5.4), sodium diacetate was shown to be inhibitory to *L. monocytogenes*, *E. coli*, *Pseudomonas fluorescens*, *S.* Enteritidis and *Shewanella putrefaciens* but not *S. aureus*, *Yersinia enterocolitica*, *Pseudomonas fragi*, *Enterococcus faecalis*, or *Lactobacillus fermentis* after 48 hr at 35°C (Shelef and Addala, 1994). In addition, 21–28 mM sodium diacetate suppressed growth by the natural microflora of ground beef after storage at 5°C for up to 8 days (Shelef and Addala, 1994). Degnan et al. (1994) found a 2.6 log decrease in viable *L. monocytogenes* in blue crab meat washed with 2M sodium diacetate after 6 days at 4°C. Dehydroacetic acid has a high pK_a of 5.27 and is therefore active at higher pH values. It is inhibitory to bacteria at 0.1–0.4% and fungi at 0.005–0.1% (Doores, 1993).

Acetic acid is used commercially in baked goods, cheeses, condiments and relishes, dairy product analogues, fats and oils, gravies and sauces, and meats. The sodium and calcium salts are used in breakfast cereals, candy, cheeses, fats and oils, gelatin, jams and jellies, meats, snack foods, sauces and soup mixes (Doores, 1993). Sodium diacetate is used in baked goods, candy, cheese spreads, gravies, meats, sauces and soup mixes (Doores, 1993).

IV. BENZOIC ACID

Benzoic acid (MW 122.12 Da) and sodium benzoate were the first antimicrobial compounds permitted in foods by FDA. Benzoic acid occurs naturally in cranberries, plums, prunes, apples, strawberries, cinnamon, cloves, and most berries. As with other weak acid antimicrobials, the undissociated form of benzoic acid ($pK_a = 4.19$) is the most effective antimicrobial agent. Therefore, the most effective pH range for use of the compound is 2.5–4.5.

Because the compound functions best at low pH, and yeasts and molds are the primary organisms of concern in acid foods, benzoic acid and sodium benzoate are used primarily as antifungal agents. The inhibitory concentration of benzoic acid at pH < 5.0 against most yeasts ranges from 20–700 μg/ml, while for molds it is 20–2000 μg/ml (Baird-Parker, 1980; Chipley, 1993). Some fungi, including *Byssochlamys nivea, Pichia membranaefaciens*, *Talaromyces flavus*, and *Zygosaccharomyces bailii* are resistant to benzoic acid (Wind and Restaino, 1995; Sofos et al., 1998). In combination with heat, 0.1% sodium benzoate reduced the time for a 3 log reduction of the ascospores of the heat resistant mold, *Neosartorya fischeri*, from 205–210 minutes at 85°C to 85–123 minutes in mango and grape juice at pH 3.5 (Rajashekhara et al., 1998).

While bacteria associated with food poisoning including *B. cereus*, *L. monocytogenes*, *S. aureus*, and *Vibrio parahaemolyticus* are inhibited by 1000–2000 μg/ml undissociated acid, the control of many spoilage bacteria requires much higher concentrations (Baird-Parker, 1980). The concentration of sodium benzoate required to inhibit growth of *L. monocytogenes* at 4 or 13°C in tryptose broth is 0.05–0.1% (500–1000 μg/ml) at pH 5.0 (El-Shenawy and Marth, 1988). At pH 5.6, the microorganism is not capable of growth at 4°C with 0.05% sodium benzoate, while 0.2% is required to inhibit the microorganism for 9 days at the same pH at 13°C. At pH 5.6 and 21°C, the minimum inhibitory concentration (defined as < 1 log increase in 5 days) is 0.25% (El-Shenawy and Marth, 1988). Benzoic acid at 0.1% is effective in reducing viable *E. coli* O157:H7 in apple cider (pH 3.6–4.0) by 3–5 logs in 7 days at 8°C (Zhao et al., 1993). Sodium benzoate in combination with propionic acid delayed growth of *E. coli* O157:H7 and *Salmonella* in the soft cheese, queso fresco (Kasrazadeh and Genigeorgis, 1994, 1995). Lopez et al. (1998) demonstrated that 0.1% sodium benzoate at pH 5.0 had a significant inhibitory effect on heat-damaged *Bacillus stearothermophilus* spores.

Sodium benzoate is used as an antimicrobial at up to 0.1% in carbonated and still beverages, syrups, cider, margarine, olives, pickles, relishes, soy sauce jams, jellies, preserves, pie and pastry fillings, fruit salads, and salad dressings and in the storage of vegetables (Chipley, 1993). An undesirable "peppery" or bitter taste is imparted by benzoate when added to fruit juices in excess of 0.1%

(Jay, 1996). A benzoic acid-based polymer coating was developed to prevent spoilage of apples (Chipley, 1993).

V. CITRIC ACID

While citric acid generally is not used as an antimicrobial, it has activity against some molds and bacteria. Citric acid retards growth and toxin production by *Aspergillus parasiticus* and *A. versicolor*, but not *Penicillium expansum* (Reiss, 1976). It is inhibitory to *Salmonella* in media on poultry carcasses and in mayonnaise, to growth and toxin production by *C. botulinum* in shrimp and tomato products, to *S. aureus* in microbiological medium and to flat sour bacteria isolated from tomato juice (Murdock, 1950; Thomson et al., 1967; Subramanian and Marth, 1968; Minor and Marth, 1970; Post et al., 1985; Xiong et al., 1999). Citric acid (0.156 % s/v) in mango juice (pH 3.5) reduced the heat resistance of *Neosartorya fischeri* spores 2.3 times at 85°C (Rajashekhara et al., 1998). The mechanism of citrate inhibition is theorized to be related to its ability to chelate metal ions. Branen and Keenan (1970) were the first to suggest that inhibition may be due to chelation, in studies with citrate against *Lactobacillus casei*. However, Buchanan and Golden (1994) found that, while undissociated citric acid is inhibitory against *L. monocytogenes*, the dissociated molecule protects the microorganism. They theorized that this protection is due to chelation by the anion.

VI. DIMETHYL DICARBONATE (DMDC)

Dimethyl dicarbonate is a colorless liquid which is slightly soluble in water. The compound is very reactive with many substances including water, ethanol, alkyl and aromatic amines, and sulfhydryl groups (Ough, 1993a). The primary target microorganisms for DMDC are yeasts including *Saccharomyces*, *Zygosaccharomyces*, *Rhodotorula*, *Candida*, *Pichia*, *Torulopsis*, *Torula*, *Endomyces*, *Kloeckera* and *Hansenula*. The compound is also bactericidal at 30–400 μg/ml to a number of species including *Acetobacter pasteurianus*, *E. coli*, *Pseudomonas aeruginosa*, *S. aureus*, several *Lactobacillus* species and *Pediococcus cerevisiae* (Ough, 1993a). DMDC (0.8 mM) was more effective than sulfur dioxide or sorbic acid at suppression of fermentation at 21°C or 31°C of grape juice inoculated with up to 4.3 log CFU/ml of yeast (Terrell et al., 1993). The compound was shown to be bactericidal at 0.025% and more effective than either sodium bisulfite or sodium benzoate against *E. coli* O157:H7 in apple cider at 4°C (Fisher and Golden, 1998). Molds are generally more resistant to DMDC than yeasts or bacteria.

VII. FUMARIC ACID

Fumaric acid (COOH—CH=CH—COOH) has been used as an antimicrobial agent in wines (Ough and Kunkee, 1974; Pilone, 1975). The compound is lethal to heat resistant ascospores of *Talaromyces flavus* and *N. fischeri* (Doores, 1993). Esters of fumaric acid (monomethyl, dimethyl, and ethyl) at 0.15–0.2% have been tested as a substitute or adjunct for nitrate in bacon. Fumaric acid esters (0.125%) retard swelling and toxin formation in canned bacon, inoculated with *C. botulinum* for up to 56 days at 30°C (Huhtanen, 1983). Methyl, dimethyl, ethyl and diethyl fumarates inhibit fungal growth in tomato juice and on bread (Huhtanen and Guy, 1984). At 1% as a sanitizer on lean beef, fumaric acid reduced *L. monocytogenes* and *E. coli* O157:H7 by approximately 1 log after 5 sec at 55°C (Podolak et al., 1996).

VIII. LACTIC ACID

Lactic acid (CH_3—CHOH—COOH; $pK_a = 3.79$; MW 90.08) is produced naturally during fermentation of foods by lactic acid bacteria including *Lactococcus*, *Lactobacillus*, *Streptococcus*, *Pediococcus*, *Leuconostoc*, and *Carnobacterium*. As an end-product in natural fermentations, lactic acid is probably one of the oldest known antimicrobial agents. Lactic acid inhibits *C. botulinum*, *Clostridium perfringens*, *C. sporogenes*, *L. monocytogenes*, *Salmonella*, *S. aureus*, and *Yersinia enterocolitica* (Minor and Marth, 1970; Brackett, 1987; Doores, 1993). While the acid itself has potential as a direct food additive, most recent interest has focused on addition of the sodium salt or use of the acid as a sanitizer or sanitizer adjunct on meat and poultry.

Sodium lactate (2.5–5.0%) inhibits *C. botulinum*, *C. sporogenes*, *L. monocytogenes* and spoilage bacteria in various meat products (Maas et al., 1989; Shelef and Yang, 1991; Unda et al., 1991; Chen and Shelef, 1992; Meng and Genigeorgis, 1993; Weaver and Shelef, 1993; Pelroy et al., 1994a). Houtsma et al. (1994) reported that toxin production by proteolytic *C. botulinum* was delayed at 15 and 20°C by sodium lactate concentrations of 2 and 2.5%, respectively and complete inhibition of toxin production at 15, 20 and 30°C occurred at 3, 4 and >4%, respectively. There have been various theories as to how the lactate salts function as antimicrobials. The salts have minimal effects on pH, thus most of the lactate remains in the less effective anionic form. Chen and Shelef (1992) and Weaver and Shelef (1993) demonstrated that the water activity in cooked meat model systems and liver sausage containing lactate salts up to 4%, was not reduced sufficiently to inhibit *L. monocytogenes*. Therefore, it is possible that, at the high concentrations of lactate used, sufficient undissociated lactic acid is present in combination with a slightly reduced pH and/or water activity, to inhibit microorganisms.

Much research has been done using lactic acid as a sanitizer on meat and poultry carcasses to reduce or eliminate pathogens. In most cases, lactic acid sprays or dips at 0.2–2.5% have shown to be effective in reducing contamination on beef, veal, pork, and poultry and, in some cases, improving shelf life (Snijders et al., 1985; Smulders et al., 1986; Visser et al., 1988; Dickson and Anderson, 1992; Anderson et al., 1992; Anderson et al., 1992; Bautista et al., 1997). Low molecular weight polylactic acid (PLA) at 2% has been shown to increase the shelf life of vacuum-packaged fresh beef by inhibiting spoilage bacteria. However the effectiveness is similar to that of 2% lactic acid (Ariyapitipun et al., 1999).

IX. NITRITES

Meat curing is a method of food preservation that utilizes salt, sugar, spices, and ascorbate or erythorbate in addition to sodium ($NaNO_2$) or potassium (KNO_2) nitrite. The primary use for nitrites as antimicrobials is to inhibit *C. botulinum* growth and toxin production in cured meats. In addition to serving as an antimicrobial, nitrite has many functions in cured meats. As nitric oxide, nitrite reacts with the meat pigment, myoglobin, to form the characteristic cured meat color, nitrosomyoglobin. It also contributes to the flavor and texture of cured meats and serves as an antioxidant. The specific contribution of nitrite to the antimicrobial effects of curing salt was not recognized until the late 1920s (NAS/NRC, 1981; Roberts et al., 1991). Nitrite inhibits bacterial sporeformers by inhibiting outgrowth of the germinated spore (Duncan and Foster, 1968; Genigeorgis and Riemann, 1979; Tompkin, 1978; Cook and Pierson, 1983; Tompkin, 1993). The effectiveness of nitrite depends on reduced pH, salt concentration, presence of ascorbate and isoascorbate, storage and processing temperatures, and initial inoculum size. The antimicrobial activity of nitrite is enhanced at low pH, under anaerobic conditions and with increased salt. Ascorbate and isoascorbate enhance the antibotulinal action of nitrite, probably by acting as reducing agents (Roberts et al., 1991). Storage and processing temperatures and initial inoculum size also significantly influence the antimicrobial effectiveness of nitrite.

Nitrite has variable effects on microorganisms other than *C. botulinum. Clostridium perfringens* is inhibited by 200 μg/ml nitrite with 3% salt or 50 μg/ml nitrite with 4% salt at pH 6.2 in microbiological medium (Gibson and Roberts, 1986). Sodium nitrite (200 mg/kg) with 5% NaCl inhibited *L. monocytogenes* growth for 40 days at 5°C in vacuum packaged and film-wrapped smoked salmon (Pelroy et al., 1994b). Strains of *Salmonella*, *Lactobacillus*, *C. perfringens*, and *Bacillus* are resistant to nitrite (Castellani and Niven, 1955; Spencer, 1971;

Grever, 1974; Rice and Pierson, 1982). Nitrite at 200 μg/ml and pH 5.0 completely inhibited growth of *E. coli* O157:H7 at 37°C (Tsai and Chou, 1996).

Meat curing may be combined with drying, heating, smoking and fermentation to produce a variety of products. For example, products that may contain nitrites include bacon, bologna, corned beef, frankfurters, luncheon meats, ham, fermented sausages, shelf-stable canned cured meats, and perishable canned cured meat (e.g., ham). Nitrite is also used in a variety of fish and poultry products. The concentration used in these products is specified by governmental regulations but is generally limited to 156 ppm (mg/kg) for most products and 100–120 ppm in bacon. Sodium nitrate is used in certain cheeses to prevent spoilage by *Clostridium tyrobutyricum* or *C. butyricum* (Tompkin, 1993).

The mechanism of nitrite inhibition of has been studied many years but the likely targets of the compound have only been identified in the past few years (Roberts and Gibson, 1986; Woods et al., 1989; Tompkin, 1993). Two enzymes in the phosphoroclastic system of clostridia, pyruvate-ferredoxin oxidoreductase (PFR) and ferridoxin have been suspected to be susceptible to nitrite (Woods and Wood, 1982; Carpenter et al., 1987; Roberts et al., 1991; Thompkin, 1993). These observations are supported by the fact that addition of iron to meats containing nitrite reduces the inhibitory effect of the compound (Thompkin et al., 1978).

X. PARABENS (*p*-HYDROXYBENZOIC ACID ESTERS)

Esterification of the carboxyl group of *p*-hydroxybenzoic acid (parabens) allows the molecule to remain undissociated up to pH 8.5 which overcomes the limitation of low pH usage of the parent benzoic acid. Parabens have a greatly expanded effective pH range of 3.0–8.0 compared to benzoic acid which is best used at pH < 4.5 (Aalto et al., 1953). In most countries, the methyl, propyl, and heptyl esters of *p*-hydroxybenzoic acid are allowed for direct addition to foods as antimicrobials while ethyl and butyl esters are approved in a limited number of countries.

The antimicrobial activity of the parabens is generally inversely proportional to the chain length of the alkyl ester (Aalto et al., 1953; Shibasaki, 1969; Fukahori et al., 1996). As the alkyl chain length increases, polarity decreases and inhibitory activity increases. This relationship is true primarily with gram positive bacteria as the lipopolysaccharides in the outer membrane of gram negative bacteria are apparently capable of screening out some non-polar antimicrobials (Freese et al., 1973; Eklund, 1980). Methyl and propyl paraben have range of minimum inhibitory concentrations against the foodborne pathogens, *A. hydrophilia*, *B. cereus*, *C. botulinum*, *E. coli*, *L. monocytogenes*, *Salmonella*, *S. aureus*, *V. parahaemolyticus*, and *Y. enterocolitica* of 1000–4000 μg/ml and 50–1000 μg/ml, respectively (Davidson, 1997). Parabens are generally more active against fungi than against bacteria. The antifungal effectiveness of the parabens has been evaluated against several food-related fungi including *Aspergillus*, *Byssochlamys*, *Candida*, *Fusarium*, *Penicillium*, *Rhizopus*, *Saccharomyces*, and *Zygosaccharomyces* (Davidson, 1997). As with bacteria, inhibition of fungi increases as the alkyl chain length of the parabens increases. The minimum inhibitory concentrations against fungi of methyl and propyl paraben are 500–1000 μg/ml and 63–360 μg/ml, respectively. Thompson (1994) reported that butyl and propyl paraben were most effective against mycotoxigenic strains of *Aspergillus*, *Fusarium* and *Penicillium* and the minimum inhibitory concentrations ranged from 1.0–2.0 mM. Methyl paraben was least effective on a concentration basis with an MIC of 3.5 to > 4.0 mM. Heptyl paraben has not been tested extensively most likely due to its limited solubility in water. It has been shown to have a minimum inhibitory concentration of 12 μg/ml against *B. cereus*, *Lactococcus lactis*, and *S. aureus* (Davidson, 1997).

Methyl and propyl parabens are normally used in combination of 2–3 : 1 (methyl : propyl) due to the increased water solubility of the former and the greater antimicrobial activity of the latter. The compounds may be incorporated into foods to a maximum of 0.1% by dissolving in water, ethanol, propylene glycol, or the food product itself. The n-heptyl ester is used in fermented malt beverages (beers) at 12 ppm or less and noncarbonated soft drinks and fruit-based beverages at 20 ppm or less. Parabens may be used in a variety of foods including baked goods, beverages, fruit products, jams and jellies, fermented foods, syrups, salad dressings, wine, and fillings.

Since parabens are phenolic derivatives, their mechanism is likely related to other phenolic compounds. A majority of the research on the mechanisms of phenolic compounds has focused on their effect on cellular membranes. Freese et al. (1973) found that the parabens inhibit serine uptake as well as the oxidation of α-glycerol phosphate and NADH in membrane vesicles of *Bacillus subtilis* and concluded that the compounds inhibit both membrane transport and the electron transport system. Eklund (1980) first demonstrated that parabens are capable of neutralizing chemical and electrical forces which establish a normal membrane gradient. He later showed (Eklund, 1985a) that parabens eliminated the ΔpH of the cytoplasmic membrane but did not significantly affect the Δψ (membrane potential) component of the proton motive force. He concluded that neutralization of the proton motive force and subsequent transport inhibition could not be the only mechanism of inhibition for the parabens. Bargiota et al. (1987) examined the relationship between lipid composition of *S. aureus* and resistance to parabens. Paraben-resistant strains were found to have a higher percentage total lipid, higher relative percentage of phosphatidylglycerol and decreased cyclopropane fatty acids than sensitive strains. These changes could influence membrane fluidity and therefore adsorption of the parabens to the membrane.

XI. PHOSPHATES

The primary functions of phosphates in food processing include acidification, alkalization, buffering, emulsification, leavening, sanitization and sequestration (Ellinger, 1981). While phosphates have not been thought of as active antimicrobials, some forms do have activity against selected bacteria and fungi. Phosphate compounds include, sodium acid pyrophosphate (SAPP; $Na_2H_2P_2O_7$), tetrasodium pyrophosphate (TSPP; $Na_4P_2O_7$), sodium tripolyphosphate (STPP; $Na_5P_3O_{10}$), sodium tetrametaphosphate (STMP; $(NaPO_3)_4$), sodium hexametaphosphate (SHMP; $(NaPO_3)_n = 13, 15, 21$) and trisodium phosphate (TSP), have variable levels of antimicrobial activity in foods (Shelef and Seiter, 1993).

Gram-positive bacteria are generally more susceptible to phosphates than gram-negative bacteria. Gould (1964) found that 0.2–1.0% SHMP permitted germination of *Bacillus* spores but prevented outgrowth. Jen and Shelef (1986) tested phosphate derivatives against the growth of *S. aureus* 196E and only 0.3% SHMP ($n = 21$) and 0.5% STPP or SHMP ($n = 13, 15$) were effective growth inhibitors. Magnesium reversed the growth-inhibiting effect. The mimimum inhibitory concentration of long chain food grade phosphates added to early-exponential-phase cells of *S. aureus ISP40* 8325 in a synthetic medium were determined to be 0.1% for sodium ultraphosphate and sodium polyphosphate glassy and 0.5% for SAPP, STPP (Lee et al. 1994). Several studies demonstrated that polyphosphates enhance the effect of nitrite, pH, and salt against *C. botulinum* in cured meats (Ivey and Robach, 1978; Nelson et al., 1980; Roberts et al., 1981; Wagner and Busta, 1983). In pasteurized process cheese, phosphate along with sodium chloride, water activity, water content, pH, and lactic acid interact to prevent the outgrowth of *C. botulinum* (Tanaka, 1982). Wagner and Busta (1985) found that SAPP alone had no effect on the growth of *C. botulinum* but delayed or prevented toxicity to mice. They theorized that SAPP was binding the toxin molecule or inactivating the protease responsible for protoxin activation. Zaika and Kim (1993) found that 1% sodium polyphosphate inhibited lag and generation times of *L. monocytogenes* in BHI broth, especially in the presence of NaCl. Various phosphate salts also have antimicrobial activity against rope-forming *Bacillus* in bread and *Salmonella* in pasteurized egg whites (Tompkin, 1983). Phosphates may also be used in retarding growth of fungi. Post et al. (1968) preserved cherries against the fungal growth by *Penicillium, Rhizopus*, and *Botrytis* with a 10% sodium tetrapolyphosphate dip.

Trisodium phosphate (TSP) is used as a sanitizer in chill water for raw poultry and other raw foods at 8–12%. It has been shown that TSP reduces viable pathogen numbers, especially *Salmonella* on poultry (Giese, 1992). Lillard (1994) reported a 2 log reduction of *S.* Typhimurium by immersing pre-chill chicken carcasses for 15 min in a 10% TSP solution. It was reported that activation was a function of the high pH of the (11–12) of the system. Kim et al. (1994) found a 1.6–1.8 log reduction in *S.* Typhimurium on post-chill chicken carcasses using a 10% TSP dipping treatment. Slavik et al. (1994) evaluated dipping of chicken carcasses inoculated with *C. jejuni* in 10% TSP and reported a 1.5 log reduction of the pathogen. Wang et al. (1997) reported that 10% TSP reduced *S.* Typhimurium attached to chicken skin by 1.6–2.3 logs. In contrast, Waldroup (1995) and Waldroup et al. (1995) concluded that TSP had no effect on pathogens or indicator microorganisms on poultry and actually may allow greater survival of *L. monocytogenes* than non-TSP processes. Fresh-laid eggs washed with solutions of TSP at 0.5, 1.0, or 5.0% resulted in changes in the shell surface as examined using scanning electron microscopy which might allow for increased bacterial penetration (Kim and Slavik, 1996). Zhuang and Beuchat (1996) found that dipping tomatoes for 15 seconds in 12% TSP reduced the population of *Salmonella* Montevideo by 4 logs on the surface and ca. 2 logs in the core. Color of the tomatoes was unaffected by TSP.

Several mechanisms have been suggested for bacterial inhibition by polyphosphates. The ability of polyphosphates to chelate metal ions, such as magnesium, appears to play an important role in their antimicrobial activity (Shelef and Seiter, 1993; Knabel et al., 1991).

XII. PROPIONIC ACID

Up to 1% propionic acid (pK_a = 4.87) is produced naturally in Swiss cheese by *Propionibacterium freudenreichii* spp. *shermanii*. Propionic acid and sodium, potassium and calcium propionates are used primarily against molds, however, some yeasts and bacteria are also inhibited. Propionates (0.1–5.0%) retard the growth of *E. coli*, *S. aureus*, *Sarcina lutea*, *Salmonella*, *Proteus vulgaris*, *Lactobacillus plantarum*, and *L. monocytogenes*, and the yeasts *Candida* and *Saccharomyces cerevisiae* (Doores, 1993). In BHI broth, 0.2% at pH 5.4, 0.15% at pH 5.1 or 0.1% at pH 4.8 inhibited rope-forming strains of *B. subtilis* and *B. licheniformis* for > 6 days at 30°C (Rosenquist and Hansen, 1998).

Propionic acid and propionates are used as antimicrobials in baked goods and cheeses. Propionic acid added at 0.1% to bread dough inhibited growth of 6 log CFU/g rope-forming *Bacillus subtilis* in wheat bread (pH 5.30) stored at 30°C for > 6 days (Rosenquist and Hansen, 1998). Even at 0.05%, the microorganisms were prevented from rope formation for 4 days. Propionates may be added directly to bread dough because they have no effect on the activity of baker's yeast (Doores, 1993). The shelflife of commercial corn tortillas was increased substantially by addition of calcium priopionate alone and in combination with potassium sorbate (Tellez-Giron et al., 1988). There is no limit to the concentration of propionates allowed in foods, but amounts used are generally less than 0.4% due to sensory changes that might occur (Doores, 1993). The primary mode of propionic acid action is probably similar to that of other organic acids, i.e., weak acid inhibition and/or interference with the cytoplasmic membrane.

XIII. SORBIC ACID

Sorbic acid was first isolated from the oil of unripened rowanberries of the mountain ash tree (Sofos and Busta, 1993). The compound is now available commercially as the free acid ($CH_3{-}CH{=}CH{-}CH{=}CH{-}COOH$; MW 112.3) or as the calcium, sodium, or potassium salt. The solubility of sorbic acid is 0.15 g/100 ml versus the calcium, sodium or potassium salts which have 1.2%, 32%, or > 50% (w/v), respectively. As with other organic acids the antimicrobial activity of sorbic acid is greatest when the compound is in the undissociated state. Sorbic acid has a pK_a of 4.75 but it does have some activity up to pH 6.0–6.5.

The primary use of sorbates as antimicrobials is as antifungal agents. Food-related yeasts and molds inhibited by sorbates include species of *Brettanomyces*, *Byssochlamys*, *Candida*, *Cryptococcus*, *Debaromyces*, *Hansenula*, *Pichia*, *Rhodotorula*, *Saccharomyces*, *Torulaspora*, and *Zygosaccharomyces* and *Alternaria*, *Aspergillus*, *Botrytis*, *Fusarium*, *Geotrichum*, *Mucor*, *Penicillin*, *Sporotrichum*, and *Trichoderma*, resectively (Sofos and Busta, 1993). Sorbates inhibit the growth of yeasts and molds in microbiological media, cheeses, fruits, vegetables and vegetable fermentations, sauces and meats (Smith and Rollin, 1954; Nury et al., 1960; Pederson et al., 1961; Harada et al., 1968; Deak et al., 1970; Baldock et al., 1979; Flores et al., 1988). Sorbate was more effective than benzoate against *Zygosaccharomyces baili* spoilage in salsa mayonnaise but the compound did not completely inhibit growth of the yeast at 23 to 25°C even at 0.3% w/w (Wind and Restaino, 1995). In combination with heat, 0.1% potassium sorbate reduced the time for a 3 log reduction of the heat resistant mold, *Neosartorya fischeri*, from 205–210 minutes at 85°C to 77–85 minutes in mango and grape juice at pH 3.5 (Rajashekhara et al., 1998). Combinations of potassium sorbate and vanillin were synergistic in their antimicrobial activity against *Penicillium digitatum, P. glabrum*, and *P. italicum* in potato dextrose agar adjusted to pH 3.5 and a_w of 0.98 (Matamoros-León et al., 1999). Sorbates inhibit growth and mycotoxin-production by *Aspergillus flavus*, *A. parasiticus*, *Byssochlamys nivea*, *Penicillium expansum* and *P. patulum* (Sofos and Busta, 1993).

Bacteria inhibited by sorbate include *Acinetobacter*, *Aeromonas*, *Bacillus*, *Campylobacter*, *Clostridium*, *Escherichia* O157:H7, *Lactobacillus*, *L. monocytogenes*, *Pseudomonas*, *Salmonella*, *Staphylococcus*, *Vibrio*, *Y. enterocolitica* (Sofos and Busta, 1993; ICMSF, 1996; Tsai and Chou, 1996). Sorbic acid inhibits primarily catalase producing bacteria (Phillips and Mundt, 1950; York and Vaughn, 1954, 1955). Catalase-negative lactic acid acteria are generally resistant to sorbates.

Sorbate inhibits the growth of many pathogenic bacteria in or on foods including *Aeromonas* spp. on sun dried fish with 1.5% salt, *Salmonella* and *S. aureus* in sausage, *S. aureus* in bacon, *V. parahaemolyticus* in seafood, *Salmonella*, *S. aureus* and *E. coli* in poultry, *Y. enterocolitica* in pork, *S.* Typhimurium in milk and cheese and *E. coli* O157:H7 and *Salmonella* in queso fresco cheese (Robach, 1980; Gram, 1991; Sofos and Busta, 1993; Kasrazadeh and Genigeorgis, 1994, 1995). Uljas and Ingham (1999) utilized 0.1% sorbic acid in combination with freeze-thawing and storage at 25°C for 12 hr or 4 hr at 35°C and storage without freeze-thawing at 35°C for 6 hr to achieve a 5 log reduction of *E. coli* O157:H7 in apple cider at pH 4.1. In addition the compound inhibits growth of the spoilage bacteria, *Pseudomonas putrefaciens* and *P. fluorescens*, histamine production by *Proteus morgani* and *K. pneumoniae* and listeriolysin O production by *L. monocytogenes* (Sofos and Busta, 1993; McKellar, 1993). Sorbates are effective anticlostridial agents in cured meats and other meat and seafood products. The compound prevents spores of *C. botulinum* from germinating and forming toxin in beef, pork, poultry and soy protein frankfurters and emulsions and bacon (Sofos and Busta, 1993). Lopez et al. (1998) showed 0.1% potassium sorbate inhibited unheated strains of *Bacillus stearothermophilus* spores by ca. 50% and prevented growth of spores heated at 121°C for 1 min. Against other strains, even 0.1% potassium sorbate had little effect on heated spores.

Certain *Penicillium*, *Saccharomyces* and *Zygosaccharomyces* species can grow in the presence of and degrade potassium sorbate (Sofos and Busta, 1993). Sorbates may be degraded through a decarboxylation reaction resulting in the formation of 1,3-pentadiene, a compound having a kerosene-like or hydrocarbon-like odor (Marth et al., 1966; Liewen and Marth, 1985).

Sorbates are applied to foods by direct addition, dipping, spraying, dusting, or incorporation into packaging. Baked goods, icing, fruit and cream fillings can be protected from yeast and molds through the use of 0.05–0.10% potassium sorbate applied either as a spray after baking or by direct addition (Sofos and Busta, 1993). Sorbates may be used in or on beverages, jams, jellies, preserves, margarine, chocolate syrup, salads, dried fruits, dry sausages, salted and smoked fish, cheeses and in various lactic acid fermentations (Sofos and Busta, 1993).

The mechanism of inhibition of vegetative cells by sorbic acid is not completely known. Ronning and Frank (1987) showed that sorbic acid reduces the cytoplasmic membrane electrochemical gradient and consequently the PMF. They concluded that loss of PMF by sorbic acid inhibited amino acid transport which could eventually result in the inhibition of many cellular enzyme systems. However, Eklund (1985b) showed that while low concentrations of sorbic acid reduce the ΔpH of the proton motive force of *E. coli* vesicles, concentrations much greater than those required for inhibition reduce, but do not eliminate the $\Delta\psi$ component. Since the $\Delta\psi$ component alone could energize active uptake of amino acids, the amino acid uptake inhibition theory does not entirely explain the mechanism of inhibition by sorbic acid (Eklund, 1989). The mechanism by which sorbic acid inhibits microbial growth may also be partially due to its effect on enzymes. For example, sorbate inhibits activation of the hemolytic activity of listeriolysis O of *L. monocytogens* by reacting with cysteine (Kouassi and Shelef, 1995).

The mechanism of sorbate action against bacterial spores has been studied extensively. In some of the only reports involving effects on the cell wall, Gould (1964) and Seward et al. (1982) demonstrated that sorbic acid inhibits cell division of germinated spores of *Bacillus* and *C. botulinum* Type E. Sorbic acid at pH 5.7 competitively inhibits L-alanine and L-a-NH_2-*n*-butyric acid induced germination of *B. cereus* T spores and L-alanine and L-cysteine induced germination of *C. botulinum* 62A (Smoot and Pierson, 1981).

XIV. SULFITES

Sulfur dioxide (SO_2) and its derivatives are added to foods as antimicrobials, antioxidants, to inhibit enzyme catalyzed reactions and to inhibit non-enzymatic browning (Lindsay, 1996). The approved applications of sulfites have been recently curtailed and they are subject to restrictive labeling regulations due to reported severe reactions to the compounds by sensitive persons with asthma (Lindsay, 1996). In addition, thiamine is cleaved by sulfite to pyrimidine sulfonate and a thiazole residue-containing product (Rose and Pilkington, 1989). Therefore sulfites are not allowed to be used in foods considered to be good sources of the vitamin. The salts of sulfur dioxide approved for use include potassium sulfite (K_2SO_3), sodium sulfite (Na_2SO_3), potassium bisulfite ($KHSO_3$), sodium bisulfite ($NaHSO_3$), potassium metabisulfite ($K_2S_2O_5$), and sodium metabisulfite (NaS_2O_5). Their primary application is in fruits and vegetable products but they are also used to a limited extent in meats in some countries.

Sulfur compounds have been used as disinfectants since the time of the ancient Greeks and Romans (Ough, 1993b). As antimicrobials, sulfites are used primarily in fruit and vegetable products to control three groups of microorganisms: spoilage and fermentative yeasts, and molds on fruits and fruit products (e.g., wine), acetic acid bacteria and malolactic bacteria (Ough, 1993b).

The most important factor impacting the antimicrobial activity of sulfites is pH. Aqueous solutions of sulfur dioxide theoretically yield $SO_2 \cdot H_2O$ (Gould and Russell, 1991). As the pH decreases, the proportion of $SO_2 \cdot H_2O$ increases and the bisulfite (HSO_3^-) ion concentration decreases. The pK_a values for sulfur dioxide, depending upon temperature are 1.76–1.90 and 7.18–7.20 (Rose and Pilkington, 1989; Gould and Russell, 1991; Ough, 1993b). The inhibitory effect of sulfites is most pronounced when the acid of $SO_2 \cdot H_2O$ is in the undissociated form and, therefore, the most effective pH is less than 4.0. Both the HSO_3^- and SO_3^{2-} ions have been shown to have significantly less antimicrobial activity than undissociated $SO_2{\cdot}H_2O$ (Rehm and Wittman, 1962; King et al., 1981). Sulfites bind carbonyl compounds present in foods which also reduces the antimicrobial effectiveness of the sulfites (Rose and Pilkington, 1989).

The antifungal inhibitory concentration range for sulfur dioxide is 0.1–20.2 μg/ml for *Saccharomyces*, *Zygosaccharomyces*, *Pichia*, *Hansenula*, and *Candida* species (Rehm and Wittmann, 1962). Roland et al. (1984) and Roland and Beuchat (1984) found that 25–100 μg/ml sulfur dioxide inhibits *Byssochlamys nivea* growth and patulin production in grape and apple juices. No synergistic antimicrobial activity was demonstrated with sulfur dioxide in combination with benzoate, sorbate or butylated hydroxyanisole against *Saccharomyces cerevisiae* (Parish and Carroll, 1988; Knox et al., 1984).

The concentration of sulfites required to inhibit bacteria is higher than that for fungi. At 100–200 μg/ml, sulfites inhibit *Acetobacter* sp. that cause wine spoilage (Rehm and Wittman, 1962; Juven and Shoman, 1985; Ough, 1993b). The concentration of sulfur dioxide required to inhibit lactic acid bacteria varies significantly depending upon conditions but can be 1–10 μg/ml in fruit products at pH 3.5 or less (Wibowo et al., 1985). Sulfur dioxide is more inhibitory to gram-negative rods than to gram-positive rods (Roberts and McWeeny, 1972). Banks and Board (1982) tested several genera of *Enterobacteriaceae* isolated from sausage for their metabisulfite sensitivity. The microorganisms tested and the concentration of free sulfite (μg/ml) necessary to inhibit their growth at pH 7.0 were: *Salmonella*, 15–109; *E. coli*, 50–195; *Citrobacter freundii*, 65–136; *Y. enterocolitica*, 67–98; *Enterobacter agglomerans*, 83–142; *Serratia marcescens*, 190–241; and *Hafnia alvei*, 200–241.

Sulfur dioxide is used to control the growth of undesirable microorganisms in fruits, fruit juices, wines, sausages, fresh shrimp, and acid pickles, and during extraction of starches. It is added at 50–100 μg/ml to expressed grape juices and used for making wines to inhibit molds, bacteria, and undesirable yeasts (Amerine and Joslyn, 1960). At appropriate concentrations, sulfur dioxide does not interfere with wine yeasts or with the flavor of wine. During fermentation, sulfur dioxide also serves as an antioxidant, clarifier, and dissolving agent. The optimum level of sulfur dioxide (50 to 75 μg/ml) is maintained to prevent post fermentation changes by microorganisms. In some countries, sulfites may be used to inhibit the growth of microorganisms on fresh meat and meat products (Kidney, 1974). Sulfite or metabisulfite added in sausages is effective in delaying the growth of molds, yeast, and salmonellae during storage at refrigerated or room temperature (Ingram et al., 1956; Banks and Board, 1982). Sulfur dioxide restores a bright color but may give a false impression of freshness.

Due to their extreme reactivity, it is difficult to pinpoint the exact antimicrobial mechanism for sulfites. The most likely targets for inhibition by sulfites include disruption of the cytoplasmic membrane, inactivation of DNA replication, protein synthesis, inactivation of membrane-bound or cytoplasmic enzymes or reaction with individual components in metabolic pathways. Sulfites dissipate the PMF and inhibit solute active transport (Rose and Pilkington, 1989).

XV. FUTURE OF CHEMICAL ANTIMICROBIALS

The fastest growing area of research on antimicrobials is their use in combinations. This is due to relatively good information on inhibition by individual compounds and the potential to achieve synergistic activity between two or more compounds. Researchers have examined combinations using only traditional antimicrobials and combinations with traditional antimicrobials and naturally occurring antimicrobials. Food processors are in favor of this because it helps them reduce individual antimicrobial concentrations and achieve equivalent or increased antimicrobial activity. The most effective method for applying antimicrobial combinations would be to utilize compounds with different mechanisms. Unfortunately not many food antimicrobials have well characterized mechanisms. Therefore, more research needs to be done to better define mechanisms so that research on combinations is more than a ''trial and error'' exercise.

Another novel area in food antimicrobials involves their incorporation into or in conjunction with packaging. Incorporation into packaging is not new as sorbic acid was incorporated into packaging as an antifungal agent as early as 1945 (Sofos and Busta, 1993). One of the limits of utilizing antimicrobials in packaging is a lack of exposure of microorganisms to food antimicrobials. There have been recent reports on packaging that releases selected antimicrobials during the course of storage. However, the microorganisms affected would be at the surface of the food product.

REFERENCES

TR Aalto, MC Firman, NE Rigier. *p*-Hydroxybenzoic acid esters as preservatives. I. Uses, antibacterial and antifungal studies, properties and determination. J Am Pharm Assoc 42:449, 1953.

MM Al-Dagal, WA Bazaraa. Extension of shelf life of whole and peeled shrimp with organic acid salts and bifidobacteria. J Food Prot 62:51–56, 1999.

UM Abdul-Raouf, LR Beuchat, MS Ammar. Survival and growth of *Escherichia coli* O157:H7 in ground, roasted beef as affected by pH, acidulants, and temperature. Appl Environ Microbiol 59:2364–2368, 1993.

MA Amerine, MA Joslyn. Commercial production of table wines. Calif Agric Exp Sta Bull 639:143, 1960.

ME Anderson, HE Huff, HD Naumann, RT Marshall. Counts of six types of bacteria on lamb carcasses dipped or sprayed with acetic acid at 25°C or 55°C and stored vacuum packaged at 0°C. J Food Prot 51:874, 1988.

ME Anderson, RT Marshall, JS Dickson. Efficacies of acetic, lactic and two mixed acids in reducing numbers of bacteria on surfaces of lean meat. J Food Safety 12:139–147, 1992.

T Ariyapitipun, A Mustapha, AD Clarke. Microbial shelf life determination of vacuum-packaged fresh beef treated with polylactic acid, lactic acid, and nisin solutions. J Food Prot 62:913–920, 1999.

AC Baird-Parker. Organic acids. Microbial ecology of foods, Vol. I. Factors affecting life and death of microorganisms. International Commission on Microbiological Specifications for Foods. New York: Academic, 1980, p. 126.

MFA Bala, DL Marshall. Organic acid dipping of catfish fillets: Effect on color, microbial load and *Listeria monocytogenes*. J Food Prot 61:1470–1474, 1998.

JD Baldock, PP Frank, PP Graham, FJ Ivey. Potassium sorbate as a fungistatic agent in country ham processing. J Food Prot 42:780, 1979.

JG Banks, RG Board. Sulfite-inhibition of *Enterobacteriaceae* including *Salmonella* in British fresh sausage and in culture systems. J Food Prot 45:1292, 1982.

EE Bargiota, E Rico-Muñoz, PM Davidson. Lethal effect of methyl and propyl parabens as related to *Staphylococcus aureus* lipid composition. Int J Food Microbiol 4:257, 1987.

DA Bautista, N Sylvester, S Barbut, MW Griffiths. The determination of efficacy of antimicrobial rinses on turkey carcasses using response surface designs. Int J Food Microbiol 34:279–292, 1997.

RE Brackett. Effect of various acids on growth and survival of *Yersinia enterocolitica*, J Food Prot 50:598, 1987.

AL Branen, TW Keenan. Growth stimulation of *Lactobacillus casei* by sodium citrate. J Dairy Sci 53:593, 1970.

RL Buchanan, SG Edelson. pH-dependent stationary-phase acid resistance response of enterohemorrhagic *Escherichia coli* in the presence of various acidulants. J Food Prot 62:211–218, 1999.

RL Buchanan, MH Golden. Interaction of citric acid concentration and pH on the kinetics of *Listeria monocytogenes* inactivation. J Food Prot 57:567–570, 1994.

CE Carpenter, DSA Reddy, DP Cornforth. Inactivation of clostridial ferredoxin and pyruvate-ferredoxin oxidoreductase by sodium nitrite. Appl Environ Microbiol 53:549, 1987.

AG Castellani, CF Niven. Factors affecting the bacteriostatic action of sodium nitrite. Appl Microbiol 3:154, 1955.

N Chen, LA Shelef. Relationship between water activity, salts of lactic acid, and growth of *Listeria monocytogenes* in a meat model system. J Food Prot 55:574–578, 1992.

JR Chipley. Sodium benzoate and benzoic acid. In: PM Davidson, AL Branen, eds. Antimicrobials in Foods, 2nd ed. New York: Marcel Dekker, Inc., 1993, pp. 11–48.

FK Cook, MD Pierson. Inhibition of bacterial spores by antimicrobials. Food Technol 37: 115, 1983.

PM Davidson. Chemical Preservatives and Natural Antimicrobial Compounds. In: MP

Doyle, LR Beuchat, TJ Montville, eds. Food Microbiology: Fundamentals and Frontiers. Washington, DC: American Society for Microbiology, 1997, pp. 520–556.

AJ Degnan, CW Kasper, WS Otwell, ML Tamplin, JB Luchansky. Evaluation of lactic acid bacterium fermentation products and food-grade chemicals to control *Listeria monocytogenes* in blue crab (*Callinectes sapidus*) meat. Appl Environ Microbiol 60:3198–3203, 1994.

T Deak, Tliske, EK Novak. Effects of sorbic acid on the growth of some species of yeast. Acta Microbiol Acad Sci Hung 17:237, 1970.

JS Dickson, ME Anderson. Microbiological decontamination of animal carcasses by washing and sanitizing systems: A review. J Food Prot 55:13–140, 1992.

S Doores. Organic acids. In: PM Davidson, AL Branen, eds. Antimicrobials in Foods, 2nd ed. New York: Marcel Dekker, Inc. 1993, pp. 95–136.

CL Duncan, EM Foster. Effect of sodium nitrite, sodium chloride, and sodium nitrate on germination and outgrowth of anaerobic spores. Appl Microbiol 16:406–411, 1968.

T Eklund. Inhibition of growth and uptake processes in bacteria by some chemical food preservatives. J Appl Bacteriol 48:423, 1980.

T Eklund. Inhibition of microbial growth at different pH levels by benzoic and propionic acids and esters of p-hydroxybenzoic acid. Int J Food Microbiol 2:159, 1985a.

T Eklund. The effect of sorbic acid and esters of p-hydroxybenzoic acid on the protonmotive force in *Escherichia coli* membrane vesicles. J Gen Microbiol 131:73, 1985b.

T Eklund. Organic acids and esters. In: GW Gould, ed. Mechanisms of Action of Food Preservation Procedures. 1989, London: Elsevier Appl Sci, 1989, pp. 161–200.

MA El-Shenawy, EH Marth. Sodium benzoate inhibits growth of or inactivates *Listeria monocytogenes. J Food Prot* 51:525–530, 1988.

RH Ellinger. Phosphates in food processing. In: TE Furia, ed. Handbook of Food Additives, 2nd ed. Cleveland: CRC Press, 1981, pp. 617–744.

E Entani, M Asai, S Tsujihata, Y Tsukamoto, M Ohta. Antibacterial action of vinegar against food-borne pathogenic bacteria including *Escherichia coli* O157:H7. J Food Prot 61:953–959, 1998.

TL Fisher, DA Golden. Survival of *Escherichia coli* O157:H7 in apple cider as affected by dimethyl dicarbonate, sodium bisulfite, and sodium benzoate. J Food Sci 63: 904–906, 1998.

LN Flores, LS Palomar, PA Roh, B Bullerman. Effect of potassium sorbate and other treatments on the microbial content and keeping quality of a restaurant-type Mexican hot sauce. J Food Prot 51:4, 1988.

JW Foster. Low pH adaptation and the acid tolerance response of *Salmonella typhimurium*. Crit Rev Microbiol 21:215–237, 1995.

E Freese, CW Sheu, E Galliers. Function of lipophilic acids as antimicrobial food additives. Nature 241:321, 1973.

M Fukahori, S Akatsu, H Sato, T Yotsuyanagi. Relationship between uptake of p-hydroxybenzoic acid esters by *Escherichia coli* and antibacterial activity. Chem Pharm Bull (Tokyo) 44:1567–1570, 1996.

C Genigeorgis, H Riemann. Food processing and hygiene. In: H Riemann, FL Bryan, eds. Food-borne Infections and Intoxications, 2nd ed. New York: Academic Press, 1979, p. 613.

AM Gibson, TA Roberts. The effect of pH, sodium chloride, sodium nitrite and storage temperature on the growth of *Clostridium perfringens* and faecal streptococci in laboratory medium. Int J Food Microbiol 3:195, 1986.

J Giese. Experimental process reduces *Salmonella* on poultry. Food Technol 46(4):112, 1992.

EF Glabe, JK Maryanski. Sodium diacetate: an effective mold inhibitor. Cereal Foods World 26:285, 1981.

GW Gould. Effect of food preservatives on the growth of bacteria from spores. In: N Molin, ed. Microbial Inhibitors in Food. Stockholm: Almqvist and Miksell, 1964, p. 17.

GW Gould, ed. Mechanisms of Action of Food Preservation Procedures. London: Elsevier Appl Sci, 1989.

GW Gould, NJ Russell. Sulphite. In: NJ Russell, GW Gould, eds. Food Preservatives. Glasgow: Blackie and Son Ltd., 1991, pp. 72–88.

L Gram. Inhibition of mesophilic spoilage *Aeromonas* spp. on fish by salt, potassium sorbate, liquid smoke, and chilling. J Food Prot 54:436–442, 1991.

ABG Grever. Minimum *nitrite* concentrations for inhibition of clostridia in cooked meat products. In: B Krol, BJ Tinbergen, eds. proceedings of the International Symposium on Nitrite in Meat Products. Wageningen, The Netherlands: Pudoc, 1974, p. 103.

K Harada, R Hizuchin, I Utsumi. Studies on sorbic acid. IV. Inhibition of the respiration in yeasts. Agric Biol Chem 32:940, 1968.

PC Houtsma, A Heuvelink, J Dufrenne, S Notermans. Effect of sodium lactate on toxin production, spore germination and heat resistance of proteolytic *C. botulinum* strains. J Food Prot 57:327–330, 1994.

CM Huhtanen. Antibotulinal activity of methyl and ethyl fumarates in comminuted nitrite-free bacon. J Food Sci 48:1574, 1983.

CM Huhtanen, EJ Guy. Antifungal properties of esters of alkenoic and alkynoic acids. J Food Sci 49:281, 1984.

ICMSF (International Commission on the Microbiological Specifications for Foods). Microorganisms in Foods 5. Microbiological Specifications of Food Pathogens. London: Blackie Academic & Professional, 1996.

M Ingram, FJH Ottoway, JBM Coppock. The preservative action of acid substances in food. Chem Ind (London) 42:1154, 1956.

FJ Ivey, MC Robach. Effect of sorbic acid and sodium nitrite on *Clostridium botulinum* outgrowth and toxin production in canned comminuted pork. J Food Sci 43:782, 1978.

JM Jay. Modern Food Microbiology, 5th ed. New York Van Nostrand Reinhold, 1996.

CMC Jen, LA Shelef. Factors affecting sensitivity of *Staphylococcus aureus* 196E to polyphosphate. Appl Environ Microbiol 52:842, 1986.

SL Jordan, J Glover, L Malcolm FM Thomson-Carter, IR Booth, SF Park. Augmentation of killing of *Escherichia coli* O157 by combinations of lactate, ethanol, and low-pH conditions. Appl Environ Microbiol 65:1308–1311, 1999.

BJ Juven, I Shomen. Spoilage of soft drinks caused by bacterial flocculation. J Food Prot 48:52, 1985.

M Kasrazadeh, C Genigeorgis. Potential growth and control of *Salmonella* in hispanic type soft cheese. Int J Food Microbiol 22:127–140, 1994.

M Kasrazadeh, C Genigeorgis. Potential growth and control of *Escherichia coli* O157: H7 in soft hispanic type cheese. Int J Food Microbiol 25:289–300, 1995.

AJ Kidney. The use of sulfite in meat processing. Chem Ind (London) 1974:717, 1974.

JW Kim, MF Slavik. Changes in eggshell surface microstructure after washing with cetylpyridinium chloride or trisodium phosphate. J Food Prot 59:859–863, 1996.

JW Kim, MF Slavik, MD Pharr, DP Raben, CM Lobsinger, S Tsai. Reduction of *Salmonella* on post-chill chicken carcasses by trisodium phosphate (Na_3PO_4) treatment. J Food Safety 14:9–17, 1994.

AD King, JD Ponting, DW Sanshuck, R Jackson, K Mihara. Factors affecting death of yeast by sulfur dioxide. J Food Prot 44:92, 1981.

SJ Knabel, HW Walker, PA Hartman. Inhibition of *Aspergillus flavus* and selected gram-positive bacteria by chelation of essential metal cations by polyphosphates. J Food Prot 54:260–365, 1991.

TL Knox, PM Davidson, JR Mount. Evaluation of selected antimicrobials in fruit juices as sodium metabisulfite replacements or adjuncts. 44th Annual Meeting, Institute of Food Technologists, Anaheim, CA, 1984.

Y Kouassi, LA Shelef. Listeriolysin O secretion by *Listeria monocytogenes* in the presence of cysteine and sorbate. Lett Appl Microbiol 20:295–299, 1995.

RM Lee, PA Hartman, DG Olson, FD Williams. Bactericidal and bacteriolytic effects of selected food-grade phosphates, using Staphylococcus aureus as a model system. J Food Prot 57:276–283, 1994.

MB Liewen, EH Marth. Growth of sorbate-resistant and -sensitive strains of *Penicillium roqueforti* in the presence of sorbate. J Food Prot 48:525, 1985.

HS Lillard. Effect of trisodium phosphate on salmonellae attached to chicken skin. J Food Prot 57:465–469, 1994.

RC Lindsay. Food additives. In: OR Fennema, ed. Food Chemistry, 3rd ed. New York: Marcel Dekker, 1996, pp. 767–823.

M Lopez, S Martinez, J Gonzalez, R Martin, Bernardo. Sensitization of thermally injured spores of *Bacillus stearothermophilus* to sodium benzoate and potassium sorbate. Lett Appl Microbiol 27:331–335, 1998.

MR Maas, KA Glass, MP Doyle. Sodium lactate delays toxin production by *Clostridium botulinum* in cook-in-bag turkey products. Appl Env Microbiol 55:2226–2229, 1989.

EH Marth, CM Capp, L Hasenzah, HW Jackson, RV hussong. Degradation of potassium sorbate by penicillium species. J Dairy Sci 49:1197, 1966.

B Matamoros-León, A Argaiz, A López-Malo. Individual and combined effects of vanillin and potassium sorbate on *Penicillium digitatum, Penicillium glabrum*, and *Penicillium italicum* growth. J Food Prot 62:540–542, 1999.

RC McKellar. Effect of preservatives and growth factors on secretion of listeriolysin O by *Listeria monocytogenes*. J Food Prot 56:380–384, 1993.

JH Meng, CA Genigeorgis. Modeling the lag phase of nonproteolytic *Clostridium botulinum* toxigenesis in cooked turkey and chicken breast as affected by temperature, sodium lactate, sodium chloride and spore inoculum. Int J Food Microbiol 19:109–122, 1993.

TE Minor, EH Marth. Growth of *Staphylococcus aureus* in acidified pasteurized milk. J Milk Food Technol 33:516, 1970.

GJ Mountney, J O'Malley. Acids as poultry meat preservatives. Poultry Sci. 44:582, 1985.

DI Murdock. Inhibitory action of citric acid on tomato juice flat sour organisms. Food Res 15:107, 1950.

NAS/NRC. The health effects of nitrate, nitrite and n-nitroso compounds. Committee on Nitrite and Alternative Curing Agents, National Research Council, National Academy Press, Washington, DC, 1981.

KA Nelson, FF Busta, JN Sofos, CE Allen. Effect of product pH and ingredient forms in chicken frankfurter emulsions on *Clostridium botulinum* growth and toxin production. 40th Annual Meeting, Institute of Food Technologists, New Orleans, LA, 1980.

FS Nury, MW Miller, JE Brekke. Preservative effect of some antimicrobial agents on high moisture dried fruits. Food Technol 14:113, 1960.

AJ Okrend, RW Johnson, AB Moran. Effect of acetic acid on the death rates at 52°C of *Salmonella newport*, *Salmonella typhimurium*, and *Campylobacter jejuni* in poultry scald water. J Food Prot 49:500, 1986.

CS Ough. Dimethyl dicarbonate and diethyl dicarbonate. In: PM Davidson, AL Branen, eds. Antimicrobials in Foods, 2nd ed. New York: Marcel Dekker, 1993a, pp. 343–368.

CS Ough. Sulfur dioxide and sulfites. In: PM Davidson, AL Branen, eds. Antimicrobials in Foods, 2nd ed. New York: Marcel Dekker, 1993b, pp. 137–190.

CS Ough, RE Kunkee. The effect of fumaric acid on malolactic fermentation in wines from warm areas. Am J Enol Vitic 25:188, 1974.

ME Parish, DE Carroll. Effects of combined antimicrobial agents on fermentation initiation by *Saccharomyces cerevisiae* in a model broth system. J Food Sci 53:240, 1988.

M Pederson, N Albury, MD Christensen. The growth of yeasts in grape juice stored at low temperature. IV. Fungistatic effect of organic acids. Appl Microbiol 9:162, 1961.

GA Pelroy, ME Peterson, PJ Holland, MW Eklund. Inhibition of *Listeria monocytogenes* in cold-process (smoked) salmon by sodium lactate. J Food Prot 57:108–113, 1994a.

GA Pelroy, ME Peterson, R Paranjpye, J Almond, MW Eklund. Inhibition of *Listeria monocytogenes* in cold-process (smoked) salmon by sodium nitrite and packaging method. J Food Prot 57:114–119, 1994b.

GF Phillips, JO Mundt. Sorbic acid as an inhibitor of scum yeast in cucumber fermentations. Food Technol 4:291, 1950.

GJ Pilone. Control of malo-lactic fermentation in table wines by addition of fumaric acid. In: JG Carr, CV Cutting, GC Whiting, eds. Lactic Acid Bacteria in Beverages and Foods. London: Academic, 1975, p. 121.

RK Podolak, JF Zayas, CL Kastner, DYC Fung. Inhibition of *Listeria monocytogenes* and *Escherichia coli* O157:H7 on beef by application of organic acids. J Food Prot 59: 370–373, 1996.

LS Post, TL Amoroso, M Solberg. Inhibition of *Clostridium botulinum* type E in model acidified food systems. J Food Sci 50:966, 1985.

FJ Post, WS Coblentz, TW Chou, DK Salunhke. Influence of phosphate compounds on certain fungi and their preservative effect on fresh cherry fruit (*Prunus cerasus* L.) Appl Microbiol 16:138, 1968.

E Rajashekhara, ER Suresh, S Ethiraj. Thermal death rate of ascospores of *Neosartorya fisheri* ATCC 200957 in the presence of organic acids and preservatives in fruit juices. J Food Prot 61:1358–1362, 1998.

HJ Rehm, H, Wittman. Beitrag zur Kenntnis der antimikrobiellen Wirkung der schwefligen Säure I. Übersicht über einflussnehmende Factoren auf die antimikrobielle Wirkung der schwefligen Säure. Z Lebensm Untersuch Forsch 118:413, 1962.

J Reiss. Prevention of the formation of mycotoxins in whole wheat bread by citric acid and lactic acid. Experientia 32:168, 1976.

KM Rice, MD Pierson. Inhibition of *Salmonella* by sodium nitrite and potassium sorbate in frankfurters. J Food Sci 47:1615–1617, 1982.

E Rico-Muñoz, PM Davidson. The effect of corn oil and casein on the antimicrobial activity of phenolic antioxidants. J Food Sci 48:1284–1288, 1983.

MC Robach. Use of preservatives to control microorganisms in food. Food Technol 34: 81, 1980.

TA Roberts, AM Gibson. Chemical methods for controlling *Clostridium botulinum* in processed meats. Food Technol 40:163, 1986.

AC Roberts, DJ McWeeny. The uses of sulfur dioxide in the food industry. A review. J Food Technol 7:221, 1972.

TA Roberts, AM Gibson, A Robinson. Factors Controlling the growth of *Clostridium botulinum* types A and B in pasteurized cured meats. II. Growth in pork slurries prepared from high pH meat (pH ranges 6.3–6.8). J Food Technol 16:267, 1981.

TA Roberts, LFJ Woods, MJ Payne, R Cammack. Nitrite. In: NJ Russell, GW Gould, eds. Food Preservatives. Glasgow: Blackie and Son Ltd., 1991, pp. 89–111.

JO Roland, LR Beuchat. Biomass and patulin production by *Byssochlamys nivea* in apple juice as affected by sorbate, benzoate, SO_2 and temperature. J Food Sci 49:402, 1984.

JO Roland, LR Beuchat, RE Worthington, HL Hitchcock. Effects of sorbate, benzoate, sulfur dioxide and temperature on growth and patulin production by *Byssochlamys nivea* in grape juice. J Food Prot 47:237, 1984.

IE Ronning, HA Frank. Growth inhibition of putrefactive anaerobe 3679 caused by stringent-type response induced by protonophoric activity of sorbic acid. Appl Environ Microbiol 53:1020, 1987.

AH Rose, BJ Pilkington. Sulphite. In: GW Gould, ed. Mechanisms of Action of Food Preservation Procedures. London: Elsevier Appl Sci, 1989, pp. 201–224.

H Rosenquist, Å. Hansen. The antimicrobial effect of organic acids, sour dough and nisin against *Bacillus subtilis* and *B. licheniformis* isolated from wheat bread. J Appl Microbiol 85:621–631, 1998.

RA Seward, RH Dielbel, RC Lindsay. Effects of potassium sorbate and other antibotulinal agents on germination and outgrowth of *Clostridium butulinum* Type E spores in microculture. Appl Env Microbiol 44:1212–1221, 1982.

LA Shelef, JA Seiter. Indirect antimicrobials. In: PM Davidson, AL Branen, eds. Antimicrobials in Foods, 2nd ed. New York: Marcel Dekker, Inc. 1993, pp. 539–570.

LA Shelef, L. Addala. Inhibition of *Listeria monocytogenes* and other bacteria by sodium diacetate. J Food Safety 14:103–115.

LA Shelef, Q Yang. Growth supression of *Listeria monocytogenes* by lactates in broth, chicken, and beef. J Food Prot 54:283–387, 1991.

I Shibasaki. Antimicrobial activity of alkyl esters of *p*-hydroxybenzoic acid. J Ferment Technol 47:167, 1969.

GR Siragusa, JS Dickson. Inhibition of *Listeria monocytogenes*, *Salmonella typhimurium* and *Escherichia coli* O157:H7 on beef muscle tissue by lactic or acetic acid contained in calcium alginate gells. *J Food Safety* 13:147–158, 1993.

MF Slavik, JW Kim, MD Pharr, DP Raben, S Tsai, CM Lobsinger. Effect of trisodium phosphate on *Campylobacter* attached to post-chill chicken carcasses. J Food Prot 57:324–26, 1994.

DP Smith, NJ Rollin. Sorbic Acid as a fungistatic agent for foods. VII. Effectiveness of sorbic acid in protecting cheese. Food Res 19:59, 1954.

La Smoot, MD Pierson. Mechanisms of sorbate inhibition of *Bacillus cereus* T and *Clostridium botulinum* 62A spore germination. Appl Environ Microbiol 42:477–483, 1981.

FJM Smulders, P Barendsen, JG van Logtestjin, DAA Mossel, GM Van der Marel. Review: lactic acid: considerations in favour of its acceptance as a meat decontaminant. J Food Technol 21:419, 1986.

JMA Snijders, JG van Logtestjin, DAA Mossel, FJM Smulders. Lactic acid as a decontaminant in slaughter and processing procedures. Vet Quart 7:277, 1985.

JN Sofos, LR Beuchat, PM Davidson, EA Johnson. Naturally Occurring Antimicrobials in Food. Task Force Report No. 132, Council for Agric Sci and Technol, Ames, IA, 1998.

JN Sofos, FF Busta. Sorbic acid and sorbates. In: PM Davidson, AL Branen, eds. Antimicrobials in Foods, 2nd ed. New York: Marcel Dekker, 1993, pp. 49–94.

R Spencer. Nitrite in curing: microbiological implications. Proc 17th European Meeting of Meat Research Workers Conference, Bristol, England, 1971.

CS Subramanian, EH Marth. Multiplication of *Salmonella typhimurium* in skim milk with and without added hydrochloric, lactic and citric acids. J Milk Food Technol 31: 323, 1968.

N Tanaka. Challenge of pasteurized process cheese spreads with *Clostridium butulinum* using in-process and post-process inoculation. J Food Prot 45:1044, 1982.

A Tellez-Giron, GR Acuff, C Vanderzant, LW Rooney, RD Waniska. Microbiological characteristics and shelf life of corn tortillas with and without antimicrobial agents. J Food Prot 51:945–948, 1988.

FR Terrell, JR Morris, MG Johnson, EE Gbur, DJ Makus. Yeast inhibition in grape juice containing sulfur dioxide, sorbic acid, and dimethyl dicarbonate. J Food Sci 58: 1132–1134, 1993.

DP Thompson. Minimum inhibitory concentrations of esters of p-hyroxybenzoic acid (paraben) combinations against toxigenic fungi. J Food Prot 57:133–135, 1994.

JE Thompson, GJ Banwart, DH Sanders, AJ Mercuri. Effect of chlorine, antibiotics, β-propiolactone, acids and washing on *Salmonella typhimurium* on eviscerated fryer chickens. Poult Sci 46:146, 1967.

RB Tompkin. The role and mechanism of the inhibition of *C. botulinum* by nitrite—is a replacement available? Proc. 31st Ann. Reciprocal Meats Conference, Storrs, CT, 1978, p. 135.

RB Tompkin. Indirect antimicrobial effects in foods: phosphates. J Food Safety 6:13, 1983.

RB Tompkin. Nitrite. In: PM Davidson, AL Branen, eds. Antimicrobials if Foods, 2nd ed. New York: Marcel Deker, 1993, pp. 191–262.

RB Tompkin, LN Christiansen, AB Shaparis. The effect of iron on botulinal inhibition in perishable canned cured meat. J Food Technol 13:521, 1978.

S Tsai, C Chou. Injury, inhibition and inactivation of *Escherichia coli* O157:H7 by potassium sorbate and sodium nitrite as affected by pH and temperature. J Sci Food Agric 71:10–12, 1996.

HE Uljas, SC Ingham. Combinations of intervention treatments resulting in a 5-$\log_{10}$-unit reductions in numbers of *Escherichia coli* O157:H7 and *Salmonella typhimurium* DT104 organisms in apple cider. Appl Environ Microbiol 65:1924–1929, 1999.

JR Unda, RA Mollins, HW Walker. *Clostridium sporogenes* and *Listeria monocytogenes*: survival and inhibition in microwave-ready beef roasts containing selected antimicrobials. J Food Sci 56:198–205, 1991.

P Van Netten, A Valentijn, DAA Mossel, JHJ Huis in't Veld. The survival and growth of acid-adapted mesophilic pathogens that contaminate meat after lactic acid decontamination. J Appl Microbiol 84:559–567, 1998.

IJR Visser, PA Koolmees, PGH Bijker. Microbiological conditions and keeping quality of veal tongues as affected by lactic acid decontamination and vacuum packaging. J Food Prot 51:208, 1988.

MK Wagner, FF Busta. Effect of sodium acid pyrophosphate in combination with sodium nitrite or sodium nitrite/potassium sorbate on *Clostridium botulinum* growth and toxin production in beef/pork frankfurter emulsions. J Food Sci 48:990, 1983.

MK Wagner, FF Busta. Inhibition of *Clostridium botulinum* 52A toxicity and protease activity by sodium acid pyrophosphate in media systems. Appl Environ Microbiol 50:16, 1985.

A Waldroup. Evaluating reduction technologies. Meat Poult 41(8):10, 1995.

A Waldroup, J Marcy, M Doyle, M Scantling. TSP: a market survey. Meat Poult 41(12): 18–20, 1995.

WC Wang, Y Li, MF Slavik, H Xiong. Trisodium phosphate and cetylpyridinium chloride spraying on chicken skin to reduce attached *Salmonella typhimurium*. J Food Prot 60:992–994, 1997.

RA Weaver, La Shelef. Antilisterial activity of sodium, potassium or calcium lactate in pork liver sausage. J Food Safety 13:1333–146, 1993.

D Wibowo, R Eschenbruch, CR Davis, GH Fleet, TH Lee. Occurrence and growth of lactic acid bacteria in wine. A review. Am J Enol Vitic 36:302, 1985.

CE Wind, L Restaino. Antimicrobial effectiveness of potassium sorbate and sodium benzoate against *Zygosaccharomyces bailii* in salsa mayonnaise. J Food Prot 58:1257–1259, 1995.

LFJ Woods, JM Wood. The effect of nitrite inhibition on the metabolism of *Clostridium botulinum*. J Appl Bacteriol 52:109, 1982.

LFJ Woods, JM Wood, PA Gibbs. Nitrite. In: GW Gould, ed. Mechanisms of Action of Food Preservation Procedures. London: Elsevier Appl Sci, 1989, pp. 225–246.

R Xiong, G Xie, AS Edmondson. The fate of *Salmonella enteritidis* PT4 in home-made mayonnaise prepared with citric acid. Lett Appl Microbiol 28:36–40, 1999.

GK York, RH Vaughn. Use of sorbic acid enrichment media for species of *Clostridium*. J Bacteriol 68:739, 1954.

GK York, RH Vaughn. Resistance of *Clostridium parabotulinum* to sorbic acid. Food Res 20:60, 1955.

LL Zaika, AH Kim. Effect of sodium polyphosphates on growth of *Listeria monocytogenes*. J Food Prot 56:577–580, 1993.

T Zhao, MP Doyle, RE Besser. Fate of enterohemorrhagic *Escherichia coli* O157:H7 in apple cider with and without preservatives. Appl Env Microbiol 59:2526–2530, 1993.

R-Y Zhuang, LR Beuchat. Effectiveness of trisodium phosphate for killing *Salmonella montevideo* on tomatoes. Lett Appl Microbiol 22:97–100, 1996.

7
Microbial Control by Packaging

Gaurav Tewari
ESL Global Inc.
San Antonio, Texas

I. INTRODUCTION

Shelf life is the time required for a food to become unacceptable from a sensory, nutritional, microbiological, or safety perspective. Provided the mechanism that causes a food product to spoil is known and can be manipulated, shelf life of a food product can be extended. Food systems are available in three forms: processed, minimally processed, and fresh. For processed food systems, shelf life is extended by using various techniques such as thermal processing (e.g., retort and aseptic processing), freeze-drying, cryogenic freezing (e.g., individual quick frozen technique, IQF), and nonthermal processing (e.g., high-pressure processing). Food industries have commercially adapted these technologies to extend the shelf life of high-acid liquid foods (e.g., aseptic or high pressure), low-acid liquid foods with or without particulates (e.g., retorting), dehydrated foods (e.g., freeze-drying), fruits and vegetables cuts (e.g., IQF), and meat (e.g., refrigeration or freezing). These food preservation techniques control microbial growth by the virtue of "processing." Therefore, the primary factor extending the shelf life is "process"; however, "packaging" further aids in preservation. It is to be noted that these processed foods need to be coupled with an excellent package with high barrier properties to prevent any microbial growth. Usually barrier properties of interest are related to water vapor and oxygen (O_2) transmission. Due to the absence of such barrier properties, moisture and O_2 levels in the package can be enhanced and thus can favor microbial growth.

For minimally processed and/or fresh food systems, refrigeration and packaging are the major and only available hurdles. Concerted interest is growing in minimally processed and fresh food products due to their superior sensory and nutritional quality. Fresh vegetables, fruits, and meats are premium priced products in local and international markets. Only a handful of techniques are available for extending the shelf life of fresh or minimally processed foods. These commercialized techniques include irradition and high-pressure processing (a nonthermal food-processing technique in which microbial inactivation occurs due to ultra–high pressure with minimal increase in product temperature). Although in recent years the U.S. Food and Drug Administration (FDA) has approved irradiation of red meat, there is still about the availability of polymers suitable for irradiated packages. There are currently only a handful of polymers approved in the CFR (Code of Federal Regulations) title 21§149.5 for use with irradiated foods. Of these, only one, ethylene vinyl alcohol (EVA), can receive electron beam (e-beam) treatment. Most of the polymers listed in this section were approved in the 1960s. It is difficult to make a contemporary multilayer structure with currently approved polymers. From the standpoint of radiation chemistry, e-beam, gamma ray, and x-ray treatments should have the same impact on polymer packaging materials. The fact that e-beam, gamma ray, and x-ray treatments are treated differently in the CFR is due to the fact that initial requests only mentioned gamma treatment. Additionally, consumer perception had always been a concern regarding irradiation. Thus, irradiation has its limitations.

In recent years, high-pressure processing (HPP) has been extensively used in Japan, and a variety of food products like jams and fruit juices have been processed. There have been 10–15 types of pressurized foods on the Japanese market, but several have disappeared, and those that remain are so specific that they would have little interest to European or American markets. Examples of commercially pressurized products in Europe and the United States include orange juice by UltiFruit®, Pernod Richard Company, France; acidified avocado purèe (guacamole) by Avomex Company, Texas/Mexico; and sliced ham (both cured-cooked and raw-cooked) by Espuna Company, Spain. Volumes produced are still very small. Pressurized fruit preparation from yogurt should soon reach the U.S. market. The European ''Novel Foods'' Directive (May 1997) introduced regulatory problems and slowed the introduction of new pressurized products (1). Although HPP has excellent commercial potential, global acceptance of this technology is several years away. The capability and limitations of HPP have been extensively reviewed (1).

Because of certain issues related to the commercialization of various presently available techniques, ''packaging and refrigeration'' remain the only methods used to control microbial growth in most minimally processed and/or food systems. Modified atmosphere packaging (MAP), packaging with O_2 scavengers, and antimicrobial and smart packaging are examples of ''packaging'' itself acting as a hurdle to microbial growth. Modified atmosphere packaging has been in use for almost a century and has been extensively studied. The concept behind MAP is to lower the O_2 levels in the atmosphere surrounding the food products by either removing all air (vacuum) or replacing air with an appropriate mixture of gases like CO_2 and N_2. Consequently, MAP delays microbiological growth due to the inhibitory effect of CO_2 on bacteria and ensures freshness and long shelf life of fresh products. Today, MAP is used in different sectors of the food industry such as conventional and centralized meat operations, fresh cut vegetables, and bakery goods. Nevertheless, MAP should be coupled with strict temperature control to achieve maximum microbial inhibition (2). Under the scope of the present chapter, a working example of MAP in centralized meat operations, where a package with excellent barrier properties is an integral component, will be used to illustrate control of microbial growth by packaging. This should serve as a guideline for other food systems, as the rate of spoilage in muscle foods is high because they provide a protein-rich medium that encourages microbial growth. Muscle food tissue continues to metabolize just after slaughter, utilizing storage carbohydrates and fats. In the absence of further processing, microbial growth results as the energy stores are depleted and metabolic products (low molecular weight organic compounds) accumulate in the tissues. This results in off-odors and off-flavors of meat, and eventually microbiological decay occurs on the meat surface. However, MAP, if coupled with an appropriate package and refrigeration, can control microbial growth even in muscle foods. Furthermore, MAP involves proper selection of packaging film "material" with appropriate CO_2/O_2 permselectivity to allow required exchange of gases through and from the atmosphere and package. This is essential to prevent anaerobic conditions in the package due to accumulation of CO_2 and to further prevent the growth of anaerobic foodborne pathogens within the product. This chapter discusses recent implementation of MAP for long shelf life of centrally prepared retail-ready meat cuts with special emphasis on packaging film permeability. Additionally, MAP for other food systems (processed, minimally processed, and fresh) will be discussed. Application of O_2 absorbent technology is another development in packaging technology to prevent microbial growth. Oxygen is the prime factor for faster microbial growth. Reduction and/or elimination of O_2 from the packaging atmosphere enhances the shelf life of foods. Therefore, concepts and application of O_2 scavengers in different food systems are discussed. Presently, certain antimicrobial agents have been incorporated into packaging materials to control microbial growth. Developments in antimicrobial packaging are also discussed. Last but not least, recent advancements in smart packaging to control microbial growth are outlined.

II. MODIFIED ATMOSPHERE PACKAGING OF MEAT: APPLICATION IN CENTRALIZED MEAT DISTRIBUTION

In centralized packaging of retail-ready cuts, final trimming and portioning take place at the initial packaging step. The packaging film contacting the meat surface is oxygen permeable. Fabricated products must withstand 20–30 days of storage in their original packages (which are usually grouped together in a modified atmosphere) prior to display in the retail case at temperatures of about 7°C. Since there is no further manipulation of the cuts at retail other than removal of retail-ready packages from the master package and placement in the display case, meat surfaces exposed at the central preparation facility must retain their attractiveness and appeal not only during the period of storage but also during subsequent display at retail for about 2 days. Therefore, packages should have excellent barrier to O_2 to prevent growth of faster growing aerobic bacteria.

A. Modified Atmosphere Packaging and Controlled-Atmosphere Storage

Modified atmosphere packaging is defined as ''the packaging of a perishable product in an atmosphere which has been modified so that its composition is other than that of air.'' This is in contrast with controlled-atmosphere storage (CAS), which involves active and continuous control of the atmosphere surrounding the food and is prominent in warehouse storage of vegetables and fruits. The CAS systems for meat use static atmospheres containing saturated concentrations of CO_2. In MAP, the initial atmosphere surrounding a food is altered by removal of O_2 or addition of CO_2 or a combination of gases (O_2, N_2, and CO_2). The function of CO_2 is to decrease the growth rate of microorganisms by increasing their lag phase and reducing product respiration, whereas N_2 is used to displace O_2 and acts as an inert filler, which prevents the package from collapsing when some of the CO_2 is absorbed by moisture in the product. The package should have proper water vapor and O_2 barrier properties to prevent any kind of spoilage through microbial growth. Under anaerobic conditions the growth of lactic acid bacteria (LAB) is enhanced, which, however, is not of a major concern from food safety perspective.

B. Methods of Atmosphere Modification

Various MAP techniques are available, and their advantages and disadvantages should be considered before selection and adaptation for centralized meat packaging (2).

1. Vacuum-Packaging

Vacuum-packaging is the earliest form of MAP developed commercially and is still used for products such as cuts of fresh red meat, cured meats, hard cheeses, and ground coffee. The product is packed in film of low O_2 permeability, the air is evacuated, and entry of O_2 from outside is restricted. In the case of vacuum-packaged fresh meat, CO_2 increases to 10–20% within the package as the respiration of the meat quickly consumes the residual O_2 (2).

Vacuum-packaged fresh beef is not preferred by consumers in the retail market because of the change in its color from red to purple (myoglobin → deoxymyoglobin) due to the low O_2 level. This is a reversible reaction, but the formation of the oxidation product metmyoglobin (brown), which is pH dependent, does occur to a small extent and is less easily reversed as time of storage increases (2). These changes also take place in pork but are less noticeable. Also, there is accumulation of exudate during prolonged storage of meat in vacuum packs, probably related to the lack of package headspace.

2. Vacuum Skin Packaging

The vacuum skin packaging (VSP), a vacuum is used to apply a thermoformable film (softened by heating) as a skin over the meat placed on a rigid backboard. This method promotes longer storage life of meat, depending on meat species and muscle pH, because exudate development and precipitation of denatured meat pigment on the meat surface is not very pronounced as in regular vacuum packages. However, dark meat color is more noticeable in meat cuts using VSP. Its application is more popular for consumer packages of cured meat products where color differences are not as apparent (2).

3. Gas Packaging

Gas packaging can be done either by mechanical replacement of air with gas mixtures or by generating the atmosphere within the package (passively or actively by using suitable atmosphere modifiers such as O_2 absorbents). Mechanical air replacement can be done either by gas flushing or by compensated vacuum. Gas flushing is usually performed on a form-fill-seal machine by injecting a continuous stream of gas into the package to replace the air. After removal of most of the air, the package is sealed. In the compensated vacuum process, a vacuum is first created in a preformed or thermoformed container holding the food, and then the desired gas or gas mixture is introduced. Gas-flushing techniques can be very fast and lend themselves to continuous operation. Ordinarily, O_2 levels in gas-flushed packs are 2–5% and therefore are not suitable for packaging O_2-sensitive foods (2).

4. Master Packs

Master packaging is commonly adapted for centralized meat distribution systems. Under master packaging operations, permeability of packaging films employed plays a major role in enhancing the shelf life of retail-ready meats. Master packaging involves placement of MAP products (wrapped in a permeable film) into a gas-impermeable package in a manner that eliminates O_2 from the atmosphere surrounding the meat. The MAP products are placed in the impermeable package prior to shipping (the internal atmosphere can be up to 100% CO_2, minimizing the growth of microorganisms). The gas-impermeable barrier is removed at the retail level, resulting in the development of a desirable bright red color of meat due to the entry of O_2 into the permeable package. Master packs must be coupled with an excellent temperature control system during distribution (−1.5 ± 0.5°C) to achieve optimal shelf life extension of meat cuts (2).

C. Microbiological and Sensory Characteristics of Meat

Foodborne illness is caused by pathogenic agents, such as bacteria, that may be affected by a combination of intrinsic and extrinsic factors. The intrinsic factors are related to the food [chemical and physical composition of food, water activity (a_w), pH, O_2-reduction potential, and natural factors inhibitory to bacteria that are present in food]. The extrinsic factors are related to the environment surrounding the food (storage temperature, packaging, gas atmospheres, preservatives, cleaning, disinfection, and hygiene). These factors influence survival or growth of microorganisms in the food. By manipulating these factors, ''hurdles'' can be created to prevent microbial growth. For MAP foods, selection of a proper package, along with refrigeration or strict temperature control at temperatures close to those causing freezing, is an important hurdle assuring product quality. This is important because psychrotrophic bacteria can grow at temperatures as low as −2°C, where most meat freezes. Adequate refrigeration must be maintained throughout the food distribution system (2).

Several researchers have studied the effects of different compositions of gases in appropriate packages on the microbiological and sensory quality of meat during centralized meat packaging. Penny and Bell (3) used MAP for extending the retail storage life of beef and lamb. They stored meat under 80% O_2 and 20% CO_2 at 5°C and found that storage life was three times better than the storage life in conventional overwrap trays. They also noted that in most situations, storage life was limited due to poor color rather than excessive microbial growth. Also, storage life was dependent on muscle type, species, and length of time the meat had been previously stored in vacuum. Pork color was more stable than beef or lamb in response to low levels of O_2.

Scholtz et al. (4) examined the potential value of prepackaging pork centrally by comparing the retail shelf life of pork loin chops packaged under an atmosphere of 25% CO_2 plus 75% O_2 with parallel samples packaged using the VSP technique and with PVC overwrapped (essentially aerobic) samples as controls. All samples were stored at 0°C for up to 21 days. Results from this group of trials were compared with those from similarly prepared retail packaged chops placed on PVC overwrapped trays that were packaged in groups of six in a master pack sealed following application of a 100% CO_2 atmosphere. Samples stored using either VSP or 25% CO_2 plus 75% O_2 were acceptable 7 days after packaging and retail display. Microbiological results were similar to those obtained from conventionally packed PVC overwrapped trays at only 4 days of storage at 0°C. The 100% CO_2 treatment yielded the most promising results with samples stored 21 days capable of 4 days subsequent retail display. Shelf life was limited by color acceptability.

In an attempt to improve pork color stability following suggestions by Taylor (5), Buys et al. (6) included O_2 at 25% or 80% in PVC wrapped retail pork chop packages and overwrapped these with barrier bags having an O_2 transmission rate (OTR) of 39 mL/(m^2 24 h) atm at 23°C, 75% RH. These were subsequently backflushed with either 100% CO_2 or 25% CO_2:50% N_2:25% O_2. No significant differences were seen following 21 days of storage, and samples achieved 2 days subsequent acceptable retail display. A trend to greater meat pigment instability was observed in high O_2-containing PVC overwrapped packages. Buys et al. (7) then explored the use of either vacuum or 100% CO_2 to preserve unsliced pork loins for up to 21 days at 0°C or 5°C and evaluated the subsequently sliced loin chops for color and microbiological condition during retail display, again in PVC overwrapped trays. Chops prepared from loins stored at 0°C for 21 days were still acceptable 4 days after retail display. Initial packaging treatments did not yield different retail shelf life. It is unfortunate that barrier bags used to simulate master packs in the trials by Buys et al. (6,7) and Scholtz et al. (4) had OTRs significantly greater than the ≤15 mL/(m^2 24 h) atm at 23°C and 75% RH currently used to create an acceptable O_2 barrier. Results from these studies did not clearly demonstrate effects that could be attributed to differences in package atmosphere composition.

Gill and Penney (8) studied the effect of initial gas volume–to–meat mass ratio and pH on the storage life of chilled beef packaged under CO_2. They reported that Enterobacteriaceae fractions were larger in number on high-pH (>6.0) meat than on normal pH meat (5.5–5.7) during storage at 1°C. Grau (9) reported that Enterobacteriaceae were unable to grow on muscle tissue of normal pH under anaerobic conditions. Later, Grau (10) and Vanderzant et al. (11) observed that Enterobacteriaceae show slow growth on vacuum-packaged, normal pH muscle tissue in packs that were prepared using film of low but measurable O_2 permeability. Gill and Penney (8) reported that all vacuum-packaged meat was spoiled by putrid flavors (high-pH meat was spoiled at 7 weeks and normal pH meat at 12 weeks). They also reported that the increasing amount of added CO_2 progressively retarded putrid spoilage (due to slow microbial growth rate), while the relative number of LAB in the flora was enhanced. This was due to the inhibitory effect of CO_2 on gram-negative organisms on both normal and high-pH meat because CO_2 extends the lag phase of microbial growth. The LAB dominate these environments at low temperatures and form less offensive metabolic by-products than organisms of other genera. This study showed that CO_2 can greatly extend the shelf life of packaged chilled meat, but saturating levels of CO_2 are required for extended storage of these products (about 2 L CO_2/kg meat).

Gill and McGinnis (12) studied the changes in microflora on commercial beef trimming during collection, distribution, and preparation for its retail sale as ground beef. They reported that during storage for up to 18 days before grinding, meat trimmings developed a LAB flora of up to 10^7 cfu/g. Normally, these materials would spoil within 3 days if packaged in O_2-permeable film and stored under the same refrigeration conditions. They recommended centralized meat distribution systems for retail-ready cuts and strict control over storage and distribution temperature for extending the shelf life of meat.

Gill and Jones (13) compared the display life of retail-ready beef steaks using vacuum packaging (−1.5°C) or master packing (2°C) under various gas atmospheres of N_2, CO_2, or $O_2 + CO_2$. The product was assessed after a storage time of 60 days. At each assessment, a vacuum package and a master pack were withdrawn from storage. Three retail packs were prepared from vacuum-packaged meat and were displayed with retail packaged meat from a master pack in a retail cabinet at air temperatures between 3 and 5.7°C. Steaks from vacuum-packaged product were considered desirable, with little metmyoglobin in the surface pigment. Numbers of bacteria on steaks from vacuum packs and N_2, CO_2, $O_2 + CO_2$ atmospheres were, respectively, $<10^4$, $<10^6$, $<10^5$, and $<10^4$ cfu/cm^2. The flora from steaks stored under CO_2 were composed entirely of LAB. Small fractions (<5%) of Enterobacteriaceae and *Brochrothrix thermosphacta* were present in the steaks prepared from vacuum-packaged product stored for 39, 46, or 53 days. Large fractions (>20%) of Enterobacteriaceae or *B. thermosphacta* were present in the flora of steaks stored under $O_2 + CO_2$ atmospheres for 8, 12, or 20 days. They suggested that master packs would minimize the deterioration of meat. The numbers attained by spoilage flora on steaks under CO_2 were insufficient to cause organoleptic effects. They concluded that master packs under CO_2 could be an appropriate technique for extending shelf life of beef for up to 7 weeks and is useful when distribution involves <4 days.

Holley et al. (14,15) did microbiological analysis of CO_2-packaged retail-ready pork. They stored wrapped boneless pork loin roasts and slices at 4°C in bulk under constant CO_2 concentrations of 50% and 100% for 1 and 2 weeks. In both treatments, levels of psychrotrophs, mesophiles, and LAB were $\leq 10^4$ cfu/cm^2 during the initial 2 weeks of storage under CO_2. Enterobacteriaceae were $\leq 10^2$ cfu/cm^2 in all samples, but *Brochrothrix* were one log higher in 50% CO_2-stored samples than in 100% CO_2-stored samples at 14 days. They concluded that samples stored under 50% CO_2 and 100% CO_2 for 2 weeks could be aerobically displayed subsequently without being unacceptable for 3 and 6 days, respectively.

Jeremiah and Gibson (16) studied the influence of different packaging atmospheres on the shelf life of retail-ready pork cuts. They used 100% N_2, 100%

CO_2, and a mixture containing 70% O_2 and 30% CO_2. They suggested that 20 days of storage life would provide the opportunity to prepare and distribute centrally packed retail ready cuts to North American customers and yield satisfactory retail display. Tests were conducted at −1.5, 2, or 5°C storage in master packs for up to 28 days followed by 30 hours of aerobic chilled display at 6.8°C. All three packaging atmospheres and storage temperatures yielded similar results, with some differences noted in terms of texture and flavor as a result of treatments. They concluded that storage life of those products was limited by growth of LAB and implied that use of pork with low initial numbers of bacteria would yield consistent success in satisfying consumer quality requirements. They concluded that either 100% N_2, 100% CO_2, or a mixture of 70% O_2 + 30% CO_2 should be satisfactory for centralized distribution of retail-ready pork cuts. Storage temperatures of up to 5°C would be somewhat satisfactory, but consistent satisfactory performance would be achieved under 100% CO_2 with storage at −1.5°C to retard growth by LAB, which limited product shelf life.

A substantial variation in the length of shelf life among different studies, discussed above, can easily be identified. Perhaps ignorance of the packaging film permeability and low initial raw meat quality could the reasons for this. Therefore, if one employs MAP for shelf life extension of food products, one should be extremely careful about the packaging film barrier properties. Recently, Tewari (17) employed master packaging film of high barrier properties in designing a centralized meat distribution system and was able to achieve 10 weeks of storage life with a subsequent display life of 3 days for beef tenderloin steaks.

III. MAP: APPLICATION FOR DIFFERENT FOOD PRODUCTS

Several studies have been conducted to determine the efficacy of MAP in different food systems. Proper barrier properties of the package, proper gas mixture, intrinsic properties of food, microorganisms of interest, and storage temperature determine the efficacy of MAP. Due to its bacterial inhibitory effects, CO_2 is recommended for different food systems. Ogihara et al. (18) determined the effect of the ratio of CO_2 to O_2 on the growth of 16 strains of foodborne pathogens and spoilage bacteria using the smear plate method. Growth of facultative anaerobic bacteria (*Aeromonas hydrophila*, *Bacillus cereus*, *Escherichia coli*, *Enterobacter cloacae*, *Listeria monocytogenes*, *Serratia liquefaciens*, *Salmonella* Typhimurium, *Staphylococcus aureus*) was not completely inhibited by CO_2, O_2, or their mixtures. However, the growth rate of those bacteria was reduced in proportion to the increase in ratio of CO_2 in the gas mixtures. Growth of microaerophilic bacteria (*Lactobacillus viridescens*) was not affected by CO_2, O_2, or their mixtures. Growth of four species (*A. calcoaceticus*, *F. lutescens*, *Pseudomonas aeru-*

ginosa, *P. fragi*) of aerobic bacteria did not occur in the presence of 100% CO_2. *Micrococcus luteus* grew to some extent under the same conditions. Growth of *P. fragi* was inhibited in the presence of 30% or more CO_2. Growth of *F. lutescens* occurred only with 10% CO_2 and 90% O_2. The susceptibility of aerobic bacteria to CO_2 was higher than that of facultative anaerobic bacteria and microaerophilic bacteria. Growth of anaerobic bacteria (*Clostridium perfringens*) was inhibited in the presence of O_2 at all concentrations. Thus, CO_2, O_2, and their mixtures can be used effectively to inhibit or delay bacterial growth. The following sections discuss application of MAP in different food systems.

A. Dairy Products

Milk and milk products are rich sources of protein and hence are excellent microbiological substrates and potential sources of pathogens. Due to this, milk pasteurization is a must, followed by milk distribution under refrigeration. This prevents microbial growth to some extent. In some countries, aseptic packaging is employed to deliver fluid dairy products that are shelf stable under ambient conditions. However, such packaging is not common in North America due to the specific flavor associated with aseptically packaged milk. Thus, the only common means of extending shelf life of dairy products remain pasteurization, proper packaging, and refrigeration. Great interest has been shown in MAP of dairy products to extend the shelf life and attain maximum profitability.

Whitley et al. (19) studied the effects of MAP on the growth of *L. monocytogenes* in mold-ripened Stilton cheese during refrigerated storage over a 6-week period. Samples of cheese were inoculated with *L. monocytogenes* and stored under MAP at N_2:CO_2:O_2 ratios of 80:10:10, 100:0:0, and 80:20:0. A significant increase in *L. monocytogenes* count was found in samples stored at 80:10:10 atmosphere. Greater inhibitory effect was achieved when CO_2 concentration was increased to 20% than by reducing the O_2 content. Results indicated that a 80:10:10 ratio of N_2:CO_2:O_2 is not suitable for use with blue Stilton cheese when *L. monocytogenes* may be present.

Nielson and Haasum (20) determined packaging conditions hindering fungal growth on cheese. Preliminary experiments suggested that the survival of fungal spores and risk of fungal growth on cheese after opening of the package was minimized if the humidity inside the package was high and the level of CO_2 was low. The study (20) isolated a wide range of fungi from foods and cultured them on a laboratory substrate and on a semi-synthetic cheese substrate under various levels of atmospheric O_2 and CO_2, a_w, pH, and temperature. Fungi capable of growing under microaerophilic conditions (*P. roquefortii* and *Geotrichum candidum*) were strongly affected by reduced a_w, but elevated CO_2 had a negligible effect. *Pseudomonas verrucosum* and *P. nalgiovense* were less affected by a_w but more by elevated CO_2. *Pseudomonas commune* was moderately tolerant to both parameters.

Fedio et al. (21) studied the effect of MAP on the growth of microorganisms in creamed cottage cheese (1% fat), which was prepared by mixing the curd with cream dressing. In addition, batches of high-count, low-fat creamed cottage cheese were inoculated with *Listeria innocua* ATC 33090. The cheese was packaged in high-barrier pouches with different headspace gases (air, 100% CO_2, 100% N_2, and a mixture of 50% CO_2:50% N_2) and stored at 5°C for up to 28 days. In all pouches the headspace gas composition remained the same throughout the storage trial, whereas samples packaged with air showed declines in O_2 concentration with increases in CO_2 content. Inoculated *L. innocua* ATC 33090 showed growth in cheese packages containing air or 100% N_2, but not in packages containing elevated CO_2 levels. Growth of pseudomonads was observed in samples packaged under air. Growth of yeasts and molds was the most strongly affected by modifying the storage atmosphere. In cottage cheese packaged in air, considerable growth of yeasts and molds was observed, while growth was suppressed in samples packaged in N_2. The yeast and mold counts of cottage cheese packaged in 100% CO_2 and a mixture of 50% CO_2:50% N_2 declined during storage. This could be attributed to the combined effects of CO_2 and the acidic environment of the cheese. These data suggest that MAP with CO_2 can inhibit the growth of spoilage microorganisms in creamed cottage cheese in high-barrier containers. The study also indicated that cottage cheese packaged with air or N_2 in conventional containers could be a vehicle for listeriosis and that packaging cottage cheese in gaseous environments containing elevated CO_2 levels could reduce the risk of *Listeria* spp. in cottage cheese.

Chen and Hotchkiss (22) studied the growth of *L. monocytogenes* and *Clostridium sporogenes* in cottage cheese in MAP. Low-fat cottage cheese (pH 5.14) was inoculated with three strains of *L. monocytogenes* and *C. sporogenes* ATCC 3584. Cheese was packaged with or without added dissolved CO_2 in polystyrene tubs overwrapped with or without high-barrier heat-shrink film and stored at 4, 7, and 21°C for up to 63 days. Initial concentration of CO_2 in the container headspace was 35% (v/v), which declined by one third over the 63 days of storage at 4°C. *Clostridium sporogenes* failed to grow under any condition applied in this study. In conventionally packaged cottage cheese, *L. monocytogenes* increased from 10^4 to 10^7 cfu/g after lag phases of 28 and 7 days at 4 and 7°C, respectively. In contrast, *L. monocytogenes* failed to grow in cottage cheese packaged with CO_2 and stored at 4°C up to 63 days and increased from 10^4 to 10^5 cfu/g in products packaged with CO_2 at 7°C. Data suggest that addition of CO_2 to cottage cheese to extend shelf life does not represent an increased *Listeria* or botulism hazard.

B. Fresh Produce

Fruits and vegetables are considered as high-acid and low-acid foods, respectively, with the exception of tomatoes, melons, and avocados (low in acid). Physiological and microbiological spoilage are rated as the number 1 and 2 factors, respectively, governing the shelf life of fresh fruits and vegetables. The former factor is enzyme-driven and can be retarded by temperature reduction. Not only does this retard physiological spoilage, but it also reduces the respiration rate and microbiological growth. For the last 40 years, considerable interest has been shown in MAP of minimally processed and/or fresh fruits and vegetables, and extensive literature is available (17). The application of gas mixtures, proper barrier properties of packaging films, and storage/distribution temperature play a major role in the successful implementation of a MAP system. It is to be noted that fresh produce may enter into anaerobic respiration if the O_2 concentration falls to near zero. This results in the production of undesirable compounds (alcohols, aldehydes, and ketones). To prevent this, packaging systems such as high–gas-permeability plastic films, plastic films disrupted with mineral fill, and films fabricated from polymers with temperature-sensitive side chains have been experimented with or used commercially. The following section discusses such successful MAP systems used for fresh produce.

Jacxsens et al. (23) conducted storage experiments to determine the behavior of bacterial pathogens on fresh-cut vegetables (trimmed Brussels sprouts, grated carrots, shredded iceberg lettuce, and shredded chicory) packaged under a modified atmosphere of 2–3% O_2, 2–3% CO_2, and 94–96% N_2 and stored at 7°C. Microbiological and sensory (appearance, taste, and odor) qualities of the samples were monitored to determine the shelf life, which was found to be 50% longer than that of control products. In a subsequent experiment, the fresh-cut vegetables were inoculated with a cocktail of psychrotrophic pathogens (*L. monocytogenes*, *Aeromonas caviae*, and *A. bestiarum*) before packaging under MA or air, and the samples were stored at 7°C. Pathogen growth was influenced more by vegetable type than by atmosphere. No growth was detected on Brussels sprouts or carrots. Both *Aeromonas* species exhibited higher growth rates than *L. monocytogenes* on shredded chicory endives and iceberg lettuce at 7°C.

Floros et al. (24) determined the combined effect of MAP and the application of a bacterial antagonist (*Erwinia* spp.) on *Botrytis cinerea* growth on apples (Golden Delicious). Inoculated apples were stored in polyethylene bags at 5°C. Based on a central composite experimental design involving five levels of O_2 (1–15%) and CO_2, the initial gas composition in each bag was set. In the absence of antagonist, measurement of mold colony diameter over time showed that O_2 had no effect on the growth of *B. cinerea*, while increased CO_2 levels delayed its growth by 4 days. Application of an antagonist resulted in a significant interaction between O_2 and CO_2. At low O_2 levels, CO_2 had no effect on mold growth, but at high O_2, CO_2 enhanced mold growth. O_2 and the antagonist worked synergistically to reduce mold growth by 6 days at low levels of CO_2, whereas at high CO_2 levels O_2 had no effect. The strongest antagonist effect was observed under ambient conditions. Generally, high CO_2 atmospheres slowed the growth of *B.*

cinerea, and *Erwinia* spp. was an effective antagonist against *B. cinerea* growth on apples, particularly under ambient conditions.

Austin et al. (25) determined the safety of fresh-cut vegetables packaged in modified atmospheres by performing challenge studies using both nonproteolytic and proteolytic strains of *C. botulinum* with a variety of fresh-cut packaged salads and vegetables stored at different temperatures. When vegetables were inoculated with spores of *C. botulinum* and incubated in low-O_2 atmospheres, spore germination and growth and toxin production were observed. Botulinum toxin was produced by proteolytic types A and B on onion, butternut squash, rutabaga squash, and salad. Nonproteolytic *C. botulinum* was capable of producing neurotoxin at temperatures as low as 5°C, whereas proteolytic strains produced neurotoxin at ≥15°C. Although most samples were visibly spoiled before detection of botulinum toxin, samples of butternut squash and onion remained acceptable after detection of toxin. The strict maintenance of low temperature (<5°C) was strongly recommended in order to control the potential growth of *C. botulinum* on fresh-cut vegetables packaged in a modified atmosphere.

Izumi et al. (26) determined the physiology and quality of carrot slices, sticks, and shreds stored in air or a controlled atmosphere (CA) of 0.5% O_2 and 10% CO_2 at 0, 5, and 10°C. The respiration of all three types of cut tissue was reduced when stored in CA, and the reduction was greater with slices or sticks than with shreds. The respiratory quotient of sticks and shreds was higher in CA than in air at all temperatures. Ethylene production was $<0.1\ \mu/kg^{-1}h^{-1}$, and off-odor was not detected with any of the samples. Controlled atmosphere was found to be beneficial in reducing decay, weight loss, pH of sticks and shreds, white discoloration on shreds, and microbial growth on sticks.

Willocx et al. (27) modeled the growth of *Pseudomonas fluorescens* (an indicator microorganism for spoilage of ready-to-eat Grade IV leafy vegetables) as a function of temperature and CO_2. The bacterial growth curve was determined with the Logistic and Gompertz equation. To estimate direct parameter, these sigmoidal functions were written in terms of lag time and maximum absolute growth rate of *P. fluorescens* under static storage conditions. In a temperature range of 4–12°C, the lag time of *P. fluorescens* showed an exponential dependence with temperature, whereas a linear relationship of the maximum absolute growth rate with temperature was observed. CO_2 was found to be effective against the proliferation of aerobic psychrotrophic gram-negative bacteria by mainly increasing the lag phase and, to a lesser extent, the generation time. The dependence of the lag time and the maximum absolute growth rate of *P. fluorescens* with CO_2 concentration (0.03–15% with an excess of O_2) can be described using an exponential function. The effectiveness of the inhibition increased with decreasing temperatures.

Osuna et al. (28) determined the effects of storage temperature and time on the microflora of fresh, vacuum-packaged, and modified atmosphere packaged asparagus. Total mesophilic, psychrophilic, lactic acid bacteria, Enterobacteriaceae, and yeast and molds were determined as a function of time. The effect of washing on the original microflora of the asparagus and the growth of the different microbial groups during the storage time were checked. From day 8, a progressive increase in the mesophile and psychrophile counts in both types of packaging was observed. After 21 days of storage, the mesophile and psychrophile counts were 10^7 cfu/g for both types of packaging. In the vacuum packaging, the final enterobactericeae counts (2.5×10^2 cfu/g) and yeast and molds (10 cfu/g) were lower than in the polyethylene bag–packaged asparagus, which showed enterobactericeae counts of 7.3×10^4 cfu/g and yeast and mold counts of 2.3×10^4 cfu/g. The nonpackaged asparagus showed low counts from the fourth day due to heavy dehydration caused by the low relative humidity in the storage chamber.

Abdul-Raouf et al. (29) determined the influence of MAP, storage temperature, and time on survival and growth of *Escherichia coli* O157:H7 inoculated into shredded lettuce, sliced cucumber, and shredded carrot. Additionally, growth of psychrotrophic and mesophillic microorganisms and changes in pH and sensory qualities of vegetables (subjective evaluation) were also monitored. Packaging under an atmosphere containing 3% O_2 and 97% N_2 had no apparent effect on populations of *E. coli*, psychrotrophs, or mesophiles. Populations of viable *E. coli* declined on vegetables stored at 5°C and increased on vegetables stored at 12 and 21°C for up to 14 days. The most rapid increases in populations of *E. coli* occurred on lettuce and cucumbers stored at 21°C. These results suggest that an unknown factor associated with carrots may inhibit the growth of *E. coli* O157:H7. The reduction in pH of vegetables can be correlated with increases in populations of *E. coli* O157:H7 and naturally occurring microfloras. Eventual decreases in *E. coli* were reported in some samples, e.g., those stored at 21°C. This could be attributed to the toxic effect of accumulated acids. Changes in visual appearance of vegetables were not influenced substantially by growth of *E. coli* O157:H7.

Omary et al. (30) determined the effects of packaging on growth of *L. innocua* in shredded cabbage. Freshly shredded white cabbage was treated with citric acid and sodium erythorbate, inoculated with *L. innocua*, and packaged in 230 g lots in four types of retail bags with OTRs of 5.6, 1500, 4000, and 6000 cc/(m^2 per 24 h). The packages were stored at 11°C for 21 days. *Listeria innocua* decreased in cabbage stored in three films. After 21 days, *L. innocua* population increased in all packages, but the increase was significant ($p < 0.05$) for cabbage packaged in film with the highest OTR.

Bennik et al. (31) employed the concept of biopreservation in MAP by evaluating the performance of two bacteriocinogenic strains of *Pediococcus parvulus* and one bacteriocinogenic *Enterococcus mundtii* strain to control growth of *L. monocytogenes* on refrigerated, modified atmosphere–stored mung bean sprouts. Three strains, which were isolated from minimally processed vegetables, grew in culture broth at 4, 8, 15, and 30°C. However, only *E. mundtii* was capable of bacteriocin production at 4–8°C. Examination of the growth of these strains on agar under 1.5% O_2 in combination with 0, 5, 20, or 50% CO_2 revealed significantly higher maximum specific growth rates for *E. mundtii* than for *P. parvulus* at CO_2 concentration <20%, which are relevant for modified atmosphere storage of vegetables. *Enterococcus mundtii* was subsequently evaluated for its ability to control growth of *L. monocytogenes* on vegetable agar and fresh mung bean sprouts under 1.5% O_2/20% CO_2/78.5% N_2 at 8°C. Growth of *L. monocytogenes* was inhibited by *E. mundtii* on sterile vegetable medium but not on fresh produce. However, mundticin (the bacteriocin produced by *E. mundtii*) was found to have potential as a biopreservative agent for modified atmosphere–stored mung beans when used in a washing step or a coating procedure.

Amanatidou et al. (32) studied the impact of a high O_2 MAP on the spoilage microorganisms of minimally processed vegetables. Pure cultures of *P. fluorescens*, *Enterobacter agglomerans*, *Aureobacterium* strain 27, *Candida guilliermondii*, *C. sake*, *Salmonella* Typhimurium, *S.* Enteridis, *Escherichia coli*, *Listeria monocytogenes*, *Leuconostoc mesenteroides*, *Lactobacillus plantarum*, and *Lactococcus lactis* were cultured on an agar-surface model system and incubated at 8°C under a modified atmosphere of 80% or 90% O_2 and balance N_2, 10% or 20% CO_2 and balance N_2, or a combination of both gases. In general, exposure to high O_2 alone did not inhibit the microbial growth strongly, while CO_2 alone reduced growth to some extent in most cases. A strong inhibition was obtained when both gases were used in combination. Therefore, it was concluded that with minimally processed vegetables, where CO_2 levels of >20% cannot be used due to physiological damage to the produce, the combined treatment of high O_2 and 10–20% CO_2 may provide adequate control of microbial growth, thus allowing a safe, prolonged shelf life.

Bennik et al. (33) examined the influence of O_2 and CO_2 on the growth of prevalent Enterobacteriaceae and *Pseudomonas* isolated from fresh and controlled atmosphere–stored vegetables. The prevalent bacterial populations present on minimally processed chicory endive and mung bean sprouts were analyzed during storage at 8°C under controlled atmosphere conditions (1.5% O_2, 78.5% N_2) for 13 days. Enterobacteriaceae and *Pseudomonas* predominated on both vegetables before and after storage. Several *Pseudomonas* and *Rahnella aquatilis* were present on fresh chicory, but *P. fluorescens* and *Escherichia vulneris* predominated after storage. An agar model system was used to assess growth of predominant isolates at 8°C under atmospheres of 1.5 or 20% O_2 with 0, 5, 20, or 50% CO_2. Results showed that there was no significant difference in growth under 1.5 or 20% O_2. Lag times were not observed for any of the strains. Maximum specific growth rates of Enterobacteriaceae and *Pseudomonas* declined as CO_2 concentration increased. It was concluded that during storage under MAP a selective suppression of growth of epiphytes may occur, some of which may have retarded growth of pathogens, and thus may affect safety of minimally processed vegetables.

Floros et al. (24) examined the combined effect of MAP and application of bacterial antagonist (*Erwinia*) on *Botrytis cinerea* growth on apples. Inoculated samples were stored in polyethylene bags at 5°C. Five levels of O_2 (1–15%) and CO_2 (0–15%) were used as initial gas composition. Without the antagonist, measurements of fungal colony diameter over time showed that O_2 had no effect on microbial growth, while increased CO_2 levels delayed its growth by 4 days. Application of antagonist resulted in a significant interaction between O_2 and CO_2. At low O_2 levels CO_2 had no effect on fungal growth, but at high O_2, CO_2 enhanced fungal growth. Oxygen and the antagonist worked synergistically to inhibit the fungal growth by about 6 days at low CO_2 levels. However at high CO_2 levels, O_2 had no effect. The strongest antagonist effect was observed under ambient conditions. The results showed that a high-CO_2 atmosphere can slow the growth of *B. cinerea* and that *Erwinia* was an effective antagonist against *B. cinerea* growth on apples, particularly under ambient conditions. Chong and Cash (35) studied the effect of varying modified atmospheres containing 75% CO_2 with 25, 10, 5, or 0% O_2 (balance N_2) and temperatures of 5, 12.5, and 25°C on the growth of *Pseudomonas fragi*, *S.* Typhimurium, and *Clostridium sporogenes* on a model minimally processed food (cooked ground beef and peas) with an internal temperature of 55°C. An atmosphere containing 75% CO_2 and 0% O_2 was the most effective in inhibiting the growth of *P. fragi*, and *S.* Typhimurium and an atmosphere containing 75% CO_2 and 25% O_2 was most effective against *C. sporogenes*.

Solomon et al. (36) investigated the ability of *C. botulinum* type A and B spores to grow and produce toxin on fresh raw potatoes in a modified atmosphere with or without sulfite at 22°C. Fresh peeled and sliced potatoes (untreated or dipped for 2 min into 0.7% sulfite solution and drained) were surface inoculated at several levels with a mixture of *C. botulinum* spores, either type A or B. They were placed in a modified atmosphere (30% N_2, 70% CO_2) in O_2-impermeable bags (200 g/bag) and incubated at room temperature. Toxicity was tested on days 0, 3, 4, 5, 6, and 7. After incubation, the potatoes were blended and centrifuged, and the Millipore-filtered supernatant fluid was injected intraperitoneally into mice. Sensory analysis (except taste) was also performed. Potatoes inoculated with *C. botulinum* type A spores but not treated with sulfite became toxic in 4–5 days, which coincided with the sensory evaluation "unfit for human consumption." Potatoes treated with sulfite regardless of inoculum size or residual sulfite level appeared acceptable for human consumption through day 7, even though they were toxic after 4 days of incubation. Although toxicity from type B spores occurred later and in fewer test samples than toxicity from type A, some potatoes again appeared acceptable but were toxic. Thus, although sulfite remarkably extended the consumer acceptability of peeled, sliced, raw potatoes at the abuse temperature, it did not inhibit the outgrowth and toxin production of *C. botulinum* under these conditions.

Wei et al. (37) studied the effects of MAP (air or vacuum), NaCl, and storage temperature (5, 10, or 15°C) on survival and growth of *L. monocytogenes* in minimally processed green beans. Growth of *L. monocytogenes* Scott A (inoculation levels 10^2–10^3 cfu/g) was increased by increasing storage temperature (10^5 cfu/g at 5 and 10°C after 9 days storage; 10^6 cfu/g at 15°C after 8 days storage). Growth of aerobes, psychrotrophs, lactics, coliforms, yeasts, and fungi was significantly affected by temperature but unaffected by NaCl or atmosphere. *L. monocytogenes* were not detected in uninoculated samples. Beans remained bright green at 5°C under all conditions, whereas at 15°C beans turned yellow after 4 days. Generally, unsalted beans had better appearance than salted beans. It is suggested that additional barriers are required to prevent growth of *L. monocytogenes* in minimally processed beans. Diaz and Hotchkiss (38) compared the organoleptic and microbiological spoilage with the survival of *E. coli* O157:H7 in modified atmosphere–stored shredded iceberg lettuce. Samples were inoculated with a nalidixic acid–resistant *E. coli* O157:H7 (ATCC 35150) and placed in chamber, which was continuously flushed with gas mixtures of 0/10/90, 3/0/97, 5/30/65, 20/0/80 ($O_2/CO_2/N_2$) and held at 13 or 22°C. APC growth was inhibited in 5/30/65 ($O_2/CO_2/N_2$) at 13°C compared to all other atmospheres, which were not significantly different from each other. Growth rates for both *E. coli* O157:H7 and APC were greatest in air at 22°C. Carbon dioxide concentration had no significant effect on growth of *E. coli* O157:H7 at either temperature. Shelf life of shredded lettuce, as judged by appearance, was extended in atmospheres containing 30% CO_2 by approximately 300% compared to air. However, the extended shelf life provided by the modified atmosphere allowed *E. coli* O157:H7 to grow to higher numbers compared to air-held shredded lettuce.

Cartaxo et al. (39) investigated the effect of controlled-atmosphere storage on the microbial growth on fresh-cut watermelon. Seedless watermelons were cut in 2.5 cm cubes and stored at 3°C for 15 days under five different atmospheres. The concentrations of 3% O_2 + 15% CO_2 and 3% O_2 + 20% CO_2 inhibited bacterial development during the entire storage but had negative effects on the visual quality of the cubes. It is to be noted that for fresh-cut fruits, the requirement for polymer package is to have high O_2 permeability and low CO_2/O_2 permselectivity—because of the high respiratory rates of fresh cut fruits—whereas polymeric materials available today have the reverse properties. Thus, in essence MAP is not the best technique to preserve fresh-cut fruit. It should be coupled with alternate food preservation technologies such as nonthermal food-processing techniques to provide extended shelf life of fresh cut fruits. Beuchat and Brackett (40) examined the effects of shredding, chlorine treatment, and MAP on survival and growth of *L. monocytogenes*, mesophilic aerobes, psychrotrophs, and yeasts and molds on lettuce stored at 5 and 10°C. With the exception of shredded lettuce, which had not been chlorine treated, no significant changes in populations of *L. monocytogenes* were detected during the first 8 days of storage at 5°C; significant increases occurred between 8 and 15 days. Significant increases occurred within 3 days when lettuce was stored at 10°C; after 10 days, populations reached 10^8–10^9 cfu/g. Chlorine treatment, MAP (3% O_2, 97% N_2), and shredding did not influence growth of *L. monocytogenes*. It was concluded that *L. monocytogenes* is capable of growing on lettuce subjected to commonly used packaging and distribution procedures used in the food industry.

Berrang et al. (41) determined the effects of controlled-atmosphere storage (CAS) on survival and growth of *L. monocytogenes* on fresh asparagus, broccoli, and cauliflower. Vegetables were inoculated with *L. monocytogenes* strain Scott A or LCDC 81-861 (10^3–10^5 cfu/g) and stored at 4 and 15°C under CAS and air. *Listeria monocytogenes* populations were monitored over 21-day (4°C) and 10-day (15°C) periods using selective recovery media and a direct plating technique. CAS lengthened the time that all vegetables were considered acceptable for consumption by subjective inspection. Populations of *L. monocytogenes* increased during storage, but CAS did not influence the rate of growth. In another study, Berrang et al. (42) investigated the effects of CAS on the survival and growth of *Aeromonas hydrophila* on fresh asparagus, broccoli, and cauliflower. Two lots of each vegetable were inoculated with *A. hydrophila* 1653 or K144. A third lot served as an uninoculated control. Following inoculation, vegetables were stored at 4 or 15°C under a CAS system previously shown to extend the shelf life of each commodity or under ambient air. Populations of *A. hydrophila* were enumerated by direct plating with selective media on the initial day of inoculation and at various intervals for 10 days (15°C) or 21 days (4°C) of storage. The organism was detected on most lots of vegetables as they were received from a commercial produce supplier. Without exception, the CAS system lengthened the time vegetables were subjectively considered acceptable for consumption. However, CAS did not significantly affect populations of *A. hydrophila* that survived or grew on inoculated vegetables.

C. Fish Products

Among all foods, fish is the most difficult to preserve in its fresh state, because of large microbiological populations present in it [most of these organisms are psychrophilic (capable of growth at refrigerated temperatures)]. Also, fish is considered to be a potential carrier of *Clostridium botulinum* type E, a non-proteolytic, psychrotrophic anaerobic pathogen. Although MAP can be applied to fish, increased chance of growth of *C. botulinum* type E cannot be ignored under reduced O_2 conditions. Nevertheless, concerted work is being done on the design of MAP systems for fish. Kimura et al. (43) investigated the growth of *C. per-*

fringens inoculated into fish fillets of jack mackerel (*Trachurus japonicus*) and packaged under a controlled CO_2 atmosphere (40% CO_2, 60% N_2) at marginal growth (15°C) and abuse temperatures (30°C). No increase in *C. perfringens* population was obtained under either atmosphere after 15°C and 3 days. *Clostridium perfringens* rapidly increased at 30°C after 2- to 4-hour lag phase regardless of atmosphere; however, growth was significantly enhanced under CO_2 within 6 hours of storage. Holding inoculated fish fillets at 5°C and 24 hours prior to 6-hour storage at 30°C prevented *C. perfringens* growth. It is suggested from the results that controlling distribution temperature to $<$5°C would reduce the health risks associated with *C. perfringens*, which may be higher in fish fillets packaged under a controlled CO_2 atmosphere.

Pastoriza et al. (44) studied the effect of MAP in a CO_2 atmosphere on microbiological quality and the biochemical and sensory properties of stored Atlantic salmon (*Salmo salar*) slices. The slices were packaged under CO_2 and stored at 2°C for 3 weeks. Results were compared with a control stored in air. The shelf life of MAP samples was nearly twice as long as that of samples stored in air; the MAP samples had lower bacterial counts, pH, and trimethylamine and total volatile base levels, as well as improved sensory properties, after 18 days of chilled storage. After 10 and 20 days of storage, exudate values for MAP samples increased by 1 and 2%, respectively. After 10 days of chilled storage, exudation was not significantly different between control and MAP samples. It was concluded that salmon slices stored under CO_2 were of a high quality after 18 days of chilled storage.

Wannapee et al. (45) studied microbial growth in catfish fillets under a master-pack system. Catfish fillets inoculated (10^5 cfu/g) with *E. coli* O157:H7 or *L. monocytogenes* Scott A were packaged in modified atmospheres (90% CO_2/2% O_2/8% N_2) and stored at 2°C for 20 days. During this period, sample packages were opened and placed at 8°C for analysis every 2 days for 6 days for *E. coli* and aerobic plate counts, or *L. monocytogenes* and psychrotrophic plate counts. Inoculated fillets showed constant counts of *L. monocytogenes* throughout 16 days at 2°C, but increased by 1 log after 20 days at 2°C and 6 days at 8°C. Psychrotrophic plate counts of inoculated fillets increased during 12 days at 2°C and kept increasing when placed at 8°C. *Escherichia coli*–inoculated samples showed decreasing counts during 20 days at 2°C, whereas counts remained steady on transfer to 8°C.

Oka et al. (46) studied growth of histamine-producing bacteria in fish fillets under modified atmospheres. Fillets of bigeye tuna (*Thunnus obesus*) were inoculated with histamine-producing bacteria (*Morganella morganii*, *Klebsiella pneumoniae*, and *Hafnia alvei*) and packaged in aluminum laminate pouches under air or a modified atmosphere (CO_2/N_2 60:40). Packaged samples were stored for 14 days at 4 or 10°C. Bacterial growth, pH, and histamine concentration were determined. No growth of *K. pneumoniae* was observed, irrespective of atmosphere or temperature. *Morganella morganii* count after 14 days was 4.4–4.6 × 10^3/g fish at 4°C and 1.1–1.2 × 10^8/g fish at 10°C, irrespective of storage atmosphere; a slight inhibitory action of the modified atmosphere was observed over the first 3 days of storage at 10°C. *Hafnia alvei* counts at the end of storage were 1.0 × 10^4–9.2 × 10^6/g fish at 4°C and 3.0–8.2 × 10^8/g fish at 10°C for both storage atmospheres; slight inhibition by the modified atmosphere was observed at both 4 and 10°C. Histamine content in samples stored for 14 days at 10°C was 550–650 mg/100 g fish in air-packaged samples and 300–560 mg/100 g fish in modified atmosphere-stored samples, irrespective of presence of the three histamine-forming bacteria. Values of pH increased during storage in all samples, the increase being greater at 10 than at 4°C, and slightly greater in air-packaged than in modified atmosphere–packaged samples. It was concluded that modified atmosphere packaging gave only slight inhibition of histamine-forming bacteria in bigeye tuna, and histamine formation in bigeye tuna fillets was not attributable to the three bacterial species studied.

Oka et al. (47) studied the growth pattern and enterotoxin production of food-poisoning bacteria in fish fillets stored under different modified atmospheres. Bigeye tuna fillets inoculated with toxigenic food poisoning bacteria (*E. coli*, *S. aureus*, and *C. perfringens*) were packed into retort pouches under air or a modified atmosphere (CO_2/N_2, 60:40). Growth pattern and enterotoxin production of the packaged samples were examined during storage for 5 days at 25°C. After 5 days, counts of *E. coli* and *S. aureus* in air-packed samples were 1.2 × 10^9/g and 1.4 × 10^7/g, respectively; in gas-packaged samples, counts were 1.6 × 10^8/g and 1.6 × 10^5/g, respectively. The modified atmosphere exerted a slight inhibitory action on these organisms. The *C. perfringens* count was 1.7 × 10^7/g in air-packaged and 3.7 × 10^7/g in gas-packaged samples; the modified atmosphere did not inhibit this species. All nine strains each of *E. coli*, *S. aureus*, and *C. perfringens* isolated from the fish were toxigenic, forming enterotoxin concentrations of 5, 1000, and 10,000 ng/mL, respectively. No inhibitory action modified atmosphere on enterotoxin formation was observed.

Suzuki et al. (48) studied growth and polyamine production of *Aeromonas* spp. in fish meat extracts under MAP. *Aeromonas* spp. are of importance in the spoilage of fish and fish products. Studies were conducted on growth and polyamine formation by a halophilic strain (S5B) and a nonhalophilic strain (S29) of *Aeromonas* in walleye pollock (*Theragra chalcogramma*) and short-finned squid (*Illex argentinus*) extracts. The inoculated fish or squid meat extracts were incubated at 25°C under N_2 or CO_2 for up to 6 days. Growth of strain S5B was inhibited by CO_2; growth of strain S29 was not. Results for formation of polyamines (putrescine, cadaverine, agmatine, histamine, spermine) were complex; it was concluded that polyamine formation was influenced by strain, atmosphere, and culture medium.

IV. MICROBIAL GROWTH CONTROL BY OXYGEN-SCAVENGING PACKAGE TECHNOLOGY

Presence of O_2 is a prerequisite for food deterioration, both from sensory and safety perspectives, caused by oxidation of food constituents or growth of aerobic bacteria and molds. Even if anaerobic pathogens are present, their rate of growth is low under refrigeration compared to aerobic pathogens. Modified atmosphere packaging can provide reduced O_2 conditions in the package, but it is not always the best solution because some residual O_2 always remains in the package and, hence, may affect ultra–O_2-sensitive food products such as red meat (beef or pork). Moreover, permeation of O_2 through the package material is inevitable, and cannot be prevented by MAP. Thus, application of O_2 scavengers for food spoilage and food safety is necessary. Generally speaking, existing O_2-scavenging technologies utilize the following concepts: iron powder oxidation, ascorbic acid oxidation, photosensitive dye oxidation, enzymatic oxidation (e.g., glucose oxidase and alcohol oxidase), unsaturated fatty acids (e.g., oleic acid or linolenic acid), and immobilized yeast on a solid material (49). The rates of O_2 absorption vary from one scavenger to another. Tewari et al. (50) determined O_2-absorption kinetics of several commercially available O_2 scavengers. Table 1 summarizes the mechanisms and the applications of commercially available scavengers. The majority of O_2 scavengers are based on the principle of iron oxidation:

$$Fe \rightarrow Fe^{2+} + 2e^-$$

$$1/2\ O_2 + H_2O + 2e^- \rightarrow 2OH^-$$

$$Fe^{2+} + 2OH^- \rightarrow Fe(OH)_2$$

$$Fe(OH)_2 + 1/4\ O_2 + 1/2\ H_2O \rightarrow Fe(OH)_3$$

Some scavengers use an enzyme reactor surface that would react with some substrate to absorb incoming O_2. Glucose oxidase, a promising O_2-scavenging agent, is an oxidoreductase that transfers two H_2O from the -CHOH group of glucose to O_2 with the formation of glucono-delta-lactone and H_2O_2. The lactone then reacts with water to form gluconic acid (51):

$$2G + 2O_2 + 2H_2O \rightarrow 2GO + 2H_2O_2,$$

where G is the substrate. Hydrogen peroxide (H_2O_2) is not acceptable as an end product; therefore catalase is introduced to break down the peroxide:

$$2H_2O_2 + \text{catalase} \rightarrow 2H_2O + O_2$$

Ascorbic acid is another O_2-scavenging component that can be applied. The basic reaction of Darex O_2-scavenging technology, designed to be incorporated into barrier packaging such as crown caps, plastic or metal closures, is ascorbate oxidizing to dehydroascorbic acid and sulfite to sulfate (52). Another

Table 1 Description of Commercially Available O_2 Scavengers

Name	Scavenging reaction	Capacity (mL O_2)	Features and suggested applications	Atmosphere (CO_2/Air/N_2)	Company
Bioka® S-100[a]	Enzyme-mediated oxidation	100	[b]	Air/N_2	Bioka® Limited, Finland
Bioka® S-75[a]	Enzyme-mediated oxidation	75	[b]	Air/N_2	Bioka® Limited, Finland
Ageless® FX-100[a]	Iron oxidation	100	For moist food; stable in air before use; excellent water resistance	Air/N_2	Mitsubishi Gas Chemical Company, Tokyo, Japan
FreshPax® M-100[a]	Iron oxidation	100	For refrigerated applications; designed for modified atmosphere packaging of moist product	CO_2	Multisorb Technologies Inc., Buffalo, NY
FreshPax® R-300[a]	Iron oxidation	300	For refrigerated and frozen application; recommended for cases where rapid removal of O_2 is required	Air/N_2	Multisorb Technologies Inc., Buffalo, NY
FreshPax® R-2000[a]	Iron oxidation	2000	For refrigerated and frozen application; recommended for cases where rapid removal of O_2 is required	Air/N_2	Multisorb Technologies Inc., Buffalo, NY
ATCO®	Iron oxidation	[c]	[b]	Air/N_2	Standa Industries, France
Freshilizer® series	Iron oxidation	[c]	[b]	Air/N_2	Toppan Pinting Company, Japan
Vitalon®	Iron oxidation	[c]	[b]	Air/N_2	Toagosei Chemical Industry Company, Japan
Sanso-cut®	Iron oxidation	[c]	[b]	Air/N_2	Finetec Company, Japan

[a] All of the scavengers are designed for products with a water activity ≥ 0.85.

[b] These scavengers are found in packages of many foods such as fresh and precooked pasta, catering, meat products (e.g., smoked ham and salami), bakery products (e.g., bread, pizza crust, pastries, cookies, cakes), cheese, coffee, nuts, and potato chips.

[c] Available in various sizes.

commercial O_2 scavenger utilizing the same concept is Oxysorb, which uses a transition metal, preferably copper, to catalyze the oxidation reaction. This scavenging material can be either included inside a pouch or incorporated into the packaging material (53).

A small coil of an ethyl cellulose film containing a dissolved photosensitive dye and a singlet O_2 acceptor in the headspace of a transparent package can also scavenge O_2. After illumination of the film with light to the appropriate wavelength, excited dye molecules sensitize O_2 molecules, which have diffused into the polymer, to the singlet state. These singlet O_2 molecules react with acceptor molecules and are thereby consumed. The photochemical reaction is (54):

$$\text{photon} + \text{dye} \rightarrow \text{dye}^*$$

$$\text{dye}^* + O_2 \rightarrow \text{dye} + O_2^*$$

$$O_2^* + \text{acceptor} \rightarrow \text{acceptor oxide}$$

$$O_2^* \rightarrow O_2.$$

An alternative to having O_2 scavengers in sachets is the incorporation of the O_2 scavenger into the packaging structure itself. Low molecular weight ingredients may be dissolved or dispersed in a plastic, or the plastic may be made from a polymeric scavenger (51). One such example is Oxyguard (Toyo Seikan Kaisha, Japan), an iron-based absorber that can be incorporated into a laminate. The main alternative to dispersal of iron into plastics is organic reactions of plastics themselves. Oxbar™ is a system developed by Carnaud-Metal Box (UK), which involves cobalt-catalyzed oxidation of a nylon polymer blended especially in polyethylene terepthalate (PET) bottles for plastic packaging of wine, beer, sauces, and other beverages (51). Amoco Chemicals (USA) produces Amosorb®, a polymer-based absorber, which can be incorporated in various packaging structures (sidewall or lid of rigid containers, flexible films, and closure liners) (51). The rate and capacity of O_2-scavenging films are considerably lower than those of iron-based O_2 scavenger sachets. Recent developments in O_2 scavenger technology include inserts in the form of flat packets, cards or sheets, and O_2-scavenging adhesive labels like Freshmax® (Multisorb Technologies Inc., Buffalo, NY) and the ATCO® labels (Standa Industrie, France) (51).

A. Applications of O_2-Scavenging Package Technology

Oxygen scavengers have been used for food preservation for the last 20 years. The purpose of O_2 scavengers is to create low-O_2 atmosphere in sealed packs of products, thereby slowing or preventing deterioration due to oxidation of product components and/or growth of aerobic microorganisms or survival of insects. Oxygen scavengers are extensively used in Japan to prevent discoloration of cured meats and teas, rancidity problems in high-fat foods, and mold spoilage of intermediate- and high-moisture bakery products; in the United States they are used to delay oxidative flavor changes in coffee and to prevent mold growth, rancidity, and staling in bakery products. Their use for the preservation of color and flavor and the prevention of microbial spoilage in cooked, cured meats has also been reported. The current uses of O_2 scavengers generally involve packs in which the atmosphere contains some substantial fraction of O_2, if not air, at the time of pack sealing and the inhibition of chemical reactions or proliferation of microorganisms, which proceed relatively slowly. Recent application of O_2-absorbent technology has been shown by Tewari (17), who employed this concept in extending the shelf life of retail-ready beef cuts. The author utilized proper package design and barrier properties and optimized O_2-absorbent capacity to achieve extended shelf life. Ageless® FX-100 (Mitsubishi Gas Chemical Co. Inc., Tokyo, Japan) and FreshPax® R-2000 (Multisorb Technologies Inc., Buffalo, NY) O_2 scavengers were used in the study. Application of these scavengers along with the proper package also prevented transient discoloration of centrally prepared beef steaks (55). These O_2 scavengers are based on iron chemical systems and need moisture (>70% relative humidity) for activation. Detailed information about these scavengers has been reported elsewhere (see Table 1). A detailed methodology for shelf life extension of centrally prepared retail-ready beef steaks is given in the following paragraph.

Fresh beef tenderloins (*psoas major*, PM) from animals slaughtered 24 hours previously were obtained from a local beef abattoir. Eighty steaks of 2 cm thickness were prepared from these tenderloins. Each steak was placed on a 152 × 114 mm absorbent pad (MP-30620, Paper Pak® Corp., La Verne, CA) in a 216 × 133 × 25 mm (L × W × H) solid polystyrene tray (clear plastic tray #2D, Western Paper & Food Distributors Ltd., Calgary, AB), with eight Ageless® FX-100 O_2 scavengers placed underneath the absorbent pad. Each retail tray was overwrapped with a shrinkable O_2-permeable film (Vitafilm 'Choice Wrap,' Goodyear Canada Ltd., Calgary, AB) with an O_2 transmission rate of 8000 mL/(m^2 24 h) at 1 atm, 23°C, and 70% RH. After sealing, the film was shrunk to the tray using a hot air gun. Then two 3 mm holes were made at the corners of the tray to allow free exchange of atmospheres during gas flushing. Four such retail trays were placed in an EVA co-extruded master pack (WINPAK ESO-PAEV2E 121575 R, Winnipeg, MB) with O_2 transmission rate of 0.55 mL/(m^2 24 h) at 1 atm, 23°C, and 70% RH. The bags were evacuated, filled with 4.5 L of N_2, and sealed using a CAP machine (CAPTRON®, SecureFresh Pacific Ltd., Auckland, New Zealand). The appropriate usage of proper barrier properties of the package in this study should be noticed.

In this study (17), despite significant ($p < 0.05$) differences between storage intervals on microbial numbers on day 0 of retail display for each storage interval, steaks had $<10^2$ cfu/cm^2 of total aerobic microbial population (Fig. 1). No differences of practical importance existed because such numbers do not

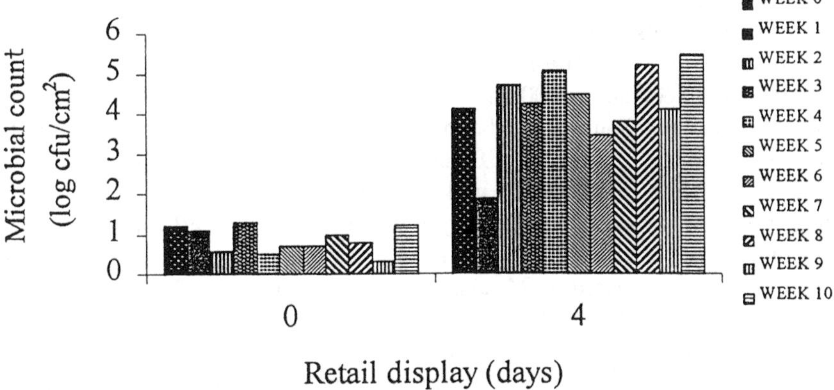

Figure 1 Mean microbial count for steaks stored under MAP along with O_2 scavengers for 10 weeks at −1.5°C.

normally cause off-odors. Microbial populations at day 0 of retail display were comparable with those of nonstored controls. On day 4 of retail display, microbial populations were $\geq 10^4$ cfu/cm^2 in most cases. Because of strict temperature control, MAP, and application of O_2 scavengers, the system resulted in a 10-week storage life and a subsequent display life of 3 days for centrally prepared retail-ready beef steaks. This presents an example of microbial growth control by employing proper package with appropriate barrier properties coupled with the concept of O_2-absorbent technology.

Patel et al. (56) compared different oxygen scavengers for their ability to enhance resuscitation of heat-injured *L. monocytogenes*. The recovery of heat-injured *L. monocytogenes* Scott A in Fraser broth (FB) supplemented with sodium thioglycolate, sodium pyruvate, L-(+)-cysteine hydrochloride, catalase or Oxyrase® was studied. After 3 hours of incubation at 30°C, recovery was enhanced by all oxygen scavengers except sodium pyruvate. Oxyrase® promoted the highest recovery (34%) compared to recovery in control broth (19%). All O_2 scavengers enhanced the recovery of injured *L. monocytogenes* in FB within 6 hours of incubation. After 6 hours at 30°C, 49 and 55% of injured cells underwent resuscitation in FB containing 2.5 mg of sodium pyruvate/mL and 400 μg of catalase/mL, respectively, compared to 24% resuscitation in FB not supplemented with oxygen scavengers. The percentage recovery was increased as the incubation time was extended to 6 and 24 hours. Nearly all injured cells were recovered within 24 hours of incubation, regardless of supplementation of FB with O_2 scavengers. Fraser broth containing 2.5 mg of sodium pyruvate/mL, 400 μg of catalase/mL, or 0.01 U of Oxyrase®/mL were tested to determine the optimal incubation time and temperature for recovering heat-injured *L. monocytogenes*. Percentage recovery of injured cells increased with an increase in temperature from 25 to 30°C and from 30 to 35°C. The highest percentage of injured cells recovered was observed in FB containing 400 μg of catalase/mL (67%) and 0.01 U of Oxyrase®/mL (68%) within 6 hours of incubation at 35°C. Catalase (400 micrograms ml^{-1}) and Oxyrase® (0.01 U/mL) in FB resulted in significantly higher recovery of injured cells from heated whole milk; however, recovery of injured cells from heated skim milk was not significantly higher. Enrichment in FB containing catalase or Oxyrase® has potential for recovering heat-injured *L. monocytogenes* cells within 6 hours compared to 24 hours required in conventional methods. Juven and Rosenthal (57) studied the effect of free-radical and O_2 scavengers on photochemically generated O_2 toxicity and on the aerotolerance of *Campylobacter jejuni*. Four strains of *C. jejuni* were spread onto Nutrient Broth No. 2 (Oxoid) agar plates, either unsupplemented or supplemented with various additives and incubated in atmospheres containing 5–21% O_2. Some plates had been previously exposed to fluorescent light and/or air. Catalase, sodium dithionite (5–10 mM), and histidine increased recoveries of *C. jejuni* on agar that had been exposed to light and air. A high inoculum of *C. jejuni* could not be recovered in an unsupplemented medium incubated in air, although 10–12% recoveries were obtained at 17 or 21% O_2 + 10% CO_2. Media supplemented with dithionite, catalase, or histidine showed some colony formation in air, but only dithionite consistently maintained viability of *C. jejuni* in air for >4 weeks. It is concluded that diverse oxygen species are involved in toxicity caused by high O_2 levels (17–21% O_2). These studies show that different O_2-scavenging systems can be incorporated into the package to inhibit the growth of undesirable microorganisms, but proper selection of such systems is required so as not to supplement growth of pathogenic microorganisms.

Oxygen scavengers have been proven to be effective in preventing growth of molds. Prevention of mold growth is important for dairy and baked products. Oxygen concentrations of 0.1% or lower are required for such prevention. Crustry rolls packaged with MAP (40% N_2/60% CO_2) and O_2 scavenger (Ageless®, Mitsubishi Gas Chemical Co., Tokyo, Japan) had the headspace O_2 concentration never increased beyond 0.05%, and the rolls remained mold-free even after 60 days. Similar mold-free shelf life was obtained in air- or N_2-packaged crusty rolls, both containing an O_2 scavenger (Ageless®, Mitsubishi Gas Chemical Co., Tokyo, Japan) (58). The mold-free shelf life of white bread packaged in polypropylene was extended from 5 to 45 days at room temperature by introducing an O_2 scavenger (Ageless®, Mitsubishi Gas Chemical Co., Tokyo, Japan). Pizza crust, which becomes moldy after 2–3 days at 30°C could be kept mold-free for more than 14 days using an appropriate O_2 scavenger (59). Powers and Berkowitz (60) determined the efficacy of an O_2 scavenger to modify the atmosphere and prevent mold growth on ready-to-eat pouched bread. An O_2-scavenging system enclosed in a pouch with ready-to-eat bread prevented growth of a mixed mold inoculum on the surface of the bread for 13 months by reducing O_2 tension inside the package. In the absence of the O_2 scavenger, growth of *Aspergillus* and *Penicillium* was visible on the bread within 14 days. Growth of molds on pouched culture media, with and without sorbate, was also prevented by O_2-scavenging sachets.

V. ANTIMICROBIAL PACKAGING

Certain antimicrobial substances can be incorporated in or coated onto food materials in order to control undesirable microorganisms on foods. The principal action of antimicrobial films is based on the release of antimicrobial components. Some of these components could pose a safety risk to consumers if their release is not tightly controlled by some mechanisms within the packaging material (51). The major potential applications for antimicrobial films include meat, fish, poultry, bread, cheese, fruits, and vegetables. Table 2 gives an overview of certain antimicrobial components that can be incorporated in the film; a few of these components are also available commercially. Most of these films have been tested for laboratory-scale efficacy; only a few are now available commercially due to regulatory concerns related to food safety.

Siragusa et al. (61) studied the incorporation of bacteriocin in plastic to retain activity and to inhibit surface growth of bacteria on meat. Sterilized beef tissue sections were inoculated with *B. thermosphacta* ATCC 11509 and variously packaged: prewrapped in nisin-impregnated plastics film; no prewrapping; and uninoculated beef with no prewrapping. These sections were then vacuum-packaged and stored at 4°C, or at 4°C for 2 days and then at 12°C. Tissue sections were analysed for *B. thermosphacta* numbers during storage. Nisin activity of impregnated and control plastic films was determined, and antimicrobial activity of nisin-impregnated plastics against *Lactobacillus helveticus* was evaluated. The effect of proteinase treatment on nisin activity was also examined. Nisin activity was retained on incorporation into plastics; conditions used to produce the film (e.g., heat, organic compounds) did not destroy antimicrobial activity. Even at 12°C, *B. thermosphacta* was inhibited by nisin-impregnated film.

Dong et al. (62) developed an antimicrobial packaging film for curled lettuce and soybean sprouts. Low-density polyethylene (LDPE) of 30 μm thickness, containing 0.1 or 1.0% grapefruit seed extract (GFSE), was fabricated by the blown-film extrusion process for developing films. The films were tested against several species of microorganisms on agar plate medium, after which they were used for packaging curled lettuce and soybean sprouts to investigate their suitability and effectiveness as antimicrobial packaging films. The films containing 1.0% GFSE showed inhibitory activity against *E. coli* and *S. aureus* in a disc test.

Table 2 Description of Antimicrobial Films for Food Packaging

Antimicrobial component	Packaging material	Mechanism of incorporation	Suggested applications	Company
Organic compound and respective acid anhydrides				
Potassium sorbate	LDPE	Wax layer, wet wax coating, or edible protein coating	Cheese	[a]
	LDPE	-do-	Culture media	[a]
	MC/Palmitic acid	-do-	Culture media	[a]
	MC/HPMC/Fatty acid	-do-	Culture media	[a]
	MC/Chitosan	-do-	Culture media	[a]
	Starch/Glycerol	-do-	Chicken breast	[a]
Calcium sorbate	CMC/Paper	-do-	Bread	[a]
Propionic acid	Chitosan	-do-	Water	[a]
Acetic acid	Chitosan	-do-	Water	[a]
Benzoic acid	PE-co-MA	-do-	Culture media	[a]
Sodium benzoate	MC/Chitosan	-do-	Culture media	[a]
Sorbic acid anhydride	PE	-do-	Culture media	[a]
Benzoic acid anhydride	PE	-do-	Fish fillet	[a]
Triclosan (chloro-organic compound)	Chitosan (Microban®)	-do-	Chopping boards, dish cloths	Microban Products Company
Fungicide/Bacteriocin				
Benomyl	Ionomer	-do-	Culture media	[a]
Imazalil	LDPE	-do-	Bell pepper	[a]
Nisin	Silicon coating SPI, corn zein films	-do-	Culture media	[a]

Peptide/Protein/Enzyme				
Lysozyme	PVOH, nylon, cellulose acetate, SPI film, corn zein films	-do-	Culture media	[a]
Glucose oxidase	Alginate	-do-	Fish	[a]
Alcohol/Thiol				
Ethanol	Silica gel sachet	In form of a sachet	Culture media	[a]
Ethanol	Silicon oxide sachet (Ethicap™): ethanol/water mixture adsorbed onto silicon dioxide powder	In form of a sachet (laminate of paper/ethyl vinyl acetate copolymer)	Bakery, apple	Freund Industrial Company (Japan)
	Negamold™: scavenges O_2 as well as generates ethanol vapors	In form of a sachet (laminate of paper/ethyl vinyl acetate copolymer)	Bakery	Freund Industrial Company (Japan)
	Oitech™	-do-	-do-	Nippon Kayaku, Japan
	Ageless® type SE	-do-	-do-	Mitsubishi Gas Chemical Company, Japan
	ET Pack	-do-	-do-	Ueno Seiyaku, Japan
Other				
Sulfur dioxide (SO_2)	Pads of sodium metabisulfite incorporated microporous material	-do-	Fruits	CSIRO (Australia)
Silver-zeolite	Plastics (LDPE)	Lamination as a thin layer (3–6 μm)	Inhibition of metabolic enzymes	[a]

Table 2 Continued

Antimicrobial component	Packaging material	Mechanism of incorporation	Suggested applications	Company
UV irradiation	Nylons	Immobilization on the surface of polymer films	Culture media	[a]
Allyl isothiocynate (essential oil of mustard or horse-radish)	Nylon	Immobilization of the surface of polymer films		[a]
Antimicrobial absorbent food pad (lower alkyls, lower alkenyls, or phenyl and substituted phenyl acid)	Absorbent pad	Dispersed within the absorbent medium of the pad	Meat and poultry	Hansen et al. (71)[b]
Antimicrobial paper (paper is slurried in water, to which a paper strengthening agent such as polyamidepolyamine-epichlorohydrin condensate, calcium carbonate or calcium phosphate treated with a silver, copper, or zinc salt such as calcium carbonate treated with copper sulfate, and zinc oxide whiskers is added	Paper	Mixture is converted to a sheet of paper	Liner for corrugated board	Yoshida et al. (72)[b]
Antimicrobial sheet (pulp is slurried in water, to which 2-(thiazolyl)benzimidazole or *N*-fluorodichloromethylthiophthalimide is added)	Sheet	Mixture is converted to a sheet of paper	Food-wrapper	Iwashiro (73)[b]

Amorphous tectosilicate (composed of silicon dioxide, aluminum oxide, sodium oxide) is slurried in water, to which silver nitrate and cetylpyridinium chloride are added	Paper	Antimicrobial composition is added to paper	Fresh food-wrapper	Suzuki et al. (74)[b]
Antimicrobial paper (paper is coated with a dispersed mixture of a silver, copper, lead, or tin salt of an *N*-(12-18-carbon)acyloyl-amino acid (0.1–1 wt%) such as silver *N*-stearoyl-L-glutamate and a binder such as PV alc. in water)	Paper	Coating	Food-wrapper	Kanie and Ichikawa (75)[b]
Antimicrobial paper (pulp is slurried in water to which radioactive mineral powder containing 0.05–2 wt% thorium oxide, silver-containing mineral powder (silver content, higher than 0.005 wt%), and/or zinc oxide powder singly or in combination are added)	Paper	Mixture is converted to a paper	Food-wrapper	Yokota (76)[b]

Table 2 Continued

Antimicrobial component	Packaging material	Mechanism of incorporation	Suggested applications	Company
Antimicrobial sheet (to a pulp slurry are added a metal salt of an oxyquinoline derivative (0.05–5 wt% of pulp) such as a copper salt of 8-oxyquinoline, a humic acid salt (0.01–1 wt% of pulp) such as a sodium or ammonium salt, and a water-soluble multivalent metal salt (1–10 wt% of pulp) such as aluminum sulfate)	Sheet	Mixture is converted to a sheet	Fruits- or seedlings-wrapper	Kato et al. (77)[b]
Antimicrobial paper (antimicrobial paper prepared by treatment with 0.005% chitosan ologosaccharide and 0.001% monolaurin is presented as speciality paper for food products to extend shelf life.)	Paper	Treatment of paper	Food-wrapper	Lee et al. (78)[b]
Hitachi Wrap®	Paper	Paper coating	Food-wrapper	Aoki (79)[b]
S-Clear KA®, a polystyrene blown film incorporating zirconium phosphate particles and silver ions to produce a bactericidal effect, Novalon, through an ion-exchange reaction	Film	Immobilized into film	Food-packaging film	Kobayashi (80)[b]

Daisangen® (made from styrol foam combined with ceramic based antimicrobial substance such as silica, alumina, and oxidized titanium)	Container	Incorporated into container	Germ protecter and temperature preserver	Toho Industry (Japan) (81)[b]
Germ Protect 365® (antimicrobial nonwoven using chitin)	Container	Incorporated into the container	Fresh food package	Shotec KK and Toyo Boseki, (Japan) (82)[b]
Take-guard® (drip-absorbing sheet containing antibiotic from a bamboo extract)	Sheet	Incorporated into the sheet	Microbial elimination on tables and kitchens (fast food industry)	Takecks Company (Japan) (83)[b]
Miracle water pack® (made from a transparent film comprising a five-layer coextrusion of nylon and polyethylene films; inner layer of film is antibiotic)	Pouch	Incorporated into the pouch	Drinking water preservation	Try Company and Taiyo Chemical Company (Japan) (84)[b]
Kapora® antimicrobial film	Film	Incorporated into the film	Inhibiting the growth of coliform bacteria and *Staphylococcus aureus*; packaging material for food and drugs	Kaito Chemical Engineering (Japan) (84)[b]
Wasa-power® (Japanese horseradish)	Sheet	PET film laminated underside	Food package	Tsuboi KK (Japan) (84)[b]
Hinokitiol® (Japanese cypress)	Film	Incorporated into the film	Preservation of freshness, mold growth; fresh food package	Nissey Kagaku Company (Japan) (84)[b]
Celagrase® (composed of silica, alumina, and titanium oxide)	Film	Incorporated into the film	Inhibition of bacterial growth; fresh food package	Tokuda (85)[b]

Table 2 Continued

Antimicrobial component	Packaging material	Mechanism of incorporation	Suggested applications	Company
Chef's Darling® (made from PET and rayon, with the PET fibers containing platinum catalyzed ceramic fibres. The ceramic gives off far-infrared radiation, which prevents cell rupture in fresh meat and fish products, preserving freshness and color. It also prevents drip loss and color deterioration and inhibits the growth of coliforms and other bacteria.)	Nonwoven fabric	Incorporated into the fabric	Fresh meat and fish products	Taiyo Kogyo (Japan) (86)[b]
BacteKiller® (composed of an alumina and silica based porous synthetic zeolite with silver and copper ions)	PET, PP and PS	Incorporated into the material	—	Kanebo Limited (Japan) (87)[b]
Bactenon 509® (manufactured by mixing 'Zeolite' inorganic based, antibacterial agent into special polyolefin resins)	Shrink-film	Incorporated into the film	Storage and distribution packaging of meat and other products	Toko Materials Industry Company Limited (Japan) (88)[b]

LDPE = Low-density polyethylene; MC = methyl cellulose; HPMC = hydroxypropyl MC; CMC = carboxyl MC; PE = polyethylene; PS = polystyrene; MA = methacrylic acid; SPI = soy protein isolate; PVOH = polyvinyl alcohol; HDPE = high-density PE.

[a] These antimicrobial films have been tested at a lab scale by several researchers and are not available commercially.

[b] Patented or commercial antimicrobial package.

Source: Partially adapted from Ref. 70.

Additionally, the growth rates of aerobic bacteria and yeast in curled lettuce packaged in the film containing 1.0% GFSE decreased, especially in cases where the bacteria count was $<10^6$/g. It was also found that within the period of shelf life for both curled lettuce and soybean sprouts, the growth rate of lactic acid bacteria could be decreased. Duck-Soon et al. (63) packaged fresh curled lettuce and cucumbers using LDPE films impregnated with antimicrobial agents. LDPE films (50 μm thickness) containing antimicrobial substances (*Rheum palmatum* extract, *Coptis chinensis* extract, sorbic acid, or Ag-substituted inorganic zirconium matrix, 1%) were manufactured. Physical properties and antimicrobial activities of the films were compared, and they were used in packaging curled lettuce and cucumbers. Films containing *R. palmatum* extract, *C. chinensis* extract, or Ag-substituted inorganic zirconium matrix showed no activity against *E. coli*, *S. aureus*, *Leuconostoc mesenteroides*, *Saccharomyces cerevisiae*, *Aspergillus niger*, *A. oryzae*, or *Penicillium chrysogenum*, while that prepared with sorbic acid inhibited *E. coli*, *S. aureus*, and *L. mesenteroides*. Addition of the antimicrobial agents changed the color and light transmittance properties of the films but did not alter mechanical tensile strength, heat shrinkage, or wettability. For curled lettuce and cucumbers packaged in the films and stored at 5 or 10°C, growth of total aerobic bacteria was reduced, but other quality attributes such as hardness and ascorbic acid content were not affected. Ling-Jane et al. (64) studied the application of antimicrobial polyethylene films and minimal microwave heating to control the microbial growth of tilapia fillets during cold storage. Benzoic anhydride–incorporated antimicrobial polyethylene (PE) films and minimal microwave heating were used to control the microbial growth of tilapia fillets during cold storage. The fillets wrapped with antimicrobial PE films (40 mg benzoic anhydride per g of PE in the initial preparation) alone had lower aerobic and anaerobic counts than the fillets wrapped with PE films without benzoic anhydride after 9 days of storage at 4°C. When the fillets were first microwaved for a short period of time (10, 20, or 30 s), PE films containing less benzoic anhydride (20 mg benzoic anhydride per g of PE in the initial preparation) also achieved the similar aerobic and anaerobic counts during a 14-day storage period at 4°C. Using benzoic anhydride–incorporated PE films, either alone or in combination with minimal microwave heating, was effective to control microbial growth of tilapia fillets during cold storage.

Cutter (65) studied the effectiveness of triclosan-incorporated plastic against bacteria on beef surfaces. Triclosan is a nonionic, broad-spectrum, antimicrobial agent with low toxicity that inhibits the growth of enteric bacteria, used in a range of personal hygiene products such as toothpastes and soaps. It is also added during the extrusion of plastic and fibers in the production of waste bags, toothbrushes, and surgical gauze. The benefits of using Triclosan-incorporated plastic (TIP) to inhibit pathogenic bacteria on meat surfaces was evaluated in plate overlay assays and in meat experiments using irradiated, vacuum-packaged beef and nonirradiated vacuum-packaged beef. The results indicated that fatty acids or components relating to adipose tissues could interfere with TIP's antimicrobial activity, and when plastic containing 1500 ppm Triclosan was combined with vacuum packaging and refrigerated storage, bacteria were not adequately inhibited on meat surfaces. Sherman (66) reported on the application of biocides in flexible foams prone to attack by bacteria and in polyvinyl chloride (PVC) products where plasticizers may be attacked by such organisms. The use of biocides to kill foodborne pathogens in products such as plastic cutting boards is now being driven by consumer hygiene expectations. Manufacturers' marketing has been scrutinized by the U.S. Environmental Protection Agency (EPA), and prosecutions for excessive claims have resulted. The author (65) reviewed individual additive developments covering Intercide ZnP from Akcros, Preventol A8 from Bayer, Metasol TK-100 from Calgon Corp, MicroFree from DuPont, Morton SB-30 and Morton 7040 from Morton International, Omacide from Olin Corp, and Troysan Polyphase from Troy Corp. It was concluded that it is only a matter of time before organic biocides will be used in food packaging. Microban® is already being used in the United Kingdom by the major supermarket chain Sainsbury, but in the United States the Food and Drug Administration is remaining cautious regarding approval of biocide additives for food contact purposes (51).

Traditional synthetic packaging provides a physical barrier against environmental contamination but does not have any intrinsic antimicrobial function, unlike edible films and coatings. It can, however, be modified to perform this function by incorporating metallic ions, which remain fixed in the material, for use with foods not high in proteins, such as mineral waters. Enzymes have an antimicrobial effect and are used in packaging for beer and fruit juices. The microbiological quality of food can be improved by edible coatings, which can also serve to restrict the diffusion of antimicrobial additives within the food product. Studies have been undertaken on the use of edible films based on wheat gluten to retain sorbic acid at the surface of moist food products (67).

Ming et al. (68) studied the application of pediocin (7.75 $\mu g/cm^2$) to food packaging films to inhibit inoculated *L. monocytogenes* on meats and poultry. The pediocin-coated bags completely inhibited growth of the *L. monocytogenes* during 12 weeks of storage at 4°C, indicating that application of bacteriocins to plastic food packaging films is effective in inhibiting growth of *L. monocytogenes* in fresh and processed meat and poultry. The inhibitory activity of pediocin was also found to be nondialyzable through cellulosic casing, meaning that it should be possible to use pediocin-coated casings as an effective antilisterial barrier on food products.

Scolaro et al. (69) described the efficacy of plastic films with zeolites added for fruit and vegetable preservation. Controlled atmosphere packaging of fruit and vegetables in cool airtight cells slows down the product's metabolism and thus extends shelf life. CAP also reduces the output of ethylene, which causes fruit decay and changes its color, texture, and taste. Stretch film wrappings hitherto used by supermarkets do not ensure adequate modified atmosphere, whereas so-called active packaging interacts with the food and its ambience: examples include oxygen/ethylene absorbers, carbon dioxide control systems, bacteriostatic or antibacterial wrapping, controlled release of additives, humidity control, etc. Especially interesting are zeolites—aluminium silicate crystals with elements such as sodium, magnesium, calcium, and (particularly useful) ions of potassium. Research by Distam (Universita degli Studi, Milan) shows that LDPE film with 5% zeolytes added had a greater permeability to various gases and could be used in fruit and vegetable active packaging to reduce carbon dioxide and ethylene levels (51).

VI. SMART PACKAGING

Active packaging, described in previous sections (packaging employing O_2 absorbent technology and antimicrobial components), originated from the term ''smart or intelligent'' packaging. Today, many active packaging systems are available, but not all of them are intelligent (i.e., they do not really change with environmental conditions but rather function in a less passive manner than conventional ''barrier'' packaging) (89). In a true sense, fresh and/or minimally processed food systems require ''smart or intelligent'' packaging systems because these food systems are *alive* and their metabolic activities vary significantly with the changes in the surrounding atmospheres. Therefore, packaging that can sense an environmental situation and then trigger a response in the packaging material functionality, such as antimicrobial activity, desiccation and moisture control, aroma emitting, oxygen or ethylene absorption, temperature control, temperature sensing, and visible indication, is required for its successful implementation in fresh and/or minimally processed food systems (89). Present active packaging systems do not meet this requirement and often result in discrepancies in their efficacies (17). Recently, Brody (89) reported a true ''smart or intelligent'' packaging system known as Intellipac™ (Landec Corp., Menlo Park, CA). This system contains polymeric materials, that are side-chain–crystallizable (SCC) polymers with the ability to effectively and reversibly melt with the increases in temperature and thus facilitate increased gas transmission through them. SCC polymers are acrylics with side chains capable of effecting characteristics irrespective of the main chain; thus, variation in the side chain length can alter the melting point from 0 to 68°C. Such a wide range is well within the extreme distribution temperature range of fresh and/or minimally processed foods (89). Additionally, the permeation properties may be modified to change the CO_2-to-O_2 permeability ratios, resulting in the lowest required O_2 concentration without going anaerobic within the package. These materials have 100 times greater O_2 permeability than the conventional polyethylene films without compromising the CO_2-to-O_2 permeability ratios. This enables the packaging system to be used for both CO_2- and O_2-sensitive fresh and/or minimally processed food systems. These types of ''smart or intelligent'' packaging systems are required for fresh and/or minimally processed food systems to control the package structure in order to deliver controlled atmosphere within the package.

VII. CONCLUSIONS

Many packaging systems have been developed over the years to enhance the shelf life of different food systems by controlling or preventing microbial growth and other undesirable factors. MAP, O_2-absorbent technology, antimicrobial, and smart packaging are examples of these developments. However, it must be emphasized that, while adapting any of these packaging systems for a food product, a systems approach needs to be taken, i.e., product/package/process (if fully or minimally processed) interactions should be determined for every possible worst-case scenario. This is necessary not only to deliver superior product quality but also to ensure food safety.

REFERENCES

1. G Tewari, DS Jayas, RA Holley. High pressure processing of foods: an overview. Sci Aliment 19:619–661, 1999.
2. G Tewari, DS Jayas, RA Holley. Centralized packaging of retail meat cuts: a review. J Food Prot 62:418–425, 1999.
3. N Penny, RG Bell. Effect of residual oxygen on the colour, odour and taste of carbon dioxide-packaged beef, lamb and pork during short term storage at chill temperatures. Meat Sci 33:245–252, 1993.
4. EM Scholtz, E Jordaan, J Kruger, GL Nortje, RT Naude. The influence of different centralised pre-packaging systems on the shelf life of fresh pork. Meat Sci 32:11–29, 1992.
5. AA Taylor. Developments in fresh meat technology. Proceedings of 36th International Congress of Meat Science and Technology, Havana, Cuba, 1990, pp. 346–365.
6. EM Buys, J Kruger, GL Nortje. The effect of centralised pre-packaging on the microbial, odor, color, and acceptability attributes of PVC-overwrapped pork loin chops. Proceedings of 39th International Congress of Meat Science and Technology, Calgary, AB, 1993.
7. EM Buys, GL Nortje, PL Steyn. The effect of wholesale vacuum and 100% CO_2 storage on the subsequent microbiological, color and acceptability attributes of PVC-overwrapped pork loin chops. Food Res Int 26:421–429, 1993.

8. CO Gill, N Penney. The effect of initial gas volume to meat weight ratio on the storage life of chilled beef packaged under CO_2. Meat Sci 22:53–63, 1988.
9. FH Grau. Role of pH, lactate, and anaerobiosis in controlling the growth of some fermentative gram-negative bacteria on beef. Applied Environ Microbiol 42:1043–1050, 1981.
10. FH Grau. Microbial growth on fat and lean surfaces of vacuum-packaged chilled beef. J Food Sci 48:326–328, 1983.
11. C Vanderzant, JW Savell, MO Hanna, V Potluri. A comparison of growth of individual meat bacteria on the lean and fatty tissues of beef, pork and lamb. J Food Sci 51:5–8, 1986.
12. CO Gill, C McGinnis. Changes in the microflora on commercial beef trimmings during their collection, distribution, and preparation for retail sale as ground beef. Int J Food Microbiol 18:321–332, 1993.
13. CO Gill, T Jones. The display life of retail-packaged beef steaks after their storage in master packs under various atmospheres. Meat Sci 38:385–396, 1994.
14. RA Holley, P Delaquis, J Gagnon, G Doyon, C Gariepy. Modified atmosphere packaging of fresh pork. Alimentech. 6(3):16–17, 1993.
15. RA Holley, P Delaquis, N Rodrigue, G Doyon, J Gagnon, C Gariepy. Controlled-atmosphere storage of pork under carbon dioxide. J Food Prot 57:1088–1093, 1994.
16. LE Jeremiah, LL Gibson. The influence of controlled atmosphere storage on the flavor and texture profiles of display-ready pork cuts. Food Res Int 30:117–129, 1997.
17. G Tewari. Centralized packaging of retail ready meat cuts. Ph.D. dissertation. University of Manitoba, Winnipeg, MB, 2000.
18. H Ogihara, M Kanie, N Yano, M Haruta. Effect of carbon dioxide, oxygen and their gas mixture on the growth of some food-borne pathogens and spoilage bacteria in modified atmosphere package of food. J Food Hygienec Soc Japan 34:283–293, 1993.
19. E Whitley, D Muir, WM Waites. The growth of *Listeria monocytogenes* in cheese packed under a modified atmosphere. J Appl Microbiol 88:52–57, 2000.
20. PV Nielson, I Haasum. Packaging conditions hindering fungal growth on cheese. Scan Dairy Inform 4:22–24, 1997.
21. WM Fedio, A Macleod, L Ozimek. The effect of modified atmosphere packaging on the growth on microorganisms in cottage cheese. Milchwissenschaft 49:622–629, 1994.
22. JH Chen, JH Hotchkiss. Growth of *Listeria monocytogenes* and *Clostridium sporogenes* in cottage cheese in modified atmosphere packaging. J Dairy Sci 76:972–977, 1993.
23. L Jacxsens, F Devlieghere, P Falcato, J Debevere. Behavior of *Listeria monocytogenes* and *Aeromonas* spp. on fresh-cut produce packaged under equilibrium-modified atmosphere. J Food Prot 62:1128–1135, 1999.
24. JD Floros, LL Dock, PV Nielson. Biological control of *Botyritis cinerea* growth on apples stored under modified atmospheres. J Food Prot 61:1661–1665, 1998.
25. JW Austin, KL Dodds, B Blanchfield, JM Farber. Growth and toxin protection by *Clostridium botulinum* on inoculated fresh-cut packaged vegetables. J Food Prot 61: 324–328, 1998.

26. H Izumi, AE Watada, NP Ko, W Douglas. Controlled atmosphere storage of carrot slices, sticks and shreds. Postharvest Biol Technol 9:165–172, 1996.
27. F Willocx, M Mercier, M Hendrickx, P Tobback. Modelling the growth of *Pseudomonas fluorescens* as a function of temperature and carbon dioxide. Proceedings of a workshop on the postharvest treatment of fruit and vegetables: modified atmosphere, Istanbul, Turkey, 1992, pp. 87–101.
28. JJ Osuna, G Zurera, RM Garcia. Microbial growth in packaged fresh asparagus. J Food Qual 18:203–214, 1995.
29. UM Abdul-Raouf, LR Beuchat, MS Ammar. Survival and growth of *Escherichia coil* O157:H7 on salad vegetables. Appl Environ Microbiol 59:1999–2006, 1993.
30. MB Omary, RF Testin, SF Barefoot, JW Rushing. Packaging effects on growth of *Listeria monocytogenes* in shredded cabbage. J Food Sci 58:623–626, 1993.
31. MHJ Bennik, W van Overbeek, EJ Smid, LGM Gorris. Biopreservation in modified atmosphere stored mungbean sprouts: the use of vegetable-associated bacteriocinogenic lactic acid bacteria to control the growth of *Listeria monocytogenes*. Lett Appl Microbiol 28:226–232, 1999.
32. A Amanatidou, EJ Smid, LGM Gorris. Effect of elevated oxygen and carbon dioxide on the surface growth of vegetable microorganisms. J Appl Microbiol 86:429–438, 1999.
33. MHJ Bennik, W Vorstman, EJ Smid, LGM Gorris. The influence of oxygen and carbon dioxide on the growth of prevalent Enterobacteriaceae and *Pseudomonas* species isolated from fresh and controlled atmosphere-stored vegetables. Food Microbiol 15:459–469, 1998.
34. Reference deleted.
35. HL Chong, JN Cash. Comparative growth rates of bacteria on minimally processed meat-vegetable product under modified atmospheres. Food Sci Biotechnol 7:6–12, 1998.
36. HM Solomon, EJ Rhodehamel, DA Kautterr. Growth and toxin production by *Clostridium botulinum* on sliced raw potatoes in a modified atmosphere with and without sulfite. J Food Prot 61:126–128, 1998.
37. T Wei, DA Grinstead, JR Mount, FA Draughon. Growth and survival of *Listeria monocytogenes* in minimally processed green beans as influenced by modified atmosphere packaging, NaCl treatment and storage temperatures. J Food Prot 58(suppl): 16, 1995.
38. C Diaz, JH Hotchkiss. Comparative growth of *Escherichia coli* O157:H7, spoilage organisms and shelf life of shredded iceberg lettuce stored under modified atmospheres. J Sci Food Agri 70:433–438, 1996.
39. CBC Cartaxo, SA Sargent, DJ Huber, L Chia-Min. Controlled atmosphere storage suppresses microbial growth on fresh-cut watermelons. Proceedings of the Florida State horticultural society, Gainesville, FL, 1998, pp. 252–257.
40. LR Beuchat, RE Brackett. Survival and growth of *Listeria monocytogenes* on lettuce as influenced by shredding, chlorine treatment, modified atmosphere packaging and temperature. J Food Sci 55:755–758, 870, 1990.
41. ME Berrang, RE Brackett, LR Beuchat. Growth of *Listeria monocytogenes* on fresh vegetables stored under controlled atmosphere. J Food Prot 52:702–705, 1989.

42. ME Berrang, RE Brackett, LR Beuchat. Growth of *Aeromonas hydrophila* on fresh vegetables stored under a controlled atmosphere. Appl Environ Microbiol 55:2167–2171, 1989.
43. B Kimura, S Kuroda, M Murakami, T Fujii. Growth of *Clostridium perfringens* in fish fillets packaged with a controlled carbon dioxide atmosphere at abuse temperatures. J Food Prot 59:704–710, 1996.
44. L Pastoriza, G Sampedro, JJ Herrera, ML Cabo. Effect of carbon dioxide atmosphere on microbial growth and quality of salmon slices. J Sci Food Agri 72:348–352, 1996.
45. B Wannapee, JL Silva, C Handumrongkul. Microbial growth and certain pathogens in catfish fillets under a master-pack system. Institute of Food Technologists (USA) meeting, USA, 1996, pp. 1082–1236.
46. S Oka, K Fukunaga, H Ito, K Takama. Growth of histamine producing bacteria in fish fillets under modified atmospheres. Bulletin Faculty Fisheries, Hokkaido Univ (Jpn) 44:46–54, 1993.
47. S Oka, H Ito, K Takama. Growth pattern and enterotoxin production of food poisoning bacteria in the fish fillets stored under different modified atmospheres. Bulletin Faculty Fisheries, Hokkaido Univ (Jpn) 43:105–114, 1992.
48. S Suzuki, J Noda, K Takama. Growth and polyamine production of *Alteromonas* spp. in fish meat extracts under modified atmosphere. Bulletin Faculty Fisheries, Hokkaido Univ (Jpn) 41:213–220, 1991.
49. JD Floros, LL Dock, JH Han. Active packaging technologies and applications. Food Cosmetics Drug Packag 20:10–17, 1997.
50. G Tewari, DS Jayas, LE Jeremiah, RA Holley. Oxygen absorption kinetics of oxygen absorbers. Int J Food Sci Technol, 2001. (In press.)
51. L Vermeiren, F Devlieghere, M van Beest, N de Kruijf, J Debevere. Developments in the active packaging of foods. Trends Food Sci Technol 10:77–86, 1999.
52. Darex technical information. Active packaging technology. Darex Container Products, WR Grace and Co., Minneapolis, MN, 1998.
53. E Graff. Oxygen removal. U.S. Patent 5284871 (1994).
54. ML Rooney. Oxygen scavenging from air in package headspaces by singlet oxygen reactions in polymer media. J Food Sci 47:291–294, 298, 1985.
55. G Tewari, DS Jayas, LE Jeremiah, RA Holley. Prevention of transient discoloration of beef steaks. J Food Sci, 2001. (Accepted for publication.)
56. JR Patel, CA Hwang, LR Beuchat, MP Doyle, RE Brackett. Comparison of oxygen scavengers for their ability to enhance resuscitation of heat-injured *Listeria monocytogenes*. J Food Prot 58:244–250, 1995.
57. BJ Juven, I Rosenthal, I. Effect of free-radical and oxygen scavengers on photochemically generated oxygen toxicity and on the aerotolerance of *Campylobacter jejuni*. J Appl Bacteriol 59:413–419, 1985.
58. JP Smith, B Ooraikul, WJ Koersen, ED Jackson, RA Lawrence. Novel approach to oxygen control in modified atmosphere packaging of bakery products. Food Microbiol 3:315–320, 1986.
59. H Nakamura, J Hoshino. Techniques for the preservation of foods by employment of an oxygen absorber. Tech Info Mitsubishi Gas Chem Co. 1:1–45, 1983.
60. EM Powers, D Berkowitz. Efficacy of an oxygen scavenger to modify the atmo-

sphere and prevent mold growth on meal, ready-to-eat pouched bread. J Food Prot 53:767–771, 1990.
61. GR Siragusa, CN Cutter, JL Willett. Incorporation of bacteriocin in plastic retains activity and inhibits surface growth of bacteria on meat. Food Microbiol 16:229–235, 1999.
62. SL Dong, H Yong II, HC Sung. Developing antimicrobial packaging film for curled lettuce and soybean sprouts. Food Sci Biotechnol 7:117–121, 1998.
63. A Duck-Soon, H Yong II, C Sung-Hwan, L Dong-Sun. Packaging of fresh curled lettuce and cucumber by using low density polyethylene films impregnated with antimicrobial agents. J Korean Soc Food Sci Nutr 27(4):675–681, 1998.
64. H Ling-Jane, H Chen-Huei, W Yih-Ming. Using antimicrobial polyethylene films and minimal microwave heating to control the microbial growth of tilapia fillets during cold storage. Food Sci (Taiwan) 24(2):263–268, 1997.
65. CN Cutter. The effectiveness of triclosan-incorporated plastic against bacteria on beef surfaces. J Food Prot 62:474–479, 1999.
66. LM Sherman. Biocides keep the bugs off your plastics. Plastic Technol 44:45–48, 1998.
67. B Cuq. Antimicrobial packaging and coatings. Ind Aliment Agric 114:127–129, 1997.
68. X Ming, GH Weber, JW Ayres, WE Sandine. Bacteriocins applied to food packaging materials to inhibit *Listeria monocytogenes* on meats. J Food Sci 62:413–415, 1997.
69. M Scolaro, L Piergiovanni, P Fava. Plastic films with zeolites added for preserving fruit and vegetables. Rass dell'imballaggio 13:4–6, 1992.
70. JH Han. Antimicrobial food packaging. Food Technol 54(3):56–65, 2000.
71. RE Hansen, CG Rippl, DG Midkiff, JG Neuwirth. Antimicrobial absorbent food pad. U.S. Patent 4865855 (1989).
72. H Yoshida, K Matsuo, J Yagi. Antimicrobial paper. Japan Patent 0K93397093 (1993).
73. S Iwashiro. Antimicrobial sheet for food container. Japan Patent 0K105000091 (1991).
74. K Suzuki, K Hokita, T Itoh. Tectosilicate antimicrobial composition. Japan Patent 0K120204091 (1991).
75. T Kanie, K Ichikawa. Antimicrobial paper. Japan Patent 0K898091 (1991).
76. Y Yokota. Antimicrobial paper. Japan Patent 0K104798090 (1990).
77. J Kato, H Komiya, Y Hata. Antimicrobial sheet. Japan Patent 0K76599083 (1983).
78. MK Lee, SM Lee, DH Oh. Manufacture of antimicrobial paper for food products. Proceedings of recent advances in paper science and technology, Seuol, Korea, 1999, pp. 217–220.
79. S Aoki. The development of antibacterial plastic food wrap ''Hitachi Wrap.'' JPI J 35:33–38, 1997.
80. H Kobayashi. Polystyrene film (S-clear KA) for protecting foods from bacteria and moulds. JPI J 35:21–26, 1997.
81. Antimicrobial packaging materials. Packpia 536:34–44, 1997.
82. New antibiotic sheet incorporates bamboo sheet. Packag Trends Jpn 97(3):6, 1997.
83. New antibiotic pouch for long term drinking water preservation. Packag Trends Jpn 97(2):2–3, 1997.

84. A strategy for the introduction of antibacterial packaging materials and related products. Packpia 535:39–49, 1997.
85. M Tokuda. Functional packaging material Celagrase. Packpia 533:20–23, 1996.
86. Ceramic in antimicrobial non-woven fabric keeps food fresh. Jpn Packag News 19: 3, 1996.
87. PJ Louis. Antimicrobial packages. Eur Packag Newslett World Rep 26:3–4, 1993.
88. Antibacterial shrink film. Packag Jpn 11:73, 1990.
89. AL Brody. Smart packaging becomes Intellipac™. Food Technol 54(6):104–106, 2000.

8

Control of Foodborne Microorganisms with Carbon Dioxide Under Elevated Pressure

Gerhard J. Haas
Fairleigh Dickinson University
Teaneck, New Jersey

Paul D. Matthews
Lehman College of the City University of New York
Bronx, New York

I. INTRODUCTION

Carbon dioxide has been known since antiquity because of its presence in volcanoes and mineral springs (1). J van Helmont (1577–1674) found that it was produced by fermentation and combustion. J. Black in 1755 termed it ''fixed air'' because he was able to release it from carbonates (2). Lavoisier showed that CO_2 could be produced from the combustion of carbon and oxygen (3). Pasteur and Jobert described the inhibitory action of CO_2 on *Bacillus anthracis* (1), and they were the first to demonstrate its antimicrobial activity. Its role in mammalian and microbial metabolism and in plant growth and its chemical and physical parameters were gradually discovered and described as scientific research continued.

The effect of CO_2 on microbes at atmospheric pressure is discussed in Chapter 9, and we restrict ourselves to CO_2 pressures higher than atmospheric. It is not a surprise that with elevated pressure the antimicrobial activity of CO_2 increases, because higher pressure leads to greater concentration at the sites of activity.

Swearingen and Lewis published (4) in 1933 research on the "nature of the effect of CO_2 under pressure upon bacteria." They quoted work by D'Arsonval and Charrin in 1893 and in 1917 by Larson, Hartzell, and Diehl, who had experimented with the effect of CO_2 pressure on a number of pathogens: *Pseudomonas aeruginosa*, *Escherichia coli*, *Mycobacterium tuberculosis*, and others. These experiments were carried out at 50 atm for 2 hours, and considerable killing power was achieved, but quantification of microbial population or temperature during the pressure treatment was not performed. These early investigators compared the effects of acidity generated by CO_2 with that generated by adding HCl and found that CO_2 at the same pH was much more effective in killing bacteria.

Swearingen and Lewis, working with *E. coli*, demonstrated the importance of treatment duration. They also observed that rapid pressure release did not enhance the destruction of microbes. The ionic concentration was an important parameter in their experiments, and at higher salt concentrations CO_2 was more effective. They theorized that the antimicrobial activity of CO_2 was due to "a precipitation of certain colloidal systems within the cell body." Unfortunately, the temperatures at which the work was done were not mentioned.

The first reported practical application of CO_2 under pressure was described in 1912 in Switzerland by Böhi (5,6) who developed this procedure in order to improve storage of fruit juices. Because these juices are very popular in Switzerland, there is a great need for a nonthermal, inexpensive method of preservation. Seven atm at 15°C was the recommended storage condition. This process also became popular in the 1930s in Germany and in other European countries, but some modifications became necessary because the microbes were not killed—just slowed in their growth. These modifications included the use of better sanitation so as to start with a low inoculum of microbes, a brief filtration step before storage, and, in some modifications, even sterile filtration before bottling.

Today most of the major fruit juice producers do not use the Böhi procedure (7), and its use is diminishing more and more among the smaller producers (F. Escher, personal communication).

In spite of the many years of research on the use of CO_2 under pressure, there is no clear picture of the mechanisms of microbial inhibition. Much higher pressures are needed to inhibit or kill microbes when gases other than CO_2 are used; thus, there is no question that CO_2 under pressure has a special effect. CO_2 differs from these other gases by virtue of its high solubility in water and in lipids. Nitrous oxide, which is also very soluble, similarly has some antimicrobial activity (8). The capacity of CO_2 for forming acid in water [$CO_2 + H_2O \rightarrow H_2CO_3 \rightarrow (H)^+ + (HCO_3)^-$] is also important. These properties of CO_2 probably lead to inactivation of microbes by one or a combination of the mechanisms listed below. The factors important to the antimicrobial action are listed in Table 1.

Table 1 Parameters Important for Inhibition of Microbes by CO_2 Under Pressure

Magnitude of pressure
Criticality of CO_2
Time under pressure
Speed of decompression
Temperature during treatment
Moisture content
Water activity (a_w)
Type of microorganism
Spore or vegetative state
Age of cells
Other constituents of the menstruum

II. MECHANISMS OF ACTION

Several mechanisms are reported to be responsible for CO_2 antimicrobial activity.

A. Penetration and Solubility

CO_2 lowers intracellular pH, but this effect alone cannot be the whole cause for its inhibitory effect on microbes, because adjusting the pH with hydrochloric acid extracellularly to pH 3.2 is far less lethal than 900 psi CO_2 pressure, which lowers the pH of nutrient broth to pH 4.35 (9). Similar results were reported by Swearingen and Lewis (4). It is not known, however, how this extracellular pH of the broth is correlated with intracellular pH of bacteria; there is no question that this intracellular pH is an extremely important factor. Due to its penetrative capabilities, the pH may actually be driven lower by CO_2 than by exogenous HCl or by other acids.

B. Disruption of Normal Metabolism

The effect of CO_2 on enzymes was measured by King et al. (10). These authors postulate a mass action effect of CO_2 on decarboxylation in *Pseudomonas*. Even though King et al.'s experiments (10) were carried out at atmospheric pressure, their findings would be enhanced at higher pressures. Such a mass action effect is also suggested by Haas et al. (9) and by Jones and Greenfield (11).

C. Mechanical Rupture

Destruction of the integral structure by pressurization and decompression is another possibily important factor. Rupture by CO_2 pressure in the macro-structure of fruits and vegetables is evident (9), and such rupture may similarly weaken the structure of microbes. Rapid decompression may enhance rupture. Isenschmid et al. (12) showed an electron micrograph of yeast cells demonstrating that a certain number of yeast cells, but not all, suffer shrinkage of the plasma membrane. Nakamura et al. (13) also showed electron micrographs of baker's yeast after CO_2 treatment under pressure.

The extent of rupture depends on the magnitude of the pressure and the speed of decompression. Rapid decompression appears to have a lesser incremental effect than one might think. Enomoto et al. (8) showed breakage of yeast cells on pressurization with 40 atm CO_2. This may have been caused by excessive swelling of the cells causing eventual rupture. With baker's yeast explosive decompression (43 atm/min) resulted only in a marginal increase in kill rate over and above a slow decompression rate of 0.33 atm/min. Arreola et al. (3) also found that increase in death due to rapid decompression may not be a factor. On the other hand, Lin and coworkers (14) observed that pressure release followed by repressurization does increase the degree of cell inactivation. Rapid decompression from the supercritical state of CO_2 may be more lethal than decompression from the noncritical state.

Enomoto et al. (8) report experiments in which decompression increases the degree of inactivation by CO_2 and cite other authors who have observed this effect. It appears that other factors, such as the number of cells surviving treatment by pressure but before release, determine whether rapid decompression leads to additional kill. Enomoto et al. (8) calls decompression ''potentially lethal.''

At the pressure where CO_2 is lethal, all experiments with nitrogen or argon, whether rapidly decompressed or not, have shown no or minimal effect (8), probably due to the low solubility of these gases; 4000–8000 atm is needed for the destruction of microbes by hydrostatic pressure.

In spite of much work by many laboratories there is no unified, accepted mechanism for the antimicrobial action of CO_2. As mentioned, Table 1 shows the important parameters that determine the degree of inactivation of microbes. Mechanisms of lethality include cell rupture, internal acidification, changes in cell membranes, or changes in cell metabolism.

The various mechanisms are by no means mutually exclusive (15). They may act simultaneously or sequentially. The change in microbial pH and the mass action effect by CO_2 and/or other metabolic effects by lower pH 915) are not mutually exclusive but may be additive or even synergistic (15). For instance, changes in metabolism may follow an internal pH change. There can be changes in the enzyme reactions, particularly decarboxylation reactions (16), which would be affected by the high concentration of CO_2. Further, the great solubility of CO_2 in the tissues and the expansion of the cell membrane will lead to changes in metabolism (16). When exposed to CO_2 in the supercritical phase state, cell constituents and particularly lipids are extracted (17). The various factors important for the mechanism have been discussed by many authors (e.g., Refs. 9,10,12,14,18,19).

Isenschmid et al. (12) postulated that cell death was mainly due to the accumulation of CO_2 within the cell membrane, an ''anesthetic effect,'' rather than cell rupture. This anesthetic effect is discussed at length by Dixon (1) with all the important related references. Isenschmid (12) found that at near-critical pressure, CO_2 is less harmful to yeast than at supercritical pressure and that the pressure at which this ''greater harmfulness'' occurs is microorganism dependent. Cell death is not correlated to cell lysis. Even at 13°C when only 45% of the cells lyse, viability had decreased by more than 99%. At 27°C there was almost no lysis, but the degree of kill was similar.

Several authors (3,15,18,20) have found two distinct phases in survival curves of microorganisms exposed to CO_2. The earlier stage is characterized by a slow rate of inactivation, which increases sharply in the later stage. They suggest that the first stage would parallel the penetration of CO_2 into the cell wall, while with most cells the actual inactivation occurs during the second stage.

The effects of CO_2 are enhanced by increased pressure, length of pressure treatment, and temperature. This has been shown by all authors who have experimented in this area. An increase in pressure and/or temperature leads to an abbreviation of the first phase and an increase of inactivation rate during the second stage (15). A reduction of moisture content and a lower water activity (a_w) reduce the effect of CO_2 under pressure (9,13,21). Also, food ingredients such as meat and fat (18,21) reduced the lethality (20,22,23). In contrast, agitating the sample during pressurization enhances the effect. Other antimicrobial agents such as ethanol and acetic acid (24) can be added to the supercritical CO_2 so as to achieve more effective sterilization. Ethanol and acetic acid may further remove cell components (24), and this could be the reason for increased lethality.

III. SUSCEPTIBILITY SPECTRUM

Susceptibility of microbes varies. Different authors have worked with a great variety of microbes of interest to them. This variety can be seen in Table 2 (bacteria) and Table 3 (fungi). Isenschmid et al. (12) showed that among the yeasts, *Kluyveromyces fragilis* is the most resistant to CO_2 under pressure, with less resistance demonstrated by *Saccharomyces cerevisiae* and *Candida utilis*. Hays et al. (22) also showed differences in susceptibility between different types of microbes.

Table 2 Studies of Bacteria Treated with CO_2 Under Pressure

Microorganism	Ref.
Aerobacter aerogenes	22
Bacillus cereus	27
Bacillus coagulans	22
Bacillus megaterium	8, 26
Bacillus subtilis	1, 17, 21, 24, 27, 29
Clostridium botulinum	22, 27
Clostridium butyricum	22
Clostridium perfringens	27
Clostridium sporogenes	9, 27
Escherichia coli	4, 9, 15, 17, 19, 24, 30
Lactobacillus	9, 19, 22
Leuconostoc dextranicum	14
Listeria monocytogenes	20, 22, 23
Pseudomonas aeruginosa	4, 10, 17, 29
Salmonella Senftenberg	9
Salmonella Typhimurium	1, 18, 23, 30
Staphylococcus aureus	1, 9, 18, 22, 24, 30
Streptococcus faecalis	1, 18, 30
Coliphage	17

IV. ACTIVITY TOWARD SPORE-FORMERS

Clostridium botulinum is the microorganism most feared by processors of shelf-stable foods because of its highly heat-resistant spores and the lethality of its toxin. *Clostridium botulinum* occurs in the soil, and so the probability of the presence of botulinum spores in the environment and crops is high. Any method

Table 3 Studies of Fungi Treated with CO_2 Under Pressure

Microorganism	Ref.
Aspergillus niger	18, 21, 22
Bissochlamys fulva	21
Candida albicans	17
Candida utilis	12
Kluyveromyces fragilis	12
Penicillium roqueforti	9, 30
Saccharomyces cerevisiae	1, 8, 9, 12, 13, 22, 24, 28, 30

that might be useful in making foods, in which *C. botulinum* has the conditions for growth, safe would have to kill these microbes. Several investigators have determined the susceptibility of spores to CO_2 under pressure. *Clostridium botulinum* is not the only spore-former pathogenic, but others, such as *Bacillus cereus* and *Clostridium perfringens*, are also pathogenic. Other important spore-formers of the genera *Clostridium* and *Bacillus*, while nonpathogenic, can lead to food spoilage. Doyle (25) has investigated botulinum toxin production under CO_2 pressure.

Spores are not easily killed by CO_2, even at elevated pressure. Kamihira et al. (24) using supercritical CO_2 found no appreciable destruction of bacillus spores at 200 atm. Using 200, 500, and 2500 atm Stahl demonstrated a maximum 6 log reduction after 20 minutes of exposure (17); unfortunately no treatment temperature was given. When ethanol or acetic acid was added to the supercritical CO_2 (24), about 50% of the endospores of *Bacillus stearothermophilus* were inactivated at 200 atm in 2 hours at 35°C. Enomoto et al. (26) were able to reduce the viable count of *Bacillus megaterium* spores: at 58 atm and 60°C for 24 hours followed by rapid decompression, he achieved a 6 log reduction. Enomoto et al. (26) found that the survival rate of the *B. megaterium* spores decreased dramatically with temperature and that kill was optimal at 58 atm and 60°C and further pressure increase actually led to a higher survival rate. Enfors and Molin (27) reported that germination of the spores of *Clostridium sporogenes* and *C. perfringens* was inhibited at hyperbaric pressure. Haas et al. (9) showed that *C. sporogenes* could be destroyed (4 log reduction) at 53 atm of CO_2 at 80°C in 2 hours when the pH of the medium had been lowered to 2.5. At pH 4 there was no effect. Ballestra and Cuq (21), working at 50 atm, found CO_2 pressure effective against the spores of the fungus *Bissochlamys fulva* at 50°C.

How efficient CO_2 application under pressure is at killing spores and what the chances are of bringing about total kill of microbes will decide the useful range of applications. Other measures added or superimposed may hold the key to where CO_2 under pressure may be a realistic tool for food preservation.

V. RESEARCH ON PRACTICAL APPLICATIONS

Several of the researchers in the field have tested CO_2 under pressure for application to food and beverages, and Table 4 lists their publications.

Food application appears limited for the following reasons:

1. Water is necessary for good activity, so there is no applicability to dried foods or herbs (24,28).
2. Spores are resistant, so to kill these recalcitrant cells another parameter is needed, such as acid conditions, elevated temperature (9), or the addition of alcohol or acetic acid (24) or possibly salt (4).

Table 4 Particular Applications for Food Preservation

Application	Ref.
Orange juice	3
Water, dairy products	1
Milk	18, 20
Grape and fruit juice	5, 6, 7
Orange juice, apple juice, spices	9
Propellants	22
Kimchi	19
Fresh celery	31
Sea buckthorn	29
Black ground pepper, coriander powder, tuber celery	32
Chicken meat, shrimp, orange juice, egg yolk, whole egg	23

3. Color, flavor, and texture changes may occur. For example, when used on herbs, aroma and/or texture may be altered (9), and in chicken, shrimp, orange juice, and egg (23), color and texture were affected upon this treatment.
4. Some food constituents reduce the inhibitory effect and protect the microbes.

There may, however, be an opportunity for application with acid foods, in which spore-formers cannot grow, and where texture is of no concern, such as destruction of *E. coli* or *Listeria monocytogenes* in apple cider. In many foods a general reduction in total microbial load could be achieved (23). Hong et al. (19) carried out research that indicated that CO_2 under pressure might be useful to reduce the number of lactobaccilli remaining after completion of fermentation of a traditional Korean food (*kimchi*); if the fermentation is permitted to continue unchecked, overacidification would ensue, which can be prevented by reduction in the number of surviving lactobacilli, and thus the shelf life of this product is extended.

The antimicrobial effect of CO_2 under pressure is real and general, but differences in pressure, phase state, time of pressurization and decompression, and temperature used by various authors are appreciable and sometimes divergent. In many instances, these differences make general conclusions and application guidelines difficult. Thus, the reader who is interested in a particular aspect must consult pertinent references herein.

REFERENCES

1. NM Dixon, DB Kell. The inhibition by CO_2 of the growth and metabolism of microorganisms. J Appl Bacteriol 67:01–136, 1989.
2. J Read. Humor and Humanism in Chemistry. London: G Bell & Sons, 1947.
3. AG Arreola, CI Wei, A. Peplew, M Marshal, J Cornell. Effect of supercritical CO_2 on microbial populations in single strength orange juice. J Food Qual 14:275–284, 1991.
4. JS Swearingen, IM Lewis. The nature of the effect of CO_2 under pressure upon bacteria. J Bacteriol 26:201–210, 1933.
5. A von Mehlitz. Süssmost. Braunschweig: Drs. Serger & Hempel, 1938.
6. RF Fritsche, E Eberlein, H Schmid. In: Schweizer Pioniere der Wirtschaft und Technik. Zürich: Verein für Wirtschaftshistorische Studien, 1974.
7. U Schobinger. Frucht- und Gemüsesäfte. Berlin: Verlag Eugen Ulmer, 1990.
8. A Enomoto, K Nakamura, K Nagai, T Hashimoto, M Hakoda. Inactivation of food microoraganisms by high-pressure carbon dioxide treatment with or without explosive decompression. Biosci Biotech Biochem 61(7):1133–1137, 1997.
9. GJ Haas, E Dudley, R Dik, C Hintlian, L Keane. Microbial destruction by high pressure CO_2. J Food Safety 9:253–265, 1989.
10. JAD King, CW Nagel. Influence of carbon dioxide upon the metabolism of *Pseudomonas aeruginosa*. J Food Sci 40:362–366, 1975.
11. RP Jones, PF Greenfield. Effect of carbon dioxide on yeast growth and fermentation. Enzym Microbiol Tech 4, 1982.
12. A Isenschmid, IW Marison, U von Stockar. The influence of pressure and temperature of compressed CO_2 on the survival of yeast cells. J Biotech 39(3):229–237, 1995.
13. K Nakamura, A Enomoto, A Fukushima, K Nagai, M Hakoda. Disruption of microbial cells by the flash discharge of high-pressure carbon dioxide. Biosci Biotech Biochem 58(7):1297–1301, 1994.
14. H-m Lin, Z Yang, LF Chen. Inactivation of *Leuconostoc dextranicum* with carbon dioxide under pressure. Chem Eng J 52:B29–B34, 1993.
15. P Ballestra, AAD Silva, JL Cuq. Inactivation of *Escherichia coli* by carbon dioxide under pressure. J Food Sci 61(4):829–831, 1996.
16. G Kritzman, I Chet, Y Hennis. Effect of carbon dioxide on growth and carboydrate metabolism in *Sclerotium rolfsii*. J Gen Micro 100:167–175, 1997.
17. E Stahl, G Rau, H Kaltwasser. Hochdruck Behandlung von Mikroorganismen. Naturwisschaft 72:144–145, 1985.
18. O Erkman. Antimicrobial effect of pressurized carbon dioxide on *Staphylococcus aureus* in broth and milk. Food Sci Technol 30(8):826–829, 1997.
19. S-I Hong, W-S Park, W-R. Pyun. Inactivation of lactobacillus from kimchi by high pressure carbon dioxide. Lebensm Wisschaft Technol 30(7):681–685, 1997.
20. H-m Lin, N Cao, L-F Chen. Antimicrobial effect of pressurized carbon dioxide on *Listeria monocytogenes*. J Food Sci 59(3):657–659, 1994.
21. P Ballestra, JL Cuq. Influence of pressurized carbon dioxide on the thermal inactivation of bacterial and fungal spores. Lebensm Wissschaft Technol 31(1):84–88, 1998.

22. G Hays, J Burroughs, R Warner. Microbiological aspects of pressure packaged foods and the effect of various gases. Food Technol 13:567–570, 1959.
23. CI Wei, MO Balaban, SY Fernando, AJ Peplow. Bacterial effect of high pressure CO_2 treatment on foods spiked with listeria or salmonella. J Food Prot 54(3):189–193, 1991.
24. M Kamihira, M Taniguchi, T Kobayashi. Sterilization of microorganisms with supercritical carbon dioxide. Biol Chem 51:407–412, 1987.
25. MP Doyle. Effect of carbon dioxide on toxin production by *Clostridium botulinum*. Eur J Appl Micro Biotech 17:53–56, 1983.
26. A Enomoto, K Nakamura, M Hakoda, N Amaya. Lethal effect of high-pressure carbon dioxide on bacterial spores. J Ferm Bioeng 83(3):305–307, 1997.
27. S-O Enfors, G Molin. The influence of high concentrations of carbon dioxide on the germination of bacterial spores. J Appl Bact 45:197–208, 1978.
28. H Kumagi, C Hata, K Nakamura. CO_2 sorption by microbial cells and sterilization by high-pressure CO_2. Biosci Biotech Biochem 61(6):931–935, 1997.
29. P Manninen, E Haivala, S Sarino, H Kallio. Distribution of microbes in supercritical CO_2 extraction of sea buckthorn (*Hippopidac rhamnoides*). Z Lebensm Unters Forsch A204:202–205, 1997.
30. JL Cuq, H Roussel, D Vivier, JP Caron. Etude des effets des gaz sous pression. 2. Influence sur l'inactivation thermique des microorganismes. Sci Aliments 13:677–696, 1993.
31. K Kuhne, D Knorr. Effects of high pressure carbon dioxide on the reduction of microorganisms in fresh celery. Eur Food Sci 41:55–58, 1990.
32. G Modlich, H Weber. Vergleich verschiedener Verfahren zur Gewürzentkeimung. Fleischwirtschaft 73(3):337–343, 1993.

9
Inhibition of Microbial Growth by Low-Pressure and Ambient Pressure Gases

Christopher R. Loss and Joseph H. Hotchkiss
Cornell University
Ithaca, New York

I. INTRODUCTION

Many studies have demonstrated the antimicrobial activity of gases at ambient and subambient pressures on food-related microorganisms. This chapter will review the published literature related to the effects of gases on microorganisms and the factors influencing the efficacy of antimicrobial atmospheres. The effect of altered atmospheres on specific foodborne pathogens will be discussed and the strategies used to control pathogens reviewed. Common food products preserved by altered atmospheres and the composition of these atmospheres will be presented.

Gases can inhibit microorganisms in two general ways. First, they can have direct toxic effects such as inhibiting growth and proliferation. Carbon dioxide (CO_2), ozone (O_3), and oxygen (O_2) are examples of gases that are directly toxic to many microorganisms. The inhibitory mechanism is dependent upon the chemical and physical properties of the gas and its interaction with the aqueous and lipid phases of the environment. Second, altering the gas composition can have indirect inhibitory effects by altering the ecology of the microbial environment. When the atmosphere under which microorganisms grow is altered, the competitive environment is also altered. Atmospheres having a negative effect on the growth of one particular microorganism may promote the growth of another. This may have negative consequences depending upon the native pathogenic microflora and their substrate. Replacement of O_2 by nitrogen (N_2) is an example of such indirect effects.

A variety of technologies are used to establish desired atmospheric conditions to affect the growth of microorganisms. A majority of these technologies rely upon temperature to augment the inhibitory effect. Common technologies include modified atmosphere packaging (MAP), controlled-atmosphere packaging (CAP), controlled-atmosphere storage (CAS), direct addition of carbon dioxide (DAC), and hypobaric storage (HS). Most of these technologies have been defined and refined over the last century, driven by a demand for higher-quality, safer foods that are distributed over extended distances and time.

Oxygen, CO_2, O_3, and N_2 are the most commonly used gases for antimicrobial atmospheres. Oxygen, CO_2, and O_3, are utilized for their direct inhibitory effects, and N_2 is used to replace other gases, primarily O_2. Oxidizing radicals generated by O_3 and O_2 are highly toxic to anaerobic bacteria and can have an inhibitory effect on aerobes depending upon their concentrations.

Carbon dioxide is an effective antimicrobial against many gram-negative bacteria [*Pseudomonas* ssp., for example (1)] and some gram-positives [*Listeria* ssp. (2)]. When dissolved in aqueous solutions, CO_2 decreases pH, generating an environmental stress. The physical and chemical properties of CO_2 allow it to disrupt critical enzymatic activity that takes place at the bacterial cell wall and is also capable of disrupting the physical integrity of the cell membrane and altering its fatty acid composition (3).

Depending upon the target microbial population, CO_2 and O_2 can be combined at specific ratios, balanced with N_2, dispersed around the product, and contained in a specific packaging barrier. For water-containing products, CO_2 can be dissolved directly into the aqueous phase to provide an antimicrobial effect. Ozone is also dissolved directly into water, which can then be used as an antimicrobial rinse or frozen and used as antimicrobial ice (4). Some other less commonly used gases include carbon monoxide, nitrous oxide, argon, hydrogen, ethylene oxide, bromide, and ammonia. These gases are effective in many cases but are more problematic to use in food systems due to their toxicity or cost.

The antimicrobial properties of defined atmospheres have been used to preserve several food products (Table 1). Controlled-atmosphere storage has been used in the meat industry for over 50 years (5) to maintain quality and improve yields during extended shipping periods. The preservation principles of antimicrobial atmospheres have been applied to fruits, vegetables, cheeses, eggs, and prepared foods (5–7).

II. EFFECT OF ATMOSPHERE ON MICROORGANISMS

A. Aerobic vs. Anaerobic Environments

Aerobic bacteria and molds cause most organoleptic food spoilage, while anaerobic environments can promote certain pathogens. Atmospheres containing limited

Table 1 Optimum Gas Compositions for Extending the Shelf Life of Refrigerated Foods

Food	CO_2(%)	O_2(%)	N_2(%)	CO(%)	Ref.
Fresh beef	50	24	25	1	15
Retail meat	100				141
Tellagio cheese	10		90		143
Pork	100				145
Cooked beef	75	10	15		8

amounts of or no O_2 will therefore inhibit food spoilage organisms but not necessarily pathogens. Nitrogen and/or CO_2 are used to create anaerobic atmospheres. Some pathogens (e.g., *Listeria monocytogenes* and *Clostridium botulinum*) that are facultative or strictly anaerobic may inadvertently be selected for or promoted under these conditions. To avoid these hazards, the origin and composition of a food to be preserved by altered atmospheres must be carefully considered.

An atmosphere containing O_2 will deter anaerobic pathogens such as *C. botulinum*, and CO_2 will retard spoilage organisms and aerobic pathogens. Combining CO_2 with O_2 may be effective at inhibiting spoilage organisms and anaerobic pathogens. Hintlan and Hotchkiss (8) combined O_2 and CO_2 in an atmosphere composed of 75% CO_2, 15% N_2, and 10% O_2 and found it inhibited growth of the aerobic spoilage organism *Pseudomonas fragi* as well as pathogens *Samonella* Typhimurium and *Staphylococcus aureus* and the anaerobic pathogen *Clostridium perfringens*. Before utilizing a strictly aerobic or anaerobic atmosphere, the native spoilage and pathogenic flora need to be investigated to ensure a safe extended–shelf life product.

B. Nitrogen

Nitrogen is an inert, soluble gas that has no direct antimicrobial effect. Its low solubility and inert quality makes it an excellent balance gas, allowing specific atmospheric environments to be created. Although N_2 is not directly inhibitory to bacteria, yeast, or molds, it can slow bacterial growth by exclusion of O_2 and hold organisms in a quiescent metabolic state resulting in extended generation times. Aerobic microflora found on chicken drumsticks packaged under N_2 (0.75–0.80 kp/cm^2) and stored at 4.5°C were observed to have incrased generation times compared to those packaged under air (9). Aerobic populations in controls were capable of doubling in number in 0.7 days, whereas N_2 caused generation times of 9.8 days. Anaerobic generation times, however, remained unchanged compared to controls, illustrating the selective effect of N_2 on a specific portion of the bacterial population.

Beauchat (10) compared the growth of yeast and molds on cowpeas and cowpea flour stored under various environmental conditions including 100% N_2. A pure N_2 atmosphere resulted in a retention of viability for total fungi and *Aspergillus flavus* populations inoculated onto the whole cowpeas and flour. These organisms experienced a log cycle reduction on the same product stored under vacuum conditions. Contradictory to this, Lee et al. (11) found that a 100% N_2 atmosphere reduced the population of yeast and molds growing on meat loaves stored at 3 and 7°C in comparison to vacuum-packaged product under the same conditions. It is possible other extrinsic or intrinsic factors could account for this observation such as permeability of the barrier, initial microbial loads, or pH of the food.

C. Oxygen

Respiration, oxidation, and microbial activity are the three major modes of O_2 utilization in flesh foods (12). O_2 is necessary to retain oxygenation of myoglobin and the desirable red meat color. Much of the data collected on the effects of O_2 on foodborne microorganisms comes from stability studies done with red meat (13–15). Oxygen also maintains the quality of foods that continue to respire after harvest. Work with fruits and vegetables under modified and controlled atmospheres containing reduced O_2 levels to control mold growth has provided important data on how microbes respond to varying levels of O_2 (16–18).

1. Bacterial Populations Affected

In food products with pH of 4.7 or greater, anaerobic, toxin-producing sporeformers are a major concern. Oxygen can function as an inhibitor of pathogens and spoilage bacteria that have metabolic limits for O_2. Hintlan and Hotchkiss (8) demonstrated that atmospheres surrounding beef containing 5–10% O_2 were sufficient for preventing the outgrowth of *C. perfringens*. However, in mixed cultures of anaerobic and aerobic organisms, the O_2 may be rapidly consumed by the aerobic organisms allowing for rapid development of anaerobic pathogens (19–21).

Gram-positive and gram-negative bacteria grown under oxidative conditions have varying sensitivities. Killing curves for *S.* Typhimurium and *E. coli* illustrate a multihit killing effect due to oxidative stress (22). Curves for *Sarcina lutea* and *S. aureus* suggested single-hit kinetics. These results suggest that the structure of the cell wall has an effect on the extent of damage caused by O_2.

The authors propose that the lipopolysaccharide (LPS) portion of the gram-negative's outer membrane acts initially as a shield against toxic reactions of pure singlet O_2, but eventually becomes saturated and serves as a pool of secondary reactions that increase the rate of cell death. By comparing the effect of O_2 exposure on mutant (low LPS) and wild-type strains, these researchers showed that outer membrane LPS plays a significant protective role against singlet O_2.

2. Effect on Growth and Metabolism

Bacterial preferences for energy sources change with varying levels of O_2. Bacteria under reduced O_2 atmospheres passed up glucose for amino acids (in beef) (23). The microflora of meat have been shown to catabolize lactate in the presence of O_2, whereas this memtabolite was not utilized by microflora under anaerobic conditions (24). When O_2 levels are low, a changeover from homofermentation to heterofermentation can occur (25), and this will have an indirect effect on the other microbial populations present because the pH and other aspects of the chemical environment will change.

Mild oxidative stress can lead to resistant strains as has been shown with *E. coli* and *S.* Typhimurium (26). Oxy R is a gene induced by oxidative stress that controls transcription of nine genes that code for proteins that indirectly effect the cells defense systems against oxidation. *E. coli* that can express oxy R can achieve resistance to HOCl (an oxidizing agent), and cells without it show increased spontaneous mutation rates.

O'Neill and Bissonnette (27) investigated the effects of antecedent O_2 growth conditions on the recovery of heat-stressed *E. coli* and discovered that exposure to O_2 before heat stress allowed for a more rapid recovery. Such findings illustrate that O_2 concentrations in atmospheres surrounding foods can have an effect on a microorganism's response to stresses commonly encountered during food processing. This is a critical concern when considering postprocessing conditions of foods stored under controlled and modified atmospheres.

D. Carbon Dioxide

1. Factors Affecting the Inhibitory Effect of CO_2

There are many known and putative inhibitory effects of CO_2 on spoilage and pathogenic microorganisms. A number of mechanisms have been investigated: reduction of intracellular pH (28–30), physiochemical alteration and regulation of enzymes (31–36), alteration of membrane integrity (3,37), and futile cycles leading to wasted energy (38–41).

Gram-negative bacteria are more sensitive to CO_2 than gram-positives (42). The degree of inhibition will depend upon physical and biochemical characteristics of the food to be preserved and the prevalence of susceptible bacteria. Carbon dioxide is highly soluble in the aqueous phase and fat portion of foods, components that harbor bacteria and spores. Solubility is also temperature dependent, with lower temperatures allowing for increased solubility. Thus the presence of an aqueous phase is a critical component for effective CO_2 inhibition. Water activity and storage temperature are important factors because they have a direct effect upon the solubility of carbon dioxide.

The inhibitory effect of CO_2 on a specific organism in a given food will be dependent upon the water activity of the food, the concentration of CO_2 dissolved in the product, and the storage temperature (43). For example, when CO_2 is dissolved in an aqueous medium with a pH of less than 8.0, hydration reactions result in the following equilibrium (6):

$$CO_2 + H_2O \leftrightarrow H_2CO_3 \leftrightarrow HCO_3^- + H^+ \quad (1)$$

As the concentration of CO_2 increases, the pH of the aqueous medium will decrease. Decreasing the pH of the aqueous environment stresses microbial populations, which respond by purging their cytoplasm of H^+ ions (44). The acid stress and the energy it takes to increase intercellular pH via proton pumps or neutralizing hydroxyls will detract from the microorganism's ability to grow and proliferate. However, the stress caused by reduced pH does not completely explain how CO_2 inhibits bacterial growth. King and Mabbitt (45) showed that CO_2 has an inhibitory effect beyond its ability to lower pH. Total colony counts of psychrotrophic bacteria were enumerated in whole milk that had been acidified to pH 6.0 with HCl and compared to CO_2 at the same pH. Acidification with HCl allowed a 10-fold greater increase in growth at 7 days of storage compared to the CO_2 acidified milk.

When CO_2 is dissolved in the aqueous phase of a product, the amount of dissolved O_2 will decrease as a result of displacement. King and Mabbitt (45) investigated the effect on growth that is due to displacement of O_2 by CO_2. Growth of aerobic bacteria over a 6-day period in milk containing 1300 ppm CO_2 and 2.7 or 5.3 ppm O_2 was reduced compared to milk containing similar dissolved amounts of O_2 and treated with N_2 instead of CO_2. Earlier work by King and Nagel (46) investigated the growth of *Pseudomonas aeruginosa* under atmospheres of varying levels of O_2 displacement. Growth was not inhibited when 75% of the air was replaced with N_2 but was inhibited when replaced by CO_2. This suggests that displacement of O_2 by CO_2 was not responsible for the observed inhibition.

Sears and Eisenberg (37) proposed an alternative mechanism for the inhibitory effect of CO_2. They suggested that CO_2 altered the cellular membrane by changing its water miscibility and thereby exposing the cell's cytoplasm to its chemical surroundings. Research with yeast by Castelli et al. (47) and Jones and Greenfield (48) found similar connections between CO_2 levels and cellular membrane disruption. These reports indicate that the lipid bilayer is disrupted by CO_2, resulting in a change in ionic permeability. Gram-negative bacteria are more sensitive to CO_2 than gram-positives (42,49,50). For CO_2 to damage the membrane of a gram-positive bacterium, it must traverse a thick peptidoglycan barrier, whereas in gram-negative organisms there is only a thin layer of LPS.

In addition to altering membrane function, there is evidence that CO_2 alters the physiochemical properties of bacterial and fungal proteins (48). Reports also indicate that CO_2 alters gene expression. Stretton and Goodman characterized two genes in *Pseudomonas* sp. S91 that are induced when exposed to 10% CO_2. More specifically CO_2 has been shown to regulate the expression of proteins that play an integral role in capsule and toxin production by *Bacillus anthraxis* (51).

2. Microbial Population Affected

Strict aerobes that have an absolute requirement for O_2 are very sensitive to CO_2 (52). Because their growth is dependent upon the amount of dissolved O_2, the dissolved CO_2 is a major deterrent. Anderson et al. (9) observed that aerobic bacterial generation times on chicken flushed with CO_2 was 4.7 days, whereas controls in air took 0.7 days to double their population.

The inhibition of CO_2 on the growth of psychrotolerant bacteria has been well documented (45,53–58). A majority of this work demonstrates an increase in lag phase on the growth of common gram-negative spoilage bacteria.

The use of CO_2 to create anaerobic atmospheres has generated concern that the risk of botulism is increased in such MAP foods. MAP and DAC of low-acid foods may support the growth of *C. botulinum* as a result of decreased O_2 levels. Glass et al. (59) addressed the issue of increased risk of botulism in milk treated with CO_2. A 10-strain mixture of proteolytic and nonproteolytic *C. botulinum* spores was inoculated into milk treated with CO_2 (400 and 790 ppm) and stored in plastic and glass at 6.1 and 21°C. All samples stored at 21°C were spoiled on a sensory and microbial (standard plate counts reached 10^6) basis before toxin was detectable. At 6.1°C toxin was undetectable in all treatments for 60 days. These results indicated that DAC to milk does not pose a food safety hazard under these conditions.

Different spore forming bacteria respond differently to CO_2. An atmosphere containing 100% CO_2 at 1 atm caused an increase in the germination rate of *Clostridium sporogenes* and *C. perfringens* (60). Higher pressures of 100% CO_2 (10 and 25 atm) prevented sporulation completely. Chen and Hotchkiss (54) observed that *C. sporogenes* inoculated into cottage cheese that had CO_2 directly added to the cream dressing did not grow after 63 days. This inoculum, which was a mixture of spores and vegetative cells, also did not grow in cottage cheese that did not have added CO_2. Earlier work by Baker et al. (61) demonstrated that an 80% CO_2 atmosphere (balance air) resulted in no significant change in viable counts for *C. sporogenes* inoculatled onto ground chicken stored at 2, 7, and

13°C for 5 days. Under an atmosphere of 100% air at the same temperatures for the same duration, counts decreased by 4 log cycles. Bennik et al. (62) observed that the maximum specific growth rate and the maximum population density of *Bacillus cereus* decreased as concentrations of CO_2 increased, and 50% CO_2 completely inhibited growth.

Carbon dioxide has been shown to affect sporulation and toxin production by yeast and molds (63–65). Calderon (66) demonstrated that high-CO_2 and low-O_2 atmospheres inhibited mold growth and toxin production in stored grains.

Surrounding atmospheres effected mold morphology. Carbon dioxide concentrations in aqueous growth media were shown to cause an alteration of hyphal growth of *Aspergillus niger*. As CO_2 increased the degree of hyphal branching decreased and the length of the hyphae increased (67). Work with molds by Tomkins (68) showed that O_2 concentrations ranging from 5 to 20% did not alter the inhibitory effect of CO_2 on the fungi.

The effect of CO_2 on the lag and log phases are dependent upon the bacteria studied (69,70). The lag phase of gram-negative organisms, especially *Pseudomonas* species, increases as levels of atmospheric and dissolved CO_2 increase (1,43,45,53,71,72). A majority of this work has investigated *Pseudomonas fluorescens* and *Pseudomonas putida*. Hendricks (1) observed the growth of *P. fluorescens* in atmospheres containing 20% O_2 and 0, 20, or 40% CO_2 (with N_2 balance) at 7.5°C. As levels of CO_2 increased, the lag phase increased. King and Mabbitt (45) showed a similar effect of CO_2 on the lag phase growth of the same organism. Chen and Hotchkiss (53) also observed this increase in lag phase on a mixture of three *Pseudomonas* species in CO_2-treated cottage cheese.

Moir et al. (72) observed slower growth rates for *P. fluorescens* at 5°C in a 100% CO_2 atmosphere when compared to growth in air. There was a linear relationship betwen CO_2 level and generation time for *P. fluorescens* (43) and *P. aeruginosa* (46). Generation time increased for *P. aeruginosa* when grown in the presence of 70% CO_2.

Enfors and Molin (73) used the relationship between CO_2 level and growth rate to define the relative inhibitory effect (RI) of CO_2 as:

$$RI = ((r_c - r_{CO_2})/r_c) \times 100$$

where r_c and r_{CO_2} are the growth rates of control cultures and CO_2-inhibited cultures, respectively. Enfors and Molin (73) found that the RI for *P. fragi* and *B. cereus* under CO_2 increased as temperatures decreased for both of these bacteria. As Dixon et al. (6) points out in an extensive review on CO_2 inhibition, this enhanced inhibition at lower temperatures is due to the increased solubility of the gas and not due to increased susceptibility to CO_2. More recently, work by Devlieghere et al. (74), established a predictive model for the effect of temperature and dissolved CO_2 on the growth of *Lactobacillus sake*. They found a synergistic effect between dissolved CO_2 and temperature on growth that was not explained by increased solubility. As CO_2 increased the growth rate decreased, but no significant change in lag phase was observed. Perhaps in this scenario the inhibition was due mostly to the depression of pH by CO_2.

E. Carbon Monoxide

Discoloration of beef under high-CO_2 atmospheres led researchers to consider CO for a MA gas. Silliker et al. (75) showed that although CO_2 has bacteriostatic properties, concentrations of 30% or greater caused undesirable discoloration in red meat. Carbon monoxide complexes with myoglobin to form carboxymyoglobin, which is bright red and stable compared to oxymyoglobin.

Inhibitory effects of CO on spoilage and pathogenic bacteria on meat was investigated (76). Growth curves for four different bacteria under atmospheres of varying CO levels (0–30%) were observed. *Escherichia coli* and *P. fluorescens* increased doubling times as the concentration of CO increased. *Achromobacter* cultured under CO atmospheres increased lag phase duration as CO concentrations increased. *Pseudomonas aeruginosa* did not alter its growth patterns under CO atmospheres. It is noted that CO levels at or below 1% neither provided additional safety nor decreased microbial loads. This was confirmed by Luno et al. (15), who compared the effect of atmospheres containing 70, 20, and 10% O_2, CO_2, and N_2, respectively, to 70, 20, 9, and 1%, O_2, CO_2, N_2, and CO, respectively, on the growth of psychrotrophic microflora of beef. Again, no difference in counts was observed after 28 days at 1°C betwen the two variables.

F. Ozone

Ozone has been used as a disinfectant for drinking water in Europe since 1906 (77). Like CO_2, the inhibitory effect of O_3 is dependent upon solubility, which is affected by temperature and pH. When O_3 dissolves in water, it decomposes into hydroxy free radicals and other oxidizing species. High oxidation potential makes O_3 a strong and effective preservative for food systems (77). Its mechanism of inhibition is its ability to oxidize fatty acid double bonds in cell walls and plasma membranes, resulting in a change of cell permeability and lysis (78). This strong oxidizing potential can be detrimental to food quality when O_3 gas is used at high levels.

Rice et al. (7) extensively reviewed the antimicrobial effects of O_3. Much of the research indicated that O_3 can inhibit molds and fungi but has a negligible effect on vegetative bacterial cells and spores. There is evidence that headspace concentrations of 0.2 mg O_3/m^3 can even promote the growth of some bacteria (7). Concentrations of O_3 gas that are sufficient to inhibit bacterial growth oxidized foods. For this reason ozone is often dissolved in water and used as an antimicrobial rinse. This application minimizes the oxidative effects of O_3 while providing antimicrobial activity. Fruits and vegetables that have a protective barrier to O_3 ingress are capable of withstanding higher concentrations without decreased quality.

According to Fournaud and Lauret (79), gram-negative (*P. fluorescens*) and gram-positive (*Leuconostoc* strain) bacteria isolated from beef and grown on culture media were inhibited by a 30-minute exposure to 100 ppm O_3. *Lactobacillus* sp. and *Microbacterium thermosphactum* were not effected by O_3 under these same conditions. Further testing of the inhibitory effect of O_3 on the same four bacteria growing on beef revealed no effect. These experiments suggested that the physical and chemical composition of a food would affect the antimicrobial activity of O_3.

Jindal et al. (78) observed the growth of microbial populations on poultry submerged in chill water containing 0.5 ppm O_3 for 45 minutes and stored at 1–3°C. Aerobic plate counts (APC) decreased in treated samples compared to untreated controls. Shelf life, based on APC levels, increased by 2 days. Ozone-water treatment decreased coliform and gram-positive bacterial counts throughout storage. Gram-negative bacteria including *P. aeruginosa* were reduced, although the inhibition did not last as long as that seen with APC and coliforms. Ozone was effective at reducing counts in the water used to chill the poultry.

Reagan et al. (80) found that an O_3 wash did not have an effect on most of the bacterial populations found on beef carcasses. Ozone levels as high as 2.3 ppm were used, but exposure time was less compared to the study by Jindal et al. (78). *Listeria* was significantly reduced by the O_3 wash, whereas *Salmonella* and *E. coli* were not.

G. Other Gases

Less commonly used gases include nitrous oxide (N_2O), argon, and low molecular weight gases such as ammonia, ethylene oxide, and methyl bromide. Fayer et al. (81) investigated the effects of ammonia, ethylene oxide, and methyl bromide on the pathogenicity of *Crytosporidium parvum* oocysts using a neonatal BALB mouse model. It was found that exposure of the oocysts to these low molecular weight gases prevented infection of the mice. The authors suggest that these gases could be used for disinfecting food-processing environments such as poultry-processing buildings and livestock-housing facilities.

Shelf-life extension of processed fruits and vegetables using MAP containing N_2O has been investigated (82). Microorganisms found on these products were inhibited by N_2O. Growth rates of *P. fluorescens*, *B. cereus*, and a coliform inoculated onto sterile cream stored at 7, 4, and 1°C were reduced by N_2O (83).

Ethanol vapors emitted from a 2 g saturated cotton pad retarded toxin production by *C. botulinum* spores by 10 days on crumpets stored at 25°C (84).

Argon has been used in conjunction with microwave plasmas and other processes to inhibit microorganisms (85,86). However, direct application to foods has not been reported.

III. FACTORS INFLUENCING EFFICACY OF ANTIMICROBIAL ATMOSPHERES

Several intrinsic and extrinsic factors including temperature, product-to-gas volume ratio, initial microbial loads and type, package barrier, and biochemical composition of the food all interact to determine the degree to which microbial food quality and safety are enhanced.

A. Temperature

The most important factor affecting efficacy of antimicrobial atmospheres is temperature. Temperature directly affects growth rate but also indirectly affects growth by affecting gas solubility. For example, lower temperatures increase the solubility of CO_2 in the aqueous phase of foods and therefore have a significant effect on its ability to inhibit bacterial growth (Fig. 1a). Devlieghere et al. (74) characterized this interaction between temperature and dissolved CO_2 on the growth rate of *L. sake*. As expected, lower temperatures increased the solubility of CO_2, resulting in a decreased growth rate compared to untreated controls.

Hintlan and Hotchkiss (8) noted different degrees of inhibition on the growth of *S. aureus* on cooked roast beef in identical atmospheres under different temperatures. The effects of atmospheres containing 75% CO_2 and O_2 ranging from 0 to 25% (N_2 balance) on bacterial growth after 30 days at 13 and 27°C were compared to air-stored controls. The difference between MA samples and controls was greater for all atmospheres at the lower temperature. They suggested this was due to the decreased solubility of CO_2 at higher temperatures.

Growth of *Yersinia enterocolitica* was characterized under a 100% CO_2 atmosphere at 2, 6, and 20°C (87). Compared to growth in air at these same temperatures a 100, 98, and 43% reduction was observed.

Ogrydziac and Brown (88), Enfors and Molin (73), Gill and Tan (89–91), and Clark and Lentz (92) have discussed the effects of temperature on the efficacy of atmospheres containing CO_2. Several reports documented the growth of *Pseudomonas*, *Achromobacter*, *Acinetobacter*, *Bacillus*, *Yersinia*, and *Moraxella* on a variey of substrates (nutrient broths, beef, and chicken) under a range of atmospheres containing CO_2, O_2, and N_2 (or air) at different temperatures. Collectively, these data suggested that the major effect of temperature is on the solubility of CO_2, and when changes in solubility are taken into account, the effects of CO_2 are independent of temperature.

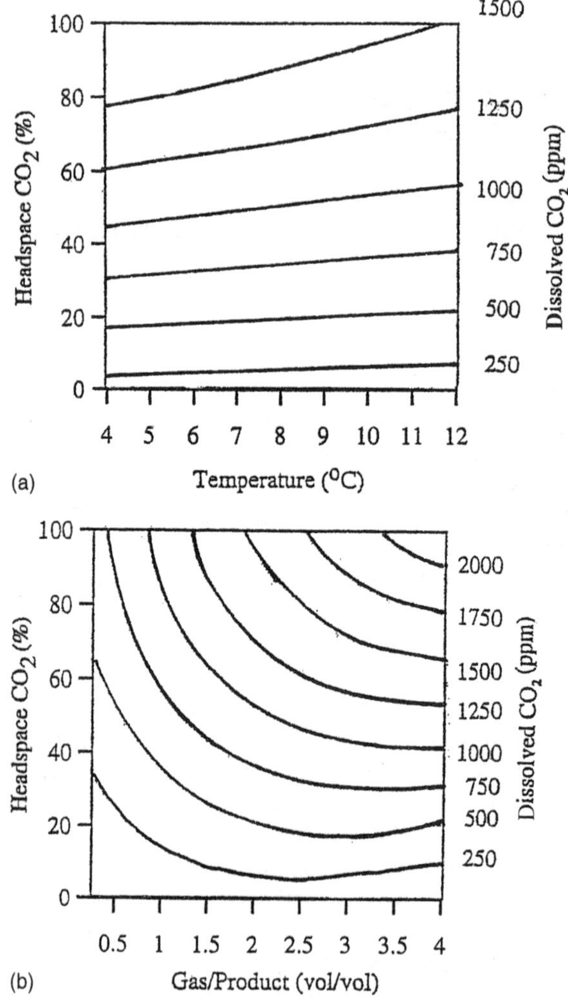

Figure 1 (a) Concentration of dissolved CO_2 in modified brain heart infusion broth as a function of initial CO_2 concentration (%) in the headspace over the broth and storage temperature (°C). (b) Concentration of dissolved CO_2 in modified brain heart infusion broth as a function of initial CO_2 concentration (%) in the headspace over the broth and gas/product (v/v) ratio. (Adapted from Ref. 43.)

B. Headspace-to-Product Volume Ratio

Devlieghere et al. (43) developed a model that determined the relationship between the gas phase concentration of CO_2 and the gas-to-product volume ratio. Gas concentration and product-to-headspace volume ratio were found to be the major factors determining the concentration of CO_2 in the aqueous phase. As the headspace of a package increased, the amount of dissolved gas in the aqueous phase increased (Fig. 1b). Henry's law was incorporated into the model to predict the amount of dissolved CO_2 in the aqueous phase:

$$(CO_2)^E \text{ aq} = K.p\ CO_2$$

where K is Henry's constant (g/L atm), pco_2 is the partial pressure of CO_2 in the gas phase at equilibrium (atm), and $(CO_2)^E$ aq is the concentration of CO_2 in the water phase at equilibrium (6). At practical food storage temperatures, the packaging configurations, particularly the product-to-headspace volume ratio, will play a major role in determining the magnitude of microbial inhibition.

C. Package Barrier Properties

Packaging barrier in MAP has a major effect on microbial growth by influencing the time in which the selected MA gases remain in contact with the product and the rate at which O_2 enters the package. Respiring produce stored under altered atmospheres will interact with the barrier and equilibrate to form new atmospheres.

Rodriguez et al. (93) suggested that barrier properties can affect safety. Indigenous *Bacillus* species may increase pH and allow for the growth of anaerobic *C. botulinum.* They observed the growth of two *Bacillus subtilis* and one *Bacillus licheniformis* strain in aseptically packaged tomato juice packaged in plastic containers of various polymeric compositions under different O_2 partial pressures. Polymers that allowed permeation of O_2 allowed the *Bacillus* to grow and increase the pH by as much as 0.8 units. They cautioned that if low-barrier containers are used, that the number of *Bacillus* spores should be kept to a minimum to prevent possible *C. botulinum* growth.

The effect of O_2 permeation rates of packaging films on the growth of beef microflora was investigated by Grau (94), who compared the growth of *Serratia liquifaciens* and *Enterobacter cloacae* under vacuum. Growths of these spoilage organisms were compared on beef vacuum-packaged in polyethylene, nylon, ethyl vinyl alcohol/polyvinylidene chloride (EVA/PVdc), and aluminum foil. In packaged beef, higher film O_2 permeation rates allowed *S. liquefaciens* counts to reach higher levels. Counts were 10^8 for polyethylene, 10^6 for nylon, and 10^6 cfu/g for EVA/PVdc packaged beef after 21 days at 5°C as would be predicted

Table 2 Days to reach 10^6 cfu/mL in Pasteurized Milk from an Initial Inoculum[b] of Approximately 10^2 cfu/mL and Days Needed to Reach Curdling at 6.1°C

Film permeability[a]	CO_2, ppm (days)			
	0	380	620	930
3801	6.4 (<17)	8.0 (<21)	8.6 (<28)	10.9 (<28)
2040	—	6.4 (<21)	9.7 (<28)	9.9 (<28)
110	—	9.1 (>28)	12.6 (>28)	13.4 (>28)
<0.5	—	9.7 (>27)	13.1 (>28)	13.4 (<28)

[a] cc/m²/24 h @7°C.
[b] Cocktail of *P. fluorescans, E. cloacae, E. aerogens.*
Source: Adapted from Ref. 95.

based on permeation rates. In contrast, no growth occurred for at least 24 days for either organism on beef packaged in the aluminum foil.

Hotchkiss et al. (95) illustrated the relationship betwen film barrier and inhibitory effect of CO_2 on bacteria in fluid milk packaged in pouches. Lower CO_2 permeability films resulted in increased lag phases for typical spoilage bacteria found in milk (*P. fluorescens, E. cloacae, Enterobacter aerogenes*). Table 2 summarizes this data.

D. Physical and Chemical Composition of Products

Water activity, salt content of the aqueous phase, pH, and fat content of foods play a role in overall inhibitory effects of antimicrobial gases. Just as with temperature, these physical and chemical characteristics of the food product have an effect on the solubility of the inhibitory gas. For example, increasing salt concentration decreases CO_2 solubility (96).

The Devlieghere et al. model (43) included the effect of pH on the solubility of CO_2. As pH increased the solubility of CO_2 increased. Therefore, foods with aqueous phases with higher pH will allow for a greater amount of CO_2 to be dissolved.

These physiochemical properties have been exploited in a method used to determine the amount of CO_2 dissolved in liquid foods (71). Sulfuric acid is added to the food to be tested and then agitated in a closed vessel, chemically and physically forcing all of the dissolved CO_2 out of the aqueous phase. A sample of headspace gas is removed and the percentage of CO_2 dissolved in the product determined by gas chromatography or other instrumentation.

Lowering pH enhanced the antimicrobial effect of anaerobic environments (100% N_2). Grau (94) studied the growth of *E. cloacae* on beef that had pH values of 6.3 and 5.4 in air and anaerobically (100% N_2) and stored at 5°C. The lower-pH beef did not support growth compared to high-pH beef under the same conditions. Growth of *S. liquefaciens* was inhibited on beef with pH 5.4 in 100% N_2 at 5°C yet grew to levels of 10^8 cfu/cm^2 in 8 days on meat with a pH of 6.3. *Yersinia enterocolitica* failed to grow on beef ranging in pH from 5.4 to 5.9 under a 100% N_2 atmosphere but grew at pH of 6.0–6.2. Under aerobic conditions pH had little effect on the growth of *Y. enterocolitica*.

The extent to which fat content affects inhibition due to CO_2 is likely dependent upon the type of fat present in the food. Devlieghere et al. (43) suggested that lipids that confer a high solid fat index absorb less CO_2 than lipids that have a low solid fat index. Carbon dioxide is more soluble in fat than in water, and therefore fat could sequester CO_2 from the aqueous phase where it has a greater bacteriostatic potential.

E. Initial Microbial Population and Type

Microbial quality of a product going into an antimicrobial atmosphere will have an effect on the extent to which the altered atmosphere can provide a safe, high-quality product for the consumer. Law and Mabbitt (97) showed that milk with initial high bacterial counts held under 1 atm pCO_2 reached unacceptable bacterial levels earlier than milk that had started with lower counts. The better the initial microbial quality of foods to be stored in inhibitory altered atmospheres is, the longer it will take for the bacteria to attain unacceptable levels.

The type and numbers of microflora can influence the effectiveness of MAP. Under a 60:40 (CO_2:N_2) atmosphere Harris and Barakat (98) found that *L. monocytogenes* and *Y. enterocolitica* growth was not affected by background microflora. Wimpfheimer et al. (99) found an increase in *L. monocytogenes* as background aerobic flora decreased under a 72.5:22.5:5 (CO_2:N_2:O_2) atmosphere.

The interaction between aerobic and anaerobic microorganisms and *C. botulinum* under modified atmosphere have been investigated (19–21,100,101). Early work (19) investigated *C. botulinum* type E toxin production together with aerobic plate counts on vacuum-packaged herring fillets stored at 20°C. Toxin was undetectable until after 24 hours, the same point at which total aerobic plate counts exceeded 10^6 CFU/g. Once aerobic growth reached these levels, it was proposed that residual O_2 had been used up and oxidation-reduction potentials reached levels that promoted *C. botulinum* growth.

Safety concerns with *C. botulinum* outgrowth on foods packaged under anaerobic atmospheres were addressed by adding low levels of O_2 to the headspace as an inhibitory mechanism (8). Subsequent research took a close look at the microbial ecology under similar conditions and found that aerobic organisms can quench the added O_2 and generate an environment that supports *C. botulinum* growth and toxin production.

Lambert et al. (21) found that MAP pork in an atmosphere containing O_2 enhanced toxin production by *C. botulinum* type A and B spores. In fresh pork inoculated with a 10-strain cocktail of spores stored at 25°C under an atmosphere containing 90% CO_2 and 10% O_2, toxin was detectable in 2 days. In pork samples stored at 15°C under an atmosphere containing 20% O_2 and 80% CO_2, toxin was detectable within 14 days. However, toxin was detectable at 21 days in samples stored at 15°C under atmospheres containing 0% initial O_2 and 100% CO_2. The atmosphere containing an initial level of 20% O_2 had 0% O_2 by 10 days of storage. This decrease in O_2 content is attributable to consumption by aerobic bacteria.

The effect of the numbers of organisms on growth is less clear. In most cases, higher numbers of spoilage organisms will reduce the likelihood of pathogen development due to competition. However, as the work of Smith et al. (20) shows, this is not necessarily the case. High levels of aerobic organisms can create anaerobic environments by consuming oxygen, creating conditions that allow rapid growth of pathogenic anaerobes. Similarly, higher numbers of pathogens do not necessarily affect growth. Wimpfheimer et al. (99) varied *L. monocytogenes* initial innoculum levels (10^1 vs. 10^3) to see how the organism would grow under a 72.5:22.5:5 (CO_2:N_2:O_2) modified atmosphere at 4°C. Within 20 days the counts were the same and the growth curves for both innoculum levels were similar, suggesting that initial bacterial counts of *L. monocytogenes* did not play a significant role in how they respond to this modified atmosphere.

Francis and O'Beirne (102) developed a model that characterized the interactions of indigenous microflora of lettuce with *L. monocytogenes* under modified atmospheres. Under modified atmospheres containing low O_2 concentrations, the growth of *L. monocytogenes* was enhanced by the presence of competitive microflora in comparison to pure culture of *L. monocytogenes* growing under similar conditions. Such interactions between pathogens and spoilage bacteria needs to be continuously assessed for new packaging conditions and new products packaged under modified atmospheres.

F. Laboratory Culture Media Versus Food Systems

Laboratory culture media do not accurately simulate the complexities of food composition, nor can the effects of production, processing, storage, and distribution be predicted for MA systems using media. Factors such as initial microbial load, water activity, fat content, and pH vary considerably between laboratory media and an actual food product. Fournaud and Lauret (79) observed an inhibition of *P. fluorescens* and a *Leuconostoc* species growing on laboratory media by O_3 but did not see the same inhibition on beef. The more dynamic system (beef) introduced factors that interacted with the gas's inhibitory effect.

IV. SAFETY

The major safety consideration in extending shelf life of foods by MAP or related technologies is the loss of organoleptic spoilage cues provided by spoilage bacteria. Without spoilage indicators it is conceivable that a food could have acceptable organoletic quality but be unsafe (103). Modified atmospheres that control spoilage bacteria potentially generate three risks: in the absence of spoilage indicators toxic food may be consumed, competitive inhibition against pathogens is decreased, and atmospheres containing increased CO_2 and/or altered O_2 levels may effect the sporulation (i.e., toxin formulation) rates of anaerobic pathogens. Fruits and vegetables that are stored and/or packaged under modified atmospheres for shelf-life extension purposes may pose a significant hazard because these foods are typically eaten without further cooking (104). Technologies that suppress senescence and extend shelf life may provide time for pathogen development that would not normally occur in products whose shelf life had not been extended.

The effect of the loss of competitive inhibition by spoilage bacteria is most pronounced on the facultative anaerobic pathogenic bacterial populations in foods under altered atmospheres (12). Growth of *Y. enterocolitica* in minced meat was inhibited by background flora under an atmosphere of 20% CO_2 and 80% O_2 compared to meat stored in air (105). At 4°C initial background flora (Pseudomonadaceae) at levels of $<10^2$ and 10^5 under the modified atmosphere inhibited growth of *Y. enterocolitica* by 1 and 4 log cycles, respectively, compared to controls stored in air.

Hintlan and Hotchkiss (8) suggested a ''safety index'' for cooked roast beef, which provides an acceptable ratio of spoilage to pathogenic bacteria under certain atmospheres and temperatures. As this ratio increases, the hazards of infection from pathogens decrease. A gas mixture of 75% CO_2, 10% O_2, and 15% N_2 was determined to be the safest and most effective (for roast beef inoculated with *S. aureus*, *S.* Typhimurium, and *C. perfringens*).

A. *Listeria monocytogenes*

The effects of modified atmospheres on *L. monocytogenes* has been investigated (2,8,54,99). Avery et al. (2) found that beef steaks sealed in stomacher bags saturated with CO_2 (2 L CO_2/500 g beef) inhibits *L. monocytogenes* by 0.6 and 1.5 log cycles compared to vacuum packaging at 5 and 10°C. They concluded that there is no increase in risk from *L. monocytogenes* under modified atmosphere. Chen and Hotchkiss (54) found that CO_2 inhibits *L. monocytogenes* inoculated into cottage cheese.

Others have found that CO_2 can enhance *L. monocytogenes* growth, presumably by inhibiting competitive organisms. For example, Wimpfheimer et al.

(99) compared the growth of *L. monocytogenes* and aerobic microflora on ground chicken under MAP in commercial packaging containing 72.5:22.5:5 (CO_2:N_2:O_2). *Listeria monocytogenes* (Scott A, Serotype 4) outgrew the aerobic spoilage flora. Aerobic plate counts under the modified atmosphere were reduced by 4 log cycles compared to growth in air, whereas *L. monocytogenes* increased by 6 log cycles under the modified atmosphere.

Zeitoun and Debevere (106) introduced a lactate buffer system to chicken held under modified atmospheres to retard growth of *L. monocytogenes* and extend shelf life. Shelf life increased by 4 days, but product quality from an organoleptic standpoint was not evaluated. Harris and Barakat (98) revealed that supplementation of MAP products with lactate buffers as a means for deterring pathogens was insufficient. Under a 40:60 (CO_2:N_2) atmosphere, ready to eat (RTE) poultry cuts were inoculated with *L. monocytogenes* or *Y. enterocoltica*. Lactate buffers were not a sufficient substitute for temperature control and proper sanitation in maintaining safety of MAPs.

Berrang et al. (107) investigated the effect of CAS on *L. monocytogenes* on asparagus, broccoli, and cauliflower stored under different atmospheres that contained 11–19% O_2 and 3–10% CO_2 (N_2 balance). At 4 and 15°C, growth rates for *L. monocytogenes* did not differ between the CAS and control (air) samples. This suggested that for these vegetables CAS did not alter the growth of *L. monocytogenes*. Francis and O'Beirne (102) investigated the effects of modified atmospheres on the growth and survival of *L. monocytogenes* and other indigenous microflora found on lettuce. Twenty percent CO_2 increased the lag phase for *L. monocytogenes* but did not alter final population density. Atmospheres of 5–10% CO_2 (with 5% O_2 and an N_2 balance) did not inhibit the growth of the pathogen in comparison to growth under air. Nitrogen (100%) did not prevent *L. monocytogenes* from surviving, although it did cease to grow. When O_2 was present in low concentrations (3%) *L. monocytogenes* grew in the presence of typical microflora found on lettuce but grew less vigorously when present in pure cultures. Francis and O'Beirne (102) concluded that *L. monocytogenes* growth could be enhanced by modified atmospheres used for minimally processed vegetables.

B. *Aeromonas hydrophila*

Aeromonas hydrophila is inhibited by CO_2 (108,109). An atmosphere of 36% CO_2, 13% O_2, and 51% N_2 reduced the growth of *A. hydrophila* on surimi stored at 5°C for 8 days by 3 log cycles compared to growth in air (110). Subsequent research indicated that this inhibition has a minimum requirement of CO_2. Growth of *A. hydrophila* on fresh vegetables (asparagus, broccoli, and cauliflower) was unhindered by CAS (111). Each vegetable was stored at 4 and 15°C under slightly different atmospheres with gas ranges of 18–11% O_2, 3–10% CO_2, and 79% N_2.

C. *Bacillus cereus*

Bacillus cereus spores and vegetative cells are sensitive to CO_2 atmospheres (5). Enfors and Molin (60) found 100% CO_2 reduced *B. cereus* spore germination by 70–90% compared to that in 100% N_2. Molin (112) found 100% CO_2 resulted in a 67 and 83% reduction in growth compared to aerobic and anaerobic (5% CO_2, 95% N_2) conditions, respectively.

D. *Yersinia enterocolitica*

Ecklund and Jarmund (87) reported 100% inhibition of growth of *Y. enterocolitica* on BHI under an atmosphere of 100% CO_2 at 2°C. However, changing substrate and storage temperature to high pH beef stored at 5 and 10°C, 100% CO_2 did not prevent *Y. enterocolitica* from growing (109).

E. *Campylobacter jejuni*

Hanninen et al. (113) observed that *C. jejuni* was capable of growing on beef under an atmosphere of 85% N_2, 10% CO_2, and 5% O_2 at 37°C. Two of the three strains investigated were capable of growing beyond 2 log cycles within 48 hours under vacuum and in 80% N_2, 20% CO_2 atmosphere. At 20 and 4°C these three atmospheres did not support growth of any of the strains. Severe temperature abuse of beef stored under ideal atmospheric conditions [80% N_2, 20% CO_2 (114)] could allow for growth.

Kiggins and Plastridge (115) found that 5% O_2, 10% CO_2, and 85% N_2 enhanced growth of *C. jejuni* compared to 100% N_2 and 80% CO_2 and 20% N_2 (116). Beef packaged under 100% N_2 and stored at 4°C was less inhibitory to growth of *C. jejuni* than beef stored under an 80% CO_2, 20% N_2 atmosphere. The lack of difference in survival of this pathogen under these CO_2-containing atmospheres led Stern et al. (116) to conclude that there is little need for concern about *C. jejuni* growth in standard MAP parameters.

F. *Clostridium botulinum*

MAP in certain cases can affect production of botulinum toxin in foods. Kautter et al. (117) found differences in toxin production among different strains of *C. botulinum* spores inoculated onto RTE sandwiches packed in N_2-flushed jars stored at 8 and 12°C. Over a 30-day storage period type A and B strains did not produce detectable amounts of toxin at 12°C but type E did. Sandwiches inoculated with spores from type A and B strains were also held at abusive temperature (26°C) under N_2 atmosphres and became toxic within 4 days although still considered organoleptically acceptable. Aerobic controls had spoiled within 3 days, yet toxin production was not detectable until 7 days.

Daifas et al. (100) compared the amount of time required for *C. botulinum* to produce toxin in crumpets under atmospheres containing 100% CO_2 and 60% CO_2 (N_2) balance stored at 25°C. The pure CO_2 atmosphere delayed toxin production as much as 3 days compared to the 60% CO_2 atmosphere.

V. TECHNOLOGIES

A. Controlled-Atmosphere Storage

The composition of the atmosphere is continuously regulated within a relatively narrow range in CAS. A CAS system, as defined by Robertson (118), consists of a refrigerated storage room in which the gas composition is controlled by vents (desired gases added) and chemical scrubbers (unwanted gases removed). This (CAS) technology is used for storage of fruits and vegetables (119) or for transport of beef over extended distances (120) and has not been widely applied to packaged products.

B. Modified Atmosphere Packaging

MAP is the packaging of a perishable product in an atmosphere that has been modified so that its composition is other than that of air. It differs from CAS in that the composition of the atmosphere changes over the storage time due to product and microbial respiration and package permeability. Temperature, native microbial population, and barrier composition affect atmosphere composition and efficacy.

MAP utilizes CO_2, O_2, and N_2. Permeability of the barrier allows an exchange of gases into and out of the package, which will have an effect on microbial growth. By manipulating the permeability of the barriers, atmospheric conditions are somewhat under control. The use of modified barriers to maintain specific atmospheric compositions is an area of research that is of particular interest to the fresh and minimally processed produce industry.

C. Direct Addition of CO_2

DAC is the incorporation of CO_2 directly into products rather than allowing a less controlled exchange with the atmosphere. This application has been most widely applied to dairy products because existing processing equipment can be fitted with inline sparging systems. Also, gram-negative spoilage organisms commonly found in dairy-based foods are significantly inhibited by CO_2. CO_2 is incorporated into the fluid dairy product after pasteurization and before filling into packages with selected barrier properties.

Alternatively, CO_2 can be added to milk before processing (e.g., in holding tanks) as a way to reduce lipolytic and proteolytic degradation. Ruas-Madiedo et al. (57) bubbled CO_2 into raw milk and held it in closed containers at 4°C for 4 days before degassing and pasteurizing. Carbon dioxide is removed because it is believed that the reduction in pH of the milk due to the CO_2 will result in increased deposition on the heat exchange surfaces. Seven days postpasteurization the previously CO_2-treated milk had total plate counts below 10^5, whereas controls had counts above regulatory levels.

D. Hypobaric Storage

Hypobaric, or low-pressure (LP), storage is a preservation technique based on the combined effects of temperature and atmospheric pressure control. In many ways, LP storage mimics vacuum-packaging. An LP system typically consists of a vacuum tank, vacuum pump, pressure regulator, humidifier, and refrigeration unit. Pressures between 10 and 150 mmHg, temperatures of 2–4°C, and relative humidity (RH) of 85–95% can be used under normal atmospheric compositions to preserve fruits, vegetables, meat, and fish (66). Restaino and Hill (121) successfully increased the lag phase of spoilage bacteria on broiler chickens by 11 days under hypobaric storage conditions. Pork chops and bone in hams showed decreased bacterial growth under the same conditions. *Pseudomonas* spp. and *Streptococcus* ssp. were the predominant bacteria isolated from these meats held under hypobaric storage.

E. Ozone

Ozone is produced for food storage in two ways: UV bulbs (for lower ozone production) and corona discharge technique (for higher concentrations). Ozone is generally recognized as safe (GRAS) for bottled water treatment and as a sanitizer for processing equipment used for bottled water (77). Cold storage rooms equipped with UV bulbs (generating wavelengths of 200 nm or less) can generate ozone in the air at low levels, which dissolves into the aqueous phase of the foods being stored. Gaseous O_3 is GRAS at levels of 0.1 ppm in coolers used for meat aging (122). There have been many other applications of O_3 around the world including grapes in Israel (123), beef in the United States (80), and salmon farming (124). Ozonized water or ice can be used to decrease bacterial growth rates on poultry and fish, respectively (7).

VI. MODELING OF MICROBIAL GROWTH UNDER ALTERED ATMOSPHERES

Prediction of microbial growth using multifactor models is a useful technique for estimating the efficacy of antimicrobials in food systems. Due to the complexity of food systems (the varying physical and chemical parameters and the diversity of bacterial populations they support), modeling of microbial growth in food systems has been difficult. Models have been developed that incorporate several parameters including the composition of the atmosphere. Table 3 summarizes the organisms, growth media, and other factors modeled in some of these studies.

Sutherland et al. (125) established a four-factor model that accounted for the effects of CO_2, NaCl, pH, and temperature on *E. coli* grown on tryptone soya broth (TSB). Their results suggested that *E. coli* O157:H7 is relatively resistant to CO_2 (up to 80–100%). Devlieghere et al. (43) established a predictive model to determine the effects of dissolved CO_2 concentration in BHI broth. The amount of CO_2 initially present in the headspace, gas-to-product volume ratio, and temperature were found to be the most significant factors affecting dissolved CO_2 concentration. As gas-to-product volume ratios decreased in configurations containing the same initial CO_2 concentrations, the dissolved CO_2 concentrations decreased. Changes in temperature between 4 and 12°C were found to have a marginal effect on dissolved CO_2 concentrations compared to temperature above realistic refrigeration temperatures. From this the authors concluded that the microbial cell is more susceptible to CO_2 at lower temperatures. This model emphasizes the importance of optimizing packaging configurations in order to create an ideal bacteriostatic MAP.

Table 3 Organisms, Growth Media, and Factors Modeled in Studies Developing Models of Atmosphere Effects on Microbial Growth

Organism/ Factors modeled	Growth medium	Intrinsic and extrinsic factors	Ref.
Lactobacillus sake	Modified brain heart infusion	Temperature, dissolved CO_2	74
Escherichia coli	Tryptone soya broth	pH, CO_2, NaCl, temperature	125
Bacillus cereus	Tryptone soya agar	Temperature, pH, NaCl, CO_2	146
Aeromonas hydrophila, Yersinia enterocolitica, Listeria monocytogenes, B. cereus	Brain heart infusion agar	O_2, CO_2	62
A. hydrophila	Modified brain heart infusion	Temperature, water activity, dissolved CO_2	147
L. monocytogenes	Modified tryptone soya broth	CO_2, NaCl, pH, temperature	126

Devlieghere et al. (74) have expanded their model to include maximum growth rates and lag phase duration of *L. sake* using dissolved CO_2 and temperature as the major parameters. The growth rate of this organism was repressed as CO_2 concentration increased, but lag phase was not effected.

Fernandez et al. (126) developed a model to predict the growth of *L. monocytogenes* in buffered TSB in atmospheres ranging from 0 to 100% CO_2, temperatures of 4–20°C, sodium chloride (0.5–8.0%), and pH (4.5–7.0). The inhibitory effect of CO_2 was shown to increase with decreasing temperatures. Doubling times increased as a result of the increased solubility of CO_2 at lower temperatures.

Modeling microbial growth and toxin production on foods stored under modified and controlled atmospheres is a very difficult endeavor. Generating growth media that accurately simulate very dynamic food systems is a major challenge. In some cases, for example, the change in pH of a food compared to synthetic media as a result of high-CO_2 atmospheres has not been considered. Many foods are highly buffered where media are not. In addition, strain-to-strain variation and recreating physiological states of the bacteria during modeling so that they accurately depict that of those found in real systems has impeded the ability to directly apply various models.

VII. APPLICATIONS

Modified atmospheres have been commercially used to reduce microbial spoilage and senescence of fruits and vegetables (127–129), raw chicken, beef, and fish (130), RTE foods (131,132), and dairy foods.

A. Minimally Processed Vegetables

Storage life of cut lettuce for salads significantly decreases due to the cellular fluids that are released from the lettuce during processing (133). Diaz and Hotchkiss (104) were able to extend the visual shelf life of iceberg lettuce by 300% by using an atmosphere containing 30% CO_2, 5% O_2, and 65% N_2. Standard plate counts under this atmosphere were lower than under air and other modified atmospheres. Aerobic plate counts at the time of visual spoilage did not, however, differ for lettuce regardless of the atmosphere under which it was held. Safety concerns were raised regarding the growth of *E. coli* O157:H7 on extended–shelf life lettuce because this pathogen reached higher counts. Hazards of foodborne illness are compounded by the fact that lettuce is not typically heat-treated, and the conditions under which it is processed may allow for *E. coli* contamination. This is an example where environmental factors and initial bacterial load play an important role. Beuchat and Brackett (134) observed the growth of *L. monocytogenes*, mesophilic aerobes, psychrotrophs, yeast, and molds on shredded and nonshredded lettuce stored under 3% O_2 and 97% N_2. Growth of *L. monocytogenes* from 10^5 to 10^8 on both lettuce preparations in 10 days at 10°C under this modified atmosphere may increase the risk of listeriosis infection. Little difference in growth among the other microbial populations was observed between MAP and air-packaged lettuce. The significance of this work is that it demonstrates MAP conditions that are unsafe and do not extend shelf life of lettuce.

B. Grains and Legumes

Mold growth and toxin formation are the major concerns for altered atmosphere storage of grains and legumes. Long-term controlled-atmosphere storage of peanuts under CO_2 was investigated by Wilson et al. (135) to determine its effects of quality and safety due to molding and aflatoxin production. Bins of peanuts held under controlled humidity and recycled CO_2 for 1 year did not support mold growth or allow for aflatoxin production. However, minor changes in the general microflora were observed. Peanuts stored in ambient and stagnant CO_2 conditions supported growth of yeast species and *Penicillium roqueforti* Thom. in localized oxic portions of the stored nuts. This illustrates the efficacy of CO_2 inhibition on strict aerobic mycoflora.

A review by Banks (136) discusses the research that elucidates the most beneficial atmospheres for storing cereal grains. Low O_2 and increased levels of CO_2 have been shown to slow mold growth significantly but not eliminate it as well as to reduce aflatoxin formation. Less than 5% O_2 decreased sporulation, and greater than 20% CO_2 decreased production of aflatoxin (137).

C. Meat

Microbial spoilage of meat will be affected by the meat substrate, the bacteria present, and the temperature conditions (138). Most spoilage bacteria found on meat are *Pseudomonas*, *Moraxella*, and *Alteromona* (*Salmonella*) *putrefaciens* (139). These bacteria cause the most undesirable changes to the surfaces of meats exposed during packaging and processing. As described earlier, CO_2 can be a very effective inhibitor of this microbial population. *Lactobacillus* and *Brochothrix thermosphacta* are the two major representatives of the gram-positive bacteria on meat (140). These bacteria are fairly resistant to the effects of CO_2.

Tewari et al. (141) reviewed the centralized packaging of retail meats with consideration of atmospheric compositions optimal for shelf-life extension and quality retention. Atmospheres of 100% CO_2, 100% N_2, and 70% N_2, 30% CO_2 are recognized as having potential for use in centralized distribution of retail-ready meats if initial microbial loads of psychrotrophic lactic acid bacteria can be minimized. An atmosphere of 100% CO_2 under −1.5 to 0.5°C is recognized as the optimal storage condition for retail meat cuts.

Luno et al. (15) also worked on optimizing the atmosphere used to extend shelf life and safety of fresh beef. The traditional O_2, CO_2, N_2 mixture was supplemented with 1% CO, creating a modified atmosphere of 25% N_2, 24% O_2, and 50% CO_2. This gas mixture extended shelf life by decreasing microbial spoilage while minimizing the hazard of pathogen growth.

In Avery et al.'s work on CAP of ultimate pH beef (2), the growth of lactic acid psychrotrophic bacteria under vacuum was compared to a CO_2 atmosphere in CAP. At 5°C under vacuum, lactic acid bacteria attained growth rates 1.8 times that under CO_2 in CAP at that temperature. At 10°C, however, these same atmospheric conditions did not significantly effect the growth rates of these bacteria.

D. Dairy Products

King and Mabbitt (45) showed that CO_2, when added to raw milk stored at 8°C, had the effect of extending the lag phase growth of *P. fluorescens* but did not change the rate of growth during the log phase. Increasing concentrations of CO_2 increased the lag phase of growth. Law and Mabbitt (97) inhibited the growth of psychrotrophic bacteria in milk by stirring it under 1 atm pco_2. At storage temperatures of 4 and 7°C, psychrotrophic counts were held below 10^6 for an additional 2 and 3 days, respectively, compared to controls that were stirred under a headspace containing air. This documented the ability of CO_2 to extend the shelf life of milk by inhibiting gram-negative psychrotrophic spoilage organisms.

Hotckiss et al. (95) points out that two factors will determine the efficacy of DAC in order to extend shelf life of dairy products: control of CO_2 addition and barrier composition. Hotchkiss et al. (unpublished data) investigated the effects of DAC on the gram-negative and total bacterial counts in raw milk that has been concentrated using reverse osmosis (RO) and ultra-filtration (UF) techniques. Ultra-filtered whole milk concentrate containing 1115 ppm dissolved CO_2 and stored at 7°C had bacterial counts of 10^5 at 14 days of storage compared to untreated controls that supported growth up to 10^8. RO whole milk concentrate containing 1100 ppm dissolved CO_2 stored at 7°C had counts of 10^3 and 10^5 for gram-negative and total bacteria counts, respectively, after 7 days compared to 10^6 for both gram-negative and total bacteria counts in control concentrates. Decreasing the storage temperature allowed for even more pronounced inhibition. Bacterial counts in carbonated RO concentrates stored at 2°C remained unchanged (10^3 and 10^2 for total counts and gram-negatives, respectively) for 21 days. The noncarbonated RO control concentrates reached counts of 10^6 by 21 days of storage at this same temperature. This study shows that DAC has potential to extend the shelf life of fluid dairy products.

Spoilage of cottage cheese is due mainly to the growth and enzymatic activity of *Pseudmonas*, *Proteus*, *Aeromonas*, and *Alcaligenes* spp. (142) and yeast and molds such as *Geotrichum*, *Penicillium*, *Mucor*, and *Alternaria* (54). The lag phases of *Pseudomonas fluorescens* and *Pseudomonas putida* inoculated into the cream dressing of cottage cheese were extended by the addition carbon dioxide (72). DAC-treated cottage cheese showed a pronounced and uniform inhibitory effect on these bacteria, compared to simply flushing the headspace. Moire et al. (72) also observed a decrease in the growth rate of *P. fluorescens* in DAC-treated cottage cheese stored at 15°C. Chen and Hotchkiss (53) demonstrated a similar effect of DAC. Cream dressing was inoculated at levels of 10^3 with a mixture of three *Pseudomonas* species (*fluorescens*, *aeruginosa*, and *marginata*). CO_2 was incorporated into half of the dressing, and both the carbonated and control portions were mixed with curd in two different batches. Samples were stored at 4 and 7°C in glass jars and standard plate counts taken periodically over an 80-day period. *Pseudomonas* growth did not occur in the CO_2-treated samples held at 4°C for 70 days. Yeast and molds were also not detected during this period. In comparison, the controls reached counts (noted to be predominantly gram-negative psychrotrophs) of 10^7 in 17 days.

In a later work, Chen and Hotchkiss (54) conducted a more extensive experiment in which they confirmed the above-mentioned effect of DAC to cottage cheese on spoilage organisms and followed the growth of *L. monocytogenes* and *Clotridium sporogenes* inoculums. *C. sporogenes* inoculated into product packaged in polystyrene tubs and held at 4, 7, and 21°C did not grow over a 63-day period in either CO_2-treated or control samples. *L. monocytogenes* inoculated at counts of 10^4 reached levels of 10^7 in 28 and 7 days when stored at 4 and 7°C respectively, whereas CO_2 treated samples suppressed growths to 10^4 and 10^5 for 63 days when stored at 4 and 7°C, respectively. These data reinforced the inhibitory effect of CO_2 on common psychrotrophic dairy spoilage organisms and a known dairy pathogen.

The effect of varying concentrations of N_2 and CO_2 gas in the atmospheres surrounding Tellagio cheese on total plate counts (TPC) and yeast and molds was investigated by Piergiovanni et al. (143). An atmosphere containing 10% CO_2 and 90% N_2 reduced TPC and yeast and molds by one log cycle of growth each after 21 days of storage at 6°C.

VIII. CONCLUSIONS

Advantages of using ambient and subambient pressure gases to inhibit microbial growth on and within foods are many. Reducing microbial loads with minimal heat treatments is a way of improving overall quality and reducing energy costs. As modeling techniques improve, specific atmospheres will be tailored for specific foods that will inhibit target organisms related to spoillage and foodborne illness. CO_2, the major inhibitory gas used in modified and controlled atmospheres, is a natural ingredient and in cases can be considered a processing aid.

One of the limitations of utilizing gases for microbial inhibition is the need for strict temperature control. As we have seen the relationship between temperature and efficacy of an inhibitory gas is not entirely understood. In general, lower temperatures have enhanced the antimicrobial effect of CO_2. But as we have also seen, food composition and initial microbial loads can play a crucial role in extending shelf life and maintaining safety. Bacterial populations are designed to adjust to such stresses as are generated by antimicrobial atmospheres. For this reason continued vigilance of the microbial interaction with these food systems needs to be maintained. This task will become easier as HACCP plans become more prevalent and microbial models become more reliable.

Combining antimicrobial atmospheres with other bacteriostatic techniques will generate hurdle technologies that can further enhance food quality and safety. For example, combining DAC with heat treatment to decrease D-values of microorganisms in milk has been described (144). The mechanism of CO_2 inhibition needs to be fully understood in order to optimize packaging and storage conditions. More models need to be generated that pertain to a wider variety of foods and the spoilage and pathogenic organisms they support.

REFERENCES

1. TM Hendricks. Modeling *Listeria monocytogenes*, *Pseudomonas fluorescens*, and spoilage organisms growth in defined media and shredded lettuce stored under controlled atmosphere. Ithaca, NY: Cornell University, 1996, p. 87.
2. SM Avery, JA Hudson, N Penney. Inhibition of *Listeria monocytogenes* on normal ultimate pH beef (pH 5.3–5.5) at abusive storage temperatures by saturated carbon dioxide controlled atmosphere packaging. J Food Prot 57(4):331–333, 1994.
3. L Nilsson, Y Chen, ML Chikindas, HH Huss, L Gram, J Montville. Carbon dioxide and nisin act synergistically on *Listeria monocytogenes*. Appl Environ Microbiol 66:(2)769–774, 2000.
4. J Salmon, L LeGall. Application of ozone to maintain the freshness and to prolong the durability of storage of fish. Rev Gen Froid (11):317–322, 1936.
5. JM Farber. Microbiological aspects of modified-atmosphere packaging technology—a review. J Food Prot 4(1):58–70, 1991.
6. NM Dixon, DB Kell. A Review—the inhibition by CO_2 of the growth and metabolism of microorganisms. J Appl Bacteriol 67(10):109–136, 1989.
7. RG Rice, JW Farquhar, JL Bollyky. Review of the applications of ozone for increasing storage times of perishable foods. Ozone: Sci Eng 4:147–163, 1982.
8. CB Hintlan, JH Hotchkiss. Comparative growth of spoilage and pathogenic organisms on modified atmosphere-packaged cooked beef. J Food Protection 50(3):218–223, 1987.

9. K Anderson, D Fung, F Cunningham, V Proctor. Influence of modified atmosphere packaging on microbiology of broiler drumsticks. Poultry Sci 64(2):420–422, 1985.
10. LR Beauchat. Survival of *Aspergillus flavus* conidiospores and other fungi on cowpeas during long-term storage under various environmental conditions. J Stored Prod Res 20(3):119–124, 1984.
11. BH Lee, RE Simard, CL Laleye, RA Holley. Shelf life of meat loaves packaged in vacuum or nitrogen gas, effect of storage temperature, light, and time on the microflora change. J Food Prot 47(2):128–133, 1984.
12. TB Labuza, B Fu, PS Taoukis. Prediction for shelf life and safety of minimally processed CAP/MAP chilled foods: a review. J Food Prot 55(9):741–750, 1992.
13. DL Huffman, KA Davis, DN Marple, JA McGuire. Effect of gas atmospheres on microbial growth, color and pH of beef. J Food Sci 40(6):1229–1231, 1975.
14. FM Christopher, SC Seideman, ZL Carpenter, GC Smith, C Vanderzant. Microbiology of beef packaged in various gas atmospheres. J Food Prot 42(3):240–244, 1979.
15. M Luno, JA Beltran, P Roncales. Shelf-life extension and colour stabilisation of beef packaged in a low O_2 atmosphere containing CO: loin steaks and ground meat. Meat Sci 48(1):75–84, 1998.
16. RE Brackett, EK Heaton. Effects of modified atmosphere packaging on the microflora of fresh tomatoes and broccoli. Presented at (Abstr.) Proc. Southern Assoc. Agr. Sci. Food Sci. Hum. Nut. Sect. 2–5, Orlando, FL, 1986, pp. 10–11.
17. RE Brackett. Influence of modified atmosphere packaging on the miccroflora on quality of fresh bell peppers (abstr). J Food Prot 51(10):829, 1988.
18. D Zagory, AA Kader. Modified atmosphere packaging of fresh produce. Food Technol 42(9):70–77, 1988.
19. K Abrahamsson, NN DeSilva, N Molin. Toxin production by *Clostridium botulinum* Type E, in vacuum-packed, irradiated fresh fish in relation to changes of the associated microflora. Can J Microbiol 11(3):523–529, 1965.
20. JP Smith, ED Jackson, B Ooraikul. Microbiological studies on gas-packed crumpets. J Food Prot 46(4):279–283, 1983.
21. AD Lambert, JP Smith, KL Dodds. Combined effect of modified atmosphere packaging and low-dose irradiation on toxin production by *Clostridium botulinum* in fresh pork. J Food Prot 54(2):94–101, 1991.
22. TA Dahl, WR Midden, PE Hartman. Comparison of killing of gram-negative and gram-positive bacteria by pure singlet oxygen. J Bacteriol 171(4):2188–2194, 1989.
23. KG Newton, WJ Rigg. The effect of film permeability on the storage life and microbiology of vacuum packaged meat. J Appl Bacteriol 47(3);433–441, 1979.
24. CO Gill, G Newton. The development of aerobic spoilage flora on meat stored at chill temperatures. J Appl Bacteriol 43(2):189–195, 1977.
25. E Borch, G Molin. The aerobic growth and product formation of *Lactobacillus*, *Leuconostoc*, *Brochothrix*, and *Carnobacterium* in batch cultures. Appl Microbiol Biotechnol 30(1):81–88, 1989.
26. B Halliwell, JC Gutteridge. Free Radicals in Biology and Medicine, 3rd ed. Oxford: Clarendon Press, 1999, pp. 1–10, 305–306.

27. CE O'Neill, GK Bissonnette. Antecedent oxygen growth conditions and recovery of heat-stressed *Escherichia coli*. J Food Prot 54(2):90–93, 1991.
28. CC Aickin, RC Thomas. Micro-electrode measurement of the internal pH of crab muscle fibres. J Physiol 252(3):803–815, 1975.
29. L Turin, A Warner. Carbon dioxide reversibly abolishes ionic communication between cells of early amphibian embryo. Nature 270(5632):56–57, 1977.
30. SK Wolfe. Use of CO- and CO2-enriched atmospheres for meats, fish, and produce. Food Technol 34(3):55–58, 1980.
31. MA Mitz. The solubility of proteins in the presence of carbon dioxide. Biochemica et Biophys Acta 25(2):426–426, 1957.
32. MA Mitz. CO_2 biodynamics: a new concept of cellular control. J Theor Biol 80(4): 537–551, 1979.
33. M Legisa, M Mattey. Glycerol synthesis by *Aspargillus niger* under citric acid and accumulating conditions. Enzyme Microbiol Technol 8(10):607–609, 1986.
34. AD King, CW Nagel. Influence of carbon dioxide upon the metabolism of *Pseudomonas aeruginosa*. J Food Sci 40(2):362–366, 1975.
35. HD Swanson, JE Ogg. Carbon dioxide regulation of formate hydrogenylase in *Escherichia coli*. Biochem Biophys Res Com 36(4):567–575, 1969.
36. B Pichard, RE Simard, C Bonchard. Effect of nitrogen, carbon monoxide, and carbon dioxide on the activity of proteases of *pseudomonas fragi* and *Streptomyces caespitosus*. Sci Aliments 4(4):595–608, 1984.
37. DF Sears, RM Eisenberg. A model representing a physiological role of CO_2 at the cell membrane. J Gen Physiol 44(5):869–887, 1961.
38. MJ Teixeira De Mattos, PJAM Plomp, OM Neijssel, DW Tempest. Influence of metabolic end-products on the growth efficiency of *Klebsiella aerogenes* in chemostat culture. Ant van Leeu 50:461–472, 1984.
39. NM Dixon, RW Lovitt, DB Kell, JG Morris. Effects of pCO_2 on the growth and metabolism of *Clostridium sporogenes* NCIB 8053 in defined media. J Appl Bacteriol 63(2):171–182, 1987.
40. GH Lovitt, JG Morris, DB Kell. The growth and nutrition of *Clostridium sporogenes* NCIB 8053 in defined media. J Appl Bacteriol 62(1):71–80, 1987.
41. J Pennock, DW Tempest. Metabolic and energetic aspects of the growth of *Bacillus stearothermophilus* in glucose-limited and glucose-sufficient chemostat culture. Arch Micro 150(5):452–459, 1988.
42. RF Stier, L Bell, KA Ito, BD Shafer, LA Brown, ML Seeger, BH Allen, MN Porcuna, PA Lerke. Effect of modified atmosphere on *Clostridium botulinum* toxigenesis and the spoilage microflora of salmon fillets. J Food Sci 46:1639–1642, 1981.
43. F Devlieghere, J Debevere, J Van Impe. Concentration of carbon dioxide in the water-phase as a parameter to model the effect of modified atmosphere on microorganisms. Int J Food Microbiol 43(1–2):105–113, 1998.
44. AA Salyers, DW Dixie. Bacterial Pathogenesis a Molecular Approach. Washington, DC: American Society for Microbiology, 1994.
45. JS King, LA Mabbitt. Preservation of raw milk by the addition of carbon dioxide. J Dairy Res 49(3):439–447, 1982.
46. AD King, CW Nagel. Growth inhibition of a *Pseudomonas* by carbon dioxide. J Food Sci 32(5):575–579, 1967.

47. A Castelli, GP Littaru, G Barbesi. Effect of pH and carbon dioxide concentration on lipids and fatty acids of *Saccharomyces cerevisiae*. Archiv Mikrobiol 66(1):34, 1969.
48. RP Jones, PF Greenfield. Effect of carbon dioxide on yeast growth and fermentation. Enzyme Microbiol Technol 4(4):210–223, 1982.
49. JP Sutherland, JT Paterson, PA Gibbs, JG Murray. The effect of several gaseous environments on the multiplication of organisms isolated from vacuum-packaged beef. J Food Technol 12:249–255, 1977.
50. JH Silliker, SK Wolfe. Microbiological safety considerations in controlled-atmosphere storage of meats. Food Technol 34(3):59–63, 1980.
51. S Stretton, AE Goodman. Carbon dioxide as a regulator of gene expression in microorganisms. Ant van Leeu 73(1):79–85, 1998.
52. JI Pitt, AD Hocking. Fungi and Food Spoilage, 2nd ed. Cambridge: Blackie Academic and Professional, 1997, pp. 8–9.
53. JH Chen, JH Hotchkiss. Effect of dissolved carbon dioxide on the growth of psychotrophic organisms in cottage cheese. J Dairy Sci 74(9):2941–2945, 1991.
54. JH Chen, JH Hotchkiss. Growth of *Listeria monocytogenes* and *Clostridium sporogenes* in cottage cheese in modified atmosphere packaging. J Dairy Sci 76(4):972–977, 1993.
55. I Sierra, M Prodonav, M Calvo, A Olano, C Vidal-Valverde. Vitamin stability and growth of psychrotrophic bacteria in refrigerated raw milk acidified with carbon dioxide. J Food Prot 59(12):1305–1310, 1996.
56. P Ruas-Madiedo, JC Bada-Gancedo, E Fernandez-Garcia, DG DeLlano, CG Reyes-Gavilan. Preservation of the microbiological and biochemical quality of raw milk by carbon dioxide addition: a pilot-scale study. J Food Prot 59(5):502–508, 1996.
57. P Ruas-Madiedo, V Bascaran, A Brana, JC Bada-Gancedo, CG de los Reyes-Gavilan. Influence of carbon dioxide addition to raw milk on microbial levels and some fat-soluble vitamin contents of raw and pasteurized milk. J Agric Food Chem 46(7):1552–1555, 1998.
58. RF Roberts, GS Torrey. Inhibition of psychrotrophic bacterial growth in refrigerated milk by addition of carbon dioxide. J Dairy Sci 71(1):52–60, 1988.
59. KA Glass, KM Kaufman, AL Smith, EA Johnson, JH Chen, J Hotchkiss. Toxin production by *Clostridium botulinum* in pasteurized milk treated with carbon dioxide. J Food Prot 62(8):872–876, 1999.
60. SO Enfors, G Molin. The influence of high concentrations of carbon dioxide on the germination of bacterial spores. J Appl Bacteriol 45(2):279–285, 1978.
61. RC Baker, RA Qureshi, JH Hotchkiss. Effect of an elevated level of carbon dioxide on the growth spoilage and pathogenic bacteria. Poultry Sci 65(4):729–737, 1986.
62. MHJ Bennik, EJ Smid, FM Rombouts, LGM Gorris. Growth of psychotrophic foodborne pathogens in a solid surface model system under the influence of carbon dioxide and oxygen. Food Microbiol 12(6):509–519, 1995.
63. GCJ Adams, EE Butler. Environmental factors influencing the formation of basidia and basidiospores in *Thanatephorus cucumeris*. Phytopathology 73(2):152–155, 1983.
64. IJ Misaghi, RG Grogan, JM Duniway, KA Kimble. Influence of environment and

culture media on spore morphology of *Alternaria alternata*. Phytopathology 68(1): 29–34, 1978.
65. DJ Neiderpruem. Role of carbon dioxide in the control of fruiting of *Schizophyllum commune*. J Bacteriol 85:1300–1308, 1963.
66. M Calderon. Food Preservation by Modified Atmospheres. Boca Raton, FL: CRC Press, 1990.
67. M McIntyre, B McNeil. Morphogenetic and biochemical effects of dissolved carbon dioxide on filamentous fungi in submerged cultivation. Appl Microbiol Biotechnol 50(3):291–298, 1998.
68. RG Tomkins. The inhibition of the growth of meat attacking fungi by carbon dioxide. J Soc Chem Ind 51:261, 1932.
69. WS Ogilvy, JC Ayres. Post-mortem changes in stored meats. V. Effects of carbon-dioxide on microbial growth on stored frankfurters and characteristics of some microorganisms isolated from them. Food Res 18(2):121–130, 1953.
70. G Valley. The effect of carbon dioxide on bacteria. Quart Rev Biol 3:209, 1928.
71. EYC Lee. Carbon dioxide gas analysis and application in the determination of the shelf life of modified atmosphere packaged dairy products. Food Science and Technology. Ithaca, NY: Cornell University, 1996, pp. 29–36.
72. CJ Moir, MJ Eyles, JA Davey. Inhibition of pseudomonads in cottage cheese by packaging in atmospheres containing carbon dioxide. Food Microbiol 10(4):1–7, 1993.
73. SO Enfors, G Molin. The influence of temperature on the growth inhibitory effect of carbon dioxide on *Pseudomonas fragi* and *Bacillus cereus*. Can J Microbil 27(1): 15–19, 1981.
74. F Devlieghere, J Debevere, J Van Impe. Effect of dissolved carbon dioxide and temperature on the groth of *Lactobacillus sake* in modified atmospheres. Int J Food Microbiol 41(3):231–238, 1998.
75. JH Silliker, RE Woodruff, JR Lugg, SK Wolfe, WD Brown. Preservation of refrigerated meats with controlled atmospheres: treatment and post-treatment effects of carbon dioxide on pork and beef. Meat Sci 1:468–488, 1977.
76. DL Gee, WD Brown. The effect of carbon monoxide on bacterial growth. Meat Sci 5:215–222, 1980.
77. DM Graham. Use of ozone for food processing. Food Technol 51(6):72–75, 1997.
78. V Jindal, AL Waldroup, RH Forsythe. Ozone and improvement of quality and shelflife of poultry products. J Appl Poultry Res 4:239–248, 1995.
79. J Fournaud, R Lauret. Influence of zone on the surface microbial flora of frozen beef and during thawing. Ind Aliment Argic 89(5):585–589, 1972.
80. JO Reagan, GR Acuff, DR Buege, MJ Buyck, JS Dickson, CL Kastner, JL Marsden, JB Morgan, RI Nickelson, GC Smith, JN Sofos. Trimming and washing of beef carcasses as a method of improving the microbiological quality of meat. J Food Prot 59(7):751–756, 1996.
81. R Fayer, TC Graczyk, MR Cranfield, JM Trout. Gaseous disinfection of *Cryptosporidium parvum* oocysts. Appl Environ Microbiol 62(10):3908–3909, 1996.
82. E Flair-Flow. Novel 'MAP' extends produce shelf-life. Flair-Flow-Reports; F-FE 306/98, 1998.
83. HS Juffs, SRJ Smith, DC Moss. Keeping quality of whipping cream stored in

dispensers pressurized with nitrous oxide. Aus J Dairy Technol 35(4):132–136, 1980.

84. DP Daifas, JP Smith, I Tarte, B Blanchfield, JW Austin. Effect of ethanol vapor on growth and toxin production by *Clostridium botulinum* in a high moisture bakery product. J Food Safety 20(2):111–125, 2000.
85. I Nakamura, T Ohta. Sterilization of the microbial cells in an aqueous solution by contact glow discharge electrolysis. J Japan Soc Food Sci Technol 44(8):594–596, 1997.
86. CS Tennessee. Development of a flowing plasma reactor for surface biodecontamination. Raleigh: North Carolina State, 1997.
87. T Eklund, T Jarmund. Microculture model studies on the effect of various gas atmospheres on microbial growth at different temperatures. J Appl Bacteriol 55(1): 119–125, 1983.
88. DM Ogrydziac, DB Brown. Temperature effects in modified atmosphere storage of seafoods. Food Technol 36(5):86–96, 1982.
89. CO Gill, KH Tan. Effect of carbon dioxide on the growth of *Pseudomonas fluorescens*. Appl Environ Microbiol 38(2):237–240, 1979.
90. CO Gill, KH Tan. Effect of carbon dioxide on growth of meat spoilage bacteria. Appl Environ Microbiol 39(2):317–319, 1980.
91. WS Ogilvy, JC Ayres. Post mortem changes in stored meats. II. The effect of atmospheres containing carbon dioxide in prolonging the storage life of cut-up chicken. Food Technol 5(3):97–102, 1951.
92. DS Clark, CP Lentz. Use of carbon dioxide for extending shelf-life of prepackaged beef. Can Inst Food Sci Technol J 5(4):175–178, 1972.
93. JH Rodriguez, MA Cousin, PE Nelson. Oxygen requirements of *Bacillus licheniformis* and *Bacillus subtillis* in tomato juice; ability to grow in aseptic packages. J Food Sci 57(4):973–976, 1992.
94. FH Grau. Role of pH, lactate, and anaerobiosis in controlling the growth of some fermentative, gram-negative bacteria on beef. Appl Environ Microbiol 42(6):1043–1050, 1981.
95. JH Hotchkiss, JH Chen, HT Lawless. Combined effects of carbon dioxide addition and barrier films an microbial and sensory changes in pasteurized milk. J Dairy Sci 82(4):690–695, 1999.
96. HS Harned, R Davies. The ionization constant of carbonic acid in water and the solubility of carbon dioxide in water and aqueous salt solutions from 0 to 50° C. J Am Chem Soc 65:2030, 1943.
97. BA Law, LA Mabbitt. New methods for controlling the spoilage of milk and milk products. In: TA Roberts, FA Skinner, eds. Food Microbiology: Advances and Prospects. London: Academic Press, 1983, pp. 131–150.
98. LJ Harris, RK Barakat. Growth of *Listeria moncytogenes* and *Yersinia enterocolitica* on cooked poultry stored under modified atmosphere at 3.5, 6.5, and 10°C. J Food Prot 58:Suppl. 38, 1995.
99. L Wimpfheimer, NS Altman, JH Hotchkiss. Growth of *Listeria monocytogenes* Scott A serotype 4 and competitive spoilage organisms in raw chicken packaged under modified atmospheres and in air. Int J Food Microbiol 11(3):205–214, 1990.
100. DP Daifas, JP Smith, B Blanchfield, JW Austin. Growth and toxin production by

Clostridium botulinum in English style crumpets under modified atmospheres. J Food Prot 62(4):349–355, 1999.

101. AE Larson, EA Johnson. Evaluation of botulinum toxin production in packaged fresh-cut cantaloupe and honeydew melons. J Food Prot 62(8):948–952, 1999.
102. GA Francis, D O'Beirne. Effect of storage atmosphere on *Listeria monocytogenes* and competing microflora using a surface model system. Int J Food Sci Technol 33(5):465–476, 1998.
103. JH Hotchkiss. Microbiological hazards of controlled/modified atmosphere food packaging. Presented at CASA Executive Seminar, Buffalo, New York, 1989, pp. 41–49.
104. C Diaz, JH Hotchkiss. Comparative growth of *Escherichia coli* O157:H7, spoilage organisms and shelf-life of shredded iceberg lettuce stored under modified atmospheres. J Sci Food Agric 70:433–438, 1996.
105. N Kleinlein, F Untermann. Growth of pathogenic *Yersinia enterocolitica* strains in minced meat with and without protective gas with consideration of the competitive background flora. Int J Food Microbiol 10(1):65–72, 1990.
106. AAM Zeitoun, JM Debevere. Inhibition, survival and growth of *Listeria monocytogenes* on poultry as influenced by buffered lactic and treatment and modified atmosphere packaging. Int J Food Microbiol 14(2):161–169, 1991.
107. ME Berrang, RE Brackett, LR Beauchat. Growth of *Listeria monocytogenes* on fresh vegetables stored under controlled atmosphere. J Food Prot 52(10):702–705, 1989.
108. SO Enfors, G Molin, A Ternstrom. Effect of packaging under carbon dioxide, nitrogen on air on the microbial flora of pork stored at 4° C. J Appl Bacteriol 47(2): 197–208, 1979.
109. CO Gill, MP Reichel. Growth of the cold-tolerant pathogens *Yersinia enterocolitica*, *Aeromonas hydrophila* and *Listeria monocytogenes* on high-pH beef packaged under vacuum or carbon dioxide. Food Microbiol 6(6):223–230, 1989.
110. SC Ingham, NN Potter. Growth of *Aeromonas hydrophila* and *Pseudomonas fragi* on mince and surimis made from atlantic pollock and stored under air or modified atmosphere. J Food Prot 51(12):966–970, 1988.
111. ME Berrang, RE Brackett, LR Beuchat. Growth of *Aeromonas hydrophila* on fresh vegetables stored under a controlled atmosphere. Appl Environ Microbiol 55(9): 2167–2171, 1989.
112. G Molin. The resistance to carbon dioxide of some food related bacteria. Eur J Appl Microbiol Biotechnol 18(4):214–217, 1983.
113. ML Hanninen, H Korkeala, P Pakkala. Effect of various gas atmospheres on the growth and survival of *Campylobacter jejuni* on beef. J Appl Bacteriol 57(1):89–94, 1984.
114. SC Seideman, GC Smith, ZL Carpenter, TR Dutson, CW Dill. Modified gas atmosphere and changes in beef during storage. J Food Sci 44(4):1036–1040, 1979.
115. EM Kiggins, WN Plastridge. Effect of gaseous environment on growth and catalase content of *Vibrio fetus* cultures of bovine origin. J Bacteriol 71:397–400, 1956.
116. NJ Stern, MD Greenberg, DM Kinsman. Survival of *Campylobacter jejuni* in selected gaseous environments. J Food Sci 51(3):652–654, 1986.

117. DA Kautter, RK Lynt, TJ Lilly, HM Solomon. Evaluation of the botulism hazard from nitrogen-packed sandwiches. J Food Prot 44(1):59–61, 1981.
118. GL Robertson. Food Packaging Principles and Practice. New York: Marcel Dekker, 1993, pp. 676–678.
119. AA Kader. Biochemical and physiological basis for effects of controlled and modified atmospheres on fruits and vegetables. Food Technol 40(5):99–104, 1986.
120. AF Egan, BJ Shay. Long-term storage of chilled fresh meats. Presented at 34th International Congress of Meat Science and Technology, Brisbane, Australia, 1988, pp. 476–481.
121. L Restaino, WM Hill. Microbiology of meats in hypobaric storage. J Food Prot 44(7):535–538, 1981.
122. R Ronk. Status of ozone for water treatment and food processing under the federal FD&C Act. Presented at the first international symposium on ozone for water and waste water treatment, Stamford, CT, 1975, pp. 830–842.
123. P Sarig, T Zahavi, Y Zutkhi, S Yannai, N Lisher, R Ben-Arie. Ozone for control and post-harvest decay of table grapes caused by Rhizopus stolonifer. Phys Mol Plant Pathol 48(6):403–415, 1996.
124. R Jennings. Ozone water treatment to aid habitat of salmon. Contra Costa Times, Concord, CA, November 15, 1996.
125. JP Sutherland, AJ Bayliss, DS Braxton, AL Beaumont. Predictive modeling of *Escherichia coli* O157:H7: inclusion of carbon dioxide as a fourth factor in a preexisting model. Int J Food Microbiol 37(2–3):113–120, 1997.
126. PS Fernandez, S George, M., CC Sills, MW Peck. Predictive model of the effect of CO_2, pH, temperature, and NaCl on the growth of *Listeria monocytogenes*. Int J Food Microbiol 37(1):37–45, 1997.
127. NB Day, BJ Skura, WD Powrie. Modified atmosphere packaging of blueberries microbiological changes. Can Inst Food Sci Technol J 23(1):59–65, 1990.
128. SPS Guleria, PK Porsdalo. Hypobaric storage of fruits and vegetables. Scand Refrg 11(2):77–80, 1982.
129. RL Shewfelt. Postharvest treatment for extending the shelf life of fruits and vegetables. Food Technol 40(5):70–74, 80, 89, 1986.
130. F LaBell. Modified atmosphere packaging of meat, poultry, seafoods. Food Proc 47(13):135–136, 1986.
131. JM Farber, DW Warburton, P Laffey, U Purvis, L Gour. Modified atmosphere packaged pasta: a microbiological quality assessment. Ital J Food Sci 2:157–166, 1993.
132. B Fabiano, P Perego, R Pastorino, Md Borghi. The extension of the shelf-life of pesto sauce by a combination of modified atmosphere packaging and refrigeration. Int J Food Sci Technol 35(3):293–303, 2000.
133. MJ Jay. Modern Food Microbiology, 4th ed. New York: Chapman Hall, 1992.
134. LR Beauchat, RE Brackett. Survival and growth of *Listeria monocytogenes* on lettuce as influenced by shredding, chlorine treatment, modified atmosphere packaging and temperature. J Food Sci 55(3):755–758, 1990.
135. DM Wilson, E Jay, RA Hill. Microflora changes in peanuts groundnuts stored under modified atmospheres. J Food Prod Res 21(1):47–52, 1985.
136. HJ Banks. Effects of controlled atmosphere storage on grain quality: a review. Food Technol Aus 33(7):335–340, 1981.

137. TH Sanders, ND Davis, UL Diener. Effect of carbon dioxide, temperature, and relative humidity on production of falatoxin in peanuts. J Am Soc Oil Chem 45: 683–685, 1968.
138. AD Lambert, JP Smith, KL Dodds. Shelf life extension and microbiological safety of fresh meat—a review. Food Microbiol 8(4):267–297, 1991.
139. CO Gill, KG Newton. The ecology of bacterial spoilage of fresh meat at chill temperatures. Meat Sci 2:207–217, 1978.
140. RH Dainty, BG Shaw, TA Roberts. Microbial and chemical changes in chill stored red meats. New York: Academic Press, 1983, pp. 151–178.
141. G Tewari, DS Jayas, RA Holley. Centralized packaging of retail meat cuts: a review. J Food Prot 62(4):418–425, 1999.
142. TF Brockelhurst, BM Lund. Microbiological chages in cottage cheese varieties during storage at 7°C. Food Microbiol 2:207, 1985.
143. L Piergiovanni, P Fava, M Moro. Shelf-life extension of taleggio cheese by modified atmosphere packaging. Ital J Food Sci 2:115–127, 1993.
144. CR Loss, JH Hotchkiss. Effect of dissolved carbon dioxide on the thermal destruction of *Pseudomonas fluorescens* R1-232 in milk. Presented at American Dairy Science Association and American Society of Animal Science Joint Meeting and Northeast Dairy Section Meeting, Baltimore, MD, 2000, p. 133.
145. RA Holley, C Gariepy, P Delaquis, G Doyon, J Gagnon. Static, controlled (CO_2) atmosphere packaging of retail ready pork. J Food Sci 59(6):1296–1301, 1994.
146. JP Sutherland, A Aherene, AL Beaumont. Preparation and validation of a growth model for *Bacillus cereus*: the effects of temperature, pH, sodium chloride and carbon dioxide. Int J Food Microbiol 30(3):359–372, 1996.
147. F Devlieghere, I Lefevere, A Magnin, J Debevere. Growth of *Aeromonas hydrophila* in modified-atmosphere-packed cooked meats. Food Microbiol 17(2): 185–196, 2000.

10

Control with Naturally Occurring Antimicrobial Systems Including Bacteriolytic Enzymes

Grahame W. Gould*
Unilever Research Laboratory
Bedford, England

I. INTRODUCTION

Preservation techniques currently used to protect the quality of foods during storage and distribution and to ensure safety from food poisoning act in three main ways:

1. Prevention of access of microorganisms to foods
2. Inactivation of microorganisms should they nevertheless have gained access
3. Prevention or retardation of the growth of microorganisms should they have gained access and not been inactivated

Most preservation techniques act by method (3), preventing or retarding microbial growth, e.g., by freezing, chilling, drying, curing, conserving, vacuum-packing, modified-atmosphere packing, acidifying, fermenting, or adding preservatives. A much smaller number of techniques act by method (2), inactivating microorganisms, e.g., by pasteurization and sterilization by heat or, to a much lesser extent, by ionizing radiation. Complementary techniques act by method, (1), by restricting the access of microorganisms to processed products, e.g., by hygienic handling, aseptic processing, and packaging (1).

---

* Retired.

A major trend that is apparent at the moment is that new and ''emerging'' preservation techniques that are being developed or are coming into use act primarily by inactivation, e.g., new physical techniques, such as the application of high hydrostatic pressure; high-voltage electrical pulses; combined heat, ultrasonication, and slight overpressure (manothermosonication, or MTS); and high-intensity pulsed light. Further strong trends are towards the use of procedures that result in products that are less heavily preserved, have higher quality, are more natural, are free from artificial additives, and are more healthful. These trends result mainly from the changing lifestyles and needs of consumers and the food industry's reactions to those changes and needs (2) (Table 1).

Antimicrobial systems that occur naturally—produced by animals, plants, and microorganisms—mostly act by inactivation (3–5). They therefore have the potential to contribute to the development of a wider portfolio of micoorganism-inactivation procedures. They also have the potential advantage of meeting many consumer needs, usually being perceived as more natural than most of the currently employed preservation techniques. It is for these reasons that interest in the use of natural systems for food preservation is growing (3).

The number of naturally occurring antimicrobial systems that are known

Table 1 Trends in Consumer Requirements for Foods and Reactions of the Food Industries

Major trends in consumer requirements

1. Improved convenience, in storage, shelf life, and ease of preparation for consumption
2. Higher quality, in flavor, texture, and appearance
3. Fresher
4. More natural, with fewer additives
5. More nutritious
6. Minimally packaged
7. Safer

Major food industry reactions

1. Development of milder processes: minimal overheating; less intensive heating; introduction of nonthermal alternatives to heat
2. Fewer additives; less use of ''chemical'' preservatives
3. Use of combination preservation systems or ''hurdle technologies'' to minimize the extreme use of any single technique
4. Use and evaluation of natural antimicrobial systems for food preservation
5. Less use of salt, saturated fats, sugars
6. More low-calorie foods
7. Reduced, environmentally friendly packaging
8. Introduction of more microorganism-inactivation techniques, in particular to eliminate food-poisoning microorganisms from foods

Table 2 Major Naturally Occurring Antimicrobial Systems

Category	Origin	Example
Animals—constitutive systems	Phagosomes	Myeloperoxidase
	Serum	Transferrins
	Milk	Lactoperoxidase, lactoferrin
	Eggs	Lysozyme, ovotransferrin (conalbumin) avidin,
Animals—inducible systems	Immune system	Antibodies, complement
	Frogs	Magainins
	Insects	Abaecin, apidaecin, attacins, cecropins, coeoptericin, defensins, diptericin, royalisin
Plants—constitutive systems	Herbs, spices, and other plants	Eugenol (cloves)
		Allicin (garlic), allyl isothiocyanate (mustard), oleoeuropein (olives)
Plants—inducible systems	Infected or injured plants	Low MW phytoalexins
		High MW polyphenols
Microorganisms	Lactic acid bacteria	Nisin, pediocin, other bacteriocins
	Other microorganisms	Other antibiotics (natamycin/pimaricin, subtilin), bacteriophages, yeast ''killer toxins,'' organic acids and other low MW metabolites

and reasonably well understood is very large. They include some systems derived from animals, some from plants, and some from microorganisms. Many of the systems are constitutive, i.e., they are present and active at all times in the particular organisms that produce them. Others are inducible, i.e., they are generated in response to some form of stimulus, such as injury or infection, as categorized in Table 2. Although numerous, very few of the naturally occurring antimicrobial systems have been exploited deliberately so far, although some probably have a history of empirical use as long as that of any other preservation technique. For example, naturally occurring lactic and alcoholic fermentations of plant and animal materials have been employed for many thousands of years.

II. ANIMAL-DERIVED SYSTEMS

Numerous highly effective antimicrobial systems operate in animals (6) (Table 2). They include the immune system in higher animals and a number of bacteriolytic and other enzymes that are components of bactericidal systems as well as some nonenzymatic proteins with antimicrobial activity. An expanding number of small antimicrobial peptides have been discovered recently that are mostly membrane-active and lethal for a wide range of microorganisms.

Although the animal-associated antimicrobial systems are numerous, by far the most commercial use has been made of just one of them, hen egg white lysozyme, while limited applications have been pursued for lactoperoxidase and lactoferrin derived from cows' milk.

A. Lysozyme

Lysozyme is a 14,600 dalton protein that is present in many body fluids and at levels up to about 3.5% of dry weight in egg white, so that it is readily available, relatively inexpensive, and natural. It lyses many, but not all, types of gram-positive bacteria in their vegetative forms but is normally inactive against gram-negative bacteria because their outer membranes restrict access to the underlying peptidoglycan layer that is the enzyme's substrate. Lysozyme acts as a muramidase, by cleaving the β(1–4) glycosidic bond between the C_1 of *N*-acetylmuramic acid and the C_4 of *N*-acetylglucosamine in the peptidoglycan layer that makes up the important skeletal structure of bacterial cell walls. Hydrolytic cleavage of this structure destroys its mechanical strength so that a sensitive microbial cell will normally lyse unless supported in an iso- or hyperosmotic medium.

Hen egg white lysozyme is the most readily and economically available muramidase and therefore finds application to inactivate microorganisms in foods. It can be produced in high purity and yield from hen egg whites by ion exchange (7,8), and there is no toxicological problem associated with its use in foods.

Introduction of lysozyme to foods by genetically modified microorganisms was suggested by Maulla et al. (9). They showed that *Kluyveromyces lactis*, transformed with the human lysozyme gene, grew in cottage cheese whey to produce up to 125 μg/mL lysozyme, and they proposed its use as a starter culture for cheese maturation.

While being very stable to heat at low pH values, surviving brief boiling at pH 3, lysozyme is more heat sensitive at the pH values typically of most foods. However, its survival in heated foods is very dependent on the food type, so that survival of activity in foods with pH values as high as 5 can be substantial (10). This has become of concern with respect to the survival and growth of psychrotrophic strains of *Clostridium butulinum* in some mildly heated foods (see below).

Other lysozymes with slightly different molecular structures can be isolated from the eggs of other species of birds and from other animal sources. Furthermore, chitinases, which help to protect some plants against fungal attack, have lysozyme-like activity. Consequently, lysozyme-like activity is naturally present in many types of food raw materials.

The major successful use of lysozyme commercially has been to lyse cells of *Clostridium tyrobutyricum* as they outgrow from germinated spores (11), thus preventing gas formation and spoilage by "blowing" in certain types of cheeses (12,13). The concentrations employed in foods are low, approximately 20–400 ppm, such that its use is economically viable in those situations where it prevents a commercially unacceptable level of spoilage. It has been estimated that in excess of 100 tons of lysozyme are used annually for this purpose (14).

Applications in many other types of foods have been explored, particularly in far eastern countries, but with more limited commercial exploitation. Tranter (10) listed early 20 categories of foods in which lysozyme had been shown to deliver a useful antimicrobial effect, sometimes when employed with adjuncts, or additional "hurdles" (see Chapter 20), that act synergistically with it (Table 3).

Although lysozyme is normally inactive against gram-negative bacteria because their peptidoglycan layer is protected by an outer membrane, not all gram-positive bacteria are naturally lysozyme-sensitive because in some, teichoic acids and other cell wall components interfere with access of the enzyme to its substrate, the antimicrobial spectrum of the enzyme can be broadened in a number of ways, some of which are compatible with foods. For example, pretreatment with some chelating agents, such as ethylenediaminetetraacetic acid (EDTA), that leads to a loss of magnesium ions from the outer membrane of gram-negative cells sensitizes them to the action of lysozyme (15). Conjugation of lysozyme with dextran increased its activity against gram-positive (*Staphylococcus aureus*,

Table 3 Use of Lysozyme and Adjuncts as Antimicrobials in Foods

Antimicrobial system	Categories of food
Lysozyme alone	Fresh meat, sausages, bacon
	Fresh fish
	Cheese, butter
	Vegetables, fruit
	Tofu bean curd
Lysozyme + sodium chloride	Seafood, fish cakes
Lysozyme + p-hydroxybenzoate esters	Wine
Lysozyme + β-glycan pyranose	Sake
Lysozyme + acetic acid (vinegar)	Sushi
Lysozyme + amino acid	Potato salad
Lysozyme + amino acids and propylene glycol	Chinese noodles
Lysozyme + amino acids and ethanol	Creamed custard
Lysozyme + lactoferrin	Infant milk formula

Source: Adapted from Ref. 10.

Bacillus cereus) and gram-negative bacteria (*Vibrio paranaemolyticus*, *Escherichia coli*, *Aeromonas hydrophila*, *Klebsiella pneumoniae*), particularly if the temperature was raised during treatment (16). The possibility of covalently bonding small molecules to lysozyme to aid its penetration of the outer structures of bacterial walls was further suggested by the preparation of a fatty acid (palmitate) derivative that was more active against *E. coli* than was the native enzyme (17). Lysozyme and the bacteriocin nisin (see Chapter 11) act synergistically under certain conditions to inhibit the growth of, and to inactivate, *Listeria monocytogenes* (18). Freeze-thaw treatments sensitized *Escherichia coli* to lysozyme (19). Gram-negative bacteria, including *Salmonella* species, become lysozyme-sensitive following an osmotic downshift (e.g., achieved by sudden dilution of a salt solution). This has been proposed as the basis for a sequential double-dip or spray decontamination treatment for poultry and other animal carcases (20).

All these treatments and adjuncts that act synergistically or raise the activity of lysozyme to improve its efficacy against vegetative bacteria have potential value in food preservation and safety. With this in mind, the stability of lysozyme in combination with a number of potential adjuncts (including NaCl, nitrite, benzoate, sorbate, propionate, parabens, acetate, lactate) was determined by Yang and Cunningham (21). Addition of lactic acid, trypsin, proteinase K, lipase, ovotransferrin, and lactoferrin (see below) potentiated the lethal action of lysozyme on *L. monocytogenes* (22).

A major further extension of the use of lysozyme would open up if bacterial spores could be inactivated by the enzyme. With a few important exceptions, however (see below), spores are lysozyme-resistant because the peptidoglycan substrate for the enzyme is protected by surrounding protein-rich coats. However, it has been known since the 1960s that there are procedures that will sensitize spores to lysozyme. Lysozyme causes such sensitized spores to undergo a process similar to that occurring during normal germination, so that they become sensitive to mild, pasteurization-like heat processing. Unfortunately, the sensitizing treatments that are most effective involve the use of strong reducing agents (e.g., β-mercaptoethanol, dithiothreitol, thioglycolic acid) or oxidizing agents (e.g., performic acid) that rupture disulfide bonds in spore coat proteins and make the coats permeable to lysozyme (23). These reagents are not compatible with food use. Alternative mild, food-compatible spore-sensitization techniques would have great potential.

A few types of spores have sufficiently permeable coats that they are completely lysozyme-sensitive (24). For other types, a fraction of their populations are naturally lysozyme-sensitive. They include some strains of *Clostridium perfringens* (25) and psychrotrophic strains of *C. botulinum* (26–29). This has proved to be of significance with respect to *C. botulinum* and its survival in mildly heated chill-stored foods, e.g., REPFEDs [refrigerated processed foods of extended durability (30,31)] and "sous vide" foods (32), in particular. This is because spores of *C. botulinum* type E and nonproteolytic type B are regarded normally to be relatively heat-sensitive and therefore inactivated by heat treatments as low as 90°C for 10 minutes or equivalent (31,33). However, apparently inactivation occurs because heat destroys some part of the germination system in these spores rather than destroying the protected contents in the spore protoplast. In this case, lysozyme, by hydrolyzing peptidoglycan in the spore cortex, bypasses the inactivated germination system, with the result that the apparently inactivated spores are revived (34). Their apparent heat resistance is then much greater than previously assumed. This has called into question the safety of RE-PFEDs if circumstances arise in which naturally occurring lysozyme or lysozyme-like activity could bring about these changes in particular foodstuffs (34,35). Anomalously high heat resistance of *C. botulinum* type E spores, though not proven to be due to lysozyme, was recently reported in canned Dungeness crab meat (36).

B. Lactoperoxidase

Lactoperoxidase is a 78,000 dalton glycoprotein enzyme that is widely distributed in animals, e.g., in mammalian colostrum, milk, saliva, and other body fluids and tissues in which it plays a role in the natural defense mechanisms of the host (37,38). It catalyes the oxidation of thiocyanate (SCN^-) by hydrogen peroxide (H_2O_2) to form hypothiocyanite ($OSCN^-$) and some other products, which are bactericidal or bacteristatic to a wide range of vegetative bacteria, but ineffective against spore forms and only weakly active against some yeasts and molds (38). Lactoperoxidase can be isolated from milk by cation exchange chromatography and has been available commercially for about 20 years for food and animal feed preservation (39). It has been shown to be particularly effectively in improving shelf life of milk in countries in which distribution infrastructure and refrigeration are poorly developed (40,41) and in animal feeds to minimize enteric infection in young farm animals (42).

The necessary thiocyanate substrate is often present at sufficient concentrations along with the enzyme as a breakdown product of glucosinolates and other chemical presursors in the host animal diet (e.g., commonly 1–10 ppm in bovine milk). However, in order to activate the lactoperoxidase system, in foods in which it occurs naturally or in foods to which it has been artificially added, it is usually necessary to supply the low level of H_2O_2 that is needed. Many imaginative proposals have been made for doing this in a food-compatible manner (43–45) (Fig. 1). For example, H_2O_2 has been delivered by direct addition, e.g., to milk. In raw milk, where the enzyme is naturally present, efficiency is improve by supplementation with thiocyanate as well as with the H_2O_2 necessary to complete the system. Optimum concentrations of H_2O_2 and SCN^- are as low as about 8 and 12 ppm, respectively (46). This has been exploited to gain useful shelflife

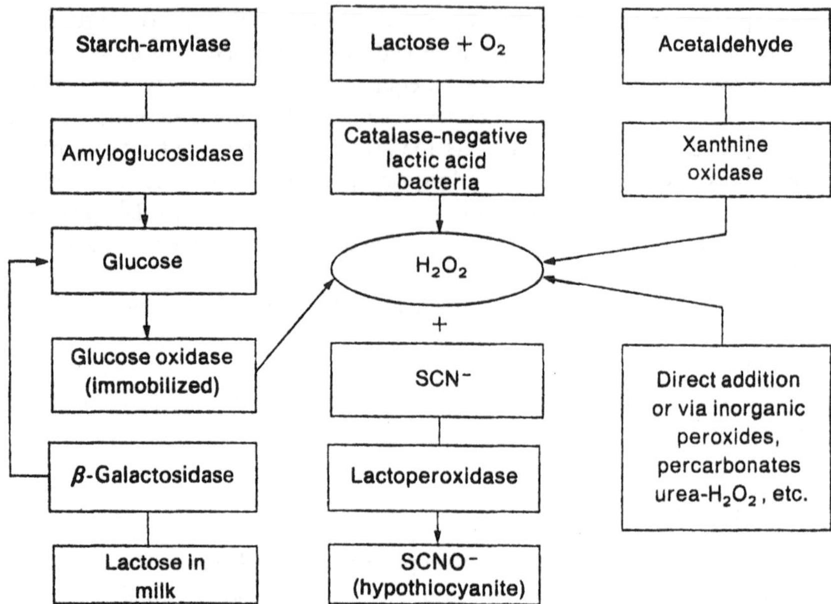

Figure 1 Schematic representation of alternative routes proposed for the generation of hydrogen peroxide to activate the milk or salivary lactoperoxidase systems and form the actimicrobial hypothiocyanite anion. (From Ref. 1.)

in a number of tropical and subtropical countries in which refrigerated transport from farm to dairy is not available (40). In manufactured human foods, an impediment to application is the need to supplement the food with thiocyanate, as well as providing a source of H_2O_2, since these are often present naturally at insufficient concentrations, and legislative constraints may prevent such additions. Alternatively to direct addition, H_2O_2 may be delivered via the addition of dry precursors, such as urea: H_2O_2 or magnesium percarbonate, that decompose to generate H_2O_2 when they become hydrated in the food or feed. Hydrogen peroxide may be generated by the action of glucose oxidase on glucose and of other direct oxidases on their substrates (e.g., xanthine oxidase in milk acting on acetaldehyde generated by lactic acid bacteria, such as in yogurt) (Fig. 1). Immobilized β-galactosidase and glucose oxidase have been demonstrated to allow ''cold pasteurization'' of milk in flow-through systems. The β-galactosidase converts a small fraction of the milk lactose into glucose and galactose; the glucose oxidase oxidizes some of the glucose so formed to generate gluconic acid and H_2O_2, which then activates the lactoperoxidase system, which is present naturally in the raw milk (45). In some animal feed applications, e.g., for weaning diets for young farm animals,

H_2O_2 may be generated naturally in the reconstituted feed by the action of endogenous or added lactic acid bacteria.

Toothpastes are on sale that contain thiocyanate along with amyloglucosidase and glucose oxidase enzymes, which, following breakdown of food starches to maltodextrins by salivary amylase, generate hydrogen peroxide (47) (Fig. 1). This activates salivary lactoperoxidase and, it is claimed, naturally promotes dental health (48).

While generally ineffective against bacterial spores and most yeasts and molds, the lactoperoxidase system has been shown to be bactericidal or bacteristatic for a wide range of important food-spoilage and food-poisoning microorganisms (49), including species of *Listeria*, *Staphylococcus*, *Campylobacter*, *Salmonella*, and *B. cereus* (50,51) and some yeasts (52). Gram-negative, catalase-positive bacteria such as *Pseudomonas*, coliforms, and salmonellae are generally inhibited by the fully activated system and may be killed if the peroxide supply is maintained (40,53). Gram-positive bacteria, e.g., the lactic acid bacteria, are generally inhibited rather than inactivated (54,55).

C. Lactoferrin and Other Iron-Binding Glycoproteins

The avian egg contains a number of antimicrobial systems apart from lysozyme that serve to keep it free of infection during the development of the young bird embryo. These systems, summarized in Table 4, include alkalinity (the pH of egg white can rise to about 9.5 soon after laying as CO_2 is lost to the atmosphere), avidin, which binds biotin, and ovoflavoprotein, which binds riboflavin suffi-

Table 4 Antimicrobial and Enzyme-Inhibitory Systems in the Avian Egg

Antimicrobial/Antienzyme system	Mechanism of action
High pH (up to about 9.5)	Inhibitory to some microorganisms and enhances the iron-binding effect of ovotransferrin
Ovotransferrin (conalbumin)	Strongly binds (chelates) Fe^{3+}, making it unavailable for microbial growth
Lysozyme	Hydrolyzes β(1-4)-glycosidic bonds in the peptidoglycan of bacterial cell walls, causing lysis
Ovoflavoprotein	Chelates riboflavin, making it unavailable as a growth factor for microbial growth
Avidin	Chelates biotin, making it unavailable as a growth factor for microbial growth
Ovomucoid	Inhibits serine proteases
Ovoinhibitor	Inhibits serine proteases
Cystatin	Inhibits cysteine proteases

ciently tightly. These systems make the vitamins unavailable to microorganisms requiring them as growth factors. A number of protease inhibitors guard against proteolytic attack and the consequent generation of amino acids and peptides that may become substrates for microbial growth. The iron-binding glycoprotein from eggs (ovotransferrin or conalbumin) and the analogous protein from milk (lactoferrin) have received attention as potential food preservatives because they have a broad spectrum of activity and are present at high concentrations. There are related glycoproteins in blood (serum transferrins). The transferrins all bind ferric iron so tightly that, in media low in free iron, such as serum and egg white, they prevent microbial growth (56).

Ovotransferrin constitutes about 12% of the dry weight of egg white. It is a 77,000 dalton glycoprotein that contains two iron-binding sites for Fe^{3+}. For each Fe^{3+} bound an anion, preferably bicarbonate, is bound as well. The analogous lactoferrin in milk is an 80,000 dalton glycoprotein but is present at lower levels (up to about 0.02% in bovine milk and about 0.2% in human milk). Both transferrins have been available commercially for many years, extracted from eggs (57) or from milk (58).

In culture media and in egg white, in which free iron levels are normally low, the antimicrobial effectiveness of the transferrins can be very strong. In cheese making, and in the manufacture of some infant feed formulae, lactoferrin is lost. However, it may be reintroduced with potential health benefits. A recently concluded European Union research program led to a patent application in Italy for the use of lactoferrin to prevent contamination of Mozarella cheese with *L. monocytogenes* (59). In many foodstuffs, however, high levels of iron are commonly present, and effectiveness is consequently reduced. Furthermore, microorganisms such as the lactic acid bacteria, which have low requirements for iron, are much less affected by the transferrins than are organisms such as the pseudomonads and coliforms, which have high iron requirements (60,61). Some of the latter microorganisms produce iron chelators of their own when in iron-depleted media, and then may compete with the transferrins for the low levels of iron that are available [e.g., enterochelin, produced by *Salmonella typhimurium* (62)]. Transferrins may therefore find more effective use in combination with other factors or procedures that limit the availability of iron. For example, Tranter (10) suggested that, like EDTA, transferrins can act to enhance the bactericidal activity of lysozyme. Transferrins have been shown to increase the activity of lysozyme against *L. monocytogenes* (63), and there may be other potentially useful exploitable combinations.

D. Lactoferricin

Lactoferricin is an antimicrobial polypeptide that can be derived from lactoferrin by limited acid hydrolysis (64) or proteolysis (65). It effectively inactivates a wide range of gram-positive and gram-negative bacteria at concentrations as low as about 1 ppm and up to about 150 ppm for different species, including *E. coli*, *Salmonella* species, *K. pneumoniae*, *Proteus vulgaris, Yersinia enterocolitica*, *Pseudomonas aeruginosa*, *Campylobacter jejuni*, *S. aureus*, *Streptococcus mutans*, *Corynebacterium diphtheriae*, and *L. monocytogenes* (66,67). Interestingly, it acts on microbial cells by means that do not depend on the binding of iron (65). It has been shown to be effective in extending the shelflife of raw milk and in reducing the microbial contamination or raw vegetables when employed as a 1% dip (68). A potential advantage of lactoferricin in food preservation is its relative heat resistance.

E. Small Membrane-Active Peptides

Small membrane-active peptides, usually containing between 20 and 40 amino acid residues, are common in animals. They include, for example, defensins in mammals, attacins and cecropins in insects, and magainins in frogs (6) (Table 2). There is considerable interest in the possibilities of their commercial application, particularly in the pharmaceutical industries and for topical application in medicine. Regulatory hurdles will probably substantially delay any application in foods.

III. PLANT-DERIVED SYSTEMS

Wilkins and Board (69) reported that more than 1340 plants were known to be potential sources of antimicrobial compounds. Many of these are elements of the natural defense mechanisms of the plants against infection (70) and include adaptive systems and constitutive ones. The former include the phytoalexins, which are low molecular mass, broad-spectrum antimicrobials, the synthesis of which is actually triggered by the initiation of infection or by damage to the plant structure (71). They include a wide variety of different chemicals (72) and are mostly antimycotic rather than antibacterial. When they do show antibacterial activity, they are more effective against gram-positive than gram-negative species (73).

Banks et al. (74) considered that the susceptibility of many animal cells to phytoalexins made their practical use as food antimicrobials unlikely. In contrast, many of the constitutive antimicrobials in plants have been explored and proposed for use as antimicrobials in foods since the late nineteenth century (75). The active compounds include many low molecular weight substances (Table 2), in particular, numerous herb and spice components, the most active of which are the phenolic compounds and essential oils [e.g., allicin in garlic, allyl isothiocyanate in mustard, oleuropein in olives, vanillin in vanilla, cinnamaldehyde in cinnamon, eugenol in cloves (76)]. While much attention has been given to some of these compounds individually, such as oleuropein, this should not draw attention from the large numbers of compounds that occur even in individual plant species. For example, Dallyn (77) listed, in addition to oleuropein, some 26 other phenolic antimicrobials found in certain olive oils and olive tissue. Often, families of related compounds are found. Plant-derived isothiocyanates with 10 different side groups were listed by Delaquis and Mazza (78). About 60 antimicrobial compounds identified in plants were listed by Nychas (79), and Walker (80) listed about 50 antibacterial and antifungal compounds that have been isolated and characterized from 20 families of plants that are normally used as sources of food. Whitehead and Threlfall (81) listed phytoalexins produced by 17 species of edible plants. Beuchat listed about 60 plants, commonly used as herbs or spices, that are also known to contain low molecular mass antimicrobial substances with activity against a wide range of bacteria, yeasts, and molds (82). The plant-derived phenolic compounds were particularly effective against fungi (83). Many herbs and spices contain essential oils that are antimicrobial. Deans and Ritchie (84) listed more than 80 plants that contain high levels of antimicrobials with potential for food use, including especially sage, rosemary, coriander, garlic, and onion.

Of course, it may not be necessary to use pure, isolated, plant-derived compounds or undesirably high concentrations of strongly favored herbs or spices or their extracts to obtain a useful preservative effect. Simple, edible plant fractions sometimes provide potentially useful antimicrobial effects. For example, even raw carrot juice has been shown to inactivate *L. monocytogenes* and other *Listeria* species (85–87).

For many years, of course, certain of the naturally occurring organic acids have been widely used as food acidulants or preservatives (88), although they are normally regarded as ''additives'' by consumers and come under the relevant legislation in most countries (e.g., citric, lactic, and acetic acid at up to percent levels, and sorbic, benzoic, and propionic acids at levels of hundreds or thousands of ppm), while the phenolic compounds many of which are related chemically to the food additives, butylated hydroxyanisol (BHA) and butylated hydroxytoluene (BHT) have been exploited in an antioxidant role, although they do show antimicrobial activity as well (89,90). The preservative acids are mostly weak lipophilic organic acids or their derivatives (e.g., *p*-hydroxybenzoic acids and their esters). They owe their efficacy to a large extent to their capacity, in their undissociated forms, to diffuse easily across cell membranes and thereby enter the microbial cytoplasm (91). In doing so they cause inhibition and, later, cell death, by collapsing the proton gradient across the membrane and allowing acidification of the cytoplasm, in addition to other more specific effects (92).

While many studies have demonstrated the effectiveness of a wide range of plant-derived antimicrobials, apart from the allowed preservatives, using laboratory media (69,93), disappointingly few studies have been undertaken using foods (94,95). Most significantly, the levels of the antimicrobials necessary to inhibit or inactivate microorganisms in foods have often been found to be considerably higher than those determined using cultures (70,96). This loss of efficacy occurs because many of the compounds bind to proteins or other food components or partition into the lipid phase of foods and so become unavailable for antimicrobial action. Other disadvantages associated with the use of many plant-derived antimicrobials include the strong flavors of some phenolic compounds and essential oils, which make them compatible with only a limited range of foods. Consequently, confident widespread use of the plant-derived materials in practical food preservation has yet to occur and probably requires better information on their effectiveness when used in combination with other permitted antimicrobials and/or procedures used in food-preservation practice in order to reduce the concentrations needed for efficacy (79).

IV. MICROORGANISM-DERIVED SYSTEMS

A. Natamycin

The tetraene antimycotic natamycin (pimaricin, tennecitin, mycoprozine), produced by *Streptomyces natalensis*, is used to prevent mold growth in some foods, e.g., on the surfaces of some cheeses and dry sausages (97), and is being considered for wider use in some countries, e.g., within the European Union (98). In these applications natamycin inhibits fungal growth at concentrations between 1 and 20 ppm and is an alternative to sorbic acids at levels about 100 times higher (99). Of the other microorganism-derived antimicrobials, by far the greatest research and application has centered on the products of the lactic acid bacteria.

B. Bacteriocins

The use of microbially derived antimicrobials for the inactivation of microorganisms in foods is dominated by the peptide ''lantibiotic'' nisin, produced by certain strains of *Lactococcus lactis*, and applied mainly in cheese products and in canned vegetables (100,101). Other applications proposed have been for meat, fish, milk, milk products, and alcoholic beverages (102).

However, the numbers of other bacteriocins that have been discovered and that may have potential in food preservation continue to grow. They have to be categorized into three classes: small, post-transcriptionally modified lantibiotics such as nisin; small, heat-stable unmodified peptides, such as pedicin PA-1; and larger, heat-labile molecules, such as helvticin J. Hill (103) listed more than 20 bacteriocins that have been well characterized, and many more have been reported but little-studied. Many are produced by strains of starter cultures (104).

While these have not been applied in food preservation in the same way as nisin, the potential for their use as additives, or by generation in situ in foods, e.g., from the addition of particular starter cultures, or in cultured food product ingredients such as Microgard™ (105) must be substantial. Bacteriocins are reviewed in Chapter 11.

C. Other Products of Lactic Acid Bacteria

The lactic acid bacteria, including the genera *Lactobacillus*, *Lactococcus* (group N streptococci), *Leuconostoc*, and *Pediococcus*, have been employed empirically for many years in most countries of the world, often in naturally occurring mixed culture fermentations with other bacteria, yeasts, and molds, for the production of a wide range of meat, dairy, and plant-derived foods and feeds (106). The main fermentation reaction that contributes to the preservation of such foods in the production of lactic acid and the resulting lowering of pH. At the same time, minor metabolic products contribute other attributes such as flavor, odor, and color, and structural changes occur that characterize particular fermented foods (107,108). Some of these minor products have antimicrobial properties, and there is increasing interest in the possibilities for optimizing their use (109,110). The minor products include acids other than lactic, such as acetic and succinic. They are normally generated in relatively small amounts, but the particular effectiveness of some of them, e.g., acetic acid, make the potential of using high-generating strains of some interest. In some Emmental-type cheese fermentations, *Propionibacterium shermanii* normally converts much of the lactic acid initially formed to propionic acid, and this greatly improves the preservation of such foods. The reaction forms the basis of the commercially available ingredients that improve the preservation of dairy-based foods (111). Diacetyl is produced from citrate by some *Lactococcus* and *Leuconostoc* species (112), and some dairy starter cultures are selected for high-diacetyl production for reasons of flavor generation. Diacetyl is also antimicrobial, however (95), and may contribute to the preservative effectiveness of some culture supernatants. The concentrations in foods are generally well below those that are inhibitory alone. However, diacetyl may act synergistically with some of the other antimicrobial factors in cultures. Many lactic acid bacteria produce hydrogen peroxide when incubated in the presence of oxygen, and, being catalase negative, the peroxide tends to persist in cultures and can even autoinhibit growth (113). As with diacetyl, production and persistence of hydrogen peroxide is generally insufficient alone to provide useful antimicrobial effects in fermented or other foods (e.g., in meat products (114)]. However, synergism may be more important, and peroxide can act as a precursor for the generation of other, more potent, antimicrobial derivatives, such as superoxide or hydroxyl free radicals, and as a substrate for the lactoperoxidase system (above). Of the other low molecular weight antimicrobials that have been detected in cultures of lactic acid bacteria, 3-hydroxypropionaldehyde (''reuterin''), produced by *Lactobacillus reuteri*, was shown to be particularly effective (115). It is generated during anaerobic growth in the presence of glycerol and may well have been present as a previously undiscovered component of the antimicrobial activity of antimicrobial culture supernatants (116).

D. Yeast "Killer Toxins"

Some yeasts secrete substances that inactivate other strains of yeasts. They used to be the cause of some mysteriously failed fermentations in the brewing and wine industries (117). The so-called ''killer toxins'' are 10,000–20,000 dalton proteins, first described for *Saccharomyces cerevisiae* (118), but since recognized in a wide range of yeast species (119). They mostly act by interacting with the cell membrane to make it permeable (120), though some act by interfering with the synthesis of DNA, e.g., that of *K. lactis* (121).

Yeast killer toxins have been detected in a number of fermented foods other than beers and wines, e.g., soy sauce, miso, and salted vegetables (120, 122), so that they may play an ecological role in these types of foods. Wide deliberate use of yeast killer toxins for food preservation is less likely because they generally have a very narrow spectrum of activity.

V. CONCLUSIONS

A large number of very effective antimicrobial systems occur in nature. However, disappointingly few of them have been made deliberate use of to activate microorganisms in foods. This may reflect the food industry's reluctance to embark on the substantial and expensive programs of toxicological testing necesssary to convince regulatory authorities of the safety of a new additive or technique. The limited application of natural antimicrobials may also reflect the absence of strong commercial, marketing, and economic incentives for their exploitation. However, this has changed in the past decade or two as consumers' desires for less processed, more natural, and safer foods has grown. As a result, it is likely that further applications of natural antimicrobials will be developed, along with other milder preservation techniques (74). In a recent comprehensive review of naturally occurring antimicrobials in food (123), Sofos et al. concluded that ''few, if any, microbials are present in foods at concentrations great enough to be antimicrobials without purification or concentration. The ultimate challenge is to find a naturally occurring antimicrobial that can be added to a 'microbiologically-sensitive' food product in a nonpurified form from another nonsensitive food.'' Developments are likely to be most rapid if natural systems are evaluated in combination with other preservation adjuncts (''hurdles''; see Chapter 20). Successful combinations would minimize the concentrations of antimicrobials needed to achieve particular desired effects. The combinations could include the more obvious potential adjuncts (e.g., lowered pH, lowered water activity, specific solutes, chelators, lowered oxygen, raised carbon dioxide, organic acids and other allowed preservatives, low temperature, mild heat-processing), but also some of the newly emerging ones (e.g., high hydrostatic pressure, electroporation, manothermosonication) (124). However, there are many more potentially useful adjuncts. In the final report of an European Union research project on preservation by combined processes, over 60 possible hurdles usable in foods, were identified (125). The potential for useful additive or synergistic interactions with natural systems is therefore substantial (5). At the same time, progress will probably be most rapid if more studies are undertaken using real foodstuffs as well as model systems and laboratory media.

REFERENCES

1. GW Gould. Ecosystem approaches to food preservation. J Appl Bacteriol Symp Suppl 73:58S–68S, 1992.
2. GW Gould Industry perspectives on the use of natural antimicrobials and inhibitors for food applications. J Food Prot (suppl):82–86, 1996.
3. VM Dillon, RG Board. Future prospects for natural antimicrobial food preservation systems. In: VM Dillon, RG Board, eds. Natural Antimicrobial Systems and Food Preservation. Wallingford, Oxon: CAB International, 1994, pp. 297–305.
4. DE Conner. Naturally occurring compounds. In: PM Davidson, AL Branen, eds. Antimicrobials in Foods. New York: Marcel Dekker, 1993, pp. 441–468.
5. RG Board, GW Gould. Future prospects. In: NJ Russell, GW Gould, eds. Food Preservatives. Glasgow: Blackie Academic and Professional, 1991, pp. 267–284.
6. RG Board. Natural antimicrobials from animals. In: GW Gould, ed. New Methods of Food Preservation. Glasgow: Blackie Academic and Professional, 1995, pp. 40–57.
7. EA Johnson. Egg white lysozyme as a preservative for use in foods. In: JS Sim, S Nakai, eds. Egg Uses and Processing Technologies. Wallingford, Oxon: CAB International, 1994, pp. 171–191.
8. FE Cunningham, VA Proctor, SJ Goettsch. Egg white lysozyme as a preservative: an overview. World Poult Sci J 47:141–163, 1991.
9. C Maullu, G Lampis, T Basile, A Ingianni, GM Rossolini, R. Pompei. Production of lysozyme-enriched biomass from cheese industry byproducts. J Appl Microbiol 86:182–186, 1999.
10. HS Tranter. Lysozyme, ovotransferrin and lysozyme. In: VM Dillon, RG Board, eds. Natural Antimicrobial Systems and Food Preservation. Wallingford, Oxon: CAB International, 1994, pp. 65–97.
11. D Carminati, E. Nevianti, G Murchetti. Activity of lysozyme on vegetative cells of *Clostridium tyrobutyricum*. Latte 10:194–198, 1985.

12. S Carini, R Lodi. Inhibition of germination of clostridial spores by lysozyme. Ind Latte 18:35–48, 1982.
13. F Wasserfall, E Voss, D Prokopek. Experiments on cheese ripening: the use of lysozyme instead of nitrite to inhibit late blowing of cheese. Kiel Milchwirtsch Forschungsber 28:3–16, 1976.
14. D Scott, FE Hammer, TJ Szalkucki. Bioconversions: enzyme technology. In: D Knorr, ed. Food Biotechnology. New York: Marcel Dekker, 1987, pp. 124–143.
15. KJ Samuelson, JH Rupnow, GW Froning. The effect of lysozyme and ethylenediaminetetraacetic acid on *Salmonella* on broiler parts. Poultry Sci 64:1488–1490, 1987.
16. R. Nakamura, A Kato, K Kobayashi. Novel bifunctional lysozyme-dextran conjugate that acts on both Gram-negative and Gram-positive bacteria. Agric Biol Chem 54:3057–3059, 1990.
17. HR Ibrahim, A Kato, K Kobayashi. Antimicrobial effects of lysozyme against Gram-negative bacteria due to covalent binding of palmitic acid. J Agric Food Chem 39:2077–2082, 1991.
18. DJ Monticello. Control of microbial growth with nisin/lysozyme formulations. Eur Pat Appl 89123445.2 (1989).
19. B Ray, C Johnson, B Wanismail. Factors influencing lysis of frozen *Escherichia coli* cells by lysozyme. Cryo Lett 5:183–190, 1984.
20. A Chatzolopou, RJ Miles, G Anagnostopoulos. Destruction of Gram-negative bacteria. Int Pat Appl WO 93/00822 (1993).
21. TS Yang, FE Cunningham. Stability of egg white lysozyme in combination with other antimicrobial substances. J Food Prot 56:153–156, 1993.
22. SE El-Kest, EH Marth. Transmission electron microscopy of unfrozen and frozen-thawed cells of *Listeria monocytogenes* treated with lipase and lysozyme. J Food Prot 55:687–696, 1992.
23. GW Gould, AD Hitchins. Sensitization of spores to lysozyme and hydrogen peroxide with agents which rupture disulphide bonds. J Gen Microbiol 33:413–422, 1963.
24. Y Suzuki, LJ Rode. Germination of bacterial spores with lysozyme. Bact Proc 22, 1967.
25. M Cassier, M Sebald. Germination lysozyme-dependence des spores de *Clostridium perfringens* ATCC 3624 apres traitment thermique. Ann Inst Pasteur Paris 117: 314–324, 1969.
26. M Sebald, H Ionesco. Germination IzP-dependente des spores de *Clostridium botulinum* type E. Comput Rend Acad Sci Paris 275:2175–2177, 1972.
27. G Alderton, JK Chen, KA Ito. Effect of lysozyme on the recovery of heated *Clostridium botulinum* spores. Appl Microbiol 17:613–615, 1974.
28. VN Scott, DT Bernard. The effect of lysozyme on the apparent heat resistance of non-proteolytic type B *Clostridium botulinum*. J Food Saf 7:145–154, 1985.
29. MW Peck, DA Fairbairn, BM Lund. The effect of recovery medium on the estimated heat-inactivation of spores of non-proteolytic *Clostridium botulinum*. Lett Appl Microbiol 15:146–151, 1992.
30. DAA Mossel, P van Netten, I Perales. Human listeriosis transmitted by food in a general medical-microbiological perspective. J Food Prot 50:894–895, 1987.

31. S Notermans, J Dufrenne, BM Lund. Botulism risk of refrigerated processed foods of extended durability. J Food Prot 53:1020–1024, 1990.
32. GE Livingston. Extended shelflife called prepared foods. J Foodserv Syst 3:221–230, 1985.
33. BM Lund, MW Peck. Heat resistance and recovery of spores of non-proteolytic *Clostridium botulinum* in relation to rerigerated, processed foods with an extended shelf life. J Appl Bacteriol Suppl 76:115S–128S, 1994.
34. MW Peck, PS Fernandez. Effect of lysozyme concentration, heating at 90°C, and then incubation at chilled temperatures on growth from non-proteolytic *Clostridium botulinum*. Lett Appl Microbiol 21:50–54, 1995.
35. GW Gould. Sous vide foods: conclusion of an ECFF botulinum working party. Food Control 10:47–51, 1999.
36. ME Peterson, GA Pelroy, FT Poysky, RN Paranjpye, FM Dong, FT Piggott, MW Eklund. Heat pasteurization process for inactivation of non-proteolytic types of *Clostridium botulinum* in picked dungeness crabmeat. J Food Prot 60:928–924, 1997.
37. KG Paul, PI Ohlsson, A Hendriksson. The isolation and some liganding properties of lactoperoxidase. FEBS Lett 100:200–204, 1980.
38. B Reiter, G Harnulv. Lactoperoxidase natural antimicrobial system: natural occurence, biological functions and practical applications. J Food Prot 47:724–732, 1984.
39. B Reiter, RJ Fulford, VM Marshall, N Yarrow, MJ Ducker, M Knutsson. An evaluation of the growth-promoting effect of the lactoperoxidase system in newborn calves. Anim Prod 32:297–306, 1981.
40. L Bjorck, Q Claesson, W Schultness. The lactoperoxidase/thiocyanate/hydrogen peroxide system as a temporary preservative for raw milk in developing countries. Milchwissenschaft 34:726–729, 1979.
41. BG Harnulv, C Kandusamy. Increasing the keeping quality of raw milk by activation of the lactoperoxidase system. Results from Sri Lanka. Milchwissenschaft 37: 454–457, 1982.
42. B Reiter. The lactoperoxidase system of bovine milk. In: KM Pruitt, J Tenovuo, eds. The Lactoperoxidase System: Chemistry and Biological Significance. New York: Marcel Dekker, 1985, pp. 123–142.
43. KM Pruitt, B Reiter. Biochemistry of peroxidase system: antimicrobial effects. In: KM Pruitt, J Tenovuo, eds. The Lactoperoxidase System. Chemistry and Biological Significance. New York: Marcel Dekker, 1985, pp. 143–178.
44. B Reiter. Protective proteins in milk—biological significance and exploitation. Int Dairy Fed Bull 191:2–35, 1985.
45. L Bjorck. The lactoperoxidase system. In: GW Gould, ME Rhodes-Roberts, AK Charnley, RM Cooper, RG Board, eds. Natural Antimicrobial Systems. Bath: Bath University Press, 1986, pp. 297–308.
46. International Dairy Federation. Code of practice for preservation of raw milk by lactoperoxidase. Bull Int Dairy Fed No. 234. Brussels: Int Dairy Fed, 1988.
47. El Thomas, KA Pera, KW Smith, AK Chwang. Inhibition of *Streptococcus mutans* by the lactoperoxidase antimicrobial system. Infect Immun 39:767–778, 1983.
48. J Tenovuo, B Mansson-Rahemtulla, KM Pruitt, R Arnold. Inhibition of dental

plaque acid production by the salivary lactoperoxidase system. Infect Immun 34: 208–214, 1981.

49. H Korhonen. A new method for preserving raw milk: the lactoperoxidase antibacterial system. World Animal Rev 35:23–29, 1980.
50. B Ekstrand. Lactoperoxidase and lactoferrin. In: VM Dillon, RG Board, eds. Natural Antimicrobial Systems and Food Preservation. Wallingford, Oxon: CAB International, 1994, pp. 15–63.
51. E Borsch, L Wallerston, M. Rosen, L Bjorck. Antibacterial effect of the lactoperoxidase/thiocynate/hydrogen peroxide system against strains of *Campylobacter* isolated from poultry. J Food Prot 52:638–641, 1989.
52. KM Pruitt, B Reiter. Biochemistry of the lactoperoxidase system: antimicrobial effects. In: K Pruitt, J Tenovuo, eds. The Lactoperoxidase System. New York: Marcel Dekker, 1985, pp. 143–178.
53. B Reiter, VM Marshall, L Bjorck, CG Rosen. Nonspecific bactericidal activity of the lactoperoxidase-thiocyanate-hydrogen peroxide system of milk against *Escherichia coli* and some Gram-negative pathogens. Infect Immun 13:800–807, 1976.
54. B Reiter, A Pickering, JD Oram. An inhibitory system—lactoperoxidase/thiocyanate/peroxide in raw milk. In: N Molin, ed. Microbial Inhibitors in Foods. Stockholm: Almqvist and Wiksell, 1964, pp. 297–305
55. JM Oram, B Reiter. The inhibition of streptococci by lactoperoxidase, thiocyanate and hydrogen peroxide. 1. The effect of the inhibitory system on susceptible and resistant strains of group N streptococci. Biochem J 100:373–381, 1966.
56. HS Tranter, RG Board. Review: the antimicrobial defense of avian eggs, Biological perspective and chemical basis. J Appl Biochem 4:295–338, 1982.
57. P Azari, R Baugh. A simple and rapid procedure for preparation of large quantities of pure ovotransferrin. Arch Biochem Biophys 118:138–144, 1967.
58. L Hambraeus. The role of lactoferrin in neonatal feeding. In: Int Dairy Fed, ed. Antimicrobial Systems in Milk, 1986, pp. 49–58.
59. Success stories from the agro-industrial research programmes. Luxembourg: EUR 18227, 1998.
60. B Reiter, JD Oram. Iron and vanadium requirements of lactic streptococci. J Dairy Res 35:67–69, 1968.
61. P Rainard. Bacteriostatic activity of bovine milk lactoferrin against mastitic bacteria. Vet Microbiol 11:387–392, 1986.
62. JA Garibaldi. Influence of temperature on the biosynthesis of iron transport compounds by *Salmonella typhimurium*. J. Bacteriol 110:262–265, 1972.
63. EA Johnson. The potential application of antimicrobial proteins in food preservation. J Dairy Sci 72:123–124, 1989.
64. H Saito, H Miyakawa, Y Tamura, Y Shinamura, M Tomita. Potent bactericidal activity of bovine lactoferrin hydrolysate produced by heat treatment at acidic pH. J Dairy Sci 74:3724–2730, 1991.
65. M Tomita, W Bellamy, M Takase K Yamauchi, H Wakabayashi, K Kawase. Potent antibacterial peptides generated by pepsin digestion of bovine lactoferrin. J Dairy Sci 74:4137–4142, 1991.
66. KS Hoeke, JM Milne, PA Grieve, DA Dionysius, R Smith. Antibacterial activity in bovine lactoferrin-derived peptides. Antimic Agents Chemother 41:54–59, 1997.

67. H Wakabayashi, W Bellamy, M Takase, M Tomita. Inactivation of *Listeria monocytogenes* by lactoferricin, a potent antimicrobial peptide isolated from cow's milk. J Food Prot 55:238–240, 1992.
68. S Kobayashi, M. Tomita, K Kawase, M Takase, H Miyakawa, K Yamauchi, H Saito, H Abe, S Shimamura. Lactoferrin hydrolysate for use as an antibacterial agent and as a tyrosinase inhibition agent. Eur Pat Appl 90125035.7 (1990).
69. KM Wilkins, RG Board. Natural antimicrobial systems. In: GW Gould, ed. Mechanisms of Action of Food Preservation Procedures. Barking, Essex: Elsevier Applied Science, 1989, pp. 285–362.
70. LR Beuchat. Antimicrobial properties of spices and their essential oils. In: VM Dillon, RG Board, eds. Natural Antimicrobial Systems and Food Preservation. Wallingford, Oxon: CAB International, 1994, pp. 167–179.
71. RA Dixon. The phytoalexin response: elicitation, signalling and control of host gene expression. Biol Rev 61:239–291, 1986.
72. RA Dixon, PM Dey, CJ Lamb. Phytoalexins: enzymology and molecular biology. Adv Enzymol 55:1–136, 1983.
73. DA Smith, SW Banks. Biosynthesis, elicitation and biological activity of isoflavenoid phytoalexins. Phytochem 25:979–995, 1986.
74. JG Banks, RG Board, NHC Sparks. Natural antimicrobial systems and their potential in food preservation of the future. Biotech Appl Biochem 8:103–147, 1986.
75. W Boyle. Spices and essential oils as preservatives. Am Perfum Essent Oils Rev 66:25–28, 1955.
76. L Hargreaves, B Jarvis, AP Rawlinson, JM Wood. The antimicrobial effects of spices, herbs and extracts from these and other food plants. London: British Food Manufacturing Industries Research Association, Scientific and Technical Services Publication No. 88, 1975.
77. H. Dallyn. Antimicrobial properties of vegetable and fish oils. In: VM Dillon, RG Board, eds. Natural Antimicrobial Systems and Food Preservation. Wallingford, Oxon: CAB International, 1994, pp. 205–221.
78. PJ Delaquis, G Mazza. Antimicrobial properties of isothiocyanates and food preservation. Food Technol 44(11):100–117, 1995.
79. GJE Nychas. Natural antimicrobials from plants. In: GW Gould, ed. New Methods of Food Preservation. Glasgow: Blackie Academic and Professional, 1995, pp. 58–89.
80. JRL Walker. Antimicrobial compounds in food plants. In: VM Dillon, RG Board, eds. Natural Antimicrobial Systems and Food Preservation. Wallingford, Oxon: CAB International, 1994, pp. 181–204.
81. IM Whitehead, DR Threlfall. Production of phytoalexins by plant tissue cultures. J Biotechnol 26:63 81, 1992.
82. L Beuchat. Antimicrobial properties of spices and their essential oils. In: VM Dillon, RG Board, eds. Natural Antimicrobial Systems and Food Preservation. Wallingford, Oxon: CAB International, 1994, 167–179.
83. VV Lantanzio, V de Cicco, D Di Venere, G Lima, M Salermo. Antifungal activity of phenolics against fungi commonly encountered during storage. Ital J Food Sci 6:23–60, 1994.

84. SG Deans, G Ritchie. Antibacterial properties of plant essential oils. Int J Food Microbiol 5:165–180, 1987.
85. LR Beauchat, RE Brackett. Inhibitory effects of raw carrots on *Listeria monocytogenes*. Appl Environ Microbiol 56:1734–1742, 1990.
86. C Nguyen-the, BM Lund. The lethal effect of carrot on *Listeria* species. J Appl Bacterial 70:479–489, 1991.
87. C Nguyen-the, BM Lund. An investigation of the bacterial effect of carrot on *Listeria monocytogenes*. J Appl Bacteriol 73:23–30, 1992.
88. S Doores. Organic acids. In: PM Davidson, AL Brannen, eds. Antimicrobials in Foods New York: Marcel Dekker, 1993, pp. 95–136.
89. JJ Kabara. Phenols and Chelators. In: NJ Russell, GW Gould, eds. Food Preservatives. Glasgow: Blackie Academic and Professional, 1991, pp. 200–214.
90. JJ Kabara, T Eklund. Organic acids and esters. In: NJ Russell, GW Gould, eds. Food Preservatives. Glasgow: Blackie Academic and Professional, 1991, pp. 44–71.
91. IR Booth, RG Kroll. The preservation of foods by low pH. In: GW Gould, ed. Mechanisms of Action of Food Preservation Procedures. Barking, Essex: Elsevier Applied Science, 1989, pp. 119–160.
92. M Stratford, PA Anslow, Evidence that sorbic acid does not inhibit yeast as a classic "weak acid preservative." Lett Appl Microbiol 27:203–206, 1998.
93. LA Shelef, EK Jyothi, M Bulgarelli. Sensitivity of some common food-borne bacteria to the spices sage, rosemary and allspice. J Food Sci 45(4):1042–1044, 1980.
94. LA Shelef. Antimicrobial effects of spices. J Food Saf 6:29–44, 1983.
95. JM Jay, GM Rivers. Antimicrobial activity of some food flavoring compounds. J Food Saf 6:129–139, 1984.
96. MI Farbood, JH Macneil, K Ostovar. Effect of rosemary spice extract on growth of microorganisms in meats. J Milk Food Technol 39:675–679, 1976.
97. WS Klis. Effectiveness of pimaricin as an alternative to sorbate for the inhibition of yeasts and moulds in foods. Food Technol 13:124–132. 1960.
98. Comprehensive EEC Directive on Food Additives. Document III-3761-rev. 2. Brussels: EU, 1989.
99. J Stark, HS Tan. Natamycin. In: NJ Russell, GW Gould, eds. Food Preservatives, 2nd ed. Gaithersburg, MD: Aspen Publishers Inc., 2001.
100. J Delves-Broughton. Nisin and its application as a food preservative. J Soc Dairy Technol 43:73–76, 1990.
101. GG Fowler, MJ Gasson. Antibiotics—nisin. In: NJ Russell, GW Gould, eds. Food Preservatives. Glasgow: Blackie Academic and Professional, 1991, pp. 135–152.
102. J Delves-Broughton, MJ Gasson. Nisin. In: VM Dillon, RG Board, eds. Natural Antimicrobial Systems and Food Preservation. Wallingford, Oxon: CAB International, 1994, pp. 99–131.
103. C Hill. Bacteriocins: natural antimicrobials from microorganisms. In: GW Gould, ed. New Methods of Food Preservation. Glasgow: Blackie Academic and Professional, 1995, pp. 22–39.
104. B Ray, MA Daeschel. Bacteriocins of starter cultures. In: VM Dillon, RG Board, eds. Natural Antimicrobial Systems and Food Preservation. Wallingford, Oxon: CAB International, 1994, pp. 133–166.

105. N Al-Zoreky, JW, Ayres, WE Sandine. Antimicrobial action of Microgard™ against food spoilage and pathogenic microorganisms. J Dairy Sci 74:748–763, 1991.
106. BJB Wood, ed. Microbiology of Fermented Foods. Barking, Essex: Elsevier Applied Science, 1985.
107. CS Pederson. Microbiology of Food Fermentations. Westport, CT: AVI Publishing, 1979.
108. AH Rose. Fermented Foods, Economic Microbiology Series. London: Academic Press, 1982.
109. MA Daeschel. Antimicrobial substances from lactic acid bacteria for use as food preservatives. Food Technol 43(1):164–167, 1989.
110. FK Lucke, RG Earnshaw. Starter cultures. In: NJ Russell, GW Gould, eds. Food Preservatives. Glasgow: Blackie Academic and Professional, 1991, pp. 215–234.
111. GH Weber, WA Broich. Shelflife extension of cultured dairy foods. Cult Dairy Prod J 21(4):19–23, 1986.
112. TM Cogan, M O'Dowd, D Mellerick. Effects of pH and sugar on acetoin production from citrate by *Leuconostoc lactis.* Appl Environ Microbiol 41:1–8, 1981.
113. El Thomas, KA Pera. Oxygen metabolism of *Streptococcus mutans*: uptake of oxygen and release of superoxide and hydrogen peroxide. J Bacteriol 154:1236–1244, 1983.
114. M Raccach, RC Baker. Formation of hydrogen peroxide by meat starter cultures. J Food Prot 41:798–799, 1978.
115. LT Axelsson, TC Chung, WJ Dobrogosz, SE Lindgren. Production of a broad spectrum antimicrobial substance by *Lactobacillus reuteri*. Microb Ecol Health Dis 2: 131–136, 1989.
116. TL Talarico, LT Axelsson, J Novotny, M Fiuzat, WJ Dobrogosz. Utilization of glycerol as a hydrogen acceptor by *Lactobacillus reuteri*: purification of 1,3 propanediol: NAD^+ oxidoreductase. Appl Environ Microbiol 56:943–948, 1990.
117. CJ Jacobs, JJJ van Vuuren. Effects of killer yeasts on wine fermentations. Am J Enol Vitic 42:295, 1991.
118. EA Bevan, M Makower. The physiological basis of the killer character in yeast. Proc XI Int Congr Genet 1:202–203, 1963.
119. F Radler, S Herzberger, I Schonig, P Schwarz. Investigation of a killer strain of *Zygosccharomyces bailii*. J Gen Microbiol 139:495–500, 1993.
120. VM Dillon, PE Cook. Biocontrol of undesirable microorganisms in food. In: VM Dillon, RG Board, eds. Natural Antimicrobial Systems and Food Preservation. Wallingford, Oxon: CAB International, 1994, pp. 255–296.
121. B Martinac, H Zhu, A Kubalski, XL, Zhou, M Cuthbertson, H Bussey, C Kung. Yeast K1 killer toxin forms ion channels in sensitive yeast spheroplasts and in artificial liposomes. Proc Natl Acad Sci USA 87:6228–6232, 1990.
122. C Suzuki, K Yamada, N Okada, S Nikkuni. Isolation and characterization of halotolerant killer yeasts from fermented foods. Agric Biol Chem 53:2593, 1989.
123. JN Sofos, LR Beuchat, PM Davidson, EA Johnson. Naturally Occuring Antimicrobials in Food. Ames, IA: Council for Agricultural Science and Technology, 1998.
124. GW Gould, ed. New Methods of Food Preservation. Glasgow: Blackie Academic and Professional, 1995.
125. L Leistner, LGM Gorris, ed. Food Preservation by Combined Processes. Brussels: EUR 15776 EN, 1994.

11

Perspectives for Application of Bacteriocins as Food Preservatives

Michael L. Chikindas and Thomas J. Montville
Rutgers University
New Brunswick, New Jersey

I. INTRODUCTION

The food industry is under continuous consumer pressure to produce fresh, unprocessed (or minimally processed) natural, healthy, and safe foods. Preservatives such as nitrites and propionates are either no longer used or are added at very low concentrations. Sterilization and pasteurization also prevent food spoilage but employ heat treatment of the food product. Unfortunately, consumers often equate ''processed'' with ''not fresh'' rather than ''safe.'' Thus, there is a need for a novel food preservative, which should originate from a natural source and deliver its antimicrobial activity without being labeled as a chemical preservative (1).

Most microorganisms are capable of producing a large variety of molecules. These molecules may be inhibitory either to the producing cell or to other bacteria and include organic acids, toxins, lytic enzymes, antibiotics, bacteriocins, etc. Virtually all microorganisms produce bacteriocins, which can be defined as proteinaceous compounds that kill closely related bacteria (2). Some authors put bacteriocins into the group of antibiotics (3), but due to their mechanism of synthesis, structure, and function, bacteriocins belong to a separate and distinct group of antimicrobial substances (4). Bacteriocins are heterogeneous with respect to inhibitory spectrum, molecular size, stability, physical and chemical properties, mode of action, etc. It should be remembered that what is known about nisin or pediocin (the most studied bacteriocins) may not be true of other bacteriocins and that most bacteriocins have not been studied extensively.

Lactic and bacteria (LAB) of the genera *Lactococcus*, *Lactobacillus*, *Leuconostoc*, *Pediococcus*, and *Carnobacterium* have been associated with food, especially dairy and meat products, for centuries, and selected strains of these microorganisms are used as starter cultures. Apart from the antimicrobial properties of lactic acid, hyrogen peroxide, and diacetyl produced by these bacteria, many species also produce bacteriocins. These bacteriocins inhibit growth of foodborne pathogens like *Listeria monocytogenes* and *Clostridium botulinum* (Table 1). By using bacteriocin-producing strains as starter cultures, or by adding purified bacteriocins to food, it is possible to repress the growth of certain food spoilage and/or pathogenic bacteria and to extend the shelf life of food products (5). The potential value of bacteriocins in the preservation of food products has been recognized and has prompted many research groups to direct their studies to investigations of these antimicrobial substances. The use of bacteriocin-producing starter cultures as ingredients may not require special consideration if the culture (microorganism) is considered GRAS (generally recognized as safe) because of its history of safe use by food industries prior to the 1958 Food Additives Amendment (6). If a purified bacteriocin is used as a food preservative, the substance might be self-affirmed as GRAS by the company according to the Code of Federal Regulations (7), but the Food and Drug Administration (FDA) may require justification.

According to FDA regulations, natural antimicrobials used as food preservatives have to be produced by GRAS microorganisms (7). Since many species of LAB have GRAS status, they have been extensively studied for the production of food-grade bacteriocins (Table 1). In addition to FDA, the Food Safety and Inspection Service (FSIS) of the U.S. Department of Agriculture (USDA) confirms the safety and evaluates efficiency of novel food preservatives. The use and application of novel bacteriocins require an integrated review by both FSIS and FDA before any applications are possible (8).

II. CLASSIFICATION

Based on progress in biochemical and genetic characterization of bacteriocins, Klaenhammer (9) defined four classes of these molecules (Table 2). Alternatively, bacteriocins produced by LAB can be divided into two groups with respect to their inhibitory spectrum (10). One group includes bacteriocins with a narrow inhibitory spectrum, active either only against closely related bacteria belonging to the same genus, e.g., diplococcin, lactocin 27, lactocin B, helveticin J (11–13), or against other bacterial genera as well, e.g., lactacin F (14,15). The second group consists of bacteriocins with a relatively broad spectrum of activity, e.g., nisin, pediocin, and leuconocin S (16–18). No activity against vegetative cells of gram-negative bacteria by gram-positive bacteriocin producers has been reported. This may be explained by their mode of action.

Table 1 Inhibition of Undesirable Microorganisms by Bacteriocin-Producing Bacteria Isolated from or Associated with Foods

Source	Strain	Activity	Ref.
Commercial probiotic product	*Streptococcus* sp. CNCM I-841	*Clostridium* sp., *L. monocytogenes*	41
Bulgarian yellow cheese	*Lactobacillus delbrueckii*	*L. monocytogenes*, *S. aureus*, *E. faecalis*, *E. coli*, *Yersinia enterocolitica*, *Y. pseudotuberculosis*	79
Vegetables	*Enterococcus mundtii*	*L. monocytogenes*, *C. botulinum*	80
Radish	*Lactococcus. lactis* subsp. *cremoris* R	*Clostridium*, *Staphylococcus*, *Listeria*, and *Leuconostoc* spp.	81
"Waldorf" salad	*Lactobacillus plantarum* BFE905	*L. monocytogenes*	82
French mold-ripened soft cheese	*Carnobacterium piscicola* CP5	*Carnobacterium*, *Listeria*, and *Enterococcus* spp.	83
Bean sprouts	*Lactococcus lactis* subsp. *lactis* ($NisZ^+$)	*L. monocytogenes* Scott A	84
Munster cheese	*L. plantarum* WHE92 ($PedAcH^+$)	*L. monocytogenes*	85
Spoiled ham	*C. piscicola* JG126	*L. monocytogenes*	86
Traditional French cheese	*Enterococcus faecalis* EFS2	*Listeria inocua*	87
Dry sausage	*L. plantarum* UG1	*L. monocytogenes*, *B. cereus*, *C. perfringens*, *C. sporogenes*	88
Irish kefir grain	*L. lactis* DPC3147	*Clostridium*, *Enterococcus*, *Listeria*, *Leuconostoc* spp.	89
Dry fermented sausage	*L. lactis* ($NisA^+$)	*L. monocytogenes*	90
Fermented sausage	*L. plantarum* SA6	*Lactobacillus* spp.	91
Red smear cheese	*Brevibacterium linens* M18	*Listeria* and *Corynebacterium* spp.	92
Meat	*Leuconostock carnosum* Ta11A ($LeuA^+$)	*L. monocytogenes*	93
Sour doughs	*L. bavaricus* ($bavA^+$)	*L. monocytogenes*	94
Wine	*Pediococcus pentosaceus* N5p	Lactic acid bacteria	95
Whey	*E. faecalis* 226	*L. monocytogenes*	96
Goat's milk	*Leuconostock mesenteroides* Y105	*L. monocytogenes*	20
Sauerkraut	*L. lactis* subsp. *lactis* (Nis^+)	*L. monocytogenes*	97

Table 2 Classification of Bacteriocins by Klaenhammer

Group		Features	Bacteriocins (group representatives)
I		Lantibiotics, small (<5 kDa) peptides containing lanthionine and β-methyl lanthionine	Nisin, lacticin 481, carnocin UI49, lactocin S, etc.
II	IIa	Small heat-stable peptides, sysnthesized in a form of precursor processed after two glycine residues, active against *Listeria*, have a consensus sequence of YGNGV-C in the N-terminal	Pediocin PA-1, sakacins A and P, curvacin A, leucocin A, etc.
	IIb	Two-component systems: two different peptides required to form an active poration complex	Lactococcins G and F, lactacin F
	IIc	Peptides that require their cysteine residues to be present in reduced for in order for the molecule to be biologically active	Lactococcin B
III		Large molecules sensitive to heat	Helveticins J and V-1829, acidophilucin A, lactacins A and B
IV		Complex molecules consisting of proteinaceous and one or more chemical moieties such as lipid or carbohydrate	Plantaricin S, leuconocin S, lactocin 27, pediocin SJ-1

III. MODE OF ACTION

A commonly accepted hypothesis of the mode of action of bacteriocins is that the interaction of the bacteriocin with a sensitive cell occurs in two stages, involving specific or nonspecific adsorption of the bacteriocin on the cell surface resulting in cell death (2,4). Bacteriocins have a bactericidal effect on sensitive cells, which in some cases results in cell lysis (19). A few bacteriocins (lactocin 27, leucosin A-UAL 187, leuconocin S) have been reported to act bacteriostatically (13,18,20). However, the cidal versus static action may be more related to the conditions of use and pathogen than specific characteristics of the molecule.

The primary target of bacteriocins is the cytoplasmic membrane. To perturb bacterial membrane, the bacteriocins often display an amphipathic α-helix (21,22). The bacteriocin molecules initiate rearrangements in the membrane structure, which alter the membrane permeability. The bacteriocin molecules make pores in membranes. This disturbs membrane transport, dissipates the proton motive force, and thus leads to inhibition of energy production and biosynthesis of proteins and nucleic acids (23,24).

Nisin is one of the most studied bacteriocins (4). It forms ion-permeable channels in the cytoplasmic membrane, resulting in an increase in the membrane permeability. This causes dissipation of the membrane potential and efflux of ATP, amino acids, and essential ions (25). These destroy energy production and inhibit biosynthesis of macromolecules, resulting in cell death. The primary disturbances may then trigger various other, secondary metabolic disorders such as cell lysis. Nisin does not require a membrane receptor, but it works better if the membrane is energized (26).

Nisin generates channels in membranes. The amount of negatively charged lipids in the membrane is likely to be a major criterion for the sensitivity of the organism to nisin. Nisin Z induces leakage of the anionic carboxyfluorescein more efficiently than the leakage of the potassium cation, suggesting that that nisin makes anion-selective pores (27).

There is some evidence that nisin inhibition of outgrowing bacterial spores involves modification of membrane sulfhydryl groups (28). The unsaturated side chains provided by the dehydro-residues may participate in these interactions, but no covalent linkages have been demonstrated. Wild-type and nisin-resistant spores of *C. botulinum* 169B studied by Mazzotta and Montville (29) contain similar amounts of unsaturated fatty acids, but the saturated straight-chain/branched-chain ratio is significantly higher in nisin-resistant (Nis^r) spores than in spores of the wild-type cells. These fatty acid differences suggest that Nis^r spore membranes may be more rigid, a characteristic that would interfere with the pore-forming ability of nisin.

Structure-function studies with subtilin (another lantibiotic-type bacteriocin) have shown that modification of the dehydroalanine residue at position 5, either chemically or by protein engineering techniques, is accompanied by the loss of its antimicrobial activity against outgrowing spores (30,31). These results indicate that the dehydroalanine-5 plays an important functional role in this type of inhibition. This mechanism has been confirmed for nisin as well (32). It has been proposed that the inhibition of spore outgrowth involves the covalent attachment of the dehydroalanine residue to a sensitive cellular target. This interaction would not require the disruption of cell membranes via channel formation and hence would not require a dehydro-residue for activity. When used in milk, nisin sensitizes *Bacillus* spores to heat and prevents recovery of survivors (33). Most likely, modifications in the membrane lead to higher heat sensitivity. This increased heat sensitivity of spores in the presence of nisin may lead to development of processes with reduced heat treatment if such a treatment is used for food preservation (34).

Pediocin PA-1, another well-studied bacteriocin (4,35), acts in a voltage-independent manner (36). Although its activity does not require protein receptor (37), it can act more efficiently in the presence of proteinaceous factors in the membrane (36). Pediocin PA-1 binding to the sensitive membrane is a result of electrostatic interactions (38), and cysteine residues in structure of pediocin molecule and in the structure of the family of pediocin-like bacteriocins are essential for its specific activity (39).

In contrast, lactococcin A increases permeability of lactococcal cytoplasmic membranes in a voltage-independent protein-mediated manner (19). The voltage-independent activity of lactococcin B, however, is dependent on the reduced state of its only cysteine residue (40).

IV. APPLICATIONS

Many newly isolated bacteriocins are attracting attention of researchers and industries (Table 3) as potent food preservatives that can be used as purified substances, crude extracts, or as a substance synthesized by food-grade microorganisms during food processing/fermentation (Table 4). Use of bacteriocin-producing starter cultures may control microorganisms in cheese (41,42) or in naturally fermented sausages (43,44). It may also help to control microorganisms during olive fermentation (45) and in many other technological processes involving fermentation. Bacteriocins may also be potent agents for decontamination of meat, acting more efficiently than water or organic acids (46).

So far nisin is the only bacteriocin approved by FDA as a GRAS antimicrobial peptide for use as a food additive (47). It was discovered as early as 1928, and its use as a food preservative began in 1951 when it was demonstrated that nisin-producing microorganisms prevent cheese spoilage from clostridial gas formation (48). It is now used for preservation of dairy products (e.g., desserts, processed cheese, cheese spreads) and canned foods all over the world (49). Toxicity studies conducted earlier clearly indicate that nisin is safe for applications as a food preservative (50,51). It has been approved for application as a food preservative in many countries (Table 5). [For a review on applications, see Delves-Broughton et al. (49).] Nisin is mainly used as an agent against *L. monocytogenes* and against spore-forming bacteria. However, its fool potential as a food preservative needs to be explored. Recently, nisin was studied as dietary additive in a mice model and was reported as having some immunomodulatory effect, causing transient change in both T-lymphocyte and B-lymphocyte counts (52). This finding may open a completely new window of opportunity for nisin's application.

Although bacteriocins have relatively narrow inhibition spectra compared to some chemical preservatives, when used as a part of hurdle technology they can work synergistically with other substances or even possess enhanced inhibition activity against a greater variety of microorganisms (Table 6). Nisin, used in combination with lactic acid, acts synergistically against *Pseudmonas fluorescens* and *Staphylococcus hominis* strains isolated from fish (53). Thomas and

Table 3 Examples of patented food applications of bacteriocins

U.S. Patent (date)	Patent title	Use	Ref.
5,817,362 (10.06.98)	Method for inhibiting bacteria using a novel lactococcal bateriocin	A method for inhibiting gram-positive bacteria in foods by using a novel bacteriocin produced by *Lactococcus lactis* NRRL-B-18535	98
5,753,614 (05.19.98)	Nisin composition for use as enhanced, broad-range bactericides	Combination of nisin, a chelating agent, and a surfactant to inhibit both gram-positive and gram-negative microorganisms in meat, eggs, cheese, and fish, used as food preservative	99
5,573,801 (11.12.96)	Surface treatment of foodstuffs with antimicrobial compositions	Use of *Streptococcus*-derived or *Pediococcus*-derived bacteriocins in combination with a chelating agent to protect food against *Listeria*	100
5,445,835 (08.29.95)	Method of producing a yogurt product containing bacteriocin PA-1	A yogurt product with increased shelf life containing a bacteriocin derived from *Pediococcus acidilactici*	101
5,219,603 (06.15.93)	Composition for extending the shelf life of processed meats	Use of a bacteriocin from *P. acidilactici* and a propionate salt to inhibit bacterial growth and to extend shelf life of raw and processed meat	102
5,186,962 (02.16.93)	Composition and method for inhibiting pathogens and spoilage organisms in foods	Use of bacteriocin-producing lactic acid bacteria to inhibit growth of foodborne pathogens	103
5,015,487 (05.14.91)	Use of lanthionines for control of postprocessing contamination in processed meat	Inhibiting the contamination of processed meat products by pathogenic or spoilage microorganisms by treating the surface of the meat product with a lantibiotic	104
4,929,445 (05.29.90)	Method for inhibiting *Listeria monocytogenes* using a bacteriocin	Inhibition of *L. monocytogenes* by a bacteriocin produced by *P. acidilactici*	105
4,883,673 (11.28.89)	Method for inhibiting bacterial spoilage and resulting compositions	Inhibition of food spoilage microorganisms in salads and salad dressings by a bacteriocin from *P. acidilactici*	106
4,790,994 (12.13.88)	Method for inhibiting psychrotrophic bacteria in cream or milk based products using a pediococcus	Inhibition of bacterial growth in cottage cheese by a bacteriocin-producing *Pediococcus pentosaceus* cells	107

Table 4 Bacteriocins as Food Preservatives: Suggested Applications

Bacteriocin	Application	Conclusion	Ref.
Nisin A	Incorporation of nisin into a meat binding system (Fibrimex)	Addition of nisin can reduce undesirable bacteria in restructured meat products	61
Pediocin AcH	Use of a pediocin AcH producer *Lactobacillus plantarum* WHE 92 to spray on the Münster cheese surface at the beginning of the ripening period	Spray prevents outgrowth of *L. monocytogenes* and can be used as an antilisterial treatment	108
Enterocin 4	Use of enterocin producer *E. faecalis* INIA4 as a starter culture for production of Manchego cheese	Use of *E. faecails* INIA4 starter allows to inhibit *L. monocytogenes* Ohio, but not *L. monosytogenes* Scott A	109
Linocin M-18	Use of *Brevibacterium* lines as a starter culture for production of red smear cheese	Causes 2 log reduction of *L. ivanovi* and *L. monocytogenes*	110
Nisin A	Use of nisin to control *L. monocytogenes* in ricotta cheese	Nisin efectively inhibits *L. monocytogenes* for 8 weeks	111
Piscicolin 126	Use of piscicolin 126 to control *L. monocytogenes* in deviled ham paste	More effective than commercially available bacteriocins	86
Leucocin A	Use of a leucocine-producing *Leuconostoc gelidum* UAL187 to control meat spoilage	Inoculation of a vacuum packed beef with the bacteriocin-producer delays the spoilage by *L. sake* for up to 8 weeks	112
Lactocin 705	Use of lactocin 705 to reduce growth of *L. monocytogenes* in ground beef	Lactocin 705 inhibits growth of *L. monocytogenes* in ground beef	113
Pediocin AcH	Use of the pediocin producer *Pediococcus acidilactici* to inhibit *L. monocytogenes*	*P. acidilactici* (Ped^+) starter culture contributes to effective reduction of *L. monocytogenes* during manufacture of chicken summer sausage	114
Pediocin PA-1	Use of *P. acidilactici* (Ped^+) strain as a starter culture in sausage fermentation	Pediocin effectively contributes to inhibition of *L. monocytogenes*	115

Table 5 Some Countries Permitting Use of Nisin

Country	Foods in which nisin is permitted	Maximum level (IU/g)
Argentina	Processed cheese	500
Australia	Cheese, processed cheese, canned tomatoes	No limit
Belgium	Cheese	100
Cyprus	Cheese, clotted cheese, canned vegetables	No limit
France	Processed cheese	No limit
Italy	Cheese	500
Mexico	Nisin is a permitted additive	500
Netherlands	Factory cheese, quard, processed cheese, cheese powder	800
Peru	Nisin is a permitted additive	No limit
Russia	Dietetic processed cheese, canned vegetables	8,000
United Kingdom	Cheese, canned foods, clotted cream	No limit
United States	Pasteurized processed cheese spreads	10,000

Wimpenny (54) showed that nisin acts synergistically with NaCl when used against *Staphylococcus aureus* and that lowering of pH improved its activity. When used in combination with sucrose fatty acid esters, nisin has higher bactericidal activity against certain strains of *L. monocytogenes*, *Lactobacillus plantarum*, and *S. aureus* and against cells and spores of *Bacillus cereus*. This activity was even greater when nisin was combined with NaCl at lower pH (55). However, it is well documented that *L. monocytogenes* can tolerate rather low pH by triggering ''the acid-tolerance response'' (56). The interaction of nisin and acid tolerance may be bidirectional. Nisin may regulate acid tolerance in *L. monocytogenes* (57), and some acid-tolerant *L. monocytogenes* cells are more resistant to nisin (58). Therefore, when nisin is used as a part of hurdle technology, pH, salt concentration, temperature, and other parameters of the food product and its storage conditions should be taken into consideration. For example, milk fat can dramatically decrease nisin's activity against *L. monocytogenes*, and its antilisteria action can be improved in the presence of emulsifier Tween 80 (59). Taking these factors into account will achieve the best antimicrobial effect and minimize inhibition of preservatives either by each other or by compound-induced antimicrobial resistance in the microorganisms.

In addition to being used in a free form, nisin can be also immobilized in gels (60) and commercial meat-binding systems (61), or it can be adsorbed on specific surfaces (62,63). This may open a new window of opportunity for the use of bacteriocins in packaging material. The controlled release of bacteriocins may prolong their action against undesired microflora, thus extending the shelf life of a food product.

Table 6 Increased Activity of Bacteriocins Used as Part of Hurdle Technology

Bacteriocin	Other Factors	Effect	Ref.
Nisin A	N_2; CO_2; low temperature	Effect on *L. monocytogenes*: increase in the lag phase (400 IU/ml); inhibition of growth (1250 IU/ml)	71
Pediocin AcH	Hydrostatic pressure and high temperature	Combination of pressure (345 MPa), temperature (50°C) and bacteriocin acts synergistically causing reduction of viability of *S. aureus*, *L. monocytogenes*, *E. coli* O157:H7, *L. sake*, *Leu. mesenteroides*	117
Nisin A	Milk lactoperoxidase (LP) and low temperature	Nisin-producing *L. lactis* acts synergistically with LP in reduction of *L. monocytogenes*	45
Nisin A	Calcium alginate gel	Gel-immobilized nisin is delivered more effectively than pure nisin and suppresses growth of *Brochothrix thermosphacta* on beed carcases	60
Pediocin AcH	Sodium diacetate	Combination of pediocin and sodium diacetate works synergistically against *L. monocytogenes* at both room and low temperature	51
Pediocin AcH	Emulsifier (Tween 80) or encapsulation of the pediocin within liposomes	Pediocin AcH possesses higher listericidal activity in slurries of nonfat milk, butterfat, or meat when present in encapsulated form or acts in the presence of Tween 80	49

Although nisin and other bacteriocins from gram-positive bacteria are not active against gram-negative microorganisms, when used in combination with certain compounds, bacteriocins can act against gram-negative organisms as well. The outer membrane of gram-negative cells is the main hurdle for bacteriocins, but when combined with chelating agents, such as disodium EDTA, nisin acts against cells of *Salmonella* (64,65). When used after treatment with sublethal concentrations of trisodium phosphate (TSP), nisin successfully inhibited *S. aureus*, *Salmonella* Enteritidis, *Campylobacter jejuni*, *Escherichia coli*, and *P. fluorescens*, which had been dried on the surface of chicken skin cells (66). This suggests that a much lower (sublethal) concentration of TSP can be used for decontamination of gram-negative bacteria on the surface of food products if treatment with TSP is followed by treatment with nisin. Boziaris and coworkers reported that heat-injured gram-negative cells are becoming sensitive to nisin (67). This could allow reduction of pasteurization time for certain food products.

Although most bacteriocins do not require specific targets in the membranes of sensitive cells, when used together they can work synergistically or possess antagonism against each other (68). Microbial resistance to bacteriocins may become an issue as their food-preservative use becomes more wide spread. In many cases bacteriocins do not induce cross-resistance (69). However, other reports indicate that microorganisms treated with certain bacteriocins become resistant to not only the bacteriocin they were in contact with, but even toward unrelated bacteriocins never used against these cells (70). This observation again suggests that bacteriocins should be used as a part of hurdle technology, which allows them to act synergistically with other food preservatives to prevent appearance of resistant bacterial forms. Previously reported results (71) indicate that nisin works synergistically with other preservatives, also in a food model system (72,73). Combining preservation techniques is a powerful tool for extending shelf life, especially of minimally processed foods (74). Hurdle technology permits use of less severe preservative levels and of techniques that are less damaging to the quality of the final product (75).

Since the spread of multidrug-resistant pathogenic bacteria has became a serious issue, it is important to emphasize that drug-resistant microorganisms are as susceptible to bacteriocins as their drug-sensitive forms (76) and the bacteriocin-resistant bacteria are as susceptible to antibiotics as their bacteriocin-sensitive forms (77), i.e., bacteriocin resistance does not confer antibiotic resistance.

V. CONCLUSIONS

In the last decade, hundreds of bacteriocins have been discovered and scores of applications have been proposed, but only nisin is in commercial use (Table 5). Euphoric enthusiasm for bacteriocins has abated as it has become clear that they are not antimicrobial ''magic bullets.'' However, the elucidation of their mechanistic action has made it clear that bacteriocins can be important new components of multiple hurdle applications (78). Research on bacteriocin resistance suggests that it does not confer resistance to antibiotics, heat, or other antimicrobials and that in multiple hurdle applications, the development of resistance is rare. There is widespread expert agreement that bacteriocins are safe, natural, and effective food preservatives. If the full potential of bacteriocins is to be realized, mechanisms of action must be traced back to amino acid sequences and protein structure and then applied to specific preservation usages in foods.

ACKNOWLEDGMENTS

Preparation of this manuscript was supported by State Appropriations, U.S. Hatch Act Funds, and the New Jersey Agricultural Experiment Station. Research in the authors' laboratory is also supported by the USDA NRI Competitive Grants Program in Food Safety.

REFERENCES

1. DL Zink. The impact of consumer demands and trends on food processing. Emerg Infect Dis 3:467–469, 1997.
2. JR Tagg, AS Dajani, LW Wannamaker. Bacteriocins of gram-positive bacteria. Bacteriol Rev 40:722–756, 1976.
3. MA Riley. Molecular mechanisms of bacteriocin evolution. Annu Rev Genet 32: 255–278, 1998.
4. TJ Montville, Y Chen. Mechanistic action of pediocin and nisin: recent progress and unresolved questions. Appl Microbiol Bitechnol 50:511–519, 1998.
5. WP Hammes, PS Tichaczek. The potential of lactic acid bacteria for the production of safe and wholesome food. Z Lebensm Unters Forsch 198:193–201, 1994.
6. PM Muriana. Bacteriocins for control of *Listeria* spp. in food. J Food Prot (suppl): 54–63, 1996.
7. FDA. Food additives. Code of Federal Regulations. Title 21, part 170. Washington, DC: U.S. Government Printing Office, 1990, pp. 5–23.
8. RC Post. Regulatory perspective of the USDA on the use of antimicrobials and inhibitors in foods. J Food Prot (suppl):78–81, 1996.
9. TR Klaenhammer. Genetics of bacteriocins produced by lactic acid bacteria. FEMS Microbiol Rev 12:39–85, 1993.
10. TR Klaenhammer. Bacteriocins of lactic acid bacteria. Biochimie 70:337–349, 1988.
11. SF Barefoot, TR Klaenhammer. Detection and activity of lactacin B, a bacteriocin produced by *Lactobacillus acidophilus*. Appl Environ Microbiol 45:1808–1815, 1983.

12. MC Joerger, TR Klaenhammer. Characterization and purification of helveticin J and evidence for a chromosomally determined bacteriocin produced by *Lactobacillus helveticus* 481. J Bacteriol 167:439–446, 1986.
13. GC Upreti, RD Hinsdill. Production and mode of action of lactocin 27: bacteriocin from a homofermentative *Lactobacillus*. Antimicrob Agents Chemother 7:139–145, 1975.
14. T. Abee, TR Klaenhammer, L Letellier. Kinetic studies of the action of lactacin F, a bacteriocin produced by *Lactobacillus johnsonii* that forms poration complexes in the cytoplasmic membrane. Appl Environ Microbiol 60:1006–1013, 1994.
15. PM Muriana, TR Klaenhammer. Purification and partial characterization of lactacin F, a bacteriocin produced by *Lactobacillus acidophilus* 11088. Appl Environ Microbiol 57:114–121, 1991.
16. VG Eijsink, M Skeie, PH Middelhoven, MB Brurberg, IF Nes. Comparative studies of class IIa bacteriocins of lactic acid bacteria. Appl Environ Microbiol 64:3275–3281, 1998.
17. JN Hansen. Nisin as a model food preservative. Crit Rev Food Sci Nutr 34:69–93, 1994.
18. ME Stiles. Bacteriocins produced by *Leuconostoc* species. J Dairy Sci 77:2718–2724, 1994.
19. MJ van Belkum, J Kok, G Venema, H Holo, IF Nes, WN Konings, T Abee. The bacteriocin lactococcin A specifically increases permeability of lactococcal cytoplasmic membranes in a voltage-independent, protein-mediated manner. J Bacteriol 173:7934–7941, 1991.
20. Y Hechard, B Derijard, F Letellier, Y Cenatiempo. Characterization and purification of mesentericin Y105, an anti-listeria bacteriocin from *Leuconostoc mesenteroides*. J Gen Microbiol 138:2725–2731, 1992.
21. Y Fleury, MA Dayem, JJ Montagne, E Chaboisseau, JP Le Caer, P Nicolas, A Delfour. Covalent structure, synthesis, and structure-function studies of mesentericin Y 105(37), a defensive peptide from gram-positive bacteria *Leuconostock mesenteroides*. J Biol Chem 271:14421–14429, 1996.
22. J Nissen-Meyer, H Holo, LS Havarstein, K Sletten, IF Nes. A novel lactococcal bacteriocin whose activity depends on the complementary action of two peptides. J Bacteriol 174:5686–5692, 1992.
23. TJ Montville, ME Bruno. Evidence that dissipation of proton motive force is a common mechanism of action for bacteriocins and other antimicrobial proteins. Int J Food Microbiol 24:53–74, 1994.
24. TJ Montville, Winkowski, K, RD Ludescher. Models and mechanisms for bacteriocin action and application. Int Dairy J 5:797–814, 1995.
25. HG Sahl, M Kordel, R Benz. Voltage-dependent depolarization of bacterial membranes and artificial lipid bilayers by the peptide antibiotic nisin. Arch Microbiol 149:120–124, 1987.
26. MJ Garcera, MG Elferink, AJ Driessen, WN Konings. In vitro pore-forming activity of othe lantibiotic nisin. Role of protonmotive force and lipid composition. Eur J Biochem 212:417–422, 1993.
27. E Breukink, C Van Kraaij, RA Demel, RJ Siezen, OP Kuipers, B De Kruijff. The

C-terminal region of nisin is responsible for the initial interaction of nisin with the target membrane. Biochemistry 36:6968–6976, 1997.

28. SL Morris, RC Walsh, JN Hansen. Identification and characterization of some bacterial membrane sulfhydryl groups which are targets of bacteriostatic and antibiotic action. J Biol Chem 259:13590–13594, 1984.
29. AS Mazzotta, TJ Montville. Nisin induces changes in membrane fatty acid composition of *Listeria monocytogenes* nisin-resistant strains at 10 °C and 30 °C. J Appl Microbiol 82:32–38, 1997.
30. W Liu, JN Hansen. Enhancement of the chemical and antimicrobial properties of subtilin by site-direct mutagenesis. J Biol Chem 267:25078–25085, 1992.
31. K Winkowski, ME Bruno, TJ Montville. Correlation of bioenergetic parameters with cell death in *Listeria monocytogenes* cells exposed to nisin. Appl Environ Microbiol 60:4186–4188, 1994.
32. WC Chan, HM Dodd, N Horn, K Maclean, LY Lian, BW Bycroft, MJ Gasson, GC Roberts. Structure-activity relationships in the peptide antibiotic nisin: role of dehydroalanine 5. Appl Environ Microbiol 62:2966–2969, 1996.
33. LR Wandling, BW Sheldon, PM Foegeding. Nisin in milk sensitizes *Bacillus* spores to heat and prevents recovery of survivors. J Food Prot 62:492–498, 1999.
34. BM Beard, BW Sheldon, PM Foegeding. Thermal resistance of bacterial spores in milk-based beverages supplemented with nisin. J Food Prot 62:484–491, 1999.
35. JD Marugg, CF Gonzalez, BS Kunka, AM Ledeboer, MJ Pucci, MY Toonen, SA Walker, LC Zoetmulder, PA Vandenbergh. Cloning, expression, and nucleotide sequence of genes involved in production of pediocin PA-1, and bacteriocin from *Pediococcus acidilactici* PAC1.0. Appl Environ Microbiol 58:2360–2367, 1992.
36. ML Chikindas, MJ Garcia-Garcera, AJ Driessen, AM Ledeboer, J Nissen-Meyer, IF Nes, T Abee, WN Konings, G Venema. Pediocin PA-1, a bacteriocin from *Pediococcus acidilactici* PAC1.0, forms hydrophilic pores in the cytoplasmic membrane of target cells. Appl Environ Microbiol 59:3577–3584, 1993.
37. Y Chen, R Shapira, M Eisenstein, TJ Montville. Functional characterization of pediocin PA-1 binding to liposomes in the absence of a protein receptor and its relationship to a predicted tertiary structure. Appl Environ Microbiol 63:524–531, 1997.
38. Y Chen, RD Ludescher, TJ Montville. Electrostatic interactions, but not the YGNGV consensus motif, govern the binding of pediocin PA-1 and its fragments to phospholipid vesicles. Appl Environ Microbiol 63:4770–4777, 1997.
39. P Bhugaloo-Vial, JP Douliez, D Moll, X Dousset, P Boyaval, D Marion. Delineation of key amino acid side chains and peptide domains for antimicrobial properties of divercin V41, a pediocin-like bacteriocin secreted by *Carnobacterium divergens* V41. Appl Environ Microbiol 65:2895–2900, 1999.
40. K Venema, T Abee, AJ Haandrikman, KJ Leenhouts, J Kok, WN Konings, G Venema. Mode of action of lactococcin B, a thiol-activated bacteriocin from *Lactococcus lactis*. Appl Environ Microbiol 59:104–1048, 1993.
41. G Giraffa, N Picchioni, E Neviani, D Carminati. Production and stability of an *Enterococcus faecium* bacteriocin during taleggio cheesemaking and ripening. Food Microbil 12:301–307, 1995.
42. HJ Joosten, P Gaya, M Nunez. Isolation of tyrosine decarboxylaseless mutants of

a bacteriocin-producing enterococcus faecalis strain and their application in cheese. J Food Prot 58:1222–1226, 1995.
43. M Hugas, M Garriga, MT Aymerich, JM Monfort. Inhibition of listeria in dry fermented sausages by the bacteriocinogenic Lactobacillus sake CTC494. J Appl Bacteriol 79:322–330, 1995.
44. M Hugas, B Neumeyer, F Pages, M. Garriga, and W.P. Hammes. Antimicrobial activity of bacteriocin-producing cultures in meat products. 2. Comparison of the antilisterial potential of bacteriocin-producing lactobacilli in fermenting sausages. Fleischwirtschaft 76:649–652, 1996.
45. JL Ruizbarba, DP Cathcart, PJ Warner, R Jimenezdiaz. Use of Lactobacillus plantarum LPCO10, a bacteriocin producer, as a starter culture in spanish-style green olive fermentations. Appl Environ Microbiol 60:2059–2064, 1994.
46. GR Siragusa. The effectiveness of carcass decontamination systems for controlling the presence of pathogens on the surfaces of meat animal carcasses. J Food Safety 15:229–238, 1995.
47. Anonymous. Nisin preparation: affirmation of GRAS status as a direct human food ingredient. Fed Reg 53:12247–11251, 1988.
48. A Hirsch, E Grinsted, HR Chapman, A Mattick. A note on the inhibition of an anaerobic sporeformer in Swiss-type cheese by a nisin-producing *Streptococcus*. J Dairy Res 18:205–207, 1951.
49. J Delves-Broughton, P Blackburn, RJ Evans, J Hugenholtz. Applications of the bacteriocin, nisin. Antonie Van Leeuwenhoek 69:193–202, 1996.
50. GG Fowler. Toxicology of nisin. Food Cosmet Toxicol 11:351–352, 1973.
51. AI Shtenberg. Toxicity of nisin. Food Cosmet Toxicol 11:352, 1973.
52. MA de Pablo, JJ Gaforio, AM Gallego, E Ortega, AM Galvez, A de Cienfuegos Lopez. Evaluation of imnmunomodulatory effects of nisin-containing diets on mice. FEMS Immunol Med Microbiol 24:35–42, 1999.
53. A Nykanen, S Vesanen, H Kallio. Synergistic antimicrobial effect of nisin whey permeate and lactic acid on microbes isolated from fish. Lett Appl Microbiol 27: 345–348, 1998.
54. LV Thomas, JW Wimpenny. Investigation of the effect of combined variations in temperature, pH, and NaCl concentration on nisin inhibition of *Listeria monocytogenes* and *Staphylococcus aureus*. Appl Environ Microbiol 62:2006–2012, 1996.
55. LV Thomas, EA Davies, J Delves-Broughton, JW Wimpenny. Synergistic effect of sucrose fatty acid esters on nisin inhibition of gram-positive bacteria. J Appl Microbiol 85:1013–1022, 1998.
56. MJ Davis, PJ Coote, CP O'Byrne. Acid tolerance in *Listeria monocytogenes*: the adaptive acid tolerance response (ATR) and growth-phase-dependent acid resistance. Microbiology 142:2975–2982, 1996.
57. AR Datta, MM Benjamin. Factors controlling acid tolerance of *Listeria monocytogenes*: effects of nisin and other ionophores. Appl Environ Microbiol 63:4123–4126, 1997.
58. W van Schaik, CG Gahan, C Hill. Acid-adapted *Listeria monocytogenes* displays enhanced tolerance against the lantibiotics nisin and lacticin 3147. J Food Prot 62: 536–539, 1999.

59. DS Jung, FW Bodyfelt, MA Daeschel. Influence of fat and emulsifiers on the efficacy of nisin in inhibiting *Listeria monocytogenes* in fluid milk. J Dairy Sci 75: 387–393, 1992.
60. CN Cutter, G.R. Siragusa. Reduction of *Brochothrix thermosphacta* on beef surfaces following immobilization of nisin in calcium alginate gels. Lett Appl Microbiol 23:9–12, 1996.
61. CN Cutter, GR Siragusa. Incorporation of nisin into a meat binding system to inhibit bacteria on beef surfaces. Lett Appl Microbiol 27:19–23, 1998.
62. CK Bower, J McGuire, MA Daeschel. Influences on the antimicrobial activity of surface-adsorbed nisin. J Ind Microbiol 15:227–233, 1995.
63. CK Bower, J McGuire, MA Daeschel. Suppression of *Listeria monocytogens* colonization following adsorption of nisin onto silica surfaces. Appl Environ Microbiol 61:992–997, 1995.
64. CN Cutter, GR Siragusa. Population reductions of gram-negative pathogens following treatments with nisin and chelators under various conditions. J Food Prot 58: 977–983, 1995.
65. KA Stevens, BW Sheldon, NA Klapes, TR Klaenhammer. Nisin treatment for inactivation of *Salmonella* species and other gram-negative bacteria. Appl Environ Microbiol 57:3613–3615, 1991.
66. DMA Carneiro, CA Cassar, RJ Miles. Trisodium phosphate increases sensitivity of gram-negative bacteria to lysozyme and nisin. J Food Prot 61:839–843, 1998.
67. IS Boziaris, L Humpheson, MR Adams. Effect of nisin on heat injury and inactivation of *Salmonella enteritidis* PT4. Int J Food Microbiol 43:7–13, 1998.
68. N Mulet-Powell, AM Lacoste-Armynot, M Vinas, MS Buochberg. Interactions between pairs of bacteriocins from lactic bacteria. J Food Prot 61:1210–1212, 1998.
69. M Rasch, S Knochel. Variations in tolerance of *Listeria monocytogenes* to nisin, pediocin PA- 1 and bavaricin A. Lett Appl Microbiol 27:275–278, 1998.
70. HJ Song, J Richard. Antilisterial activity of three bacteriocins used at sub minimal inhibitory concentrations and cross-resistance of the survivors. Int J Food Microbiol 36:155–161, 1997.
71. EA Szabo, ME Cahill. The combined affects of modified atmosphere, temperature, nisin and ALTA 2341 on the growth of *Listeria monocytogenes*. Int J Food Microbiol 43:21–31, 1998.
72. L Nilsson, HH Huss, L Gram. Inhibition of *Listeria monocytogenes* on cold-smoked salmon by nisin and carbon dioxide atmosphere. Int J Food Microbiol 38:217–227, 1997.
73. EA Szabo, ME Cahill. Nisin and ALTA 2341 inhibit the growth of *Listeria monocytogenes* on smoked salmon packaged under vacuum or 100% CO_2. Lett Appl Microbiol 28:373–377, 1999.
74. DM de Tapia, SM Alzamora, JW Chanes. Combination of preservation factors applied to minimal processing of foods. Crit Rev Food Sci Nutr 36:629–659, 1996.
75. GW Gould. Methods for preservation and extention of shelf life. Int J Food Microbiol 33:51–64, 1996.
76. EA Severina, A Severin, A Tomasz. Antibacterial efficacy of nisin against multidrug-resistant gram-positive pathogens. J Antimicrob Chemother 41:341–347, 1998.

77. AD Crandall, TJ Montville. Nisin resistance in *Listeria monocytogenes* ATCC 700302 is a complex phenotype. Appl Environ Microbiol 64:231–237, 1998.
78. J Richard. Use of bacteriocins for a surer production of food products—myth or reality. Lait 76:179–189, 1996.
79. V Miteva, I Ivanova, I Budakov, A Pantev, T Stefanova, S Danova, P Moncheva, V Mitev, X Dousset, P Boyaval. Detection and characterization of a novel antibacterial substance produced by a *Lactobacillus delbrueckii* strain 1043. J Appl Microbiol 85:603–614, 1998.
80. MH Bennik, B Vanloo, R Brasseur, LG Gorris, EJ Smid. A novel bacteriocin with a YGNGV motif from vegetable-associated *Enterococcus mundtii*: full characterization and interaction with target organisms. Biochim Biophys Acta 1373:47–58, 1998.
81. Z Yildirim, MG Johnson. Detection and characterization of a bacteriocin produced by *Lactococcus lactis* subsp. *cremoris* R isolated from radish. Lett Appl Microbiol 26:297–304, 1998.
82. CM Franz, TM Du, NA Olasupo, U Schillinger, WH Holzapfel. Plantaricin D, a bacteriocin produced by *Lactobacillus plantarum* BFE 905 ready-to-eat salad. Lett Appl Microbiol 26:231–235, 1998.
83. S Herbin, F Mathieu, F Brule, C Branlant, G Lefebvre, A Lebrihi. Characteristics and genetic determinants of bacteriocin activities produced by *Carnobacterium piscicola* CP5 isolatled from cheese. Curr Microbiol 35:319–326, 1997.
84. Y Cai, LK Ng, JM Farber. Isolation and characterization of nisin-producing *Lactococcus lactis* subsp. *lactis* from bean-sprouts. J Appl Microbiol 83:499–507, 1997.
85. S Ennahar, D Aoude-Werner, O Sorokine, A Van Dorsselaer, F Bringel, JC Hubert, C Hasselmann. Production of pediocin AcH by *Lactobacillus plantarum* WHE 92 isolated from cheese. Appl Environ Microbiol 62:4381–4387, 1996.
86. RW Jack, J Wan, J Gordon, K Harmark, BE Davidson, AJ Hillier, RE Wettenhall, MW Hickey, MJ Coventry. Characterization of the chemical and antimicrobial properties of piscicolin 126, a bacteriocin produced by *Carnobacterium piscicola* JG126. Appl Environ Microbiol 62:2897–2903, 1996.
87. S Maisnier-Patin, E Forni, J Richard. Purification, partial characterization and mode of action of enterococcin EFS2, an antilisterial bacteriocin produced by a strain of *Enterococcus faecalis* isolated from a cheese. Int J Food Microbiol 30:255–270, 1996.
88. G Enan, AA el-Essawy, M Uyttendaele, J Debevere. Antibacterial activity of *Lactobacillus plantarum* UG1 isolated from dry sausage: characterization, production and bactericidal action of plantaricin UG1. Int J Food Microbiol 30:189–215, 1996.
89. MP Ryan, MC Rea, C Hill, RP Ross. An application in cheddar cheese manufacture for a strain of *Lactococcus lactis* producing a novel broad-spectrum bacteriocin, lacticin 3147. Appl Environ Microbiol 62:612–619, 1996.
90. JM Rodriguez, LM Cintas, P Casaus, N Horn, HM Dodd, PE Hernandez, MJ Gasson. Isolation of nisin-producing *Lactococcus lactis* strains from dry fermented sausages. J Appl Bacteriol 78:109–115, 1995.
91. N Rekhif, A Atrih, G Lefebvre. Activity of plantaricin SA6, a bacteriocin produced by *Lactobacillus plantarum* SA6 isolated from fermented sausage. J Appl Bacteriol 78:349–358, 1995.

92. N Valdes-Stauber, S Scherer. Isolation and characterization of linocin M18, a bacteriocin produced by *Brevibacterium linens*. Appl Environ Microbiol 60:3809–3814, 1994.
93. JV Felix, MA Papathanasopoulos, AA Smith, A von Holy, JW Hastings. Characterization of leucocin B-Talla: a bacteriocin from *Leuconostoc carnosum* Tal 1a isolatled from meat. Curr Microbiol 29:207–212, 1994.
94. AG Larsen, FK Vogensen, J Josephsen. Antimicrobial activity of lactic acid bacteria isolated from sour doughs: purification and characterization of bavaricin A, bacteriocin produced by *Lactobacillus bavaricus* MI401. J Appl Bacteriol 75:113–122, 1993.
95. AM Strasser de Saad, MC Manca de Nadra. Characterization of bacteriocin produced by *Pediococcus pentosaceus* from wine. J Appl Bacteriol 74:406–410, 1993.
96. F Villani, G Salzano, E Sorrentino, O Pepe, P Marino, S Coppola. Enterocin 226NWC, a bacteriocin produced by *Enterococcus faecalis* 226, active against *Listeria monocytogenes*. J Appl Bacteriol 74:380–387, 1993.
97. LJ Harris, HP Fleming, TP Klaenhammer. Characterization of two nisin-producing *Lactococcus lactis* subsp. *lactis* strains isolated from a commercial sauerkraut fermentation. Appl Environ Microbiol 58:1477–1483, 1992.
98. PA Vandenbergh, SA Walker, BS Kunka. Method for inhibiting bacteria using a novel lactococcal bateriocin. U.S. Patent 5,817,362 (1998).
99. P Blackburn, J. Polak, S-A Gusik, SD Rubino. Nisin compositions for use as enhanced, broad range bactericides. U.S. Patent 5,753,614 (1998).
100. DL Wilhoit. Surface treatment of foodstuffs with antimicrobial compositions. U.S. Patent 5,573,801 (1996).
101. ER Vedamuthu. Method of producing a yogurt product containing bacteriocin PA-1. U.S. Patent 5,445,835 (1995).
102. DP Boudreaux, MA Matrozza. Composition for extending the shelf life of processed meats. U.S. Patent 5,219,603 (1993).
103. RW Hutkins, ED Berry, MB Liewen. Composition and method for inhibiting pathogens and spoilage organisms in foods. U.S. Patent 5,186,962 (1993).
104. MW Collison, TF Farver, CA McDonald, PJ Herald, DJ Monticello. Use of lanthionines for control of post-processing contamination in processed meat. U.S. Patent 5,015,487 (1991).
105. PA Vandenbergh, MJ Pucci, BS Kunka, ER Vedamuthu. Method for inhibiting *Listeria monocytogenes* using a bacteriocin. U.S. Patent 4,929,445 (1990).
106. CF Gonzalez. Method for inhibiting bacterial spoilage and resulting compositions. U.S. Patent 4,883,673 (1989).
107. MA Matrozza, MF Leverone, DP Boudreaux. Method for inhibiting psychrotrophic bacteria in cream or milk based products using a pediococcus.; U.S. Patent 4,790,994 (1988).
108. S Ennahar, O Assobhel, C Hasselmann. Inhibition of *Listeria monocytogenes* in a smear-surface soft cheese by *Lactobacillus plantarum* WHE 92, a pediocin AcH producer. J Food Prot 61:186–191, 1998.
109. M Nunez, JL Rodriguez, E Garcia, P Gaya, M Medina. Inhibition of *Listeria monocytogenes* by enterocin 4 during the manufacture and ripening of Manchego cheese. J Appl Microbiol 83:671–677, 1997.

110. I Eppert, N Valdes-Stauber, H Gotz, M Busse, S Scherer. Growth reduction of *Listeria* spp. caused by undefined industrial red smear cheese cultures and bacteriocin-producing *Brevibacterium* lines as evaluated in situ on soft cheese. Appl Environ Microbiol 63:4812–4817, 1997.
111. EA Davies, HE Bevis, J Delves-Broughton. The use of the bacteriocin, nisin, as a preservative in ricotta-type cheeses to control the food-borne pathogen *Listeria monocytogenes*. Lett Appl Microbiol 24:343–346, 1997.
112. JJ Leisner, GG Greer, ME Stiles. Control of beef spoilage by a sulfide-producing *Lactobacillus sake* strain with bacteriocinogenic *Leuconostoc gelidum* UAL187 during anaerobic storage at 2 °C. Appl Environ Microbiol 62:2610–2614, 1996.
113. G Vignolo, S Fadda, MN de Kairuz, H de Ruiz, G Oliver. Control of *Listeria monocytogenes* in ground beef by lactocin 705, a bacteriocin produced by *Lactobacillus casei* CRL 705. Int J Food Microbiol 29:397–402, 1996.
114. G Baccus-Taylor, KA Glass, JB Luchansky, AJ Maurer. Fate of *Listeria monocytogenes* and pediococcal starter cultures during the manufacture of chicken summer sausage. Poult Sci 72:1772–1778, 1993.
115. PM Foegeding, AB Thomas, DH Pilkington, TR Klaenhammer. Enhanced control of *Listeria monocytogenes* by in situ-produced pediocin during dry fermented sausage production. Appl Environ Microbiol 58:884–890, 1992.
116. Technical Information Sheet, Aplin & Barrett, Beaminster, Dorset, United Kingdom, 1989.
117. N Kalchayanand, A Sikes, CP Dunne, B Ray. Interaction of hydrostatic pressure, time and temperature of pressurization and pediocin AcH on inactivation of foodborne bacteria. J Food Prot 61:425–431, 1998.

12

Formulating Low-Acid Foods for Botulinal Safety

Kathleen A. Glass and Eric A. Johnson
University of Wisconsin–Madison
Madison, Wisconsin

I. INTRODUCTION

Historically, humans have relied on preservation methods such as drying, salting, smoking, and acidification by naturally occurring microflora to ensure the microbiological safety of foods. As our understanding of food preservation developed, a variety of chemical antimicrobials and processing techniques became available to control growth and toxin production by foodborne pathogens. Today's consumers desire minimally processed, convenient, fresh-tasting, healthful foods. These requirements challenge manufacturers as they formulate safe food products. Low-acid foods—those having an equilibrium pH > 4.6, a water activity (a_w) < 0.85, and receive a minimal treatment to reduce or eliminate microorganisms—create a special concern for growth of certain pathogens and spoilage organisms. Of particular concern, and the central topic of this chapter, is control of the spore-former *Clostridium botulinum*, which produces the most poisonous toxin known (1). The primary means for controlling *C. botulinum* in low-acid foods is to inactivate the spores by heat or to prevent their growth, primarily by acidification and control of a_w. Additionally, many low-acid refrigerated foods depend on secondary barriers to ensure their safety in the event that strict refrigeration is interrupted. Similarly, shelf-stable low-acid foods packaged in hermetically sealed containers often rely on combinations of factors such as mild heat, pH, a_w, and chemical preservatives for their safety. This chapter will focus on preservation of foods to control growth and toxin production by *C. botulinum*. Although *C. botulinum* is highlighted in this chapter, similar principles can be applied to control the growth of other spore-forming and vegetative pathogens in low-acid foods.

II. DESCRIPTION AND DIAGNOSIS OF HUMAN BOTULISM

Foodborne botulism is a rare disease, but it has been associated with high morbidity, long hospital stays, and historically with a high fatality rate (1–5). Foodborne botulism is a true intoxication caused by the consumption of foods contaminated with preformed botulinal neurotoxin, while infant and wound botulism are caused by opportunistic infections and toxin formation at the infected site (1,3,5). Human botulism is caused predominantly by neurotoxins formed by Group I (proteolytic) *C. botulinum* types A and B and Group II (nonproteolytic) types B and E (1,6). Serotypes C and F have rarely been associated with human illness, and types D and G have not been reported to cause human botulism (6). Symptoms generally appear 12–36 hours after consumption of contaminated foods, but incidences with onset times of 2 hours to 14 days have also been reported (3,5). Patients may first exhibit gastrointestinal symptoms including nausea, vomiting, and diarrhea. Subsequent neurological symptoms develop, including blurred or double vision, dry mouth, difficulty in swallowing, speaking, and breathing, and peripheral muscle weakness. Severe cases require assisted ventilation for several weeks or months (7). Death may occur due to respiratory failure. Fatality due to botulism has significantly decreased during the latter part of the twentieth century as a result of prompt administration of antitoxin and good supportive care, particularly by respiratory intubation and assisted feeding (7). Mortality due to botulism occurred in 5–10% of the cases for the period 1980–1996, compared with approximately 60% in the years prior to 1949 (7).

Diagnosis of botulism is often not obvious since the symptoms can mimic other diseases and many physicians do not have experience in diagnosing the disease. Botulism symptoms can resemble those exhibited by Guillain-Barré syndrome, myasthenia gravis, stroke, ingestion of chemical poisons, or certain other syndromes (2). Cases that present mild symptoms may be unrecognized unless they are clustered with more severe cases, suggesting that botulism may be underreported (8–10). Exposure to low levels of toxin may result in late onset (1–2 weeks) of symptoms and is more frequently associated with mild symptoms, including dry mouth, muscle weakness, dysphagia, and constipation, which may not require hospitalization (2,11).

III. RISK OF BOTULISM ASSOCIATED WITH FOOD

From 1950 to 1996, 65% of the 444 confirmed outbreaks of botulism in the United States were associated with foods prepared in the home compared with 7%

associated with commercially prepared foods (7). The sources for the remaining outbreaks were unknown. Although home-prepared foods (fermented fish products and marine animal parts such as seal flippers, home-canned meats and vegetables) were responsible for the majority of the outbreaks, recent outbreaks have been associated with restaurant/food service foods that were temperature abused (foil-wrapped baked potatoes, sautéed onions, garlic in oil, recontaminated process cheese sauce), inadequate thermal processing or other deficiencies in commercial canning (mushrooms, canned tuna fish, beef stew, hazelnut yogurt), and temperature abuse of commercial products that did not contain a secondary barrier to botulinal growth (garlic in oil, mascarpone cheese, clam chowder, black bean dip) (Table 1). The largest recorded outbreak in the United Kingdom was caused by type B botulinal toxin with 27 cases and one death (12). The outbreak was caused by yogurt containing toxin-contaminated hazelnut conserve. Heat-processing of the hazelnut preserve was inadequate to destroy *C. botulinum* spores; botulinal toxin was produced in the can during storage and prior to addition to the yogurt. The largest restaurant-associated botulism outbreak in the past 25 years in the United States caused 30 cases including 4 with severe symptoms that required mechanical ventilation (13). The culprit food was a potato-based dip that was prepared using potatoes that were baked in aluminum foil and subsequently temperature abused prior to preparation of the dish. The Pasteur Institute in Algeria recently confirmed a large botulism outbreak during July 1998. Several spoiled meat and poultry products, including paté and a processed meat called "casher," were implicated in 1400 cases of botulism including 19 deaths (14). Poor hygiene in the processing plant and inadequate refrigeration in distribution contributed to the outbreak. In March and April 1997, an outbreak of foodborne botulism affecting 27 patients with one death was caused by locally made cheese in Iran (15).

The current good safety record for commercial foods is due, in large part, to the diligence of food manufacturers in formulating, processing, and controlling temperature during distribution of foods. Other contributing factors include the low incidence of botulinal spores in prepared foods (16–18), competition with spoilage organisms, and consuming foods before toxin production can occur.

IV. DISTRIBUTION OF SPORES IN NATURE AND FOODS

Distribution and prevalence of botulinal spore types vary by geographical region (19,20). Regions with high levels of spores are associated with a higher frequency of botulism than regions containing fewer spores. Type A botulism is the predominant toxin type found in the western continental United States, China, and Argentina and is associated mainly with vegetables. Outbreaks in Europe are generally associated with type B. *C. botulinum*; meat, yogurt, vegetables, and other foods have been involved as vehicles (21). *Clostridium botulinum* type B is also preva-

Table 1 Outbreaks of Foodborne Botulism Associated with Commercial or Restaurant-Prepared Foods

Year	Location	Product	Toxin type	Cases (deaths)	Factors	Ref.
1973	Switzerland, France	Brie	B	75	Ripening of cheese on contaminated straw	22,132
1974	Argentina	Process cheese spread	A	6 (3)	Improper formulation	38
1974	Georgia	Commercial canned beef stew	A	2 (1)	Underprocessed	133
1977	Michigan	Restaurant home-canned peppers	B	59	Underprocessed	10
1978	Colorado	Potato salad	A	7	Baked potatoes held at room temp. up to 5 days	134
1978	New Mexico	Restaurant potato salad		34	Leftover baked potatoes; temperature abuse	135,136
1982	California	Commercial pot pie	A	1	Heated, then temperature abuse	137
1983	Illinois	Restaurant sautéed onions	A	28 (1)	Covered with oil; temperature abuse	138,139
1985	Canada	Commercial garlic-in-oil	B	36	Bottled; no preservatives, temperature abuse	9
1988	Florida	Coleslaw	A	4	MAP shredded cabbage	111
1989	New York	Commercial garlic-in-oil	A	3	Bottled; no preservatives; temperature abuse	140
1989	United Kingdom	Commercial hazelnut yogurt	B	27 (1)	Hazelnut conserve underprocessed	12
1993	Georgia	Restaurant commercial process cheese sauce	A	8 (1)	Recontamination; temperature abuse	37
1993	Italy	Commercial canned roasted eggplant in oil	B	7	Bottled; improper acidification·pH > 4.6;	142
1994	Texas	Restaurant; potato based dip (skordalia)	A	30	Potatoes held at room temperature	13,143
1994	California	Commercial clam chowder	A	2	No secondary barrier; temperature abuse	124,144
1994	California	Commercial black bean dip	A	1	No secondary barrier, temperature abuse	124,144
1996	Italy	Commercial mascarpone cheese	A	7	No competitive microflora; pH 6–6.25; temperature abuse	145–147

lent in eastern U.S. soils (19,20). Type E, associated with marine and freshwater foods, is the predominant cause of botulism in cooler aquatic regions, including coastal regions of Canada, Alaska, and northern Japan (19,21). Twenty-four percent of the soils tested in the United States were found to harbor botulinal spores (20). The incidence of spores was significantly greater in sediments and soils in and adjacent to Lake Michigan and along the North American Pacific coast. Surveys revealed that 30–95% of these samples were positive, predominantly for type E (20). The relative incidence of botulinal spores in soils and sediments are similar to analogous regions of Canada, Central and South America, Europe, and Asia. Surveys in Australia and New Zealand reported a low incidence of spores for inland soil samples, although sediments in southern Australia were heavily contaminated with type C spores, regions where outbreaks of avian botulism have been reported.

Clostridium botulinum spores have also been associated with a variety of foods (16,19). The incidence of spores in various raw foods often reflects the prevalence of the spores in the geographical origin of the foods. For example, fish are often contaminated with *C. botulinum* spores due to the prevalence of spores in many marine and freshwater coastal environments. Foods harvested from continental soils such as vegetables and fruits often contain type A and B spores. Certain foods are rarely contaminated with *C. botulinum* spores including many types of meats such as raw poultry, beef, pork, and dairy products (16,19,22). However, spores can be inadvertently added, particularly in dry vegetable or dry ingredients such as spices. Processing conditions and plant hygiene may also affect the contamination of foods by *C. botulinum* spores. New process techniques must be scrutinized to verify that they will not increase the risk of botulinal growth in foods. In addition to processing considerations, the formulation of foods taking into consideration intrinsic and extrinsic factors (including packaging) is critical for controlling *C. botulinum* growth and toxin production in raw and minimally processed low acid foods.

V. FORMULATING SHELF-STABLE FOODS TO INHIBIT THE GROWTH AND TOXIN PRODUCTION BY *C. BOTULINUM*

Low-acid canned foods (LACF) are defined as those foods having an equilibrium pH > 4.6 (excluding tomatoes having an equilibrium finished pH < 4.7) and a water activity > 0.85, and packaged in a hermetically sealed container (23). The common method of preserving LACF for shelf stability is by thermal processing at 121°C, or equivalent time-temperature processes, for a sufficient time to achieve commercial sterility (19,24). The minimal thermal process for a given food and size container must be sufficient to produce a hypothetical 12-log decrease in botulinal spores (12*D* botulinal cook) (19,24,25). The minimum process used for commercially canned foods is 12*D* values, which is equivalent to F_0 = 2.4 min at 121°C. However, longer heating times are employed by the industry to assure commercial sterility. The thermal destruction of *C. botulinum* spores depends on many factors, particularly pH, a_w, the nutrient composition of the food, and the presence of added solutes or gases. The thermal process must be validated by a competent process authority. The 12*D* process has had a remarkable record in ensuring the safety of canned foods (26,27) and represents one of the first predictive models employed by the food industry (28,29).

Alternatively, a less severe heat treatment can be utilized to inactivate vegetative pathogens and spoilage organisms in conjunction with exploiting barrier technology to inhibit growth and toxin production of *C. botulinum*. Also critical to the safety of these products are good manufacturing practices and well-designed and executed hazard analysis critical control point (HACCP) plans. As with thermally processed LACF, it is incumbent upon the manufacturer to verify the safety of formulation-dependent LACF before introducing the product into commerce. Recent recalls of processed cheese products and modified-atmosphere packaged baked goods illustrate the necessity to verify that the products do not support growth of *C. botulinum* (30,31).

Several commercial products apply minimal processing with the formulation-safe approach to achieve shelf stability. Well-known examples include shelf-stable cured luncheon meat products and many pasteurized process cheese spreads and sauces. More recently, certain modified-atmosphere packaged baked goods and cream cheese–based spreads have appeared on the market that ensure safety based on minimal processing and formulation.

The safety of shelf-stable canned cured meat products depends on multiple factors. Investigators formulated inoculated ground pork luncheon meat with varying concentrations of sodium chloride and nitrite, heated the product to F_0 = 0.15–3.70, and stored the cans at 30°C for up to 18 months (32). An F_0 = 0.7, equivalent to 70 min at 113°C, results in ca. 6-log reduction in viable proteolytic botulinal spores when heated in buffer. When inoculated with 10^4 spores/g, toxin was detected in meat formulated with a 5.5% brine (NaCl in aqueous phase), 146 ppm nitrite, and heated to F_o = 0.15, but no toxin was detected in the product when heated to F_o = 0.30. The F_o = 0.30 process was insufficient to inhibit toxigenesis when meat was inoculated with 10^6 spores/g, but F_o = 0.60 prevented botulinal growth. Without nitrite, toxin was detected in meat formulated with 4.8% brine, inoculated with 10^4 spores/g, and heated to F_o = 0.60–0.64; the addition of 15 ppm nitrite was sufficient to inhibit toxin production. Heat damages the spores, making them more susceptible to the effects of the antimicrobials in the product. At the same time, nitrite enhances heat injury (33). These results revealed that low levels of spores in raw materials, initial sodium nitrite and sodium chloride levels, and the amount of thermal processing are critical factors in determining the safety of these products (32,34).

VI. PASTEURIZED PROCESS CHEESE AS MODEL FOOD FOR ASSESSING BOTULINAL SAFETY

Pasteurized process cheese, and related products, which appeared in the early twentieth century (35) are among the most studied substrates for model formulation-safe foods. These products are not amenable to retort processing and are heated to 85 to 100°C. Stability of process cheese products relies on pasteurization to inactivate vegetative microorganisms and on combinations of moisture, salts, pH, and other antimicrobials to prevent the outgrowth of *C. botulinum*.

Investigators at the Food Research Institute, University of Wisconsin–Madison, developed a user-friendly formulation model that enables manufacturers to predict the microbial stability of formulated process cheese products (36). Preliminary studies suggested that levels of moisture, pH, and sodium chloride and the emulsifier disodium phosphate (DSP) were the primary determinants of cheese spread safety. In subsequent studies, over 300 cheese spread formulations were evaluated for the ability to support botulinal toxin production. Statistical analysis revealed that sodium chloride and disodium phosphate levels were similar and additive in their antibotulinal effect. Total salts, along with pH and moisture content, were a primary factor in controlling botulinal toxin formation. The resulting predictive model was reported as a series of graphical representations, which can be easily used to determine the safety of new process cheese spread formulations.

The effect of a_w was also evaluated. Tanaka et al. (36) observed that although water activity functions in determining safety, it is not always an accurate indicator of safety for cheese spreads. When water activity was at or below 0.944, no toxin was detected, whereas all formulations with $a_w > 0.957$ supported toxin production. In the a_w range between 0.944 and 0.957, botulinal safety was best predicted by the moisture, pH, NaCl, and DSP levels as specified by the model.

Process cheese products have an excellent history of safety, but the few outbreaks that were associated with high-pH/high-a_w process cheese products highlight the importance of formulation (22,37,38). The cheese sauce implicated in the 1993 Georgia outbreak was thermally processed to eliminate *C. botulinum* spores; therefore, it was not formulated to prevent botulinal growth (pH 5.8, a_w 0.96). The product was recontaminated in the restaurant, and was stored at ambient temperature for several days before use, permitting *C. botulinum* to grow and produce toxin (37). The cheese spread implicated in the 1974 Argentina outbreak, however, was not commercially sterilized. Laboratory experiments revealed botulinal toxin was produced in cheese spread samples with a similar formulation (pH 5.7, a_w 0.97) after 30–70 days of storage at 30°C.

To date, the University of Wisconsin model remains the ''gold standard'' in evaluating the safety of standard-of-identity process cheese spread formulations. However, as formulations have evolved in recent years, trends have been to reduce cheese content, fat and sodium levels, and increase moisture. Alternative emulsifiers, salts, and antimicrobials have been assessed for their effect on safety.

In the United States sodium monophosphates (orthophosphates) have been widely used as emulsifiers in process cheese spreads. Other commonly used emulsifiers include citrates and polymeric or condensed phosphates. Several studies reported greater antibotulinal effect of phosphate based emulsifiers compared with sodium citrate emulsifiers (39–41). Experiments in our laboratory have confirmed these findings (42) and further suggest that sodium citrate does not contribute to the antibotulinal properties of process cheese spreads. However, products can be safely formulated with citrate-based emulsifiers provided the moisture/pH/NaCl+DSP levels comply with the model described by Tanaka et al. (36).

Few studies have compared the antibotulinal effects of mono- and polyphosphate emulsifiers. The molecular weights of commercially manufactured polyphosphates are not defined. Therefore, comparison of efficacies of phosphates can only be made on a percentage basis rather than molar equivalents. Trials in liquid media and high-moisture process cheese spread suggest that several polyphosphates added at the 2% level exhibit equivalent or slightly greater antibotulinal activity than 2% DSP (43).

Because most polyphosphates will hydrolyze over time to disodium orthophosphate (35), the pH of process cheese spreads containing polyphosphates may take as many as 3 days to equilibrate (A. Larson, personal communication, 1997). Standardization of pH is essential to accurately assess equivalence in antibotulinal effect. Eckner et al. (44) reported greater inhibition of botulinal growth by certain polyphosphates. However, the pH values of the products varied among the formulations by as much as 0.35 pH units. It is recommended that formulations be evaluated for botulinal stability only after the pH has been allowed to stabilize to 0.1 pH unit.

Sodium-reduction in process cheese products is restrained because total salt levels (defined as sodium chloride plus sodium phosphate) are essential elements in determining safety of the products. Potassium salts were considered as alternatives to sodium salts. A study by Karahadian et al. (39) did not provide conclusive evidence on the efficacy of potassium-based salts, although DSP used as the sole emulsifier appeared to impart greater antibotulinal activity than dipotassium phosphate (DKP). However, ionic strength was not considered in the experimental design. Hence, the product formulated with DKP had an a_w of 0.97 compared with an a_w of 0.96 for the product formulated with DSP, which would affect the botulinal safety.

Experiments at the Food Research Institute (K. Glass, J. McDermott, and J. Nelson, unpublished data, 1995) compared the antibotulinal activity of sodium- and potassium-based salts in reduced-fat process cheese products (5% fat). Empirically, the results suggested that safe reduced-fat process cheese products may be formulated with potassium salts as a partial replacement (up to 75% replacement) for sodium salts. However, the statistical significance of potassium salts on the development of botulinal toxin was considered inconclusive because of the relatively small number of toxic formulations in the experiment.

Results from this study also suggested that *C. botulinum* toxin production is delayed in reduced-fat and fat-free process cheese products compared with full-fat products with similar levels of moisture, salt, and pH. The mechanism of inhibition in reduced-fat process cheese is unknown. Investigators in The Netherlands observed a delay in clostridial growth in products formulated with 50–60% nonfat cheese compared with those formulated with comparable levels of full-fat cheese (40). They attributed the enhanced stability to the higher lactic acid content of the nonfat cheese, rather than any protective value contributed by the fat. Spore loads also affect time to toxicity. Ter Steeg et al. (40) reported that initial spore loads as low as 0.1 spore/g were sufficient for botulinal toxin production in permissive formulas. Botulinal toxin was detected by ELISA at 4 weeks when initial spore levels were 100–1000 spores/g, at 6 weeks in samples with 10 spores/g, and at 8 weeks in products with 1 and 0.1 spores/g. These findings are in contrast to those reported for shelf-stable cured canned meats, which concluded that a low spore load was essential for safety (32,34).

Several laboratories have investigated secondary barriers for preventing botulinal toxin production in process cheese and other foods. One of the more extensively studied added barriers is nisin. Nisin is a bacteriocin produced by certain strains of *Lactococcus lactis* that is effective against gram-positive bacteria. It is believed that nisin inhibits the spore germination process at the preemergent swelling stage and that heat damage can sensitize spores to nisin (45). In countries that manufacture cheese spreads with relatively low NaCl levels, nisin is routinely added to enhance the antibotulinal activity of the products (46). Usage levels in the United States are limited to 250 ppm (47). Somers and Taylor (46) reported that 100–250 ppm nisin were required to inhibit toxin production in products with 57% moisture, 1.2% NaCl, and 1.4% DSP. The pH values for these formulations were not reported. Other experiments showed that 125 ppm nisin were more effective at pH 5.5 than at pH 6.0 (48,49). Nisin did not inhibit outgrowth of *C. botulinum* in cooked meat media adjusted to pH 7–8. Levels of addition to a product should consider that the antimicrobial activity of nisin diminishes as a function of heat processing, pH, substrate, and length and temperature of storage (45).

Lactic acid has been demonstrated to prevent botulinal toxin formation beyond that attributed to pH reduction (36,40,42). Additional work in our laboratory

(K. A. Glass, K. M. Kaufman, A. L. Smith, and E. A. Johnson, unpublished results, 1998) demonstrated that 1.5% sodium lactate (calculated on anhydrous basis) inhibited botulinal toxin production in process cheese products formulated with ≤58% moisture, pH < 5.8, and >2.4% combined salts (NaCl+DSP). Sodium lactate did not delay toxin production in products with 58% moisture and pH > 5.8, but it inhibited toxin production in 55% moisture products formulated with >2.3% total salts and pH 5.9. Similar formulations without 1.5% sodium lactate supported toxin production. These results suggest that sodium lactate enhances the safety of high-moisture cheese spreads without significant pH reductions.

VII. TOXIN PRODUCTION BY *C. BOTULINUM* IN SHELF-STABLE BAKERY PRODUCTS

There is an increasing trend in the food industry to market ready-to-eat shelf-stable foods that contain no preservatives and have an extended shelf life. In these foods, the primary factors used in formulation, particularly pH and a_w, must be carefully maintained to ensure botulinal safe products. One class of foods of current interest are bakery products. Bakery products have seldom been considered a botulinal safety risk because mold growth and staling limit the shelf life. With the recent advent of modified-atmosphere packaging (MAP) for bakery goods to extend shelf life, concerns have been raised of the botulinal safety of some of these products. Canadian researchers evaluated the safety of English-style crumpets packaged in different atmospheres (50). Botulinal toxin was detected in high-moisture crumpets (a_w 0.990, pH 6.5) within 4–6 days, regardless of the initial atmosphere. Botulinal toxin was also detected in high-moisture pizza crusts (a_w 0.960) after 6 weeks of storage, whereas bagels formulated with a_w of 0.944 did not support botulinal toxin production. Water activity, in combination with antimicrobials such as potassium sorbate, could be useful to enhance the safety of these products.

VIII. GROWTH AND TOXIN FORMATION OF *C. BOTULINUM* IN MINIMALLY PROCESSED REFRIGERATED FOODS

Group II *C. botulinum* (nonproteolytic) strains are notable because of their ability to grow and produce toxin at refrigeration temperatures (≥3.3°C) after extended incubation times (18,51–53). Refrigerated, processed foods of extended durability (REPFEDs) are heat-treated or processed using other treatments to reduce competitive microflora (25,54). Several compendia have provided recommendations for manufacturers of low-acid refrigerated foods with an extended shelf life of more than 10 days: (1) thermally treat the products to achieve a 6-log kill of nonproteolytic spores, usually 90°C for 10 minutes (51); (b) increase acidity to pH below 5.0; (c) adjust the sodium chloride in aqueous phase to 3.5–5% (51, 54); or (d) use combinations of water activity, pH, and antimicrobials to inhibit growth of psychrotrophic strains of *C. botulinum*.

Notermans et al. recommends all REPFEDs be stored at <3.3°C to prevent growth of nonproteolytic *C. botulinum* (53). However, surveys of U.S. refrigeration temperatures in retail operations and in homes suggest that relying on refrigeration alone is an inadequate barrier to pathogen growth. Van Garde and Woodburn reported that 21% of home refrigerators had temperatures ≥10°C (55). Over 25% of delicatessen refrigerated cases surveyed had temperatures >10°C, and 12.9% were >12.8°C (56). A 1997 survey (57) reported that refrigeration temperatures for 63% of the U.S. household refrigerators had temperatures ranging from 5.6–7.2°C (42 to 45°F) with 23% having temperatures >7.2°C (45°F). A 1999 survey demonstrated an improvement in temperature control, but 23% of U.S. households evaluated still had refrigerator temperatures of 5.6–7.2°C and an additional 9% had refrigeration temperatures >7.2°F (58).

A study in Denmark monitored the temperature of chilled products at the receiving dock of a retail chain (59). The retailer rejected loads with temperatures exceeding 7°C. During the 14-month interval from May 1989 through July 1990, more than 120 shipments (total number of shipments not reported) of processed meat products were rejected, about 10% of which were found to have temperatures in excess of 10°C. Several loads had product temperatures as high as 16–17°C. As one might expect, the highest rate of rejections occurred during summer months, with few rejections recorded during November through March. Strict enforcement of acceptable product temperature (<7°C) at the receiving dock resulted in higher compliance by shippers.

Although laboratory investigations demonstrate that Group II strains of *C. botulinum* grow at refrigeration temperatures, few outbreaks of botulism have been associated with foods that were refrigerated at ≤4°C (12,13). Investigation of outbreaks of ''refrigerated'' foods suggested that ingredients (baked potato, hazelnut conserve) used in the products were stored at ambient temperature, during which time botulinal growth and toxin production occurred. The preformed toxin in the ingredients remained stable during subsequent storage at refrigeration temperatures and under acid conditions; the skordalia potato dip and yogurt cited in these studies had equilibrium pHs of 3.7 and 4.4, respectively (12,13). In most of these outbreaks, Group I *C. botulinum* grew and produced toxin in the food ingredient at the elevated temperature. Botulinal toxin is quite stable at acidic pH but is markedly sensitive to heating. The addition of the ingredient to the acidic refrigerated food without subsequent heating prior to consumption contributed to the botulism outbreak. It is also noteworthy that *C. botulinum* has been reported to grow at pH values < 4.6 in media or foods containing high concentrations of certain proteins (60).

IX. USE OF ANTIMICROBIALS TO PREVENT BOTULINAL TOXIN PRODUCTION IN FOODS

Several intrinsic and extrinsic factors have been evaluated to control growth and toxin production by *C. botulinum* in various foods (Table 2). The primary intrinsic factors controlling *C. botulinum* growth and used in formulation of foods are the equilibrium pH and a_w. In addition, numerous antimicrobials as well as competitive microflora have been demonstrated to influence toxin production, several of which are discussed in the following section. The technology for controlling *C. botulinum* and other pathogens is based on the concept of hurdle technology, which has been extensively reviewed in this book (Chapter 20) and elsewhere (61–63).

Of the many secondary barriers for potential control of botulinal toxin production as well as other foodborne pathogens, organic acids have been most extensively evaluated. When added to foods as acidulants, it should be kept in mind that it may require several hours to days for the acid to equilibrate within a food, during which time the food should be kept refrigerated to prevent *C. botulinum* growth and toxin production in nonacidified microenvironments.

Several investigations have demonstrated that sodium lactate may serve as a secondary barrier to prevent botulinal toxin production in culture media, beef, poultry, fish, refrigerated pasta, and sauces (64–67; K. Glass, J. McDermott and

Table 2 Primary Formulation Components to Assure the Safety of Minimally Processed Low-Acid Foods

pH and acidity
a_w and solute composition
Presence of antimicrobials
Organic acids
Nitrites, sulfites, phenolic compounds
Polyphosphates
Fatty acids and esters
Gas composition
Naturally occurring antimicrobials
Competitive microflora
Indirect antimicrobials

J. H. Nelson, unpublished data, 1995). Toxin production by proteolytic *C. botulinum* was delayed in vacuum packaged cooked turkey formulated with 2–3.5% sodium lactate and stored at 27°C (64,66,68). Reduced temperatures, ranging from 4 to 20°C, increase the inhibition against both proteolytic and nonproteolytic botulinal spores (67,69,70).

Sodium diacetate (0.25%) used in combination with 1.5% sodium lactate inhibits growth of a nonpathogenic, psychotrophic spoilage *Clostridium* sp. in cook-in-bag turkey breast stored at 4°C for 22 weeks. Spoilage occurred at 7 weeks without antimicrobials (defined as off odor and 7-log increase in clostridial counts) compared with spoilage at 12 weeks for either the diacetate or lactate alone (71). No obvious synergistic effect was observed between sodium lactate and sodium chloride (69).

Experiments in our laboratory (K. A. Glass, J. McDermott, and J. H. Nelson, unpublished data, 1995) demonstrated that pH reduction potentiates the effect of sodium lactate. Refrigerated entrees consisting of chicken, broccoli, and a cream-based sauce were formulated with pH 6.3 or 5.1 with or without 2% lactate, inoculated with proteolytic or nonproteolytic *C. botulinum*, and stored at 27 or 12.8°C, respectively. Botulinal toxin was detected in both entrees with pH 6.3 and in the pH 5.1 entree without sodium lactate after 7 days storage at 27°C. No toxin was detected in the product formulated with pH 5.1 and 2% lactate through 105 days of storage at 27°C. For products inoculated with nonproteolytic spores and stored at 12.8°C, both pH 6.3 and 5.1 supported toxin production at day 7. Toxin was detected in the pH 5.1 product formulated without lactate at day 35, whereas no toxin was detected through 105 days in the entree formulated to pH 5.1 and 2% lactate.

Miller et al. (72) compared the antibotulinal effect of several organic acid salts in uncured cooked turkey stored at 28°C. The study confirmed results reported by Maas et al. (66) that 2% sodium lactate delayed botulinal toxin production for several days compared with the control. Product formulated with 2% acetate and propionate delayed toxin production 5 days compared with a 4-day delay for lactate, but no delay in toxin production was found in product formulated with 2% pyruvate or citrate. Botulinal toxin was detected in turkey formulated with 6% citrate at 18 days, but no toxin was detected during the 18-day incubation period for products formulated with similar levels of acetate, propionate, pyruvate, or lactate. As expected, the efficacy of the monocarboxylic acid salts were related to both concentration used and their respective pK_a. Their antibotulinal effects are enhanced as the pH of the substrate decreases, shifting to the equilibrium to the undissociated form. Organic acid salts with higher pK_a values, such as acetate and propionate, may be more effective in low-acid foods.

Potassium sorbate and sorbic acid, traditionally added to foods as antimycotic agents, have been reported to inhibit bacterial growth (73). Early research reported that 0.1% sorbic acid could be used as a selective agent for catalase-negative lactic acid bacteria and clostridia in media by inhibiting catalase-positive microorganisms (74). York and Vaughn (75) claimed that 0.12% sorbic acid was metabolized as a carbon source and later reported clostridial growth in the presence of 2% sorbate in liver infusion media adjusted to pH 6.7 (76).

In contrast, numerous studies reported that 0.13–0.26% potassium sorbate delayed botulinal growth and toxin production in cured and uncured meat and poultry products (77–82). Unlike the media studies, the pH of the processed meat products generally ranged between 5.8 to 6.4, with sorbic acid treatments exhibiting lower pH values. As with organic acid salts, the effect of sorbic acid and sorbate salts increases as the substrate pH decreases to below 6 as a function of the concentration of undissociated sorbic acid and pH (83–85).

Potassium sorbate, which has a water solubility of over 50%, has greater antimicrobial activity than sorbic acid. The water solubility of sorbic acid is only 0.15%. Therefore, sorbic acid is less effective in higher-fat products because of its tendency to partition into the fat phase rather than the aqueous phase (86).

The antimicrobial properties of phosphates have been documented in several refrigerated food substrates (87). Phosphates function in food processing by their ability to enhance moisture retention, inhibit oxidation, stabilize emulsions, and modify pH. Their role as antimicrobials is not known but may be related to their ability to sequester metal ions essential for metabolism, to inhibit enzymes involved in cell division, or to physically interact with cells forming channels and promoting leakage and cell lysis (88–90). Current U.S. regulations limit phosphate usage to the amount needed for functionality. As a result, many refrigerated foods, such as processed meats, include much lower levels of phosphate (<1%) than shelf-stable process cheese (as much as 3% phosphate) (35).

In refrigerated foods, the antibotulinal effect varies according to polyphosphate species, food substrate, and the presence of adjunct antimicrobials. Barbut et al. (91) reported that 0.4% sodium acid pyrophosphate (SAPP) but not sodium hexametaphosphate or sodium tripolyphosphate delayed botulinal toxin production in turkey frankfurters when used in conjunction with 150 ppm sodium nitrite and >1.5% NaCl. The addition of SAPP to nitrite-free chicken frankfurter emulsions did not delay botulinal toxin production, where the combination of 0.4% SAPP, 0.26% potassium sorbate, and 40 ppm sodium nitrite delayed toxin production fourfold (91). Wagner and Busta (92) confirmed that SAPP delayed botulinal toxin formation in beef/pork frankfurter emulsions. Although sodium polyphosphates have been demonstrated to inhibit various clostridia including *C. botulinum* in media and in certain foods (88,90,93), more work is necessary to define the chemical nature of the inhibitory components in the polyphosphate mixtures and to evaluate their interactions with other antimicrobials.

Nisin, approved for use in process cheese spreads, and other bacteriocins have been proposed to inhibit botulinal toxin production in refrigerated foods. Toxin production by type E *C. botulinum* was delayed in fish treated with nisin and packaged in 100% carbon dioxide and stored at 10 and 26°C while having no effect on time to spoilage (94). Nisin was considered effective in delaying botulinal toxin production in cured chicken frankfurter emulsions (95). However, other investigations report that the antibotulinal effect of nisin may be limited by a number of factors. Somers and Taylor reported that binding of nisin to meat particles decreased its efficacy (48), and later experiments revealed limited antibotulinal effect in bacon attributed to low solubility in the brine (96). The efficacy of nisin was also reduced as a function of high phospholipid or protein concentration, pH $\geq$ 6.0, and at abuse storage temperatures (49,97). Development of resistance in botulinal spores and vegetative cells to nisin and other bacteriocins should also be considered (98).

Recently, naturally occurring antimicrobials primarily from plants and lactic acid bacteria have attracted interest as secondary barriers against pathogenic bacteria (99,100). These include phenolic compounds, terpenoids and essential oils, flavonoids, coumarins, alkaloids, lipid compounds, lectins and polypeptides, and other uncharacterized inhibitors. Although many of these compounds have been demonstrated to inhibit pathogenic bacteria, spoilage fungi, and human viruses in vitro, relatively few trials have been performed in foods. This is expected to be an active area of food safety research in the future as the public demands natural foods.

X. EFFECT OF MODIFIED ATMOSPHERE ON BOTULINAL TOXIN PRODUCTION IN FOODS

MAP and high-barrier films are frequently used to extend the shelf life of foods. The reduced oxygen levels and increased concentrations of carbon dioxide reduce oxidative and chemical deterioration and inhibit aerobic spoilage organisms. Concerns have been raised that these conditions may also select for anaerobic and facultative anaerobic microorganisms and increase the risk for psychrotrophic pathogen growth in refrigerated foods (101), particularly if the food has been processed to inactivate vegetative cells but leaving the spores viable.

Whiting and Naftulin (102) reported that the critical level of oxygen for germination and growth by *C. botulinum* is approximately 1–2%. Other studies revealed that atmospheres consisting predominantly of carbon dioxide enhance germination of botulinal spores (103). Although reduced oxygen and increased carbon dioxide levels may promote botulinal growth and toxin production, increased safety risk depends not only on the gas environment, but also on the product, storage temperature, packaging film used, and indigenous competitive flora.

Carbon dioxide has been shown to increase the shelf life of dairy products, including cottage cheese and pasteurized milk, by inhibiting the growth of spoilage microorganisms (104,105; see also Chapter 8). Collaborative research between the University of Wisconsin–Madison and Cornell University investigated the potential for botulinal growth in pasteurized milk treated with 9.1 and 18.2 mM carbon dioxide (106). The study revealed that botulinal toxin was detected after 4 days in milk treated with 9.1 mM carbon dioxide and stored at 21°C compared with 6 days for the untreated milk samples. However, milk from all treatments were considered inedible based on odor and appearance prior to toxin detection. No toxin was detected in any milk samples stored at 6°C for 60 days. The results suggest that modified atmosphere does not increase risk of botulism in pasteurized milk under abuse or refrigerated temperatures.

Post et al. (107) evaluated the effect of modified atmosphere and vacuum-packaging on nonproteolytic type E botulinal toxin production and spoilage in several varieties of fish fillets stored at 4–26°C. Vacuum-packaging or flushing with N_2, CO_2, or gas mixtures containing 65–90% CO_2 and 2–4% O_2 (balance N_2) extended the shelf life compared with packaging in air, but it also increased the likelihood that toxin production would occur prior to spoilage. Reddy et al. (108) compared the effect of packaging tilapia fish fillets in high-barrier films in 100% air, 75% CO_2/25% N_2, and vacuum-packaging on growth of type E *C. botulinum*. In this study sensory spoilage and toxin production coincided in MAP and vacuum-packaged fish held at abuse temperatures (16°C), whereas spoilage preceded toxin production in fish packaged in air. A greater safety margin was observed at 8 and 4°C.

The combination of barrier films and respiring foods may result in decreased oxygen levels for foods packaged under ambient atmosphere. Reduced oxygen, in turn, may enhance the growth of *C. botulinum*. Sugiyama and Yang (109) evaluated fresh mushrooms packaged in semipermeable plastic film for the ability to support botulinal toxin production. Fresh mushrooms were inoculated with *C. botulinum* spores, wrapped, and stored at 20°C. Respiration by the mushrooms decreased oxygen levels to <2% within 2 hours after packaging. Toxin was detected in prepackaged mushrooms after 3 days of storage, although mushrooms were still considered organoleptically acceptable. Later studies (110) revealed that holes in the wrap provided greater air exchange. No toxin was detected in mushrooms packaged containing at least 4% oxygen. However, the investigators cautioned that oxygen content alone may not be sufficient to inhibit botulinal growth and toxin production. The effect of food components, such as sulfhydryls, may reduce oxidative-reduction potential to a level at which *C. botulinum* could grow.

The significance of barrier films with high and low oxygen transmission rates may be observed by comparing four studies on toxin production in inoculated shredded cabbage. Solomon et al. (111) packaged cabbage under a modified atmosphere of 70% CO_2/30% N_2 in high barrier bags. Botulinal toxin was detected at 4 days of storage at 22°C. Product was organoleptically acceptable at the time of toxin detection. Petran et al. (112) compared toxin production in cabbage packaged in vented and nonvented flexible pouches. No toxin was detected in either package type stored at 4.4 or 12.7°C or in vented pouches stored at 21°C. Botulinal toxin was detected in nonvented pouches after 7 days storage at 21°C, but not at 4 days. Toxic samples were considered inedible prior to toxin detection. Oxygen levels were not assayed in either study. Researchers at the University of Wisconsin–Madison and the University of Georgia each evaluated toxin production in cabbage packaged in low (3000 $cm^3/m^2/24$ h) or high oxygen (6000–8000 $cm^3/m^2/24$ h) transmission rate films and stored at 4, 12, or 13, and 21°C until spoilage (113,114). Oxygen values were generally >1% oxygen for both package types. For both studies, no botulinal toxin was detected in any of the samples tested prior to overt spoilage of the products. These results suggest that oxygen concentration and the spoilage of the products by competitive microflora during the shelf life may be critical to inhibit botulinal growth.

The effect of modified atmosphere packaging on botulinal toxin formation has been evaluated for high-moisture bakery goods (50). Crumpets were inoculated with botulinal spores, packaged in air with or without an oxygen scavenger in a carbon dioxide/nitrogen mixture, and stored at ambient temperature. Botulinal toxin was detected within 4 to 6 days in all products, regardless of the initial atmosphere. As with the vegetable studies described above, oxygen levels in all packaging treatments were <2% within 4 days, likely due to respiration of the naturally occurring flora in the product. These observations confirm that the initial presence of oxygen during packaging does not provide sufficient protection against botulinal growth.

XI. ROLE OF COMPETITIVE AND SPOILAGE MICROORGANISMS IN ENSURING SAFETY

Competitive bacterial cultures are added to fermented foods, such as fermented sausage and cheese to inhibit growth and toxin production by pathogens such as *Staphylococcus aureus* and *C. botulinum*. In particular, lactic acid starter cultures enhance the safety of fermented foods by production of lactic acid and other antimicrobials such as bacteriocins, hydrogen peroxide, and diacetyl (115).

Mesophilic lactic acid bacteria can also be utilized as a protective culture for low-acid refrigerated foods. If the product is temperature abused, the lactic acid bacteria would grow, leading to acidification of the product and formation of gas and off-odors and flavors (116). Acidification by the added lactic bacteria inhibits growth of pathogens, while the changes in sensory attributes serves as a deterrent to consumption.

Pediococcus acidilactici culture and 0.7% sucrose were used to develop a reduced nitrite bacon (Wisconsin Process) (117). Bacon prepared with culture, sucrose, and 40 or 80 ppm sodium nitrite provided greater antibotulinal activity than control bacon prepared with 120 ppm nitrite and no starter culture or carbohydrate source. Hutton and coworkers reported that the ''Wisconsin process'' can also be applied to inhibit botulinal toxin production in chicken salad (118). The pH of the salad decreased under temperature abuse because of lactic acid production from the metabolism of dextrose by *P. acidilactici*.

Research at Rutgers University revealed that inhibition of *C. botulinum* by lactic acid bacteria (*Lactococcus*, *Pediococcus*, *Lactobacillus*) in a model gravy system and in *sous vide* beef depended on whether the strain produced bacteriocins against *C. botulinum*, the presence of glucose and buffering capacity in the substrate, and storage temperature (119,120). The biopreservation systems evaluated were more effective at 15°C than they were at 4 and 10°C or 25 and 35°C. The results emphasis the need to verify food-safety systems in specific substrates under a variety of conditions.

Other innovative approaches include incorporating competitive bacteria in edible films. The presence of selected nonpathogenic, nonspoilage bacteria inhibit growth of pathogenic and spoilage bacteria on the surface of food products (121).

Food manufacturers continually strive to improve plant sanitation and reduce microbial load in incoming raw ingredients as a means to enhance food safety and improve shelf life. However, food spoilage has generally been considered as a deterrent to consumption of unsafe product. Jay (122) questioned the wisdom of eliminating or inhibiting spoilage organisms that have previously served to inhibit foodborne pathogens. For example, while MAP serves as an effective means to increase the shelf life of foods, it may affect the growth of certain groups of beneficial organisms, some of which may inhibit *C. botulinum*. Temperature abuse of MAP cabbage may have contributed to a 1987 outbreak associated with coleslaw. Laboratory experiments revealed that inoculated shredded cabbage in high-barrier bags supported botulinal toxin production at 4 days of storage at room temperature without overt spoilage (111).

Larson and Johnson (123) examined botulinal toxin production in packaged cut melons treated to reduce spoilage microorganisms. When samples were treated with 50 ppm sodium hypochlorite and incubated at 15°C, gross spoilage preceded botulinal toxin production. However, samples treated with UV light and stored at 27°C developed toxin prior to onset of gross spoilage. These results have prompted some distributors of MAP produce to consider a reduced shelf life as a means to ensure safe product (J. Rosen, Fresh Express, personal communication).

Unfortunately, anecdotal evidence suggest that even spoilage may not deter some human beings from consuming unsafe products. Temperature-abused black bean dip and clam chowder associated with separate botulism outbreaks were reported by the patients to have tested or smelled bad, yet they proceeded to eat the spoiled products (124).

XII. USE OF PREDICTIVE MODELING AND CHALLENGE STUDIES TO DEFINE CONTROL LIMITS

Among the earliest applications of predictive food microbiology was the use of thermal death time models to determine commercial sterility for canned foods (28,29). Recent predictive models are useful to product developers for estimating formulation limits of low-acid refrigerated and shelf-stable foods. Various models have been published evaluating the effects of growth parameters, such as temperature, pH, salt, or water activity on growth or toxin production by *C. botulinum* in liquid microbiological media (125–128).

Two comprehensive predictive modeling programs include the Pathogen Modeling program, available from the USDA-Agriculture Research Service, Eastern Regional Research Center, and the Food MicroModel from Leatherhead Food Research Association. These programs, also based on data derived from media experiments, model the effects of multiple variables, such as pH, water activity, aerobic-anaerobic atmosphere, and temperature, on the growth or survival of several foodborne pathogens. Although predictive models based on growth in pure culture are useful, they do not account for the presence of competitive microflora or the antimicrobial effect of food components such as the impact of carbohydrates or proteins on reducing water activity.

Predictive models that are particularly useful are those based on challenge studies of multiple formulations for a specific food rather than media. Models have been developed to predict the probability of botulinal toxin production in uncured poultry products at refrigeration and abuse temperatures (68,70). The predictive model reported by Tanaka et al. (36) has been used extensively by industry and regulators to estimate the botulinal safety of process cheese spreads. The cheese spread model, however, is valid only for full-fat process cheese products formulated with phosphate-based emulsifiers. A model published by ter Steeg and Cuppers attempts to account for apparent safety differences between reduced-fat and full-fat cheese spreads and citrate and phosphate emulsifiers (129). To increase the power of the model, the investigators chose to determine growth of *C. botulinum* via direct plate counts rather than using standard toxicity assays. Accurate assessment of *C. botulinum* populations in foods using direct plate counts is difficult; therefore toxicity detection is generally considered the standard procedure for evaluating botulinal growth (130). Because the ter Steeg and Cuppers model is based on *Clostridia* plate counts rather than toxin production, additional verification of this model should be performed.

Predictive models provide valuable assistance on manipulating formulation parameters to control pathogen growth in foods. However, they should be validated in specific foods with inoculated-pack studies by a laboratory with expertise in handling *C. botulinum* and its toxins. The National Advisory Committee on the Microbiological Criteria for Foods developed guidelines for challenge studies of extended shelf-life refrigerated foods (131). Similar guidelines can be applied to shelf-stable low-acid foods, except toxin production should be monitored for twice the expected shelf life rather than 1½ times the expected shelf life as recommended for refrigerated foods.

Although considerably more costly than media studies, a well-designed inoculated pack study can verify whether or not a formulation will inhibit botulinal toxin production, as well as identify a margin of safety around the formulation specifications. Assaying individual samples for botulinal toxin may reveal more information about probability of toxin production than pooling the same samples. Multiple sampling intervals during the testing period will similarly clarify the effect of formulation specifications and antimicrobials.

The impact of any formula modifications should be carefully considered. Substitution of acidulants, modification of fat, moisture, or salt levels, new processing techniques that may affect spore levels or competitive microflora, or changes in spice or other raw materials vendors should be evaluated for their effect on safety.

Finally, good manufacturing practices and a well-designed and administered HACCP program are essential for the safety of low-acid refrigerated and shelf-stable foods. ''Keep refrigerated'' or ''Refrigerate after opening'' labels should be prominent, and time-temperature indicators should be considered if risk of temperature abuse is high.

REFERENCES

1. LDS Smith, H, Sugiyama. Botulism. The Organism, Its Toxins, the Disease, 2nd ed. Springfield, IL: Charles C. Thomas, 1988.
2. M Cherington. Clinical spectrum of botulism. Muscle Nerve 21:701–710, 1998.
3. CL Hatheway. Botulism: the present status of the disease. In: C Montecucco, ed. Clostridial Neurotoxins. Berlin: Springer-Verlag, 1995, pp. 55–75.
4. CL Hatheway, EA Johnson. *Clostridium*: the spore-bearing anaerobes. In: A Balows, BI Duerden, eds. Topley and Wilson's Microbiology and Microbial Infections. Vol. 2. Systematic Bacteriology. London: Arnold, 1998, pp. 731–782.
5. EA Johnson, MC Goodnough. Botulism. In: L Collier, A Balows, M Sussman, eds. Topley and Wilson's Microbiology and Microbial Infections. Vol. 3. Bacterial Infections. London: Arnold, 1998, pp. 723–741.

6. CL Hatheway. *Clostridium botulinum* and other organisms that produce botulinum neurotoxin. In: AHW Hauschild, KL Dodds, eds. *Clostridium botulinum*. Ecology and Control in Foods. New York: Marcel Dekker, Inc., 1993, pp. 3–20.
7. Centers for Disease Control and Prevention: Botulism in the United States, 1899–1996. Handbook for Epidemiologists, Clinicians, and Laboratory Workers. Atlanta: Centers for Disease Control and Prevention, 1998.
8. KL MacDonald, ML Cohen, PA Blake. The changing epidemiology of adult botulism in the United States. Am J Epidemiol 124:794–799, 1986.
9. ME St. Louis, HS Shaun, MB Peck, D Borwering, BG Morgan, J Blatherwick, S Bnerje, GDM Kettyls, WA Black, ME Milling, AHW Hauschild, RV Tauxe, PA Blake. Botulism from chopped garlic: delayed recognition of a major outbreak. Ann Intern Med 108:363–368, 1988.
10. WA Terranova, JC Breman, RO Locey, S Speck. Botulism type B: epidemiological aspects of an extensive outbreak. Am J Epidemiol 108:150–156, 1978.
11. E Chiorboli, G Fortina, G Bona. Flaccid paralysis cause by botulinum toxin type B after pesto ingestion. Pediat Infect Dis J 16:425–426, 1997.
12. MO O'Mahoney, E Mitchel, RJ Gilbert. An outbreak of foodborne botulism associated with contaminated hazelnut yogurt. Epidemiol Infect 104:389–395, 1990.
13. FJ Angulo, J Getz, JP Taylor, KA Hendricks, CL Hatheway, SS Barth, HM Solomon, AE Larson, EA Johnson, LN Nickey, AA Ries. A large outbreak of botulism: The hazardous baked potato. J Infect Dis 178:172–177, 1998.
14. L Oubira. Casher: la psychose. Liberte (July 15): 3, 1998.
15. MR Pourshafie, M Saifie, A Shafie, P Vahdani, M Aslani, J Salemian. An outbreak of food-borne botulism associated with contaminated locally made cheese in Iran. Scand J Infect Dis 30:94, 1998.
16. K Dodds. *Clostridium botulinum* in foods. In: AHW Hauschild, KL Dodds, eds. *Clostridium botulinum*: Ecology and Control in Foods. New York: Marcel Dekker, 1993, pp. 53–68.
17. NF Insalata, SJ Witzeman, GJ Fredericks, HL Bodily. Incidence study of spores of *Clostridium botulinum* in convenience foods. Appl Microbiol 17:542–544, 1969.
18. PA Gibbs, AR Davies, RS Fletcher. Incidence and growth of psychotrophic *Clostridium botulinum* in foods. Food Control 5:5–7, 1994.
19. AGW Hauschild. *Clostridium botulinum*. In: MP Doyle, ed. Foodborne Bacterial Pathogens. New York: Marcel Dekker, 1989, pp. 111–189.
20. K Dodds. *Clostridium botulinum* in the environment. In: AHW Hauschild, KL Dodds, eds. *Clostridium botulinum*: Ecology and Control in Foods, New York: Marcel Dekker, 1993, pp. 21–51.
21. AHW Hauschild. Epidemiology of human foodborne botulism. In: AHW Hauschild, KL Dodds, eds. *Clostridium botulinum*. Ecology and Control in Foods. New York: Marcel Dekker, Inc., 1993, pp. 69–104.
22. DL Collins-Thompson, DS Wood. Control in dairy products. In: AHW Hauschild, KL Dodds, eds. *Clostridium botulinum*: Ecology and Control in Foods, New York: Marcel Dekker, 1993, pp 261–277.
23. U.S. Food and Drug Administration. Thermally process low-acid foods packaged in hermetically sealed containers. Code of Federal Regulations, Title 21, Part 113.

Washington DC: Office of the Federal Register, U.S. Government Printing Office, 1997, pp. 188–192.

24. P Setlow, EA Johnson. Spores and their significance. In: MP Doyle, LR Beuchat, TJ Montville, eds. Food Microbiology: Fundamentals and Frontiers. Washington, DC: ASM Press, 1997, pp. 30–65.
25. ICMSF. *Clostridium botulinum.* In: International Commission on Microbiological Specifications for Foods. Microorganisms in Foods 5: Microbiological Specifications for Pathogens. New York: Blackie Academic & Professional, an imprint of Chapman & Hall, 1996, pp. 66–111.
26. RK Lynt, DA Kautter, RB Read Jr. Botulism in commercially canned foods J Milk Food Technol 38:546–550, 1975.
27. NFPA/CMI Container Integrity Task Force. Botulism risk from post-processing contamination of commercially canned foods in metal containers. J Food Prot 47: 801–816, 1984.
28. JR Esty, KF Meyer. The heat resistance of the spores of *B. botulinus* and allied anaerobes. XI. J Infect Dis 31:650–653, 1922.
29. CT Townsend, JR Esty, FC Baselt. Heat resistance studies on spores of putrefactive anaerobes in relation to determination of safe processes for canned foods. Food Res 3:323–346, 1938.
30. R Gibson. ConAgra recalls nonfat cheese loaf, citing health risk. The Wall Street Journal, November 24, 1992.
31. Recalls and field corrections: food—class II. FDA Enforcement Report, April 7, 1999.
32. H Pivnick, HW Barnett, HR Nordin, LJ Rubin. Factors affecting the safety of canned, cured, shelf-stable luncheon meat inoculated with *Clostridium botulinum.* Can Inst Food Technol J 2:141–148, 1969.
33. CL Duncan, EM Foster. Role of curing agents in the preservation of shelf-stable canned meat products. Appl Microbiol 16:401–405, 1968.
34. AGW Hauschild, B Simonsen. Safety assessment for shelf-stable canned cured meats—an unconventional approach. Food Technol 40(4):155–158, 1986.
35. FV Kosikowski, VV Mistry. Process cheese and related products. In: Cheese and Fermented Milk Foods, Vol. 1. Origins and Principles, 3rd ed. Westport, CT: FV Kosikowski, LLC, 1997, pp. 328–352.
36. N Tanaka, E Traisman, P Plantinga, L Finn, W Flom, L Meske, J Guggisberg. Evaluation of factors involved in antibotulinal properties of pasteurized process cheese spreads. J Food Prot 49:526–531, 1986.
37. JM Townes, PR Cieslak. An outbreak of type A botulism associated with a commercial cheese sauce. Ann Intern Med 125:558–563, 1996.
38. J Briozzo, E Amato de LaGarge, J Chirife, JL Parada. *Clostridium botulinum* type A growth and toxin production in media and process cheese spread. Appl Environ Microbiol 45:1150–1152, 1983.
39. C Karahadian, RC Lindsay, LL Dillman, RH Deibel. Evaluation of the potential for botulinal toxigenesis in reduced-sodium process American cheese foods and spreads. J Food Prot 48:63–69, 1985.
40. PF ter Steeg, HGAM Cuppers, JC Hellemons, G Rijke. Growth of proteolytic *Clostridium botulinum* in process cheese products: I. Data acquisition for modeling the

influence of pH, sodium chloride, emulsifying salts, fat dry basis, and temperature. J Food Prot 58:1091–1099, 1995.

41. N Tanaka, JM Goepfert, E Traisman, WM Hoffbeck. A challenge of pasteurized process cheese spread with *Clostridium botulinum* spores. J Food Prot 42:787–789, 1979.
42. KA Glass, J Loeffelholz, JP Ford, J Nelson. Evaluation of the production of *Clostridium botulinum* toxin in a process cheese spread in the presence of citrate. In: 1992 Annual Report. Madison, WI: Food Research Institute, pp. 64–65, 1992.
43. KA Glass, AE Larson, AL Duerr, EA Johnson. Inhibition of *C. botulinum* by phosphate-based salts in media and process cheese spread. Abstracts 85th Annual Meeting International Association Milk Food Environmental Sanitarians. J Food Prot 61(suppl A):43, 1998.
44. KF Eckner, WA Dustman, AA Ry-Rodriguez. Contribution of composition, physicochemical characteristics and polyphosphates to the microbial safety of pasteurized cheese spreads. J Food Prot 57:295–300, 1994.
45. J Delves-Broughton. Nisin and its uses as a food preservative. Food Technol 44(11): 100, 102, 104, 106, 108, 111–112, 117, 1990.
46. EB Somers, SL Taylor. Antibotulinal effectiveness of nisin in pasteurized process cheese spreads. J Food Prot 50(10):842–848, 1987.
47. U.S. Food and Drug Administration. Pasteurized process cheese spread. Code of Federal Regulations, Title 21, Part 133.179. Washington, DC: Office of the Federal Register, U.S. Government Printing Office, 1997, pp. 329–331.
48. EB Somers, SL Taylor. Further studies on the antibotulinal effectiveness of nisin in acidic media. J Food Sci 46:1972–1973, 1981.
49. VN Scott, SL Taylor. Temperature, pH, and spore load effects on the ability of nisin to prevent the outgrowth of *Clostridium botulinum* spores. J Food Sci 46: 121–126, 1981.
50. D Phillips Daifas, JP Smith, B Blanchfield, JW Austin. Growth and toxin production by *Clostridium botulinum* in English-style crumpets packaged under modified atmospheres. J Food Prot 62:349–355, 1999.
51. GD Betts, JE Gaze. Growth and heat resistance of psychrotrophic *Clostridium botulinum* in relation to 'sous vide' products. Food Contr 6:57–63, 1995.
52. MW Eklund, DI Wieler, FT Poysky. Outgrowth and toxin production of nonproteolytic type B *Clostridium botulinum* at 3.3 to 5.6°C. J Bacteriol 93:1461–1462, 1967.
53. S Notermans, J Dufrenne, BM Lund. Botulism risk of refrigerated, processed foods of extended durability. J Food Prot 53:1020–1024, 1990.
54. Codex Committee on Food Hygiene. Draft code of hygienic practice for refrigerated packaged foods with extended shelf life. Rome: FAO/WHO, 1998.
55. SJ Van Garde, M Woodburn. Food discard practices of householders. J Am Diet Assoc 87:322–329, 1987.
56. RW Daniels. Applying HACCP to new-generation refrigerated foods at retail and beyond. Food Technol 46(6):122, 124, 1991.
57. RW Daniels. Home food safety. Food Technol 52(2):54–56, 1998.
58. RW Daniels. Audits International's Home Food Safety Survey. Http://www.audits.com/, 1999.

59. A Tolstoy. Practical monitoring of the chill chain. Int J Food Microbiol 13(3):225–230, 1991.
60. GJM Raatjes, JPPM Smelt. *Clostridium botulinum* can grow and form toxin at pH values lower than 4.6. Nature 281:398–399, 1979.
61. GW Gould, MV Jones. Combination and synergistic effects. In: GW Gould, ed. Mechanisms of Action of Food Preservation Procedures. London: Elsevier Applied Science, 1989, pp. 401–421.
62. L Leistner. Use of hurdle technology in food processing: recent advances. In: GV Barbosa-Cánovas, J Welti-Chanes, eds. Food Preservation by Moisture Control: Fundamentals and Applications. Lancaster: Technomic Publishing Co. Inc, 1995, pp. 378–396.
63. JA Troller. Combinations of factors to obtain the microbiological safety of foods. In: GV Barbosa-Canovas, J Welti-Chanes, eds. Food Preservation by Moisture Control: Fundamentals and Applications. Lancaster: Technomic Publishing Co. Inc., 1995, pp. 535–551.
64. RJ Anders, JG Cerveny, AL Milkowski. Method for delaying *Clostridium botulinum* growth in fish and poultry. US Patents 4,798,729, Jan 17, 1989 and 4,888,191, December 19 (1989).
65. S Doores. Organic Acids. In: PM Davidson, AL Branen, eds. Antimicrobials in Foods, 2nd ed. New York: Marcel Dekker, 1993, pp. 95–136.
66. MR Maas, KA Glass, MP Doyle. Sodium lactate delays toxin production by *Clostridium botulinum* in cook-in-bag turkey products. Appl Environ Microbiol 55: 2226–2229, 1989.
67. J Meng, CA Genigeorgis. Delaying toxigenesis of *Clostridium botulinum* by sodium lactate in 'sous-vide' products. Lett Appl Microbiol 19:20–23, 1994.
68. MR Maas. Development and use of probability models: the industry perspective. J Indust Micro 12:162–167, 1993.
69. PC Houtsma, A Heuvelink, J Dufrenne, S Notermans. Effect of sodium lactate on toxin production, spore germination and heat resistance of proteolytic *Clostridium botulinum* strains. J Food Prot 57:327–330, 1994.
70. J Meng, CA Genigeorgis. Modeling lag phase of nonproteolytic *Clostridium botulinum* toxigenesis in cooked turkey and chicken breast as affected by temperature, sodium lactate, sodium chloride and spore inoculum. Int J Food Microbiol 19:109–122, 1993.
71. JD Meyer, JB Luchansky, JG Cerveny. Inhibition of a psychrotrophic *Clostridium* species by sodium diacetate and sodium lactate in a cook-in-the-bag, refrigerated turkey breast product. J Food Prot 58(suppl A):34, 1995.
72. AJ Miller, JE Call, RC Whiting. Comparison for organic acid salts for *Clostridium botulinum* control in an uncured turkey product. J Food Prot 56:958–962, 1993.
73. JN Sofos, FF Busta. Sorbic acid and sorbates. In: PM Davidson, AL Branen, ed. Antimicrobials in Foods, 2nd ed. New York: Marcel Dekker, 1993, pp. 49–94.
74. LE Emard, RH Vaughn. Selectivity of sorbic acid media for the catalase negative lactic acid bacteria and clostridia. J Bacteriol 63:487–494, 1952.
75. GK York, RH Vaughn. Use of sorbic acid enrichment media for species of *Clostridium*. J Bacteriol 68:739–744, 1954.

76. GK York, RH Vaughn. Resistance of *Clostridium parabotulinum* to sorbic acid. Food Res 20:60–65, 1955.
77. FJ Ivey, KJ Shaver, LN Christiansen, and RB Tompkin. Effect of potassium sorbate on toxinogenesis by *Clostridium botulinum* in bacon. J Food Prot 41:621–626, 1978.
78. CN Huhtanen, J Feinberg. Sorbic acid inhibition of *Clostridium botulinum* in nitrite-free poultry frankfurters. J Food Sci 45:453–457, 1980.
79. CN Huhtanen, J Feinberg, H Trenchard, and JG Phillips. Acid enhancement of *Clostridium botulinum* inhibition in ham and bacon prepared with potassium sorbate and sorbic acid. J Food Prot 46(9):807–810, 1983.
80. JN Sofos, FF Busta, CE Allen. Sodium nitrite and sorbic acid effects on *Clostridium botulinum* spore germination and total microbial growth in chicken frankfurter emulsions during temperature abuse. Appl Environ Microbiol 37:1103–1109, 1979.
81. JN Sofos, FF Busta, CE Allen. Influence of pH on *Clostridium botulinum* control by sodium nitrite and sorbic acid in chicken emulsions. J Food Sci 45:7–12, 1980.
82. JN Sofos, FF Busta, K Bhothipaksa, CE Allen, MC Robach, MW Paquette. Effects of various concentrations of sodium nitrite and potassium sorbate on *Clostridium botulinum* toxin production in commercially prepared bacon. J Food Sci 45:1285–1292, 1980.
83. JC Blocher, FF Busta, JN Sofos. Influence of potassium sorbate and pH on ten strains of type A and B *Clostridium botulinum*. J Food Sci 47:2028–2032, 1982.
84. JC Blocher, FF Busta. Multiple modes of inhibition of spore germination and outgrowth by reduced pH and sorbate. J Appl Bacteriol 59:469–478, 1985.
85. BM Lund, SM George, JG Franklin. Inhibition of type A and type B (proteolytic) *Clostridium botulinum* by sorbic acid. Appl Environ Microbiol 53:935–941, 1987.
86. JN Sofos, FF Busta. Antimicrobial activity of sorbate. J Food Prot 44:614–622, 1981.
87. RB Tompkin. Indirect antimicrobial effects in foods: phosphate. J Food Safety 6:13–27, 1983.
88. RH Ellinger. Phosphates as Food Ingredients. Cleveland: CRC Press, 1972.
89. A Kornberg. Inorganic polyphosphate: a molecule of many functions. In: HC Schröder, WEG Müller, eds. Inorganic Polyphosphates. Biochemistry, Biology, Biotechnology. Berlin: Springer-Verlag, 1999, pp. 1–18.
90. RA Molins. Antimicrobial uses of phosphates. In: RA Molins. Phosphates in Foods. Boca Raton, FL: CRC Press, 1991, pp. 207–234.
91. S Barbut, N Tanaka, RG Cassens, AJ Maurer. Effects of sodium chloride reduction and polyphosphate addition on *Clostridium botulinum* toxin production in turkey frankfurters. J Food Sci 51(5):1136–1138, 1172, 1983.
92. MK Wagner, FF Busta. Effect of sodium acid pyrophosphate in combination with sodium nitrite or sodium nitrite/potassium sorbate on *Clostridium botulinum* growth and toxin production in beef/pork frankfurter emulsions. J Food Sci 48:990–991, 993, 1983.
93. MJ Loessner, SK Maier, P Schiwek, S Scherer. Long-chain polyphosphates inhibit growth of *Clostridium tyrobutyricum* in processed cheese spreads. J Food Prot 60:493–498, 1997.

94. LY Taylor, DD Cann, BJ Welch. Antibotulinal properties of nisin in fresh fish packaged in an atmosphere of carbon dioxide. J Food Prot 53:953–957, 1990.
95. SL Taylor, EB Somers, LA Krueger. Antibotulinal effectiveness of nisin-nitrite combinations in culture medium and chicken frankfurter emulsions. J Food Prot 48:234–239, 1985.
96. SL Taylor, EB Somers. Evaluation of the antibotulinal effectiveness of nisin in bacon. J Food Prot 48:949–952, 1985.
97. AM Rogers, TJ Montville. Quantification of factors which influence nisin's inhibition of *Clostridium botulinum* 56A in a model food system. J Food Sci 59:663–668, 686, 1994.
98. AS Mazzotta, AD Crandall, TJ Montville. Nisin resistance in *Clostridium botulinum* spores and vegetative cells. Appl Environ Microbiol 63:2654–2659, 1997.
99. Council for Agricultural Science and Technology (CAST). Naturally Occurring Antimicrobials in Foods. Task Force Report No. 132. Iowa: Council for Agricultural Science and Technology, pp. 1–103, 1998.
100. MM Cowan. Plant products as antimicrobial agents. Clin Microbiol Rev 12:564–582, 1999.
101. JM Farber. Microbiological aspects of modified-atmosphere packaging technology—a review. J Food Prot 54:58–70, 1991.
102. RC Whiting, KA Naftulin. Effect of headspace oxygen concentration on growth and toxin production by proteolytic strains of *Clostridium botulinum*. J Food Prot 55:23–27, 1992.
103. PM Foegeding, FF Busta. Effect of carbon dioxide, nitrogen and hydrogen gases on germination of *Clostridium botulinum* spores. J Food Prot 46:987–989, 1983.
104. JH Chen, JH Hotchkiss. Effects of dissolved carbon dioxide on the growth of psychrotrophic organisms in cottage cheese. J Dairy Sci 74:2941–2945, 1991.
105. JH Hotchkiss, JH Chen, HT Lawless. Combined effects of carbon dioxide addition and barrier films on microbial and sensory changes in pasteurized milk. J Dairy Sci 82:690–695, 1999.
106. KA Glass, KM Kaufman, AL Smith, EA Johnson, JH Chen, J Hotchkiss. Toxin production by *Clostridium botulinum* in pasteurized milk treated with carbon dioxide. J Food Prot 62:872–876, 1999.
107. LS Post, DA Lee, M Solberg, D Furgang, J Specchio, C Graham. Development of botulinal toxin and sensory deterioration during storage of vacuum and modified atmosphere packaged fish fillets. J Food Sci 50:990–996, 1985.
108. NR Reddy, A Paradis, MG Roman, HM Solomon, EJ Rhodehamel. Toxin development by *Clostridium botulinum* in modified atmosphere-packaged fresh tilapia fillets during storage. J Food Sci 61:632–635, 1996.
109. H Sugiyama, KH Yang. Growth potential of *Clostridium botulinum* in fresh mushrooms packaged in semipermeable plastic film. Appl Microbiol 30:964–969, 1975.
110. H Sugiyama, KS Rutledge. Failure of *Clostridium botulinum* to grow in fresh mushrooms packaged in plastic film overwraps with holes. J Food Prot 41:348–350, 1978.
111. HM Solomon, DA Kautter, T Lilly, EJ Rhodehamel. Outgrowth of *Clostridium botulinum* in shredded cabbage at room temperature under modified atmosphere. J Food Prot 53:831–833, 1990.

112. RL Petran, WH Sperber, AB Davis. *Clostridium botulinum* toxin formation in romaine lettuce and shredded cabbage: effect of storage and packaging conditions. J Food Prot 624–627, 1995.
113. AE Larson, EA Johnson, CR Barmore, MD Hughes. Evaluation of the botulism hazard from vegetables in modified atmosphere packaging. J Food Prot 60:1208–1214, 1997.
114. YY Haw, RE Brackett, LF Beauchat, MP Doyle. Microbiological quality and the inability of proteolytic *Clostridium botulinum* to produce toxin in film-packaged fresh-cut cabbage and lettuce. J Food Prot 61:1148–1153, 1998.
115. MA Daeschel. Antimicrobial Substances from lactic acid bacteria for use as food preservatives. Food Technol 43(1):164–167, 1989.
116. WP Hammes, PS Tichaczek. The potential of lactic acid bacteria for the production of safe and wholesome food. Z Lebensm Unters Forsch 198:193–201, 1994.
117. N Tanaka, L Meske, MP Doyle, E Traisman, DW Thayer, RW Johnston. Plant trials of bacon made with lactic acid bacteria, sucrose, and lowered sodium nitrite. J Food Prot 48:679–686, 1985.
118. MT Hutton, PA Chehak, JH Hanlin. Inhibition of botulinum toxin production by *Pediococcus acidilactici* in temperature abused refrigerated foods. J Food Safety 11:255–267, 1991.
119. AD Crandall, TJ Montville. Inhibition of *Clostridium botulinum* growth and toxigenesis in a model gravy system by coinoculation with bacteriocin-producing lactic acid bacteria. J Food Prot 56:485–488, 492, 1993.
120. AD Crandall, K Winkowski, TJ Montville. Inability of *Pediococcus pentosaceus* to inhibit *Clostridium botulinum* in sous vide beef with gravy at 4 and 10°C. J Food Prot 57:104–107, 132, 1994.
121. RP Clayton, RA Bowling. Method of preserving food products and food products made thereby. U.S. Patent No. 5,869,113 (1999).
122. JM Jay. Do background microorganisms play a role in the safety of fresh food? Trends Food Sci Technol 8:421–424, 1997.
123. AE Larson, EA Johnson. Evaluation of botulinal toxin production in packaged fresh-cut cantaloupe and honeydew melons. J Food Prot 62:948–952, 1999.
124. California State Department of Health Services. California Morbidity May 19, 1995.
125. AF Graham, DR Mason, MM Peck. Predictive model of the effect of temperature, pH, and sodium chloride on growth from spores of non-proteolytic *Clostridium botulinum.* Int J Food Microbiol 31:69–85, 1996.
126. PJ McClure, MB Cole, SPPM Smelt. Effects of water activity and pH on growth of *Clostridium botulinum*. J Appl Bacteriol Symp Suppl 76:105S–114S, 1994.
127. RC Whiting, JE Call. Time of growth model for proteolytic *Clostridium botulinum*. Food Microbiol 10:295–301, 1993.
128. RC Whiting, JC Oriente. Time-to-turbidity model for nonproteolytic type B *clostridium botulinum*. Int J Food Microbiol 35:49–60, 1997.
129. PF ter Steeg, HGAM Cuppers. Growth of proteolytic *Clostridium botulinum* in process cheese products: II. predictive modeling. J Food Prot 58:1100–1108, 1995.
130. DA Baker. Probability models to assess safety of foods with respect to *Clostridium botulinum*. J Industr Microbiol 12:156–161, 1993.

131. MP Doyle. Evaluating the potential risk from extended-shelf-life refrigerated foods by *Clostridium botulinum* inoculation studies. Food Technol 45(4):154–156, 1991.
132. M Sebald, J Jouglard, G Gilles. Botulisme humain de type B agres ingestion de fromage. Ann de Microbiol 125A:349–357, 1974.
133. PA Blake, MA Horwitz, L Hopkins, GL Lombard, JE McCroan, JC Prucha, MH Merson. Type A botulism from commercially canned beef stew. South Med J 70: 5–7, 1977.
134. JE Seals, JD Snyder, TA Eddell, CL Hatheway, CJ Johnson, RC Swanson, JM Hughes. Restaurant-associated type A botulism: Transmission by potato salad. Am J Epidemiol 113:436–444, 1981.
135. Botulism—New Mexico. MMWR 27:138, 1978.
136. Follow-up on botulism—New Mexico. MMWR 27:143, 1978.
137. Botulism and commercial pot pie—California. MMWR 32:39–40, 45, 1983.
138. KL MacDonald, RF Spengler, CL Hathaway, NT Hargrett, ML Cohen. Type A botulism from sautéed onions: clinical and epidemiologic observations. JAMA 253: 1275–1278, 1985.
139. HM Solomon, DA Kautter. Growth and toxin production by *Clostridium botulinum* in sautéed onions. J Food Prot 49:618–620, 1986.
140. DL Morse, LK Pickard, JJ Buzewich, BD Devine, M Sharyegani. Garlic-in-oil associated botulism: episode leads to product modification. Am J Public Health 80: 1372–1373, 1990.
141. HM Solomon, DA Kautter. Outgrowth and toxin production by *Clostridium botulinum* in bottled, chopped garlic. J Food Prot 51:862–865, 1988.
142. Type B botulism associated with roasted eggplant in oil—Italy, 1993. MMWR 44: 33–36, 1995.
143. Restaurant botulism incident hospitalizes 18 in El Paso. Food Chem News (April 18):69, 1994.
144. Botulism cases prompt warning from California officials. Food Chem News 36(34): 3048, 1994.
145. Food safety: outbreak of botulism. Weekly Epidemiol Rec (December 6):49, 1996.
146. P Aureli, G Franciosa, M Pourshaban. Foodborne botulism in Italy. Lancet 348: 1594, 1996.
147. G Franciosa, M Pourshaban, M Gianfranceschi, A Gattuso, L Fenicia, AM Ferrini V Mannoni, G Deluca, P Aureli. *Clostridium botulinum* spores and toxin in mascarpone cheese and other milk products. J Food Prot 62:867–871, 1999.

13
Reduction of Microbial Contaminants on Carcasses

Alejandro Castillo
University of Guadalajara
Guadalajara, Mexico

Margaret D. Hardin
Sara Lee Foods
Cordova, Tennessee

Gary R. Acuff
Texas A&M University
College Station, Texas

James S. Dickson
Iowa State University
Ames, Iowa

I. INTRODUCTION

Several methods are currently being used for carcass decontamination. These methods are intended to reduce pathogen numbers. However, reduction of total bacterial populations is often used to evaluate the impact of these methods on meat safety, and pathogen reduction may not correlate with the reduction of other organisms. In contrast, when using pathogens for testing various methods of carcass decontamination, the evaluation is frequently conducted under laboratory conditions, since food processors would never allow foodborne pathogens to be brought into the plant. These laboratory conditions may not represent the conditions under which meat is obtained during normal slaughter operations. Therefore, numbers reported for bacterial reductions by different methods of carcass decontamination must be viewed cautiously when trying to determine which methods, reportedly, achieve the best results.

The methods for bacterial reduction on carcasses that have most commonly been studied include cleaning methods such as water wash (Hardin et al., 1995; Reagan et al., 1996) or trimming (Prasai et al., 1995a,b), as well as sanitizing by organic acid rinses (Hardin et al., 1995; Dorsa et al., 1997b). Other methods such as hot water sprays, steam pasteurization, or trisodium phosphate have also been proposed (Smith and Graham, 1978; Dickson et al., 1994; Phebus et al., 1997). All of these interventions will ultimately reduce microbial contamination, but since sanitizers such as organic acids or hot water show bactericidal activity, the reductions achieved by sanitizing carcasses must be greater than those of cleaning methods. A description of each different carcass decontamination method will provide a better understanding of the benefits of these interventions.

II. BACTERIAL ATTACHMENT TO MEAT SURFACES

Meat decontamination is usually achieved by sprays or washes with water or antimicrobial solutions. Depending on the treatment applied, the reduction in microbial numbers on the meat surface will be due to physical removal of the microorganisms, a killing effect of the decontaminating solution, or a combination of both factors. Among the factors influencing the effectiveness of carcass interventions, bacterial attachment onto the meat surface has attracted the attention of many researchers. The molecular basis for bacterial attachment has been reviewed by Hardin (1995). Anderson et al. (1987a) found consistently higher bacterial counts on carcass samples collected by excising tissue when compared to bacterial counts from samples obtained by swabbing the carcass surface. This may indicate that microorganisms on meat surfaces are usually attached to the surface. If this is the case, the attached bacteria may be more difficult to remove by swabbing than nonattached bacteria after applying carcass washes. Fratamico et al. (1996) reported that, once bacteria are attached to meat, rinse solutions such as acetic acid or trisodium phosphate were not effective in removing a large part of the contaminating bacteria. In contrast, Castillo et al. (1998a) and Hardin et al. (1995) did not find any differences in counts of *Escherichia coli* O157:H7 or *Salmonella* Typhimurium on beef carcass surfaces treated by hot water or organic acids immediately or 20–30 minutes after contaminating the beef surfaces. Since bacterial attachment on carcass surfaces has been reported to occur within as little as 20 minutes of contact (Butler et al., 1979), those authors concluded that bacterial attachment did not affect the antimicrobial effect of hot water or organic acid sprays against pathogens on beef.

Other factors such as the surface fat characteristics of the carcass region may be more important for the effectiveness of antimicrobial sprays. Using transmission and scanning electron microscopy, Mattila and Frost (1988) described the attachment of *E. coli* to beef and chicken surfaces. By this method, they demonstrated that bacteria interlace with collagen fibers during growth and that, after growing on the surface, the cells form a network of fine fibrils and particles of condensed glycocalix on the cell surface, which help in cell colonization. Several investigators have observed no differences in attachment of bacterial pathogens to meat surfaces on lean or adipose tissue (Chung et al., 1989; Bouttier, 1997; Cabedo et al., 1997). The effect of the surface fat characteristics of the meat is important because fat, being hydrophobic, may interfere with the effectiveness of water washes to remove bacterial contamination. In contrast to the above reports, Dickson (1991a) found more attachment of *S.* Typhimurium and *Listeria monocytogenes* to lean tissue than to fat tissue. He also found organism-intrinsic factors such as inoculum size, temperature to which the cells had been exposed before attaching, and age of the cells to have a significant effect on the attachment of these pathogens to beef surfaces. Dickson and Koohmaraie (1989) determined the hydrophobicity and surface charge of different bacteria and their effect on their ability to attach to beef tissue. These authors found linear significant correlation between negative charge and attachment to either lean or fat beef surfaces, whereas the cell hydrophobicity correlated with attachment to adipose tissue only. Similar results were reported by Bouttier et al. (1997). These authors also studied the role of bacterial flagella on attachment to meat surfaces by determining the number of cells of *Salmonella choleraesuis* adhering to lean or fat tissue after treating with antibodies to flagellar antigens. The count of attached cells was significantly lower for antibody-treated cells than that of control cells or cells treated with antibodies to somatic antigens. Although this would indicate that flagella have an effect on bacterial attachment, other experiments involving potential chemical receptors on beef tissues saturated by a suspension of flagella showed no effect of flagella on the attachment of *S. choleraesuis*. This demonstrated that the surface of beef did not possess any receptors for the flagella of this species. Furthermore, although mechanical removal of the flagella did not alter the hydrophobicity, it did alter the electric charge on the cell surface. This indicates that, even though flagella do not directly affect the bacterial attachment to meat surfaces, they do determine the electrical charge, which, as described above, plays a role in the ability of bacteria to attach to meat surfaces. This may explain the findings of Butler et al. (1979), who reported greater attachment of gram-negative motile cells than gram-positive or nonmotile bacteria. Cells that are attached to meat surfaces have been shown to transfer between surfaces at a lower rate than unattached cells, although the highest transfer rate was observed from adipose to lean tissue (Dickson, 1990a). This is of practical importance because carcasses are often in close contact for long times in coolers, and cross-contamination may occur during storage. The presence of organic matter may be another factor affecting the extent of bacterial attachment on meat surfaces. Dickson and MacNeil (1991) reported greater attachment of *S.* Typhimurium and *L. monocytogenes* to beef carcass surfaces when the inoculum had been diluted in phosphate buffer compared to cow manure. The nature of the inoculum needs to be taken into consideration when research is conducted involving inoculation of pathogens onto meat surfaces.

Bacterial attachment to meat tissue is a complex mechanism, which may have a practical effect on the transfer of pathogens between carcasses, the effectiveness of sampling methods, and the performance of treatments for carcass decontamination.

III. CLEANING METHODS

A. Knife Trimming

Current U.S. Department of Agriculture (USDA) Food Safety and Inspection Service (FSIS) regulations require that all feces, ingesta, and milk be physically removed from beef carcasses by knife trimming or, when such contamination is less than one inch in its greatest dimension, by vacuuming with hot water or steam (Federal Register, 1996). The efficacy of the traditional approach of trimming for removing contamination from beef carcasses has been evaluated by several authors. Gorman et al. (1995b) reduced by 2.0–2.5 $\log_{10}$ ($p < 0.05$) the aerobic plate counts and *E. coli* counts on inoculated beef brisket by knife trimming without any other combined treatment. No additional reduction was observed when a water wash was applied subsequent to trimming, unless the water temperature was at least 66°C. Other authors have also reported extensive reductions of bacterial counts by trimming (Gorman et al., 1995a; Hardin et al., 1995; Prasai et al., 1995b; Reagan et al., 1996). However, in most instances the evaluations of knife-trimming decontamination have been conducted under laboratory conditions. Prasai et al. (1995b) concluded that the high bacterial reduction obtained by trimming might have been due to the artificial conditions under which this operation was accomplished. In their work, the trimming samples were collected from locations that had been completely trimmed by making one cut using a sterile knife. This may not be comparable to the trimming procedure practiced in plants during normal slaughter operations. In a study conducted in a beef packing plant, Gill et al. (1996) found no differences in total bacterial and *E. coli* counts from carcasses before or after trimming. These authors also found that a water wash treatment subsequent to trimming did not produce further reduction in bacterial counts, concluding that the reduction in bacterial numbers achieved by either trim or water wash is insufficient to enhance the safety of the meat. Prasai et al. (1995a) also observed no differences in aerobic plate count (APC) obtained from hot-fat trimmed carcasses when compared with nontrimmed carcasses. This hot-fat trimming step did not affect the microbiological quality of subprimals even after 14 days of storage. These reports may indicate that, although trimming has been reported to significantly reduce bacterial counts under laboratory conditions, the circumstances under which slaughter plants typically conduct this intervention may not favor effective carcass decontamination. In addition, the spread of microbial contamination from areas of fecal contamination to other clean areas is a concern for the beef industry (Hardin et al., 1995).

B. Water Wash

Water washing of carcasses after slaughter is a common practice during beef slaughter. This water wash is intended to eliminate visible contamination from carcasses as well as improve the visual quality of the meat. A controlled water wash treatment designed by Anderson et al. (1980) was proven to fulfill the requirements of standard acceptability tests which were used by the USDA-FSIS at the time these evaluations were conducted. Anderson et al. (1981) compared a careful hand wash to an automated water wash in a spraying unit and found no differences between the two types of water wash. From their results, these authors concluded that a cabinet water wash can be adjusted and standardized to ensure that washed carcasses will pass inspection. Further reports by these authors indicated that this type of spray chamber, if not properly designed, may produce too many water droplets, which wet surfaces in the slaughter area, as well as have the potential to accumulate soil and dirt in the interior of the chamber. Further improvements in the design of their unit (Anderson et al., 1982a,b,c) allowed for cleaning in place and preventing water droplets from escaping into the slaughtering area in addition to controlling the carcass water absorption and shrinkage.

Under the USDA-FSIS's zero-tolerance policy for fecal contamination, only knife trimming and, to some extent, steam vacuum treatments are allowed as means to eliminate visible fecal contamination from carcasses. However, many researchers have studied the ability of water washes to reduce microbial contamination. Anderson et al. (1975) studied the effect of factors such as water volume and pressure, angle of droplet impact, droplet size, spray force, and the speed at which the meat passed through the spray on the removal of the yeast *Rhodatorula rubra* on beef plate meat. According to this study, factors such as water pressure, water flow rate, and speed of movement of meat through spray had significant effect on microbial removal, whereas mean droplet size was not a significant factor. The angle of droplet impact was not significant when the pressure was 28 kg/cm^2 but became significant as the pressure was decreased. Contrasting with the findings of Anderson et al. (1975), Crouse et al. (1988) did not find any reduction in numbers of Enterobacteriaceae or APC on beef carcasses as affected by the spray pressure or the chain speed. In a similar study, DeZuniga et al. (1991) found no significant differences due to pressure for the reduction in APC or counts of Enterobacteriaceae on meat surfaces.

In their report, DeZuniga et al. (1991) also addressed the effect of high-pressure water wash on bacterial penetration into the meat. Using an insoluble dye (blue lake), the particle size of which is only slightly smaller than most bacteria, these authors found that the depth of blue lake penetration after automated water wash was directly proportional to the line pressure and that the type of nozzle used had a significant effect on penetration of blue lake at pressures above 4140 kPa. DeZuniga et al. (1991) concluded that bacteria might penetrate into the meat as a result of high-pressure cabinet water wash treatment. These bacteria may be able to grow, and if a sanitizer is used after washing for decontamination purposes, the solution might not reach the site where bacteria are implanted in the meat.

Another factor associated with water washing carcasses is the potential antimicrobial effect of the presence of chlorine in the washing water. Anderson et al. (1977) found no differences in log APC reductions on meat sampled after spraying tap water or water added with 200 ppm sodium hypochlorite at different flow rates, while spraying with 3% acetic acid showed significantly larger log count reductions. Kotula et al. (1974) also used chlorinated water (200 ppm chlorine) for reducing bacterial contamination on beef carcasses; however, these investigators did not include a nonchlorinated water control to assess the bactericidal effect of chlorine. According to these authors, the magnitude of the bacterial reduction was affected by factors such as line pressure or water wash temperature. In addition, total aerobic counts were lower when samples were collected 24 hours after washing that those from samples obtained immediately after washing. This indicates a potential continued effect of the chlorinated wash during storage of the carcasses.

Despite the several studies evaluating the potential of carcass water washes to remove bacterial contamination from carcasses, few studies have addressed the temperature of the wash. It is generally accepted that treatments with hot water (74°C) will produce a sanitizing effect rather than a simple washing effect (Federal Register, 1996). Cabedo et al. (1996) produced larger reductions of *E. coli* on beef brisket spraying with water at 74°C when compared to spraying brisket with water at 35°C. However, under sublethal temperature conditions, higher temperatures must have an effect on fat softening, which may also affect the ability to remove bacterial contamination. Gorman et al. (1995b) produced larger *E. coli* reductions on beef tissue by spraying water at 35°C than by spraying water at 16°C. However, there were no differences in *E. coli* reductions after spraying water at 35, 66, or 74°C. When sprays at 16, 35, or 74°C were followed by a second wash at 16°C, the reductions increased with the temperature of the first wash. This might indicate that, in fact, a warm carcass wash would be more helpful than a cold wash at removing microbial contamination, especially if the water wash is followed by a sanitizing step.

C. Trimming Versus Washing

Most studies of trimming as a means of reducing bacterial contamination also compare the reductions obtained by water washing. These studies indicate that trimming produces similar or larger reductions than those obtained by water wash (Gorman et al., 1995a; Hardin et al., 1995; Prasai et al., 1995b). The USDA-FSIS affirms that trimming, if performed properly, will effectively remove the visible contamination as well as any accompanying microbial contamination, whereas, if not properly conducted, it may spread the contamination to other newly exposed areas (Federal Register, 1996). However, Gill et al. (1996) reported that numbers of *E. coli*, coliforms, and aerobic bacteria that contaminate beef carcasses during dehiding and evisceration were not reduced by trimming and were halved by washing. Conversely, Reagan et al. (1996) showed a significant superiority of trimming over water wash at reducing aerobic bacteria and *E. coli* biotype I on beef carcasses.

Whether trim reduces more contamination than water wash or vice versa, both methods seem not to be effective means of decontaminating carcasses. In studies where trim and water wash are compared to sanitizing treatments for beef carcass decontamination, both water wash alone and trim alone have been reported to produce significantly smaller reductions than sanitizing agents such as hot water or organic acids (Gorman et al., 1995b; Hardin et al., 1995; Reagan et al., 1996). These studies indicate that neither trimming nor water washing of carcasses should be practiced for decontamination purposes, but only for carcass cleaning. Even a visually clean carcass may be contaminated with pathogenic bacteria at unsafe levels. In addition, both treatments are likely to spread pathogenic contamination to clean areas of the carcass. Therefore, trim, wash, or any other cleaning treatment should be followed by a subsequent sanitizing treatment.

D. Steam Vacuum

Application of hot water or steam combined with vacuuming has been included by the USDA-FSIS as an acceptable carcass cleaning process, which can be used instead of knife trimming to physically remove fecal contamination while sanitizing the contaminated area (Federal Register, 1996). A typical steam vacuum machine includes a vacuum wand with a hot water spray nozzle inside, which delivers water at 82–88°C. This internal nozzle is intended to sanitize the carcass surface as the vacuum removes the fecal material. Two external spray nozzles are positioned on the top and bottom of the wand to provide a continuous steam flow. This design would allow for steam continuously keeping the outside of the wand clean and sterile, while also helping in the carcass surface sanitation process. The steam vacuum machine has been designed to clean only small areas of contamination and is not applicable to clean and sanitize the entire carcass surface. Because of this limitation, applying steam vacuum treatment for carcass cleaning is permitted only when the extent of the contamination is no larger than 6.25 cm^2 (1 in^2) (Federal Register, 1996). Nevertheless, several investigators have studied the efficency of spot-cleaning vacuum machines for reducing bacterial contamination on carcass surfaces. Dorsa et al. (1996a) observed reductions of *E. coli* O157:H7 of 5.5 log_{10}/cm^2 on inoculated beef carcass short plates using a commercial steam vacuum system. In another study, Dorsa et al. (1996b) obtained reductions in APCs of 3.0 log_{10}/cm^2 on fecally contaminated beef carcass short plates. This reduction was not different from the reduction obtained by a hot water wash consisting of spraying water to a surface contact temperature of 72°C followed by spraying water at 30°C, or by a combination of both steam vacuum and double water wash. Phebus et al. (1997) observed no differences between a steam vacuum system and interventions such as steam pasteurization and knife trimming in reducing *E. coli* O157:H7, *S.* Typhimurium, or *L. monocytogenes* on freshly slaughtered beef surfaces. However, these authors indicate that steam vacuum is a spot-cleaning device, whereas steam pasteurization is a full-carcass treatment. These authors used small cuts for their comparison of treatments; therefore, the overall effect of each treatment on the microbial food safety and quality of beef carcasses may be different from the effect on pieces of beef tissue. During in-plant evaluations of two steam-vacuuming units, Kochevar et al. (1997) found lower APC and coliform counts on carcass surfaces that had been treated with the steam vacuum unit, when compared to surfaces that had been knife trimmed. Although no side-by-side comparison of both steam vacuum units was conducted, the reductions obtained by both units seem to be similar. According to the results of their study, steam vacuum reduced microbiological contamination and improved visual appearance of carcasses for which knife trimming would otherwise have been required by the inspector. In a review on procedures for beef carcass decontamination, Dorsa (1997) describes an in-plant testing of the steam vacuum designed to determine its efficacy under industrial use, indicating that a steam-vacuuming unit consistently reduced bacterial populations from contaminated areas of less than 2.5 cm. From these results, he concludes that a commercial steam vacuum system could outperform knife trimming for removing bacterial contamination on beef carcasses. Dorsa et al. (1997a) studied the effect of steam vacuum and hot water sprays on the activity of various pathogenic and spoiling organisms. They found a significant initial reduction in bacterial populations on inoculated beef short plates treated with either steam vacuum, hot water, or a combination of both treatments. However, no differences in growth rates were observed for most of these organisms during storage of the meat at 5°C for 21 days when compared to untreated controls. Inversely, *E. coli* O157:H7 and *Clostridium sporogenes* showed no ability to grow at 5°C. From this cold inhibition added to the initial reduction of these organisms by steam vacuum or hot water, these authors concluded that moist heat interventions might add a degree of safety to beef products when *Escherichia* spp. and vegetative *Clostridium* spp. are initially present on the carcass surface. Again, perhaps the excitement of producing large reductions in contamination on a specific area led these authors to conclude that steam vacuum systems are, in fact, an efficient decontamination method. The steam vacuum apparatus seems to be efficient at decontaminating small areas of visible contamination. If bacterial contamination is present, but the accompanying fecal contamination is not extensive enough to be detected by the inspector, then the decontamination treatment, whether steam vacuum or knife trimming, will not be applied. In contrast, other treatments applied to the entire carcass should be able to reduce contamination even when it is not detected. Another factor that remains to be studied is the time of treatment with the steam vacuum machine. Most of the previously described studies used a treatment of three even passes at a rate of ~1 s per pass. This may be too short a time to achieve real bacterial destruction. The individual effects of vacuum and hot water remain to be determined.

IV. SANITIZING TREATMENTS

A. Hot Water

Unlike regular water washes, sprays with water at temperatures above 74°C may be used as sanitizing interventions. The effect of applying hot water versus washing carcasses with warm water is clearly shown by Smith (1992), who produced an average reduction of different pathogens on fresh meat of 0.2 $\log_{10}/cm^2$ after washing with water at 40°C, while obtaining a reduction of 3.1 $\log_{10}/cm^2$ after applying water at 80°C. Different reports indicate that the application of hot water treatments can effectively reduce microbiological contamination on meat carcasses. In addition, several studies demonstrate that washing carcasses with water at temperatures greater than 80°C will not produce permanent discoloration of the carcass surface (Patterson, 1969; Smith and Graham, 1978; Barkate et al., 1993). In an early report on hot water decontamination, Patterson (1969) reported that beef carcasses treated with a steam and hot water spray (80–96°C) for 2 minutes contained significantly lower bacterial numbers than untreated carcasses. A volume of 18.9 L of water was sprayed on each carcass; however, the actual temperature at the carcass surface during the treatment was not provided in this study. Smith and Graham (1978) reported that pouring hot water (80°C) on beef and lamb samples for 10 s destroyed more than 99% of *E. coli* and *Salmonella* inoculated at levels of 6.5 $\log_{10}/cm^2$. Although the surface tissues of the beef and mutton were not permanently discolored by this treatment, the authors reported discoloration when water at 90°C for 120 s was used. In a laboratory evaluation of a hot water cabinet, Davey and Smith (1989) obtained *E. coli* reductions of

2.98 $\log_{10}/cm^2$ for artificially contaminated beef carcass sides treated with hot water that elevated the carcass surface temperature to 83.5°C for 20 s. Kelly et al. (1981) reported that lamb carcasses sprayed with hot water at temperatures above 80°C caused significant decreases (>1.0 $\log_{10}/cm^2$) in aerobic plate counts. In more recent studies where hot water treatments were evaluated, Dorsa et al. (1996b) and Gorman et al. (1995b) obtained reductions in coliform or *E. coli* counts of approximately 3.0 $\log_{10}/cm^2$.

Designing an appropriate hot water treatment is of paramount importance for obtaining effective reduction in bacterial populations. Heat losses in the spray from the nozzle to the carcass surface may produce an insufficient temperature increase at the carcass surface. Barkate et al. (1993) sprayed areas of hot beef carcass surfaces using 95°C water with the objective of raising the carcass surface temperature to 82°C for approximately 10 s. As a result of this treatment, the bacterial contamination on the carcass surface was reduced significantly. According to these authors, problems in applying hot water included obtaining a water spray that would adequately raise the surface temperature of the carcass to a bactericidal level. The volume of the spray and the size of the water droplets were found to have a profound effect on the temperature of the water after leaving the spray nozzle and before contacting the carcass surface. Using a type of nozzle that addressed the limitations reported by Barkate et al. (1993), Castillo et al. (1998a) sprayed hot water onto different hot carcass surface regions, obtaining average reductions of initial counts for *E. coli* O157:H7 and *S.* Typhimurium of 3.7 and 3.8 $\log_{10}/cm^2$, respectively. Corresponding reductions for APC and counts of coliforms and thermotolerant coliforms in their study were 2.9, 3.3, and 3.3 $\log_{10}/cm^2$, respectively. In this study, the hot water spray was combined with a previous water wash at 35°C, which significantly improved the visual quality of the carcass surfaces.

B. Organic Acid Sprays

The antimicrobial properties of lipophilic organic acids have been utilized by the food industry for food preservation, and the meat industry also has benefited from these antimicrobial compounds in decontaminating carcasses. The antibacterial mechanism of organic acids has not been completely described. It is generally accepted that the undissociated molecule of the organic acid or ester is responsible for antimicrobial activity (Baird Parker, 1980). Many weak acids, in their undissociated form, can penetrate the cell membrane and accumulate in the cytoplasm. If the intracellular pH is higher than the pKa of the acid, the protonated acid will then dissociate, releasing a proton and acidifying the cytoplasm of the microorganism (Booth, 1985). Other authors, however, indicate that the undissociated molecule is not the only toxic form of organic or other weak acids. Eklund (1983) developed a mathematical model to calculate the effect of pH on the minimum inhibitory concentration (MIC) of sorbic acid against different gram-positive and gram-negative bacteria and some yeasts. This author reported some microbial inhibition by undissociated molecules, but this was 10–600 times lower than the inhibition obtained by undissociated molecules. Ita and Hutkins (1991) studied the survival of *L. monocytogenes* exposed to different acids and the changes in the intracellular pH as affected by the type of acidulant. They observed that the internal pH decreased as the pH of the culture medium decreased. However, the internal pH was always higher than the external pH. The internal pH was maintained at levels close to 5.0 even when the cells were treated with lactic, citric, or hydrochloric acid at pH 3.5. In this study, acetic acid showed the highest inhibitory activity when compared to lactic, citric, or hydrochloric acid. The higher pKa of acetic acid (4.76) resulted in higher concentration of protonated form of this acid, which must have determined its higher toxicity. From these observations, these authors conclude that neither the external pH nor the internal pH was the determining factor causing bacterial destruction, but a specific effect of protonated acid molecules on the metabolic activity of *L. monocytogenes*. In contrast, Gill and Newton (1982) reported that, at least for lactic acid, the inhibitory effect in meat on different gram-negative psychrotrophs was mainly due to the decrease in pH and not due to the undissociated molecule. They also observed greater bacterial inhibition by acetic acid than by lactic acid. Another factor affecting the antimicrobial effect of organic acids may be the concentration of undissociated acid that can be reached at a certain pH by specific acids. McClure et al. (1989) reported stronger bacterial inhibition as the pH of the medium was lowered. Sorrells et al. (1989) reported the overall antimicrobial activity of different acids at equal pH to be acetic > lactic > citric > malic > hydrochloric, whereas the antimicrobial activity based on equal molar concentration was citric > malic > lactic > acetic > hydrochloric. They also reported that the inhibition is strongly influenced by the temperature. In their study, the greatest inhibition was obtained at 37°C compared to 25 or 10°C. In contrast, the greatest survival was observed at 10°C. Conner et al. (1990) also reported greater bacterial inhibition at 30°C than at 10°C; however, in their study, lactic acid resulted in more inhibition than acetic or citric acid.

These differences in antimicrobial activity of different acids, even at similar pH and when acids with similar pKa are compared (Farber et al., 1989; Conner et al., 1990), may indicate that different acids might have different mechanisms for bacterial toxicity. Moon (1983) determined the inhibition rates of different yeasts in the presence of acetic or propionic acids. If the antimicrobial mechanism of both acids is similar, then their inhibition rate at equal pH should be similar. However, her results show differences in the degree of inhibition by both acids. In addition, since the pKa of lactic acid is 3.86, then the concentration necessary to show the same inhibitory effect as acetic acid at equal pH values should be seven times the concentration of acetic acid. However, according to this author, for the same degree of inhibition, lactate concentration is roughly twice as much as the concentration of acetate. This may indicate that inhibition is not entirely due to the amount of acid and that the mechanism of inhibition may not be the same for all acids. Smulders et al. (1986) summarized the bactericidal and bacteriostatic effects of lactic acid as (a) pH effect, (b) extent of dissociation, and (c) specific effect related to the acid molecule. The above information could also be extended to all organic acids, although additional factors need to be considered. Most studies on bacterial inhibition by organic acids have included *L. monocytogenes* as the test organism, and other microorganisms might show differences in sensitivity to organic acids. Furthermore, the primary antimicrobial mechanism of organic acids might vary for different microorganisms.

Among organic acids, acetic and lactic acids have been more extensively used for carcass decontamination. The effectiveness of these two acids, as well as other carcass interventions, has been reviewed by Dickson and Anderson (1992). In an early study on pork carcass decontamination, Biemuller et al. (1973) reported that spraying the carcasses with acetic acid at pH 2.0 for 30 or 60 s resulted in large reductions of naturally occurring microflora and inoculated *S.* Enteritidis on pork carcasses at a slaughter plant. Ockerman et al. (1974) also reported significant reductions in numbers of naturally occurring microorganisms on lamb carcasses by acetic and lactic acids. Additionally, these authors reported a small residual effect of the acids on the microbial numbers on the carcasses during 12 days of refrigerated storage, which was affected by the concentration and type of the acid. Anderson et al. (1977) produced reductions of 2.55 log on counts of viable microorganisms on meat by spraying 3% acetic acid. These reductions were significantly greater than the reductions obtained by spraying hypochlorite solution (200–250 ppm). Quartey-Papafio et al. (1980) sprayed beef strips with acetic acid, formic acid, and a mixture of acetic, formic, and propionic acids. All treatments significantly reduced the bacterial counts with respect to untreated controls. However, the reductions were usually less than 1 log. Treatment with 1% formic acid produced the smallest reduction in viable counts (0.66 log), followed by a mixture of 0.5% acetic, 0.25% formic, and 0.25% propionic acids (0.76 log), 3% acetic acid (0.89 log), and 2% formic acid (1.56 log). After 7 days of storage at 7°C, bacterial counts on the strips increased between 0.92 and 2.24 log, whereas the counts on untreated controls increased 4.66 log. They also reported that 5% ascorbic acid sprayed to prevent browning of meat treated with formic acid also enhanced the antimicrobial effect. Dickson (1992) observed consistent reductions in populations of inoculated *S.* Typhimurium on lean and adipose beef tissue sprayed with 2% acetic acid, irrespective of the initial cell population. The acetic acid treatment had an immediate lethal effect on part of the population of *S.* Typhimurium, while another part was sublethally injured. In general, the reductions in counts of different pathogens on beef, as reported by different authors, vary between 2 and 4.3 log cycles after spraying 2% acetic acid (Dickson, 1991b; Dickson and Anderson, 1991; Hardin et al., 1995; Tinney et al., 1997). Variations in reductions obtained by different investigators may be due to differences in factors such as the temperature of the acid solution, which ranged from room temperature to 55°C in these reports. In the only report on organic acid treatments being ineffective to decontaminate beef tissue, Brackett et al. (1994) found acetic, citric, and lactic acid solutions at different concentrations to be unable to reduce *E. coli* O157:H7 on beef sirloin pieces regardless of the concentration and temperature of the acid solution. These authors explain the differences between their findings and those of most reports by the differences in their methods for application of the treatments. In several studies, the acid treatment is applied by dipping beef pieces in acid solutions, whereas these authors sprayed the acid solutions onto beef pieces. However, pH data in this study indicates that the inability of organic acid solutions to reduce counts of *E. coli* O157:H7 on beef was most likely due to their failure to reduce the beef surface pH to antimicrobial levels. In other similar papers, Anderson and Marshall (1990a) reduced the pH of beef dipped in lactic acid solutions from 5.6 (untreated meat) to 3.95, and Hardin et al. (1995) obtained surface pH values on beef carcass surfaces of 2.64–2.88 after spraying lactic acid and of 3.14–3.47 after spraying acetic acid. Even though *E. coli* O157:H7 has been reported to be resistant to low pH environments, recent studies indicate that lactic or acetic acid sprays, when applied at 55°C, can effectively reduce levels of *Salmonella* or *E. coli* O157:H7 (Hardin et al., 1995; Castillo et al., 1998b).

Studies on carcass decontamination using lactic acid indicate that this acid shows a strong antibacterial capacity. Hardin et al. (1995) reported that lactic acid was more effective than acetic in reducing *E. coli* O157:H7 and as effective as acetic acid in reducing *S.* Typhimurium on beef carcass surfaces. Woolthuis et al. (1984) found immersing porcine livers for 5 minutes in a 0.2% lactic acid solution to be significantly more effective than immersing in hot water (65°C) for 15 s in reducing total bacterial counts and lactic acid bacteria, whereas Enterobacteriaceae counts were reduced at the same rate after both treatments were applied. In another study, mean Enterobacteriaceae counts of 1.8 $\log_{10}$ CFU/cm^2 were reduced to undetectable levels on calf carcasses by spraying 1.25% L-lactic acid. This treatment also reduced the APCs by 0.8–1.3 $\log_{10}$ CFU/cm^2 depending on the carcass region treated (breast or perineum) (Woolthuis and Smulders, 1985). Prasai et al. (1991) reduced the APC of beef carcasses at two slaughter plants by ~2 $\log_{10}$/cm^2 by spraying 1% lactic acid at 55°C after dehiding and eviscerating. However, APCs of vacuum packaged loins cut from these carcasses were not different from those of loins cut from nontreated carcasses, indicating that the quality of subprimals depends, to a large extent, on the degree of recontamination after applying carcass intervention.

A continued antimicrobial effect has been observed during storage of meat after spraying lactic or acetic acid solutions. Kotula and Thelappurate (1994) reported that APCs and *E. coli* counts increased more rapidly on untreated steaks than on steaks treated with acetic or lactic acid. Similar results were reported by Dorsa et al. (1997b) for *E. coli* O157:H7, *L. innocua*, and *C. sporogenes* on beef carcass surface tissue. In a study on the effects of lactic acid on calf carcasses and primal veal cuts (Smulders and Woolthuis, 1985), the APCs of cold- or hot-boned primal cuts obtained from carcasses treated with lactic acid increased just slightly over a 14-day storage at 3°C, whereas the APCs of untreated controls increased by 2 $\log_{10}/cm^2$. Percent samples with detectable Enterobacteriaceae (positive samples) remained the same in cold- or hot-boned primal cuts. The percent positive samples of cold-boned cuts from treated carcasses gradually decreased during the storage, whereas positive samples of hot-boned cuts from treated carcasses were almost absent after 2.5 hours postmortem and during the 14-day storage period. In a study on the shelf life of strip loins sprayed with a hot (55°C) 50:50 mixture of 2% lactic and 2% acetic acids, counts of psychrotrophic, aerobic, anaerobic, and lactic acid bacteria were significantly lower than those of nonsprayed loins over 84 days of storage at −1°C (Goddard et al., 1996). Using the BioSys™ system for rapid detection of bacterial activity, Shelef et al. (1997) determined that treating meat with lactate or citrate extended the shelf life of ground beef samples by 2–4 days.

Some researchers have reported on the impact of organic sprays on the sensory characteristic of meat. Bell et al. (1986) did not observe significant ($p < 0.05$) discoloration of beef after dipping the meat in 1.2% v/v acetic acid for 1 minute. When the treatment was extended to 10 minutes, a concentration of 0.6% lactic acid was enough to produce significant discoloration when compared to untreated controls. A mixture of 0.6% acetic acid and 0.046% formic acid was not different from 1.2% acetic acid in its antibacterial activity and did not produce discoloration or noticeable flavors in the meat. Garcia-Zepeda et al. (1994a) compared the changes in psychrotrophic counts and acceptability scores of chuck subprimals obtained from carcasses treated with 3% lactic acid, 200 ppm chlorine, or water. Subprimals obtained from carcasses sprayed with lactic acid showed lower psychrotrophic counts but also lower acceptability scores than subprimals from carcasses treated with chlorine or water. In contrast, Goddard et al. (1996) found no differences in meat color, fat color, or odor in beef strip loins treated with a mixture of lactic and acetic acid when compared to untreated controls. Acuff et al. (1987) found no differences in bacterial counts and shelf life of steaks cut from beef loins sprayed with different acid solutions and steaks cut from untreated loins. Similar results were obtained by Dixon et al. (1987) in strip loins packed in either polyvinyl chloride or high-oxygen barrier films.

Different factors may impact the effectiveness of organic acid treatments for decontaminating carcasses. The ability of different organisms to acquire acid tolerance has been reviewed by Rowbury (1995). This author mentioned growth at high temperatures or acidic conditions as environmental factors inducing acid tolerance. Heat- or acid-induced acid tolerance in bacterial pathogens requires the cells remaining under specific environmental conditions long enough to synthesize the outer membrane proteins required for the acquired acid tolerance (Wang and Doyle, 1998). Since the length of all treatments in carcass decontamination is reduced to few seconds, the impact of this acquired acid tolerance on the effectiveness of different carcass interventions needs to be evaluated. According to Chung and Goepfert (1970), the minimum pH at which different serovars of *Salmonella* could initiate growth in culture media was affected by the type of acidulant, ranging from as low as 4.05 when citric acid was used to adjust the pH to 5.50 when propionic acid was used as acidulant. Other factors affecting the minimum pH for growth were the level of inoculum and the incubation temperature. Dickson and Kunduru (1995) addressed the acid adaptation in different strains of *Salmonella* as a potential factor influencing the effectiveness of organic acid rinses on beef. All strains of acid-adapted *Salmonella* were at least as sensitive to the organic acid rinses as the nonadapted parent strains. This study also addressed the effect of acid adaptation on heat resistance in *Salmonella*. Again, no effect of acid adaptation on the heat resistance was observed for any strain of *Salmonella*. These investigators concluded that acid adaptation of salmonellae in the environment, if it occurred, would not create a new hazard with the use of organic acid rinses on beef carcasses. Another factor that might affect the practical application and lethality evaluation of organic rinses against pathogens is sublethal injury. VanNetten et al. (1984) demonstrated that acid-stressed cells of various pathogenic Enterobacteriaceae can remain undetected during evaluations of organic acid rinses, so that pathogen reduction by these treatments may be overestimated.

Several studies have been conducted to determine the effect of temperature and concentration of the acid solution on the reduction of meatborne pathogens and spoilage bacteria on beef surfaces (Anderson et al., 1987b, 1988, 1992; Anderson and Marshall, 1989, 1990a,b; Greer and Dilts, 1992). In all these studies the temperature of the acid solution was found to have a profound effect on the magnitude of the reductions in bacterial counts. The concentration of the acid in the sanitizing solution has generally been determined to be of minimal importance for the effectiveness of organic rinses when it is above 1% (Anderson and Marshall, 1990a; Greer and Dilts, 1992).

Dickson (1990b) studied the effect of factors such as surface moisture and osmotic stress on the sanitizing of beef tissue surfaces. Osmotic stress by calcium chloride or sucrose solutions enhanced the effect of acetic acid on reducing the counts of *S.* Typhimurium and *L. monocytogenes* on inoculated fat tissue, whereas this effect was not observed on lean tissue. In addition, cells stressed with 20% sodium chloride before inoculating and spraying the beef tissue with acetic acid were more sensitive to the acetic acid spray than non-stressed cells. In this same study, samples allowed to dehydrate at 5°C for up to 6 hours before applying the acid treatment presented reductions of 1–3.5 log cycles compared to untreated controls.

Other factors such as bacterial attachment, type of meat surface (lean vs. fat), rigor state, inoculating menstruum, and level of inoculum have been studied in their effect on the efficacy of organic acid sprays for carcass decontamination. Dickson (1992) found no differences in populations of attached *S.* Typhimurium on pre- or postrigor, lean or fat beef tissue. Two separate reports (Dickson, 1992; Cutter et al., 1997) indicate that rigor state (prerigor and postrigor) does not affect the removal of pathogenic bacteria following treatment with 2% acetic acid. In general, the inoculating menstruum (buffer, tryptic soy broth, rumen fluid or feces) had no effect on reduction of *S.* Typhimurium or *E. coli* O157:H7 after spraying acetic acid onto the inoculated beef tissues. However, Dickson (1992) observed less reduction of *S.* Typhimurium populations on fat tissue after acid treatment when the inoculum menstruum was manure. The attachment rate and type of surface tissue do not seem to impact the effectiveness of treatments for carcass decontamination. Hardin et al. (1995) found no differences in reduction of *S.* Typhimurium and *E. coli* O157:H7 in the outside round, brisket, flank, and clod carcass surface regions treated with acetic or lactic acid immediately and 20–30 minutes after inoculation. The reductions of both pathogens were significantly smaller on the inside round region, which shows a surface mostly composed of lean muscle. However, as with the other carcass surface regions, no effect of bacterial attachment was observed on bacterial reductions for outside round. Similar results were reported by Castillo et al. (1998a) for beef carcass surfaces treated with hot water. Cutter and Siragusa (1994a) also reported greater reduction rates for gram-negative organisms on fat than on lean tissues. The inoculation menstruum has been shown to affect the attachment rate as well. Dickson and Macneil (1991) reported that *S.* Typhimurium and *L. monocytogenes* attached to beef carcass surfaces at a higher rate when the inoculum had been diluted in phosphate buffer compared to cow manure.

In addition to lactic and acetic acids, other organic acids have been tested for ability to reduce bacterial populations on beef. Podolak et al. (1996) found fumaric acid solutions to produce greater reduction in microbial populations and growth in ground beef. In another study, these same authors found fumaric acid at concentrations of 1% and 1.5% to be more effective than 1% lactic or acetic acids in reducing populations of *E. coli* O157:H7 and *L. monocytogenes* on beef lean muscle. In contrast, Anderson et al. (1992) found lactic acid to be more effective than acetic acid or a mixture of lactic, acetic, citric, and L-ascorbic acids in reducing gram-negative pathogens on lean meat. Garcia-Zepeda et al. (1994b) compared gluconic acid (1.5 and 3.0%), 1.5% lactic acid, and combinations of gluconic and lactic acid as fresh beef decontaminants. Beef samples were inoculated with *Lactobacillus fermentum*, treated with the different acid solutions, vacuum packaged, and stored at 1°C for up to 56 days. A mixture of 3% gluconic acid combined with 1.5% lactic acid produced the lowest psychrotrophic or lactobacilli counts. However, this mixture was detrimental to the color characteristics of the meat. A 50:50 mixture of 1.5% each gluconic and lactic acids appeared to be beneficial for the color characteristics of the meat at display, whereas lactic acid alone effectively reduced the bacterial counts but negatively affected the redness of the meat. Cutter and Siragusa (1994a) found no differences in log reductions of *E. coli* O157:H7 and *Pseudomonas fluorescens* on beef surfaces after spraying citric, acetic, or lactic acid at equal concentrations. Reynolds and Carpenter (1974) used a 60:40 w/w mixture of acetic and propionic acid, which is used as fungicide in cereal storage, for pork carcass decontamination. By modifying the molarities of these two acids, they reduced bacterial populations by 2.0 $\log_{10}$. In general, concentrations below 2.15 M of each acid produced little effect on the visual quality of the carcasses, while achieving bacterial reductions similar to those obtained by applying the acid mixture with higher molarities. More research on the usefulness of other organic acids with reported antibacterial activity (Richards et al., 1995) might be necessary for offering alternatives for carcass decontamination.

C. Other Chemicals

Several chemicals with antimicrobial activity have been tested for usefulness as carcass sanitizers. Not all compounds possessing antimicrobial properties will have practical application in carcass decontamination. Oh and Marshall (1993) almost completely inhibited *L. monocytogenes* in culture media by adding 5.0% ethanol (5.0% vol/vol) or glycerol monolaurate (monolaurin, 20 μg/mL). These authors discuss that, although monolaurin showed the greatest inhibitory effect, its applications as a food preservative may be limited due to its activity is antagonized by many food components.

A nonacid compound commonly used for carcass decontamination is trisodium phosphate (TSP), and treatments including TSP have been patented for poultry decontamination (Bender and Brotsky, 1992). By spraying TSP solutions at 55°C, Dickson et al. (1994) obtained reductions of *S.* Typhimurium, *L. monocytogenes*, and *E. coli* O157:H7 on lean beef muscle ranging from ~0.8 to 1.2 $\log_{10}/cm^2$, while on adipose tissue the reductions ranged from 1.2 to 2.5 $\log_{10}/cm^2$. Data from this study show that TSP concentration was not a significant factor in bacterial reduction by this chemical and that, in general, greater reductions were observed when the temperature of the TSP solution was increased from 25 to 55°C. The high pH of the TSP solution (~13) may be responsible for the bacteria reductions reported by Dickson et al. (1994). Antimicrobial effect of high-pH solutions on foodborne bacterial pathogens has been reported and is apparently due to membrane disruption of the cells and an increase in the water solubility of the DNA at high pH (Mendonca et al., 1994).

The effectiveness of bacteriocins such as nisin in beef decontamination has also been tested. Cutter and Siragusa (1994b) sprayed nisin solution (5000 activity units/mL) onto beef carcass tissue inoculated with different gram-positive organisms. Reductions in bacterial counts produced by this treatment ranged from 1.79 to 3.54 $\log_{10}/cm^2$, with variable effect of the tissue surface fat characteristics. Mixing nisin with 50 mM ethylenediaminetetraacetic acid (EDTA) significantly enhanced the reduction of *S.* Typhimurium and *E. coli* O157:H7 in different buffers (Cutter and Siragusa, 1995a). When nisin mixed with EDTA and other chelators was applied to inoculated beef tissue (Cutter and Siragusa, 1995b), only nisin + lactate and nisin + EDTA produced reductions significantly greater than those produced by other mixtures. However, the actual log reductions were so small that this statistical significance may not comprise any biological significance. Cutter and Siragusa (1996) enhanced the inhibition of *Brochothrix thermosphacta* on beef surfaces by immobilizing nisin in calcium alginate gels. Using this approach, not only produced greater bacterial reductions on the beef immediately after treatment, but the counts of this spoilage organism remained lower than those of beef treated with nisin during refrigerated storage for 7 d. After storage, the numbers of *B. thermosphacta* on the beef surfaces were 7.1 $\log_{10}$ CFU/cm^2 for the controls, 6.45 CFU/cm^2 for beef treated with calcium alginate only, 5.26 CFU/cm^2 for beef treated with nisin only, and 2.37 CFU/cm^2 for beef treated with alginate-immobilized nisin.

The antibacterial mechanism of nisin seems to involve inhibition of murein synthesis, as well as disintegration of the cytoplasmic membrane resulting on inhibition of synthesis of DNA, RNA, and proteins (Henning et al., 1986). Another toxic effect must be involved, since it has been shown that this bacteriocin can produce a deleterious effect on microorganisms attached to beef carcass surfaces as soon as within 30 seconds (Cutter and Siragusa, 1996). Other bacteriocins have also been used for controlling pathogens on meat (Goff et al., 1996). An advantage of using bacteriocins such as nisin, which is already used as a food preservative, would be that, since their mechanism of action involves cell structural damage, factors such as pH or temperature would not need to be controlled as when applying other compounds such as organic acids. On the other hand, using nisin would increase considerably the cost of this treatment, when compared with inexpensive organic acid or hot water sprays, which produce similar or greater bacterial reductions.

Several other chemicals have been proposed as carcass sanitizers. Cutter and Dorsa (1995) found chlorine dioxide at concentrations of up to 20 ppm to be no more effective than regular water in reducing bacteria of fecal origin on beef carcass tissue. Likewise, Anderson et al. (1977) found no differences in bacterial reductions on meat after spraying tap water or water added with 200 ppm sodium hypochlorite. In contrast, Kotula et al. (1974) found lower bacterial counts on beef carcasses treated with 200 ppm sodium hypochlorite compared to untreated carcasses. In addition, they reported some continued effect of chlorine during storage of the carcasses. Since treatment of carcasses with nonchlorinated water was not included in their study, the reported effect of chlorinated water on bacterial counts may have involved both washing effect and an extended antimicrobial effect. After inoculating different pathogenic and nonpathogenic organisms onto lean and adipose beef tissue, Dickson (1988) applied washes with phosphate buffer, ethanol, sodium chloride, sodium hydroxide, and potassium hydroxide. Phosphate buffer, ethanol, and sodium chloride produced reductions of less than 1 log cycle, whereas sodium and potassium hydroxide effectively reduced the populations of the inoculated bacteria by as much as 4 log cycles. Again, adipose tissue showed more reduction in bacterial counts than lean tissue. As previously discussed for TSP, high-pH solutions have been shown to possess strong antimicrobial activity (Mendonca et al., 1994). Cutter et al. (1996) have found copper sulfate or *N*-alkyldimethylbenzylammonium chloride to be ineffective in decontaminating beef. Beef carcass tissue treated with a commercial compound including copper sulfate as the active ingredient showed no more bacterial reduction than beef tissue treated with water, unless the concentration of the compound was as high as 160 ppm, with a pH as low as 2.03. A commercial compound including 40% *N*-alkyldimethylbenzylammonium chloride in 60% stabilized urea failed to produce bacterial reductions in bacterial populations on beef tissue distinguishable from the reductions produced by a simple water wash. The antimicrobial activity of these compounds may have been affected by the organic matter contents of the tissue being treated. Mullerat et al. (1995) compared the antimicrobial capacity of a commercial oxyhalogen disinfectant including sodium chlorite as the active ingredient against different pathogenic bacteria commonly associated with muscle foods. This compound showed a strong biocidal activity against *Salmonella* spp., *E. coli* O157:H7, *L. monocytogenes, S. aureus*, and *P. aeruginosa*. However, this activity was affected by the presence of organic matter in the system, which could make this compound useless for beef carcass decontamination. The biocidal activity was increased with the concentration of the disinfectant and by the addition of EDTA. The more added EDTA, the greater the bacterial inactivation. The reduction of *E. coli* O157:H7 and *S.* Typhimurium on beef carcass surfaces by applying acidified sodium chlorite (ASC) solutions was recently studied by Castillo et al. (1999). When phosphoric acid was used to acidify sodium chlorite, the resulting ASC solution reduced populations of both pathogens by 3.8–3.9 log cycles, while when ASC solutions were prepared by acidifying with citric acid, the reductions obtained ranged from 4.5 to 4.6 log cycles. ASC has recently been approved as a direct food additive for use in decontamination of poultry and red meats.

D. Steam Pasteurization

Developing steam treatments of beef or sheep carcasses has been attempted by various investigators with no clear success (Anderson et al., 1979; Dorsa et al.,

1996a). Dorsa (1997) has reviewed the studies that led to the development of a steam pasteurization treatment. Recently, a steam pasteurization process has been developed and acquired rapid popularity among meat processors (Wilson and Leising, 1994). Phebus et al. (1997) assembled an experimental pasteurization chamber consisting of an insulated stainless steel cabinet with two internal compartments separated by a sliding barrier. One compartment contained the steam reservoir, and the other contained the carcass tissue to be treated. An inlet valve at the bottom of the compartment allowed for filling with steam while the samples were placed in the other compartment. Then the sliding barrier would be removed for a specified time to allow for steam contact with the sample. During this process, the authors found that the treatment occurred at pressures above atmospheric pressure. After the treatments, the samples were immediately placed in a chamber where a cold water wash decreased the temperature to avoid carcass damage. Using this chamber, Phebus et al. (1997) reduced *E. coli* O157:H7, *S.* Typhimurium, and *L. monocytogenes* by 3.4–3.7 log cycles on surfaces of freshly slaughtered beef. However, steam pasteurization alone showed no greater reductions than other treatments such as knife trimming or steam vacuuming. Nutsch et al. (1997) conducted commercial evaluations of the steam pasteurization process in a beef-processing plant. The carcasses were subjected to a preliminary water wash and then passed through potent air blowers to eliminate excessive humidity that would favor steam condensation. After air elimination of the excessive carcass surface humidity, the carcasses were passed through a steam chamber that was supplied by steam from a tank and then passed to another section of the cabinet where cold water was applied. Applying this treatment, they reduced APCs on carcasses from 2.12–2.19 $\log_{10}$ CFU/cm^2 to 0.56–0.84 $\log_{10}$ CFU/cm^2. No differences were observed between 6 or 8 s of treatment. Counts of *E. coli* were also reduced from original counts of 0.60 to 1.53 $\log_{10}$ CFU/cm^2 to undetectable levels after 6 or 8 s of steam treatment.

The most important advantage of using steam for carcass decontamination is the ability to achieve great bacterial reductions without using any chemical that may have corrosive action against the equipment or negatively affect the carcass visual quality. On the other hand, the high cost of the equipment might make this intervention unnecessary when the same or greater reductions have been reported using other treatments such as hot water or lactic acid sprays.

E. Other Interventions

Low-voltage pulsed electricity has recently been found to have antimicrobial activity and has been proposed as a potential intervention to destroy pathogens on carcasses (Li et al., 1995; Slavik et al., 1995). Bawcom et al. (1995) and Tinney et al. (1997) reported some reduction in counts of coliforms or meatborne pathogens on beef surfaces using pulsed electricity. However, these reductions were smaller than 1 log cycle. In addition, acetic acid treatment of beef produced significantly greater reductions than pulsed-power electricity (Tinney et al., 1997). Nevertheless, these authors are optimistic about the potential benefits of pulsed-power electricity for the shelf life and perhaps the safety of meat.

Chemical dehairing is another treatment that may have potential benefits for the elimination of pathogens on meat carcasses. This process was developed by Bowling and Clayton (1992) based on the hair-removal process used in the tanning of leather. Recently, Schnell et al. (1995) adapted the chemical dehairing process to the slaughter operations of a commercial plant. These authors found no differences in APCs or coliform counts in samples from carcasses of dehaired cattle and those of conventionally slaughtered cattle. They concluded that dehairing enhanced the visual cleanliness of the carcasses but was not able to effectively reduce bacterial counts. This report did not include any determination of the extent to which bacterial counts were reduced on the hides of the slaughtered cattle by the dehairing process, therefore the impact of this treatment on the prevention of fecal contamination from the hides to the carcass surfaces could not be determined. After adapting the chemical dehairing process to laboratory conditions, Castillo et al. (1998c) found reductions in *E. coli* O157:H7, *S.* Typhimurium, *E. coli*, coliforms, and APCs on artificially contaminated bovine skin ranging from 3.4 to $>4.8 \log_{10}/cm^2$. This may indicate that dehairing might be applied successfully to control a major source of contamination during beef slaughter operations.

Other agents such as ozonated water or hydrogen peroxide have also been evaluated with variable results when compared to treatments such as trimming or washing (Gorman et al., 1995b; Reagan et al., 1996). Meat irradiation has recently been approved for decontamination of fresh frozen meat, and it has been previously approved for treating poultry and pork. Many reports indicate that this process is quite effective to eliminate pathogenic contamination from ground beef and poultry carcasses (Clavero et al., 1994; Lee et al.; 1996, Tarté et al., 1996). However, this seems to be a process that can be best applied to fabricated meat products instead of carcasses. The fact that large amounts of already packed and frozen product can be irradiated (so that meat safety is better ensured) would make carcass irradiation for decontamination purposes unnecessary.

F. Combined Interventions

Combining two or more different treatments may increase the decontaminating capability of carcass interventions. Gorman et al. (1995b) combined the application of different antimicrobial compounds with water washes at different temperatures. When a water wash was applied at temperatures of 16 or 35°C, application of sanitizers such as TSP, hydrogen peroxide, or ozonated water decreased significantly the total bacterial numbers as well as the numbers of inoculated *E. coli*. However, when hot water (74°C) was used for the preliminary wash, no extra reduction was observed after applying the other sanitizers. Hardin et al. (1995) and Castillo et al. (1998b) also reported that a 35°C water wash followed by a sanitizing treatment such as organic acid or hot water sprays produced significantly greater reductions than those produced by water wash alone. Phebus et al. (1997) combined treatments such as trimming, water washing, steam vacuuming, and spraying 2% lactic acid with steam pasteurization process. Reductions in counts of different pathogens by a 35°C water wash followed by steam pasteurization were not significantly different from those obtained by combined treatments including successive application of trim, water wash, lactic acid, and steam pasteurization, or steam vacuum, water wash, lactic acid, and steam pasteurization. This would indicate that single treatments, when having enough biocidal power, can be effectively used instead of combined sanitizing treatments. However, Castillo et al. (1998b) recently reported that a 35°C water wash followed by hot water spray (95°C at the source, 82°C on the carcass surface), then followed by 2% lactic acid spray at 55°C, reduced significantly the numbers of *E. coli* O157:H7 or *S.* Typhimurium when compared to water wash followed by single hot water treatment and, in some cases, to water wash followed by lactic acid spray. In addition, this combined intervention reduced counts of these pathogens, as well as *E. coli* that had been spread over the carcass surface by the water wash, to undetectable levels in 100% of the samples tested. These authors conclude that multiple interventions can more likely cause greater reductions than single treatments and will also provide a more "fail-safe" pathogen reduction strategy.

V. POTENTIALLY NEGATIVE ASPECTS OF CARCASS INTERVENTIONS

Recent concerns have been expressed regarding the potential for different carcass interventions to reduce not only pathogens, but also beneficial or antagonistic flora on carcass surfaces (Jay, 1995). This generalized microbial reduction would impact the numbers of organisms that, under traditional conditions, have acted on refrigerated meats as a natural system against the growth of bacterial pathogens, as reported by Goepfert and Kim (1975). Jay (1995) proposed that a carcass decontamination treatment be followed by spraying one or various gram-positive organisms that have been shown to have antagonistic effect against foodborne pathogens, such as lactic acid bacteria or *Brochothrix* spp. This author also indicated that a postdecontamination inoculation step might increase the shelf life of the carcasses. Nevertheless, the potential effect of this treatment on either the quality or the safety of the meat by including this step in the slaughter process is not known. In addition, this step may not be necessary. Goddard et al. (1996) reported that the counts of lactic acid bacteria, psychrotrophs, aerobes, and anaerobes in acid-sprayed beef strip loins increased significantly during storage at −1°C. In a similar study, Anderson et al. (1988) found significant increases in counts of *B. thermosphacta* and lactobacilli on acetic acid–treated lamb carcasses from 0.2–0.5 $\log_{10}/cm^2$ to 1.6–4.8 $\log_{10}/cm^2$ after 8 weeks of storage at 0°C and to 2.4–6.2 $\log_{10}/cm^2$ after 12 weeks of storage. In contrast, gram-negative and coliform bacteria did not increase their counts from 8 to 12 weeks of storage. This shows that, even though all flora is usually reduced significantly by decontamination treatments, the carcass is not sterilized, and the surviving gram-positive flora has the ability to compete with surviving gram-negative bacteria, which may include some foodborne pathogens.

Another concern associated with carcass decontamination, especially regarding organic acid sprays, is the reduction of antagonistic flora while selecting acid-tolerant pathogens (Anonymous, 1997). Although researchers have found that sublethal heat, acid, or freeze injury does favor a shock response that confers heat or acid resistance to the bacterial cells (Jackson et al., 1996; Garren et al., 1998; Wang and Doyle, 1998), several reports indicate that organic acid sprays, when applied at 55°C, can effectively reduce bacterial pathogens (Smulders and Woolthuis, 1985; Hardin et al., 1995), even if they have been acid adapted (Dickson and Kunduru, 1995). If the antimicrobial mechanism for all organic acids is only the extracellular pH damaging the cells, then acid adaptation of sublethally injured pathogenic cells may be a limitation for the effectiveness of organic acid sprays. If, on the other hand, other mechanisms are involved in organic acid reduction of pathogens, then acid adaptation may not be a problem. As discussed above, the most plausible explanation for the antimicrobial activity of organic acids is that if the undissociated molecule has the ability to penetrate into the cell, then dissociation will occur in the cytoplasm, liberating a proton causing an internal pH decline (Booth, 1985). Since the degree of inhibition by lactic and acetic acid has been shown to be temperature-dependent (Anderson et al., 1987b, 1988, 1992; Anderson and Marshall, 1989, 1990a,b; Conner et al., 1990; Greer and Dilts, 1992), it is possible that at high temperatures (55°C) the acid penetration of the cells is accelerated, causing immediate damage not related to any shock response leading to acid adaptation.

Production of a pathogen-free product cannot be guaranteed under current production conditions. However, the incorporation of a decontamination step during slaughter-dressing procedures can effect improvement of the microbial quality and safety of meat products. Multiple decontamination steps have been shown to be even more effective. Advantages associated with whole carcass treatments such as organic acid rinses or hot water sprays include treatment of the entire carcass surface with the ability to reduce bacterial contamination that may be present but not associated with visible fecal contamination. The ability of wash/

sanitizer treatments to reduce pathogens on the entire carcass surface, compared to trimming isolated areas of the carcass, enhances their value as potential critical control points in a HACCP system.

REFERENCES

GR Acuff, C Vanderzant, JW Savell, DK Jones, DB Griffin, JG Ehlers. Effect of acid decontamination of beef subprimal cuts on the microbiological and sensory characteristics of steaks. Meat Sci 19:217–226, 1987.

ME Anderson, RT Marshall. Interaction of concentration and temperature of acetic acid solution on reduction of various species of microorganisms on beef surfaces. J Food Prot 52:312–315, 1989.

ME Anderson, RT Marshall. Reducing microbial populations on beef tissues: concentration and temperature of lactic acid. J Food Saf 10:181–190, 1990a.

ME Anderson, RT Marshall. Reducing microbial populations on beef tissues: concentration and temperature of an acid mixture. J Food Sci 55:903–905, 1990b.

ME Anderson, RT Marshall, HD Naumann, WC Stringer. Physical factors that affect removal of yeasts from meat surfaces with water sprays. J Food Sci 40:1232–1235, 1975.

ME Anderson, RT Marshall, WC Stringer, HD Naumann. Combined and individual effects of washing and sanitizing on bacterial counts of meat—a model system. J Food Prot 40:688–670, 1977.

ME Anderson, RT Marshall, WC Stringer, HD Naumann. Microbial growth on plate beef during extended storage after washing and sanitizing. J Food Prot 42:389–392, 1979.

ME Anderson, RT Marshall, WC Stringer, HD Naumann. In-plant evaluation of a prototype carcass cleaning and sanitizing unit. J Food Prot 43:568–570, 1980.

ME Anderson, RT Marshall, WC Stringer, HD Naumann. Evaluation of a prototype beef carcass washer in a commercial plant. J Food Prot 44:35–38, 1981.

ME Anderson, RT Marshall, WC Stringer, HD Naumann. Evaluation of a unique chamber for a beef carcass cleaning unit. J Food Prot 45:19–22, 1982a.

ME Anderson, RT Marshall, WC Stringer, HD Naumann. Inexpensive device for CIP cleaning of the inside of a beef carcass spray chamber. J Food Prot 45:643–645, 1982b.

ME Anderson, RT Marshall, WC Stringer, HD Naumann. Effect of nozzle configuration and size on water absorbed by meat during cleaning. J Food Prot 45:276–278, 1982c.

ME Anderson, HE Huff, HD Naumann, RT Marshall, J Damare, R Johnston, M Pratt. Evaluation of swab and tissue excision methods for recovering microorganisms from washed and sanitized beef carcasses. J Food Prot 50:741–743, 1987a.

ME Anderson, HE Huff, HD Naumann, RT Marshall, JM Damare, M Pratt, R Johnston. Evaluation of an automated beef carcass washing and sanitizing system under production conditions. J Food Prot 50:562–566, 1987b.

ME Anderson, HE Huff, HD Naumann, RT Marshall. Count of six types of bacteria on lamb carcasses dipped or sprayed with acetic acid at 25° or 55°C and stored vacuum packaged at 0°C. J Food Prot 51:874–877, 1988.

ME Anderson, RT Marshall, JS Dickson. Efficacies of acetic, lactic, and two mixed acids in reducing numbers of bacteria on surfaces of lean meat. J Food Saf 12:139–147, 1992.

Anonymous. *E. coli* O157:H7 seen 'rewriting rulebooks' of microbiologists. Food Chem News. July 7:4–5, 1997.

AC Baird Parker. Organic acids. In: Microbial Ecology of Foods. Vol. I. International Commission on Microbiological Specifications for Foods. New York: Academic Press, 1980, pp. 126–135.

ML Barkate, GR Acuff, LM Lucia, DS Hale. Hot water decontamination of beef carcasses for reduction of initial bacterial numbers. Meat Sci 35:397–401, 1993.

DL Bawcom, LD Thompson, MF Miller, CB Ramsey. Reduction of microorganisms on beef surfaces utilizing electricity. J Food Prot 58:35–38, 1995.

MF Bell, RT Marshall, ME Anderson. Microbiological and sensory tests of beef treated with acetic and formic acids. J Food Prot 49:207–210, 1986.

FG Bender, E Brotsky. Process for treating poultry carcass to control salmonellae growth. U.S. patent 5,143,739 (1992).

GW Biemuller, JA Carpenter, AE Reynolds. Reduction of bacteria on pork carcasses. J Food Sci 38:261–263, 1973.

IR Booth. Regulation of cytoplasmic pH in bacteria. Microbiol Rev 49:359–378, 1985.

S Bouttier, C Linxe, C Ntsama, G Morgant, MN Bellon-Fontaine, J Fourniat. Attachment of *Salmonella choleraesuis* to beef muscle and adipose tissues. J Food Prot 60:16–22, 1997.

RA Bowling, RP Clayton. Method for dehairing animals. U.S. patent 5,149,295, Monfort Inc. Greeley, CO (1992).

RE Brackett, Y-Y Hao, MP Doyle. Ineffectiveness of hot acid sprays to decontaminate *Escherichia coli* O157:H7 on beef. J Food Prot 57:198–203, 1994.

JL Butler, JC Stewart, C Vanderzant ZL, Carpenter, GC Smith. Attachment of microorganisms to pork skin and surfaces of beef and lamb carcasses. J Food Prot 42:401–406, 1979.

L Cabedo, JH Sofos, GC Smith. Removal of bacteria from beef tissue by spray washing after different times of exposure to fecal material. J Food Prot 59:1284–1287, 1996.

L Cabedo, JH Sofos, GR Schmidt, GC Smith. Attachment of *Escherichia coli* O157:H7 and other bacterial cells grown in two media to beef adipose and muscle tissues. J Food Prot 60:102–106, 1997.

A Castillo, LM Lucia, KJ Goodson, JW Savell, GR Acuff. Use of hot water for beef carcass decontamination. J Food Prot 61:19–25, 1998a.

A Castillo, LM Lucia, KJ Goodson, JW Savell, GR Acuff. Comparison of water wash, trimming, and combined hot water and lactic acid treatments for reducing bacteria of fecal origin on beef carcasses. J Food Prot 61:823–828, 1998b.

A Castillo, JS Dickson, RP Clayton, LM Lucia, GR Acuff. Chemical dehairing of bovine skin to reduce pathogenic organisms and bacteria of fecal origin. J Food Prot 61: 623–625, 1998c.

A Castillo, LM Lucia, GK Kemp, GR Acuff. Reduction of *Escherichia coli* O157:H7 and

Salmonella Typhimurium on beef carcass surfaces using acidified sodium chlorite. J Food Prot 62:580–584, 1999.

KC Chung, JM Goepfert. Growth of *Salmonella* at low pH. J Food Sci 35:326–328, 1970.

KT Chung, JS Dickson, JD Crouse. Attachment and proliferation of bacteria on meat. J Food Prot 52:173–177, 1989.

MRS Clavero, JD Monk, LR Beuchat, MP Doyle, RE Brackett. Inactivation of *Escherichia coli* O157:H7, salmonellae, and *Campylobacter jejuni* in raw ground beef by gamma irradiation. Appl Environ Microbiol 60:2069–2075, 1994.

DE Conner, VN Scott, DT Bernard. Growth, inhibition, and survival of *Listeria monocytogenes* as affected by acidic conditions. J Food Prot 53:652–655, 1990.

JD Crouse, ME Anderson, HD Naumann. Microbial decontamination and weight of carcass beef as affected by automated washing pressure and length of time of spray. J Food Prot 51:471–474, 1988.

CN Cutter, WJ Dorsa. Chlorine dioxide spray washes for reducing fecal contamination on beef. J Food Prot 58:1294–1296, 1995.

CN Cutter, GR Siragusa. Efficacy of organic acids against *Escherichia coli* O157:H7 attached to beef carcass tissue using a pilot scale model carcass washer. J Food Prot 57:97–103, 1994a.

CN Cutter, GR Siragusa. Decontamination of beef carcass tissue with nisin using a pilot scale model carcass washer. Food Microbiol 11:481–489, 1994b.

CN Cutter, GR Siragusa. Population reductions of gran-negative pathogens following treatments with nisin and chelators under various conditions. J Food Prot 58:977–983, 1995a.

CN Cutter, GR Siragusa. Treatments with nisin and chelators to reduce *Salmonella* and *Escherichia coli* on beef. J Food Prot 58:1028–1030, 1995b.

CN Cutter, GR Siragusa. Reduction of *Brochothrix thermosphacta* on beef surfaces following immobilization of nisin in calcium alginate gels. Lett Appl Microbiol 23: 9–12, 1996.

CN Cutter, WJ Dorsa, GR Siragusa. Application of Carnatrol™ and Timsen™ to decontaminate beef. J Food Prot 59:1339–1342, 1996.

CN Cutter, WJ Dorsa, GR Siragusa. Parameters affecting the efficacy of spray washes against *Escherichia coli* O157:H7 and fecal contamination on beef. J Food Prot 60: 614–618, 1997.

KR Davey, MG Smith. A laboratory evaluation of a novel hot water cabinet for the decontamination of sides of beef. Int J Food Sci Tech 24:305–316, 1989.

AG Dezuniga, ME Anderson, RT Marshall, EL Iannotti. A model for studying the penetration of microorganisms into meat. J Food Prot 54:256–258, 1991.

JS Dickson. Reduction of bacteria attached to meat surfaces by washing with selected compounds. J Food Prot 51:869–873, 1988.

JS Dickson. Transfer of *Listeria monocytogenes* and *Salmonella typhimurium* between beef tissue surfaces. J Food Prot 53:51–55, 1990a.

JS Dickson. Surface moisture and osmotic stress as factors that affect the sanitizing of beef tissue surfaces. J Food Prot 53:674–679, 1990b.

JS Dickson. Attachment of *Salmonella typhimurium* and *Listeria monocytogenes* to beef tissue: effects of inoculum level, growth temperature and bacterial culture age. Food Microbiol 8:143–151, 1991a.

JS Dickson. Control of *Salmonella typhimurium, Listeria monocytogenes*, and *Escherichia coli* O157:H7 on beef in a model spray chilling system. J Food Sci 56:191–193, 1991b.

JS Dickson. Acetic acid action on beef tissue surfaces contaminated with *Salmonella typhimurium*. J Food Sci 57:297–301, 1992.

JS Dickson, ME Anderson. Control of *Salmonella* on beef tissue surfaces in a model system by pre- and post-evisceration washing and sanitizing, with or without spray chilling. J Food Prot 54:514–518, 1991.

JS Dickson, ME Anderson. Microbiological decontamination of food animal carcasses by washing and sanitizing systems. A review. J Food Prot 55:133–140, 1992.

JS Dickson, M Koohmaraie. Cell surface charge characteristics and their relationship to bacterial attachment to meat surfaces. Appl Environ Microbiol 55:832–836, 1989.

JS Dickson, MR Kunduru. Resistance of acid-adapted salmonellae to organic acid rinses on beef. J Food Prot 58:973–976, 1995.

JS Dickson, MD Macneil. Contamination of beef tissue surfaces by cattle manure inoculated with *Salmonella typhimurium* and *Listeria monocytogenes*. J Food Prot 54: 102–104, 1991.

JS Dickson, CJN Cutter, GR Siragusa. Antimicrobial effect of trisodium phosphate against bacteria attached to beef tissue. J Food Prot 57:952–955, 1994.

ZR Dixon, C Vanderzant, GR Acuff, JW Savell, DK Jones. Effect of acid treatment of beef strip loin steaks on microbiological and sensory characteristics. Int J Food Microbiol 5:181–186, 1987.

WJ Dorsa. New and established carcass decontamination procedures commonly used in the beef-processing industry. J Food Prot 60:1146–1151, 1997.

WJ Dorsa, CN Cutter, GR Siragusa. Effectiveness of a steam-vacuum sanitizer for reducing *Escherichia coli* O157:H7 inoculated to beef carcass surface tissue. Lett Appl Microbiol 23:61–63, 1996a.

WJ Dorsa, CN Cutter, GR Siragusa, M Koohmaraie. Microbial decontamination of beef and sheep carcasses by steam, hot water spray washes, and a steam-vacuum sanitizer. J Food Prot 59:127–135, 1996b.

WJ Dorsa, CN Cutter, GR Siragusa. Effects of steam-vacuuming and hot water spray wash on the microflora of refrigerated beef carcass surface tissue inoculated with *Escherichia coli* O157:H7, *Listeria innocua*, and *Clostridium sporogenes*. J Food Prot 60:114–119, 1997a.

WJ Dorsa, CN Cutter, GR Siragusa. Effects of acetic acid, lactic acid and trisodium phosphate on the microflora of refrigerated beef carcass surface tissue inoculated with *Escherichia coli* O157:H7, *Listeria innocua*, and *Clostridium sporogenes*. J Food Prot 60:619–624, 1997b.

T Eklund. The antimicrobial effect of dissociated and undissociated sorbic acid at different pH levels. J Appl Bact 54:383–389, 1983.

JM Farber, GW Sanders, S Dunfield, R Prescott. The effect of various acidulants on the growth of *Listeria monocytogenes*. Lett Appl Microbiol 9:181–183, 1989.

Federal register. Notice of policy change; achieving the zero tolerance performance standard for beef carcasses by knife trimming and vacuuming with hot water or steam; use of acceptable carcass interventions for reducing carcass contamination without prior agency approval. Food Safety and Inspection Service, USDA. Fed Reg 61: 15024–15027, 1996.

OM Fratamico, FJ Schultz, RC Benedict, RL Buchanan, PH Cooke. Factors influencing attachment of *Escherichia coli* O157:H7 to beef tissues and removal using selected sanitizing rinses. J Food Prot 59:453–459, 1996.

CM Garcia Zepeda, CL Kastner, PB Kenney, RE Campbell, JR Schwenke. Aroma profile of subprimals from beef carcasses decontaminated with chlorine and lactic acid. J Food Prot 57:674–678, 1994a.

CM García Zepeda, CL Kastner, BL Willard, RK Phebus, JR Schwenke, BA Fijal, RK Prasai. Gluconic acid as a fresh beef decontaminant. J Food Prot 57:956–962, 1994b.

DM Garren, MA Harrison, SM Russell. Retention of acid tolerance and acid shock responses of *Escherichia coli* O157:H7 and non-O157:H7 isolates. J Food Prot 60: 1478–1482, 1998.

CO Gill, KG Newton. Effect of lactic acid concentration on growth on meat of gram-negative psychrotrophs from a meatworks. Appl Environ Microbiol 43:284–288, 1982.

CO Gill, M Badoni, T Jones. Hygienic effects of trimming and washing operations in a beef carcass dressing process. J Food Prot 59:666–669, 1996.

BL Goddard, WB Mikel, DE Conner, WR Jones. Use of organic acids to improve the chemical, physical, and microbial attributes of beef strip loins stored at −1°C for 112 days. J Food Prot 59:849–953, 1996.

JM Goepfert, HU Kim. Behavior of selected food-borne pathogens in raw ground beef. J Milk Food Technol 38:449–452, 1975.

JH Goff, AK Bhunia, MG Johnson. Complete inhibition of *Listeria monocytogenes* on refrigerated chicken meat with pediocin AcH bound to heat-killed *Pediococcus acidilactici* cells. J Food Prot 59:1187–1192, 1996.

BM Gorman, JB Morgan, JN Sofos, GC Smith. Microbiological and visual effects of trimming and/or spray washing for removal of fecal material from beef. J Food Prot 58:984–989, 1995a.

BM Gorman, JN Sofos, JB Morgan, GR Schmidt, GC Smith. Evaluation of hand-trimming, various sanitizing agents, and hot water spray-washing as decontamination interventions for beef brisket adipose tissue. J Food Prot 58:899–907, 1995b.

GG Greer, BD Dilts. Factors affecting the susceptibility of meatborne pathogens and spoilage bacteria to organic acids. Food Res Int 25:355–364, 1992.

MD Hardin. Comparison of methods for contamination removal from beef carcass surfaces. Doctoral dissertation, Texas A&M University, College Station, Texas, 1995.

MD Hardin, GR Acuff, LM Lucia, JS Oman, JW Savell. Comparison of methods for contamination removal from beef carcass surfaces. J Food Prot 58:368–374, 1995.

S Henning, R Metz, WP Hammes. Studies on the mode of action of nisin. Int J Food Microbiol 3:121–134, 1986.

PS Ita, RW Hutkins. Intracellular pH and survival of *Listeria monocytogenes* Scott A in tryptic soy broth containing acetic, lactic, citric, and hydrochloric acids. J Food Prot 54:15–19, 1991.

TJ Jackson, MD Hardin, GR Acuff. Heat resistance of *Escherichia coli* O157:H7 in a nutrient medium and in ground beef patties as influenced by storage and holding temperatures. J Food Prot 59:230–237, 1996.

JM Jay. Foods with low numbers of microorganisms may not be the safest foods or, why did human listeriosis and hemorrhagic colitis become foodborne diseases? Dairy Food Environ Sanit 15:674–677, 1995.

CA Kelly, JF Dempster, AJ McLoughlin. The effect of temperature, pressure and chlorine concentration of spray washing water on numbers of bacteria on lamb carcases. J Appl Bacteriol 51:415–424, 1981.

SL Kochevar, JH Sofos, RR Bolin, JO Reagan, GC Smith. Steam vacuuming as a pre-evisceration intervention to decontaminate beef carcasses. J Food Prot 60:107–113, 1997.

KL Kotula, R Thelappurate. Microbiological and sensory attributes of retail cuts of beef treated with acetic and lactic solutions. J Food Prot 57:665–670, 1994.

AW Kotula, WR Lusby, JD Crouse, B De Vries. Beef carcass washing to reduce bacterial contamination. J Anim Sci 39:674–679, 1974.

M Lee, JG Sebranek, DG Olson, JS Dickson. Irradiation and packaging of fresh meat and poultry. J Food Prot 59:62–72, 1996.

Y Li, JT Walker, MF Slavik, H Wang. Electrical treatment of poultry chiller water to destroy *Campylobacter jejuni*. J Food Prot 58:1330–1334, 1995.

PJ McClure, TA Roberts, PO Oguru. Comparison of the effects of sodium chloride, pH and temperature on the growth of *Listeria monocytogenes* on gradient plates and in liquid medium. Lett Appl Microbiol 9:95–99, 1989.

T Mattila, AJ Frost. Colonization of beef and chicken muscle surfaces by *Escherichia coli*. Food Microbiol 5:219–230, 1988.

AF Mendonca, TL Amoroso, SJ Knabel. Destruction of gram-negative food-borne pathogens by high pH involves disruption of the cytoplasmic membrane. Appl Environ Microbiol 60:4009–4014, 1994.

NJ Moon. Inhibition of the growth of acid tolerant yeasts by acetate, lactate, and propionate and their synergistic mixtures. J Appl Bact 55:453–460, 1983.

J Mullerat, BW Sheldon, NA Klapes. Inactivation of salmonella species and other food-borne pathogens with Salmide®, a sodium chlorite-based oxyhalogen disinfectant. J Food Prot 58:535–540, 1995.

AL Nutsch, RK Phebus, MJ Riemann, DE Schafer, JE Boyer, RC Wilson, JD Leising, CL Kastner. Evaluation of a steam pasteurization process in a commercial beef processing facility. J Food Prot 60:485–492, 1997.

D-H Oh, DL Marshall. Antimicrobial activity of ethanol, glycerol monolaurate or lactic acid against *Listeria monocytogenes*. Int J Food Microbiol 20:239–246, 1993.

HW Ockerman, RJ Borton, VR Cahill, NA Parrett, HD Hoffman. Use of acetic and lactic acid to control the quantity of microorganisms on lamb carcasses. J Milk Food Technol 37:203–204, 1974.

JT Patterson. Hygiene in meat processing plants 4. Hot-water washing of carcasses. Rec Agric Res Minist Agric NI 18:85–87, 1969.

RK Phebus, AL Nutsch, DE Schafer, RC Wilson, MJ Riemann, JD Leising, CL Kastner, JR Wolf, RK Prasai. Comparison of steam pasteurization and other methods for reduction of pathogens on surfaces of freshly slaughtered beef. J Food Prot 60: 476–484, 1997.

RK Podolak, JF Zayas, CL Kastner, DYC Fung. Reduction of bacterial populations on vacuum packaged ground beef patties with fumaric and lactic acids. J Food Prot 59:1037–1040, 1996.

RK Prasai, GR Acuff, LM Lucia, DS Hale, JW Savell, JB Morgan. Microbiological effects of acid decontamination of beef carcasses at various locations in processing. J Food Prot 54:868–872, 1991.

RK Prasai, RE Campbell, LR Vogt, CL Kastner, DYC Fung. Hot-fat trimming effects on the microbiological quality of beef carcasses and subprimals. J Food Prot 58:990–992, 1995a.

RK Prasai, RK Phebus, CM Garcia Zepeda, CL Kastner, AE Boyle, DYC Fung. Effectiveness of trimming and/or washing on microbiological quality of beef carcasses. J Food Prot 58:1114–1117, 1995b.

EA Quartey-Papafio, RT Marshall, ME Anderson. Short-chain fatty acids as sanitizers for beef. J Food Prot 43:168–171, 1980.

JO Reagan, GR Acuff, DR Buege, MJ Buyck, JS Dickson, CL Kastner, JL Marsden, JB Morgan, R Nickelson, CG Smith, JN Sofos. Trimming and washing of beef carcasses as a method of improving the microbiological quality of meat. J Food Prot 59:751–756, 1996.

AE Reynolds, JA Carpenter. Bactericidal properties of acetic and propionic acids on pork carcasses. J Animal Sci 38:515–519, 1974.

RME Richards, DKL Xing, TP King. Activity of *p*-aminobenzoic acid compared with other organic acids against selected bacteria. J Appl Bact 78:209–215, 1995.

RJ Rowbury. An assessment of environmental factors influencing acid tolerance and sensitivity in *Escherichia coli, Salmonella* spp. and other enterobacteria. Lett Appl Microbiol 20:333–337, 1995.

TD Schnell, JN Sofos, VG Littlefield, JB Morgan, BM Gorman, RP Clayton, GC Smith. Effects of postexsanguination dehairing on the microbial load and visual cleanliness of beef carcasses. J Food Prot 58:1297–1302, 1995.

LA Shelef, S Mohammed, W Tan, ML Webber. Rapid optical measurements of microbial contamination in raw ground beef and effects of citrate and lactate. J Food Prot 60:673–676, 1997.

MF Slavik, JW Kim, Y Li, JT Walker, H Wang. Morphological changes of *Salmonella typhimurium* caused by electrical stimulation in various salt conditions. J Food Prot 58:375–380, 1995.

MG Smith. Destruction of bacteria on fresh meat by hot water. Epidemiol Infect 109: 491–496, 1992.

MG Smith, A Graham. Destruction of *Escherichia coli* and salmonellae on mutton carcasses by treatment with hot water. Meat Sci 2:119–128, 1978.

FJM Smulders, CHJ Woolthuis. Immediate and delayed microbiological effects of lactic acid decontamination of calf carcasses—influence on conventionally boned versus hot-boned and vacuum-packaged cuts. J Food Prot 48:838–847, 1985.

FJM Smulders, P Barendsen, JG Van Logtestijn, DAA Mossel, GM Van Der Marel. Lactic acid: considerations in favour of its acceptance as meat decontaminant. J Food Technol 21:419–436, 1986.

KM Sorrells, DC Enigl, JH Hatfeld. Effect of pH, acidulant, time, and temperature on the growth and survival of *Listeria monocytogenes*. J Food Prot 52:571–573, 1989.

RR Tarté, EA Murano, DG Olson. Survival and injury of *Listeria monocytogenes, Listeria innocua* and *Listeria ivanovii* in ground pork following electron beam irradiation. J Food Prot 59:596–600, 1996.

KS Tinney, MF Miller, CB Ramsey, LD Thompson, MA Carr. Reduction of microorganisms on beef surfaces with electricity and acetic acid. J Food Prot 60:625–628, 1997.

P Van Netten, H Van Der Zee, DAA Mossel. A note on catalase enhanced recovery of acid injured cells of gram negative bacteria and its consequences for the assessment of the lethality of L-lactic acid decontamination of raw meat surfaces. J Appl Bact 57:169–173, 1984.

G Wang, MP Doyle. Heat shock response enhances acid tolerance of *Escherichia coli* O157:H7. Lett Appl Microbiol 26:31–34, 1998.

RC Wilson, JD Leising, J Strong, J Hocker, J O'Connor. Method for steam pasteurization of meat. US Patent 5,976,005 (1999).

RC Wilson, J Strong, J Hocker, J O'Connor, JD Leising. Apparatus for steam pasteurization of meat. US Patent 5,711,981 (1998).

CHJ Woolthuis, FJM Smulders. Microbial decontamination of calf carcasses by lactic acid sprays. J Food Prot 48:832–837, 1985.

CHJ Woolthuis, DAA Mossel, JG Van Logtestijn, JM Dekruijf, FJM Smulders. Microbial decontamination of porcine liver with lactic acid and hot water. J Food Prot 47: 220–226, 1984.

14

Inactivation by High-Intensity Pulsed Electric Fields

Juan Fernández Molina, Gustavo V. Barbosa-Cánovas, Barry G. Swanson, and Stephanie Clark
Washington State University
Pullman, Washington

I. INTRODUCTION

Pulses of high-voltage electrical fields (PEF) are effective in the destruction of bacteria, yeast, and molds (1–12). An electric field of high strength (kV), applied as short-time pulses (μs) to aqueous suspensions of living cells has tremendous effects on the cell membrane and in some cases can even kill the organism (13). Mittal (14) also pointed out that energy conservation is the greatest advantage of processing food with high-intensity pulsed electric fields. Adequate pasteurization of 1 L of milk requires more than 300 kJ (or 71,000 calories) of energy with a conventional heating process, whereas the new technology requires 200–240 kJ. Unlike pasteurization, the process is successful in destroying spores (14).

Conventional heat treatment inactivates both microorganisms and enzymes, but the organoleptic and nutritional properties of the food often suffer because of loss or alteration of volatile flavors, lipid oxidation, and protein denaturation (15,17). PEF in principle does not alter the quality of foods because pasteurization occurs at ambient or refrigerated temperatures for seconds or less, preserving the fresh-like qualities (7,16).

Treatment of raw skim milk (2% fat) with PEF enhanced the shelf life of the milk up to 2 weeks with no significant changes in the physical and chemical properties of the milk (7). A sensory panel found no differences between PEF and thermally pasteurized skim milk (17). Castro (18) partially inactivated the enzyme alkaline phosphate from bovine milk with PEF treatment electric field intensity of 22.3 kV/cm (70 pulses) for 0.7–0.8 μs. Alkaline phosphatase is used as an indicator of adequate thermal pasteurization of milk Ho et al. (9) applied a PEF process of 13–87 kV/cm electric field intensity with a treatment time of 2 μs and 20°C and reported that the activities of some enzymes were reduced after PEF treatments. Lipase, glucose oxidase, and heat-stable α-amilase exhibited reduction of 70–85%; peroxidase and polyphenol oxidase exhibited a moderate decrease of 30–40%; alkaline phosphatase only exhibited a reduction of 5% under the condition employed. Vega-Mercado et al. (20) achieved 90% inactivation of plasmin using a converged electric field intensity of 30–45 kV/cm (50 pulses) at a temperature of 15°C. The inactivation of enzymes is dependent on the type of enzyme, the intensity of electric field, and the number of pulses applied.

II. ELECTRO-PURE PROCESS

The ohmic heating electro-pure process is one of the available PEF technologies to pasteurized liquid foods. This process was one of the first attempts to use electrical fields to inactivate microorganisms in milk. Fetterman (21) chose electrical conductivity to increase the temperature of the milk to 165°C in 15 seconds to inactivate *Tubercle bacilli* and *E. coli* present in milk. The electro-pure system consists of a vertical rectangular chamber with two sides being carbon electrodes and two sides an insulating glass. The milk is pumped upward through the chamber while the conduction of electric current between the two carbon electrodes results in a rapid heating of the milk (18).

III. ELSTERIL AND ELCRAK

The ELSTERIL and Elcrak PEF processes were developed by Krupp Maschinentechnik GmbH (Hamburg, Germany) for inactivating microorganisms in fluid media using a high-voltage generator charging capacitor of 5 μF to discharge 5–15 kV through a treatment chamber (ESTERIL). A second goal of the electrical process was the breakdown of vegetable and animal cells (Elcrak). Successful preliminary inactivation of microorganisms present in orange juice and milk were achieved with the application of these electrical processes (22).

Qin et al. (17) reported the development of a continuous PEF system to preserve pumpable fluids by Krupp in a collaborative agreement with the Technical University of Berlin and NaFuTec. The collaborative efforts successfully inactivated selected microorganisms in fluid foods such as orange juice and milk. Reported results were obtained in a batch system consisting of a impulse generator, two parallel plane carbon electrodes, and a capacitor discharging electric pulses of 28 kV/cm through the fluid food in the treatment chamber (11).

IV. PURE PULSE TECHNOLOGY, INC.

Maxwell Laboratories (Food Corporation, San Diego, California) holds the U.S. patents to preserve liquid foods such as milk, fruit juices, and liquid egg with PEF (17). The fluids are treated with electric field intensities between 12 and 25 kV/cm with treatment times of 1–100 μs (18). The patents describe both the batch and continuous systems, with research reported with the batch system (17). Researchers at various institutions and around the world are currently working toward the validation of the PEF system as an alternative to conventional thermal processing of liquid and semiliquid foods: Ohio State University; Unilever Research Laboratory, The Netherlands; North Carolina State University; Batelle Memorial Institute; and the U.S. Army Natick RD&E Laboratories are a few examples (23).

V. WASHINGTON STATE UNIVERSITY

Vega-Mercado (24) disclosed two patents for the design and development of static and continuous PEF chambers intended for processing of liquid foods with high-intensity PEF. The continuous chamber was designed by adding a laboratory scale a flow channel inside a static chamber. The treatment volume is 8 cm^3, and the gap between the electrodes 0.5 cm. The design criteria include: flow rate, 100–500 mL/min; PEF intensity maximum of 80 kV/cm; pulse width, 0.5–5 s; and pulse repetition rate of 0.10–10 Hz. Cooling the chamber is accomplished by circulating water through the jackets built between the two stainless steel electrodes. Fluid products such as fruit juices, raw skim milk, pea soup, and beaten eggs were successfully pasteurized with PEF technology at Washington State University (WSU) using peak electric fields ranging from 35 to 50 kV/cm with a pulse duration of 2μs (17). The temperature during the process is maintained between 45 and 53°C, low enough to retain fresh-like physical, chemical, and sensory properties of the fluid foods.

PEF pasteurization of raw skim milk at WSU extended the shelf life compared to commercial milk by reducing the total plate count and keeping the number of coliforms within acceptable limits for 10–14 days (23). PEF pasteurization of liquid whole eggs in a pilot plant system yielded a shelf life of more than 4 weeks with no coagulation of proteins. PEF pasteurization of liquid whole eggs is a potential alternative to conventional thermal processing.

Mathematical models were developed at WSU to simulate movement of charges along the membrane of microorganisms in a continuous PEF system to operate in a single pass or in recirculation mode and a heat conduction model to simulate heat transfer in microorganisms suspended in liquid foods. The inactivation of microorganisms with PEF is due in part to electrical forces generated along the cell membrane during PEF treatment (25,26). The inactivation of microorganisms in a single pass operation provided a good first approximation of the results obtained in the recirculation mode. For both single pass and recirculation operating conditions, microbial inactivation is dependent on the electrical conductivity of the food. Whenever the field intensity or electrical conductivity of the food increased, the temperature of the food leaving the treatment chamber increased (23). A rise in temperature increases the inactivation of microorganisms due to a synergistic effect of temperature of the test medium and the applied high-voltage pulse. The accumulation of charges adjacent to the cell membrane will also increase the electric field adjacent to the cell membranes and result in a more rapid increase in temperature across the cell membrane (23).

VI. MICROBIAL INACTIVATION WITH PEF

Many experiments validate the concept that high-intensity PEF provides appropriate microbial inactivation for consideration as a nonthermal pasteurization process (1–3,6–8,10,17,27–29). A series of degradative changes in blood, algae, bacteria, and yeast cells are attributed to high-intensity pulsed electric field treatments (27). The degradative changes include electroporation and disruption of semipermeable membranes, which lead to cell swelling or shrinking and finally to breakdown of organelles within the cell or of the cell itself (18).

Hypothesized mechanisms for the inactivation of microorganisms include electric breakdown, ionic punch-through effect, and electroporation of cell membranes (30,31). The inactivation of microorganisms is attributable primarily to an increase in membrane permeability due to compression and poration (2).

Castro et al. (27) reported a 5 log reduction in bacteria, yeast, and mold populations suspended in milk, yogurt, orange juice, and liquid whole eggs treated with PEF. Zhang et al. (6) achieved a 9 log reduction in *E. coli* suspended in simulated milk ultrafiltrate (SMUF) and treated with PEF by applying a converged electric field intensity of 70 kV/cm and a treatment time of 160 μs. A 5–7 log reduction of specific target microorganisms is required for safe and wholesome commercial food pasteurization (6).

Table 1 summarizes the inactivation of microorganisms achieved with PEF. Martin et al. (9) inactivated *E. coli* in skim milk by one cycle by applying 20–40 kV/cm and 64 pulses in a static or continuous chamber. Qin et al. (29) achieved more than 6 log reduction cycles in *E. coli* suspended in SMUF using an electric field intensity of 36 kV/cm with a five-step (50 pulses) PEF treatment. The temperature in the chamber was maintained below 40°C during the PEF treatment, which is lower than the temperature of commercial pasteurization (70–90°C) for milk. Hülsheger et al. (13) reported a 4 log reduction of *E. coli* with

Table 1 Summary of Microbial Inactivation Achieved by Pulsed Electric Fields

Microorganism and initial inoculation	Suspension media	Log reduction in viability (cfu/mL)	Temperature (°C)	Peak electric field strength (kV/cm)	Treatment time (μs)	Number of pulse treatments	Ref.
E. coli (10^9 cfu/mL)	Skim milk	1–3	15	20–45	0.7–1.8	64	9
B. subtilis 3×10^3 spores/mL	0.15% NaCl	3.4	25	50	2–3	30	10
B. cereus 10^5 spores/mL	0.15% NaCl	5	25	50	5–6	50	
E. coli (10^9 cfu/mL)	Modified SMUF	9	20	70	160	80	7
E. coli	Phosphate buffer	4	20	20	1,080	30	13
E. coli (10^7)	Milk	3	—	22	200	5	22
B. subtilis	Modified SMUF	5	60	60 + HHP*	—	75	5
E. coli	Modified SMUF	3–4	37	16	200–300	60	3
S. aureus	Modified SMUF	3	37	16	200–300	60	
B. subtilis and *L. delbrueckii*	Modified SMUF	4–5	30	16	200–300	40–50	4
E. coli (10^7)	Modified SMUF	2.2	10	40–50	—	8	24
S. cerevisiae	Apple juice	>6	<40	36	—	—	29
Zygosaccharomyces bailii	Orange, apple, grape, pineapple, and cranberry juice	5	20	32–36.5	2–3.5	—	8
Total coliforms, *E. coli*, yeast, mold	Spices	1	—	20–80	1–99	11–28	1
L. monocytogenes	Milk	1–4	24–50	25–35	100–600	—	42
Salmonella Dublin (10^5)	Skim Milk, KCl	3	10–50	15–40	12–127	—	32

HHP: High hydrostatic pressure.

an electric field intensity of 40 kV/cm accompanied by a long treatment time of 1080 μs.

Pothakamury et al. (3) validated PEF with 4 log reduction in *E. coli* and *Staphylococcus aureus* suspended in SMUF using a peak electric field strength of 16 kV/cm for 60 pulses with a treatment time ranging between 200 and 300 μs. The extent of microbial inactivation increases when the applied electric field intensity increases. Inactivation of *E. coli* and *Bacillus subtilis* suspended in pea soup increased with an increase in the intensity of the electric field, number of pulses, and pulsing frequency (28). Pagán et al. (5) found that spores of *B. subtilis* were not inactivated when PEF (60 kV/cm, 75 pulses) was used in combination with high hydrostatic pressure (HHP) (1500 atm, 30 min, 40°C). However, these treatments induced the germination of the spores of *B. subtilis* of more than 5 log cycles, making them sensitive to subsequent pasteurization heat treatment. Thus, combinations of HHP and PEF treatments constitute an alternative to the stabilization of food products by heat to inactivate spores. Vega-Mercado et al. (2) concluded that combining pH, ionic strength, and PEF intensity markedly improved the inactivation of microorganisms such as *E. coli*. Two log reductions in plate counts were observed when both pH and electric field were modified from pH of 6.8–5.7 and electric field intensities of 20–55 kV/cm. Similar results were achieved when ionic strength was reduced from 168 to 28 mM^2. Raso et al. (8) achieved a 5 log reduction of *Zygosaccharomyces bailii* suspended in orange, apple, grape, pineapple, and cranberry juices by applying an electric field intensity of 32–36.5 kV/cm for 20 pulses with a treatment time of 2–3.5 μs. Sensoy et al. (32) developed a kinetic model for the inactivation of *Salmonella dublin* suspended in skim milk with a co-field flow PEF treatment system. A 50% reduction of viable cells of *S. dublin* were obtained using a critical electric field potential (E_c) ranging between 12.64 to 16.24 kV/cm with a treatment of 25–100 μs.

VII. MECHANISMS FOR MICROBIAL INACTIVATION WITH PEF

Several hypotheses are proposed as mechanisms leading to the inactivation of microorganisms in high-intensity PEF. The most commonly proposed mechanisms are electroporation and electropermeabilization (2,27,28,31,33,34). The application of electrical fields to biological cells in an aqueous medium such as food results in accumulation of electrical charges to cell membranes (25,35). Membrane permeabilization and deterioration occurs when the induced membrane potential exceeds a critical value of 1 V in many cellular systems, which corresponds to an external field of about 10 kV/cm for *E. coli* (27).

A. Electrical Breakdown

Zimmermann (30) explained schematically (Fig. 1) the steps in the electrical breakdown or permeabilization of cell membranes. The membranes are considered as a capacitor filled with a dielectric (Fig. 1a). The normal potential difference across the membrane $V'm$ is about 10 mV. Exposure of the membranes to a high external electrical field for a short time period (μs) (Fig. 1b) leads to an increase in the potential difference (V) across the membranes. The potential difference (V) is proportional to the field intensity (E) and radius of the cell or organelle. An increase in the membrane potential leads to reduction of membrane thickness. Breakdown of the membrane occurs if the critical breakdown voltage (Vc) of larger than 1 V is reached though an increase in the external field intensity (Fig. 1c). Membrane permeabilization and breakdown results from the formation of channels filled with conductive solution in the membranes. The evolution of channels in the membranes is attributed to the intensity of the electric field exceeding the critical breakdown voltage, which leads to an immediate discharge and irreversible permeabilization of the membranes. Breakdowns of the membranes are reversible if the channels in the membranes are small in relation to the total membrane surface. In exceedingly intense electric fields for long exposure times, larger weak areas of the membrane are subjected to channel formation and breakdown (Fig. 1d). If the size and number of channels become large in relation to the total membrane surface, reversible breakdown and recovery becomes irreversible and is associated with physical dysfunction, permeabilization, and lysis of the cell membranes.

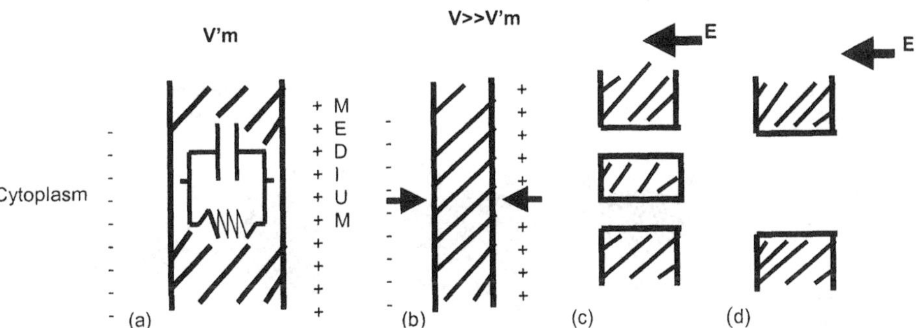

Figure 1 Schematic diagram of reversible and irreversible breakdown. (a) Cell membrane with potential V′m, (b) membrane compression, (c) pore formation with reversible breakdown, (d) large area of the membrane subjected to irreversible breakdown with large pores. (From Ref. 30.)

The critical electric field is defined as $E_{critical} = V_{critical}/f_a$, where a is the radius of the cell and f is a form constant dependent on the shape of the cell (35). For a spherical cell, f is 1.5; for cylindrical cells of length l and hemispheres of diameter d at each end, the form factor is $f = l(l - \text{d})/3$. Typical values of $V_{critical}$ required for the irreversible permeabilization of *E. coli* is approximately 1 V. The critical field intensity for the irreversible permeabilization of spherical microbial cells with a radius of approximately 1 μm and critical voltage of 1 V across the cell membrane is approximately 10 kV/cm for pulses of 10 μs duration (35).

B. Electroporation

Electroporation is the phenomenon in which the application of a high-voltage electric field of 10 kV/cm or higher to a cell temporarily destabilizes the lipid bilayer and proteins of cell membranes (27). The plasma membranes of cells become permeable to small molecules after being exposed to a high-intensity electric field. Membrane permeabilization alters the osmotic equilibrium, which leads to swelling and eventual rupture of the cell membranes (Fig. 2) (2). The main effect of an electric field on microbial cells or organelles is to increase membrane permeability attributed to membrane compression and electroporation (2). Kinosita and Tsong (36,37) induced pores of approximately 1 nm in diameter in human erythrocytes with an electric field of 2.2 kV/cm (16). Kinosita and Tsong (36) suggested a two-step mechanism for pore formation in which the initial perforation is a response to an electrical suprathreshold potential followed by a time-dependent expansion of the pore size (Fig. 2). Large pores are created by increasing the intensity of the electric field, number of pulses, or pulse dura-

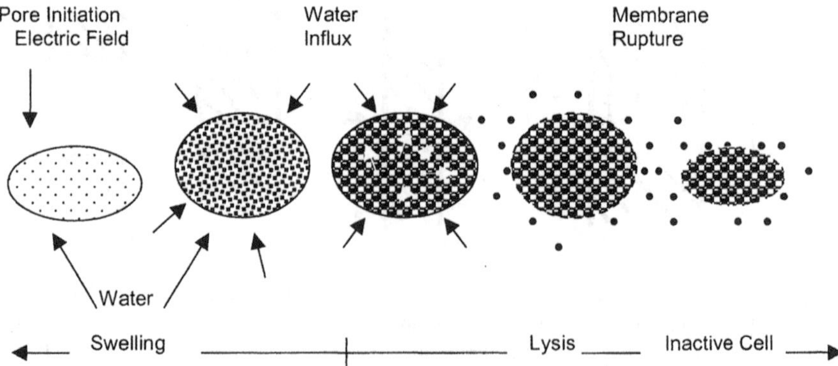

Figure 2 Electroporation of cell membrane during PEF processing.

tion, or by reducing the ionic strength of the medium. In the cell membranes, charges from lipid, protein, carbohydrate, and ion dipoles and the polarization of dipolar molecules make up the electric field. Therefore, electroporation occurs in membranes of organelles within the cell membranes (38).

The electroporation movement of ions and water dipoles through hydrophobic pores in the membranes of cell organelles is the first event of electroporation, after which molecules in the membrane bilayer rearrange to form more stable hydrophilic pores. In a cell membrane, initiation of electroporation may also result from the formation of hydrophobic pores. However, hydrophilic protein channels, pores, and pumps may also be present. Cell membranes with characteristic bilayer structure are extremely sensitive to transmembrane electric field potential and become initiation sites for the electroporation (38,39). The gating potentials for electroporation or formation of hydrophilic protein channels are approximately in the 50 mV range (27).

Miller et al. (40) reported that the insertion of DNA into mammalian cells and plant protoplasts is attributed to electroporation. Electroporation induces transient permeability to cell membranes. The utility of high-intensity electric field electroporation to initiate transformation of genetic material into intact bacterial cells was demonstrated with the enteric pathogen *Campylobacter jejuni*. Reversible electroporation involved the exposure of a *C. jejuni* cell suspension to a high-intensity electric field decay discharge of 5–13 kV/cm with a short treatment time ranging from 2.4 to 2.6 μs in the presence of plasmid DNA. Reversible electroporation and transformation of *C. jejuni* resulted in frequencies as high as 1.2×10^6 transformats per μg of DNA.

VIII. PEF GENERATOR AND FACILITY

Liquid food materials are generally electrical conductors because of the presence of large concentrations of ions as electrical charge carriers (9). Zhang et al. (7) reported that generation of high-intensity PEF with foods requires a large flux of electric current to flow through a piece of food in a treatment chamber for a very short period of time (e.g., μs). The construction of a typical pulse generator and a PEF commercial facility, as proposed by Vega-Mercado (24), is presented in Figs. 3 and 4. The pulser configuration consists of a high-voltage power supply to charge the capacitor and a discharge switch that releases the stored electric energy from the capacitor (s) in the form of an electric field through the food material (Fig. 3). Inductors modify the shape and polarity of the pulse into exponential or square waveforms.

Exponential decay waveforms exhibit a long tail following peak intensity. During exponential decay of the electric field, excess heat is generated in the food with little electroporation. Square or quadratic waveforms retain peak inten-

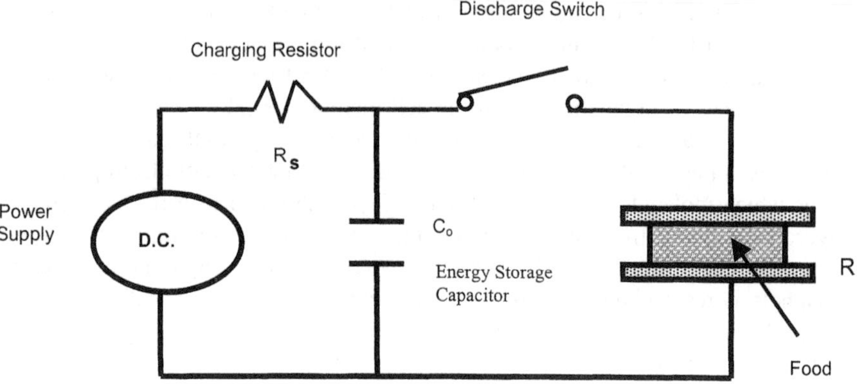

Figure 3 Typical pulser configuration for high-voltage pulsed electric fields.

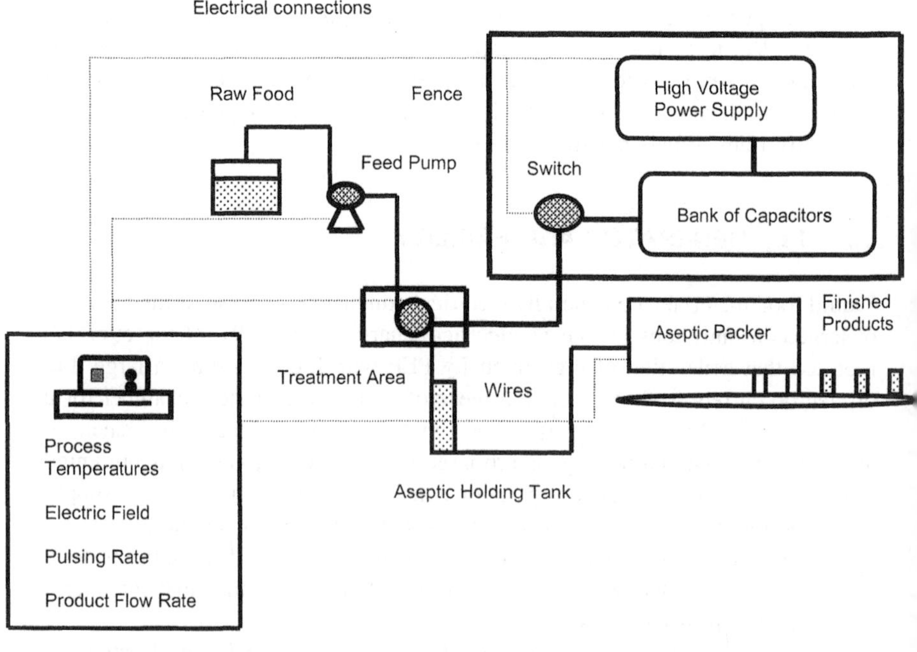

Figure 4 PEF facility for food processing. (Adapted from Ref. 24.)

sity for a longer time than exponential waveforms (9). Since both waveforms inactivate microorganisms, square waveforms conserve energy and require less cooling effort (7).

A commercial PEF facility is depicted in Figure 4. The power supply, the capacitor, and treatment chamber must be confined in a security area with interlocked gates because of the high-voltage hazard. Electrical connections are isolated and the pipes carrying material to the treatment chamber are connected to a heavy ground to prevent energy leakage through the fluid food or refrigerant in contact with the treatment chamber (24).

IX. FUNDAMENTAL MATHEMATICAL EQUATIONS RELATED TO PEF PASTEURIZATION

The intensity of an electric field (E) is defined as the difference in electrical potential (V) between two given points divided by the distance (d) between these two points (16):

$$E = \frac{V}{d} \tag{1}$$

In the case of a charged capacitor (C_o) discharging through a resistance (R), the voltage through a piece of food in the treatment chamber follows an exponential decay with a pulse width d[illegible]s:

$$\tau = RC_o \tag{2}$$

The treatment time is gen[illegible]y defined as $t = \tau\, n$, where n = the number of pulses and τ = pulse width.

The microbial surviving velocity (S) is related to the treatment time (t) and to the electric field intensity as follows:

$$S = \left(\frac{t}{t_c}\right)^{-\frac{(E - E_c)}{K}} \tag{3}$$

where k is a constant dependent on the type of microorganism, t_c the critical treatment time, and E and E_c the actual maximum and critical electric field intensities (41).

The energy supplied for one exponential decay pulse is approximately equal to:

$$Q = \left[\frac{V_o^2 C_o n}{2v}\right] = \frac{[V_o^2 t]}{2Rv} \tag{4}$$

where V_o = initial discharge voltage, C_o = capacitance of energy storing capacitor, n = number of applied pulses, t = treatment time, v = volume of the treatment chamber, and R = effective resistance.

For square pulses, the supplied energy is given by:

$$Q = \frac{[VI\tau n}{V} = \frac{[V^2\ \tau n]}{RV} = \frac{V^2 t}{RV} \tag{5}$$

where V, I, and τ are the potential difference, current intensity, and square pulse width, respectively.

Martin et al. (16) found that the effective resistance (R) and effective capacitance (C) in fluid foods could be measured, providing the electric and dielectric properties of the food are homogeneous:

$$C = \frac{[\varepsilon_0 \varepsilon_r]A}{d} \tag{6}$$

and

$$R = \frac{d}{RA} = \frac{\rho d}{A} \tag{7}$$

where ε_0 = vacuum constant = 8.8×10^{-12} (F/m), ε_r = relative dielectric constant, A = electrode area (m^2), d = distance or gap between the two parallel electrodes (m), σ = conductivity of food material (s m^{-1}), and ρ = the resistivity of the food material (Ωm). In general, the electric and dielectric properties of foods increase with water content (ε_r = 80 for pure water), and decrease with increasing temperature (14).

X. CONCLUSIONS

High-intensity pulsed electric fields are an effective nonthermal process for the inactivation of pathogenic and spoilage microorganisms in liquid foods. The application of PEF (10–100 kV/cm) and relatively short treatment times on the order of micro- or milliseconds induces electroporation and electropermeabilization of the cell or organelle membranes of microorganisms suspended in solid and fluid foods. Inactivation of microorganisms is attributed to irreversible electropermeabilization of the microbial membranes due to channeling and electroporation during exposure of microorganisms to pulses of a high-intensity electric field.

The temperature of the food and treatment chamber during PEF pasteurization is maintained below 53°C, a temperature less than required for adequate thermal pasteurization (70–90°C). Therefore, the fresh-like qualities and nutritive value of the treated foods are not altered and the energy savings are significant. Approximately 200–240 kJ are required to pasteurize 1 L of milk with PEF compared to more than 300 kJ for conventional thermal pasteurization. The shelf life of refrigerated skim milk, fruit juices, and whole liquid eggs are extended compared to raw products after following PEF treatment. High-intensity pulsed electric field technology is being used for insertion of selected plasmids and other generic material into microbial cells and inactivation of selected enzymes in model food systems.

Combination of PEF with high hydrostatic pressure, acidification, ionic strength, processing temperature, and/or antimicrobial agents markedly improves the inactivation of microorganisms. Inactivation of 1–9 log cycles of target pathogenic microorganisms suggests that PEF has potential benefits for commercial pasteurization of fluid acid foods. Pasteurization with high-intensity PEF is best accomplished with a single pass through sequential continuous flow treatment chambers.

ACKNOWLEDGMENTS

This work was funded by AVISTA Utilities, Inc., Spokane, WA, and the International Program for Agricultural Commodities and Trade (IMPACT), Washington State University, Pullman, WA.

REFERENCES

1. W Keith, L Harris, L Hudson, M Griffiths. Pulsed electric fields as a processing alternative for microbial reduction in spice. J Food Res Int 30(3/4):185–191, 1997.
2. H Vega-Mercado, U Pothakamury, F Chang, G Barbosa-Cánovas, B Swanson. Inactivation of *Escherichia coli* by combining pH, ionic strength and pulsed electric fields. J Food Res Int 29(2):117–121, 1996.
3. U Pothakamury, A Monsalve-González, G Barbosa-Cánovas, B Swanson. Inactivation of *Escherichia coli* and *Staphylococcus aureus* in model foods by pulsed electric field technology. J Food Res Int 28(2):167–171, 1995.
4. U Pothakamury, A Monsalve-González, G Barbosa-Cánovas, B Swanson. High voltage pulsed electric field inactivation of *Bacillus subtilis* and *Lactobacillus delbrueckii*. Rev Esp Cienc Tecnol Aliment 35(1):101–107, 1995.
5. R Pagán, S Espulgas, M Góngora-Nieto, G Barbosa-Cánovas, B Swanson. Inactivation of *Bacillus subtilis* spores using high intensity pulsed electric fields in combination with other food conservation technologies. J Food Sci Technol Int 4:33–44, 1997.
6. Q Zhang, B Qin, G Barbosa-Cánovas, B Swanson. Inactivation of *E. coli* for food pasteurization by high-strength pulsed electric fields. J Food Proc Pres 19:103–118, 1994.

7. Q Zhang, G Barbosa-Cánovas, B Swanson. Engineering aspects of pulsed electric field pasteurization. J Food Eng 25:261–281, 1995.
8. J Raso, M Calderon, M Gongora-Nieto, G Barbosa-Cánovas, B Swanson. Inactivation of *Zygosaccharomyces bailii* in fruit juices, high hydrostatic pressure and pulsed electric fields. J Food Sci 63(6):1042–1044, 1998.
9. O Martin, B Qin, F Chang, G Barbosa-Cánovas, B Swanson. Inactivation of *Escherichia coli* in skim milk by high intensity pulsed electric fields. J Food Proc Eng 20:317–336, 1997.
10. V Marquez, G Mittal, M Griffiths. Destruction and inhibition of bacterial spores by high voltage pulsed electric fields. J Food Sci 62(2):399–401, 1997.
11. B Qin, U Pothakamury, G Barbosa-Canovas, B Swanson. Nonthermal pasteurization of liquid foods using high intensity pulsed electric fields. Crit Rev Food Sci Nutr 36(6):603–627, 1996.
12. G Barbosa-Canovas, M Gongora-Nieto, B Swanson. Nonthermal electrical methods in food preservation. Food Sci Int 4(5):363–370, 1998.
13. H Hülsheger, J Pottel, E Niemann. Electric field effects on bacteria and yeast cells. Radiat Environ Biophys 22:149–162, 1983.
14. G Mittal. New energy-efficient process could replace pasteurization. External communication, University of Guelpg, Ontario, Canada, 1996.
15. D Quass. Pulsed electric field processing in the food industry. A status report. CR-109742. Electric Power Institute, Palo Alto, CA, 1997.
16. O Martin, Q Zhang, A Castro, G Barbosa-Cánovas, B Swanson. Empleo de pulsos eléctricos de alto voltaje para la conservación de alimentos. Microbiología e ingeniería del proceso. Rev Esp Cienc Tecnol Aliment 34(1):1–34, 1994.
17. B Qin, U Pothakamury, H Vega, O Martin, G Barbosa-Cánovas, B Swanson. Food pasteurization using high intensity pulsed electric fields. J Food Technol 49(12):55–60, 1995.
18. A Castro. Pulsed electrical field modification of activity and denaturation of alkaline phosphatase. Ph.D. dissertation. Washington State University Pullman, WA, 1994.
19. S Ho, G Mittal, J Cross. Effects of high field electric pulses on the activity of selected enzymes. J Food Eng 31:69–85, 1997.
20. H Vega-Mercado, R Powers, G Barbosa-Cánovas, B Swanson. Plasmin inactivation with pulsed electric fields. J Food Sci 60:1143–1146, 1995.
21. J Fetterman. The electric conductivity method of processing milk. Agric Eng 9:107–108, 1928.
22. T Grahl, W Sitzmann, H Markel. Killing of microorganisms in fluid media by high-voltage pulses. DECHEMA Biotechnol. Conference Series. 5B:675–678, 1992.
23. N Mermelstein. Processing papers cover wide range of topics. J Food Technol 52(7): 50–54, 1998.
24. H Vega-Mercado. Inactivation of proteolytic enzymes and selected microorganisms in foods using pulsed electric fields. Ph.D. dissertation. Washington State University, Pullman, WA, 1996.
25. R Bruhn, P Pedrow, R Olsen, G Barbosa-Canovas, B Swanson. Electrical environment surrounding microbes exposed to pulsed electric fields. IEEE Trans Dielectric Electrical Insulation 4(6):806–812, 1997.
26. R Bruhn, P Pedrow, R Olsen, G Barbosa-Cánovas, B Swanson. Electrical environ-

ment surrounding microbes exposed to pulsed electric fields. IEEE Trans Dielectric Electrical Insulation 5(6):878–885, 1998.

27. A Castro, G Barbosa-Cánovas, B Swanson. Microbial inactivation of foods by pulsed electric fields. J Food Proc Pres 17:47–73, 1993.
28. H Vega-Mercado, O Martin-Belloso, F Chang, G Barbosa-Cánovas, B Swanson. Inactivation of *Escherichia coli* and *Bacillus subtilis* in pea soup using pulsed electric fields. J Food Proc Pres 20:501–510, 1996.
29. B Qin, G Barbosa-Cánovas, B Swanson, P Pedrow. Inactivating microorganism using a pulsed electric field continuous treatment system. IEEE Trans Indus Applic 34(1):43–49, 1998.
30. U Zimmermann. Electric breakdown, electropermeabilization and electrofusion. Rev Physiol Biochem Pharmacol 105:175–256, 1986.
31. G Barbosa-Canovas, M Gongora-Nieto, U Pothakamury, B Swanson. Preservation of Foods with Pulsed Electric Fields. New York: Academic Press, 1999.
32. I Sensoy, Q Zhang, S Sastry. Inactivation kinetic of *Salmonella dublin* by pulsed electric fields. J Food Proc Eng 20:367–381, 1997.
33. A Sale, W Hamilton. Effect of high electric fields on microorganisms. I. Killing of bacteria and yeast. Biochim Biophys Acta 148:781–788, 1967.
34. U Zimmermann, R Benz. Dependence of the electrical breakdown voltage on the charging time in *Valonia utricularis*. J Membr Biol 53:33–43, 1980.
35. K Schoenbach, F Peterkin, R Alden, S Beebe. The effect of pulsed electric fields on biological cells: experiments and applications. IEEE Trans Plasma Sci 25(2): 284–292, 1997.
36. K Kinosita, T Tsong. Voltage induced pore formation and haemolysis erythrocytes. Biochim Biophys Acta 471:227–242, 1977.
37. K Kinosita, T Tsong. Voltage-induced conductance in human erythrocyte membranes. Biochim Biophys Acta 554:479–497, 1979.
38. T Tsong. Electrical modulation of membrane proteins: enforced conformational oscillations and biological energy signals. Annu Rev Biophys Chem 19:83–106, 1990.
39. M Calderon-Miranada. Inactivation of *Listeria innocua* by pulsed electric fields and nisin. M.S. thesis, Engineering, Washington State University, Pullman, WA, 1998.
40. J Miller, W Dower, L Tompkins. High-voltage electroporation of bacteria: generic transformation of *Campylobacter jejuni* with plasmid DNA. Proc Natl Acad Sci 85: 856–860, 1988.
41. H Hülsheger, J Pottel, E Niemann. Killing of bacteria with electric pulses of high field strength. Radiat Environ Biophys 20:53–65, 1981.
42. LD Reina, TZ Jin, QH Zhang, AE Yousef. Inactivation of *Listeria monocytogenes* in milk by pulsed electric fields. J Food Prot 61:1203–1206, 1998.

15

Magnetic Fields as a Potential Nonthermal Technology for the Inactivation of Microorganisms

Gustavo V. Barbosa-Cánovas, M. Fernanda San Martín, Federico M. Harte, and Barry G. Swanson
Washington State University
Pullman, Washington

I. INTRODUCTION

Several attempts have been made to elucidate whether exposure of living organisms to magnetic fields (MF) is harmful, beneficial, or neutral. Concern appears as a consequence of human exposure to electromagnetic fields (EMF) generated by household appliances and from the hypothesis that if MF were able to modify biological processes, they could be advantageously used in diverse areas. Areas of specific focus include those in which microorganisms play an important role, such as food preservation and pharmaceutical and fermentative processes.

The increase in demand for foods with improved and/or fresh-like quality and little heat-induced degradation of nutritional and organoleptic properties gave rise to the development of nonthermal processes for the preservation of food. These include utilization of electric fields (EF) or MF, ionizing radiation, light pulses, high hydrostatic pressure, bacteriocins, and antimicrobial enzymes (1). The primary advantages of MF as a potential tool in the inactivation of microorganisms have been cited as (a) minimal thermal denaturation of nutritional and organoleptic properties, (b) reduced energy requirement for adequate processing, and (c) potential treatment of foods inside a flexible film package. However, from a realistic standpoint the latter two benefits are still far from reach since contradictory results have been obtained and an extensive part of the field remains to be explored.

It is the purpose of this chapter to present an overview of some of the proposed interaction mechanisms as well as some of the results that have been obtained by exposing living systems to different varieties of MF (i.e., low-frequency, low-intensity fields and high-frequency, high-intensity fields). Previous works can be found in some other extensive reviews (1–5).

II. MAGNETIC FIELDS AND SOME RELATED CONCEPTS

Among the different types of MF, static fields are those that have a constant strength over time. These fields can be produced with permanent magnets or electromagnets of direct current. Oscillating magnetic fields (OMF) are generated with electromagnets of alternating current and the intensity varies periodically according to the frequency and type of wave in the magnet. OMF that are generated by pulses (pulsed MF) are electromagnetic by nature and have an associated electric field component capable of inducing electrical currents in stationary biological systems (5). Homogeneous fields have constant strength over space, and heterogeneous MF exhibit a gradient depending on the nature of the magnet. MF intensity (H) is measured in Oersted, while magnetic flux density (B) is measured in gauss or Tesla (1 Tesla = 10,000 gauss). One Oersted is defined as one line of strength by square centimeter. The relationship between MF intensity and MF density is given by $B = \mu H$, where μ is the magnetic permeability of the media, which for vacuum and—for practical purposes—for air, is approximately equal to 1. Therefore, intensity and density are used interchangeably. According to their relative strength, weak MF have intensities in tens of gauss, while high-intensity MF rank between thousands of gauss and over.

Particles that have equal susceptibility to magnetization in the three orthogonal axes are called isotropic, and those that exhibit unequal susceptibility are called anisotropic. While carbon atoms are isotropic, two carbon atoms bonded by single, double, or triple bonds exhibit anisotropy (1,5).

III. INTRODUCTION TO EXTREMELY LOW-FREQUENCY MAGNETIC FIELDS

Interaction between weak oscillating at extremely low-frequency (ELF) MF and biological systems has received special attention from various groups of investigators. The mechanism for explaining the action of ELF AC fields coupled with low-intensity DC fields (e.g., geomagnetic fields) could form the basis for the harmful action of electric transmission lines on surrounding inhabitants (6), and since many home electrical devices generate such types of MF, the search for possible effects on cell physiology has been of continuous interest.

Different attempts have been made to explain the effect of ELF MF on living systems, including models to illustrate the observed results. However, constraints exist that in many cases make it difficult to interpret such results. Mittenzwey et al. (7) stated that contradictory results and lack of reproducibility are typical problems in ELF research and attempts should therefore be made to improve experimental conditions. Furthermore, Saffer and Phillips (8) suggested that sources of variability are not only biological but also technical. Response to ELF fields may depend on physiological and genetic characteristics of cells. In addition, cells possess a variety of control mechanisms such as signaling pathways, transcription regulatory elements, and feedback loops that initiate adaptation mechanisms to adverse conditions. Moreover, some cells have back-up systems that can be activated when a given function is altered, and this may mask a magnetic effect. ELF EMF may act on bacteria as a co-stressing factor, which activates a process or reaction already initiated by other stresses (7). Cell density (an important factor in many EMF experiments) is another consideration because it is often related to biological function in that both cell-cell contacts and the concentration of cell-produced agents influence cell behavior (8). Furthermore, Fitzsimmons et al. (9) reported that the greater a cell's dimension deviates from spherical, the greater the sensitivity to an external EMF.

Since ELF EMF differ in wave forms, frequency, and strength, it is possible that a sharp "window" (i.e., discrete combination of frequency and strength) is necessary to render a visible effect (7). At a molecular level, electric and MF can accelerate ions and charged proteins. However, collisions with other molecules tend to impede acceleration so that—at least for weak MF—there is not a significant change in motion compared to thermal kinetic energy.

Since living cell environments are electrically noisy, for an EMF to present a signal detectable by cells, it must be distinguishable in magnitude, frequency, and coherence from the intrinsic noise. Therefore, the signal-to-noise (S/N) ratio is an important factor when evaluating the bioeffects of EMF (10). Because of this noisy environment within a cell, it has been proposed that for an electromagnetic effect to be identified by a cell, the induced potential gradient must be greater than the thermal noise limit. However, cellular responses have been identified at induced fields lower than the thermal noise limit (9).

Based on this constriction, Barnes (11) identified two questions to be elucidated concerning the interaction of external signals and ELF MF on biological systems:

1. How large does an external signal have to be in order to perturb the ongoing natural signal that is being used to communicate or control some biological process?

2. How much of the signal field typically leaks away to form a background noise environment for surrounding processes?

Answers to these questions are currently being sought, and a generally accepted theory has not yet been formed.

Finally, temperature must also be considered as playing an important role in the combined action of MF and biological system EMF (7) because interaction between MF and temperature was found by different authors (12–14) in cellular response to applied MF.

IV. MECHANISM OF MAGNETIC FIELD INTERACTION WITH BIOLOGICAL SYSTEMS

Several mathematical and physical models have been proposed by different authors (15–17) to elucidate how MF interact with biological entities. However, none of these models has been widely accepted.

A. Ion Cyclotron Resonance Model

One of the first attempts to explain the effect of OMF on living systems was presented by Liboff in 1985 (15). According to his model a charged particle entering a static and uniform MF experiences Lorenz force at right angles to the particle velocity $\vec{v}$ and the MF strength $\vec{B}$. Due to the MF the particle is forced to execute a curved path with an angle between 0 and 90°. By balancing the centripetal and Lorenz forces, it is stated that the gyrofrequency ν of the ion depends only on the strength of the MF (B) and the charge to mass ratio (q/m) yielding the condition for cyclotron resonance. Externally applied electromagnetic signals at frequencies close to a given resonance and parallel to a static MF (Fig. 1) may couple to the corresponding ionic species in such a way as to selectively transfer energy to these ions and thus indirectly to the metabolic activities in which they are involved. The earth's total field ranges from 25 to 70 μT, and most of the slightly and doubly charged ions of biological interest have corresponding gyrofrequencies in the ELF range 10–100 Hz for this field strength.

However, Liboff (15) noted some restrictions: (a) ions are rarely found "naked" on biological systems, but are more often hydrated which reduces their cyclotron resonance frequency; (b) ions need $2\pi r$ free space in order to maintain circular or helical paths that are determined by the ratio of gyrofrequency to the collision frequency, although this is highly unlikely to occur in cells unless they have very intense fields (10^9 Tesla); (c) thermal noise fluctuation prevents coherent currents to form (i.e., the radius of the path of an ion at an energy comparable to the thermal background would be nearly three meters); and (d) the ion selectiv-

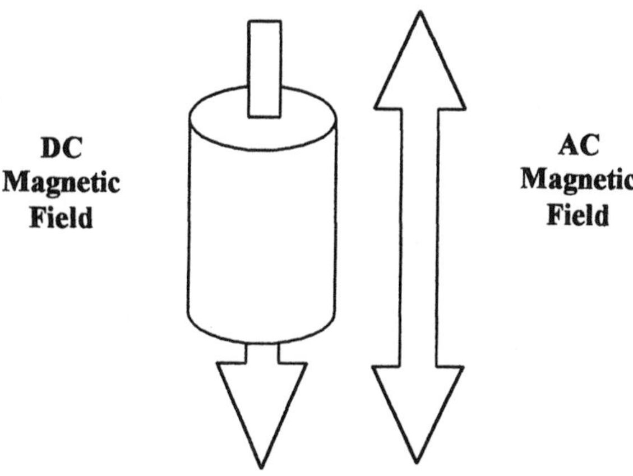

Figure 1 Required AC and DC MF orientation to achieve ion cyclotron resonance.

ity of membrane channels tends to reduce the scattering by other ionic species, enzymes, and proteins. These restrictions do not apply when considering a constrained preexisting path through a membrane since such helical pathways through membrane-bound proteins are suitable for a cyclotron process. In this case, a membrane mechanism at the molecular level is suggested in which ions move in helical, ion-selective channels (15).

Various experiments were done to verify the ion cyclotron resonance (ICR) model and are well documented elsewhere (18–21). However, research in this area has been blocked due to the inability of experiment replication by different investigators (5,18). EMF experiments adjusted for ICR conditions have produced diverse and even opposing results. While Galt et al. found negative effects for Ca flux changes through an artificial lipid bilayer (19) and Ca uptake by lymphocytes (18), increased ^{45}Ca uptake by human bone cells exposed to a combined MF was reported by Fitzsimmons et al. (9). Smith (22) observed the response in diatoms subjected to MF at ICR for Ca to be strain dependent.

Galt et al. (19) also studied the influence of MF at ICR conditions for K^+, H^+, and Ca^{2+} ions on their passage through gramicidin A channels embedded in an artificial lipid bilayer of glyceryl monooleate dissolved in decane or hexadecane. Calculated ICR frequencies were 20, 40, and 760 Hz for K^+, Ca^{2+}, and H^+, respectively, at a DC MF intensity of 50 μT and AC MF intensity of 50 μT_{RMS}. If ion cyclotron effects were present, a change in ion transport properties and pore-opening kinetics would be expected. However, no evidence of ICR in transmembrane channels was observed.

According to Zhadin (6), a mechanism in which a DC MF causes Ca ion rotation around a field line and an AC field accelerates the ion at the cyclotron frequency could not work in a liquid medium due to its high density and resistance to ion motion. The assumption that the ICR model can explain the penetration of Ca ions through spiral channels (assuming conditions close to vacuum) from extracellular medium into the cell has been criticized because the rotation radius of an ion moving with a thermal speed has been stated as being several meters at a DC field comparable to the geomagnetic field (6,18). Other research, as reported in Zhadin (6), considered the action of a DC field on ion motion and rejected any possibility for resonance phenomena at joint action of DC and AC fields. Edmonds (23) focused attention on mutually perpendicular AC and DC fields, suggesting that parallel fields cannot cause resonance.

B. Ion Parametric Resonance Models

Based on the premise that ICR cannot take place in liquids even under high-intensity MF, Lednev (16) proposed a new interpretation for cyclotron resonance in biosystems. This model considers the splitting of quantum energy levels in a molecular ion under the influence of a DC MF (Zeeman effect). The absorption of quanta of an AC field at cyclotron frequency is able to change the quantum state of the calcium ion in a calcium-binding protein molecule and thus change the free calcium concentration in the intracellular medium. Parametric resonance provides the basis for this effect. However, Zhadin (6) states that considering a background of constantly acting thermal noise with energy on the order of 10^{-2} eV, it seems unlikely that the absorption of separate quanta on the order of 10^{-13} eV (corresponding to a frequency of 30 Hz) is able to affect the quantum state of the molecular ion.

The ion parametric resonance (IPR) model has been reviewed from its physical basis by Adair (24), who does not think it can be used to explain the interaction between weak EMF and biological systems. IPR suggests that low-frequency (10–100 Hz) modulating AC fields together with DC fields could modify the probability of the radiative transition of ions in the excited state to the ground state, but Adair (24) claims that if correctly calculated, the modulating field cannot affect the transition rate.

C. Radical Recombination Model

Cozens and Scaiano (25) suggest that whatever radical-molecule processes may take place in biological systems, the radical concentration will be ultimately regulated by radical-radical processes, which remove free radicals from the system. Such radical processes should be the subject of MF effects, and small fields should be sufficient to cause significant effects. Valentinuzzi, as cited in Vaughan and Weaver (13), attempted to theoretically predict the inhibiting effect of MF on the growth of organisms. His hypothesis supported the concept that MF exert an influence on the free radicals present in enzymatic reactions in metabolic pathways because as temperature is increased, growth inhibition by MF will diminish (13). Valberg et al. (10) showed that MF have the ability to extend the lifetime of free radicals by holding the electron's magnetic moment. If the lifetime of free radicals were increased, the proportion of free radicals chemically reacting with macromolecules would increase and potential for adverse effects on cell function would exist. However, for free radicals and significant rates of triplet-singlet transitions to occur, field intensities of hundreds of gauss must be applied, and to achieve free radical recombination effects, frequencies higher than 10^5 Hz or 100 kHz should be used.

A model that considers the thermal motion of the ion in the macromolecule under the influence of AC and DC MF was proposed by Zhadin (6). According to his model, switching on the fields initiates the occurrence of a set of new resonance frequencies in the ion thermal oscillations, and the magnitude of the energetic variations produced by the weak DC and AC MF may be sufficient for triggering changes in the quantum state of the macromolecule. The most prominent effect is expected on the cyclotron frequency of the AC field and depends on many parameters: the charge to mass ratio of the ion, DC MF, frequency of the AC MF, and structure of the molecule near the ion within the molecular cleft. The theory outlined is quite applicable to the motion of the Ca^{2+} ion inside the calcium-binding proteins like calmodulin, troponin C, and others. Changing their conformation, the molecules of these proteins can capture and release the calcium ions from the cytoplasm, influencing the intracellular concentration of free calcium. In the cytoplasm, the free calcium ions act as a second messenger and take part in signal transmission from the membrane receptors to the effector mechanism of the cell.

V. SITES OF INTERACTION

Most suggestions concerning the mechanism of interaction between weak MF and cells have focused on the plasma membrane as a major target site for interaction between EMF and biological systems (26,27). Figure 2 shows a proposed cascade of responses that occurs as a consequence of a MF stimulus in the cell membrane. In a similar manner, Barnes (11) proposed that since some organisms are capable of synthesizing small particles of magnetite (Fe_3O_4) when arranged in a string, a MF can apply a torque in these particles, which is believed to be large enough to exert a force on membranes capable of opening channels in cell membranes for fields on the order of the gradients of the earth's MF.

Vaughan and Weaver (13) proposed that MF pulses may interact with biologically generated magnetite or contaminant magnetic particles to generate metastable pores in the membranes of cells. The basic idea is that rotational motion of

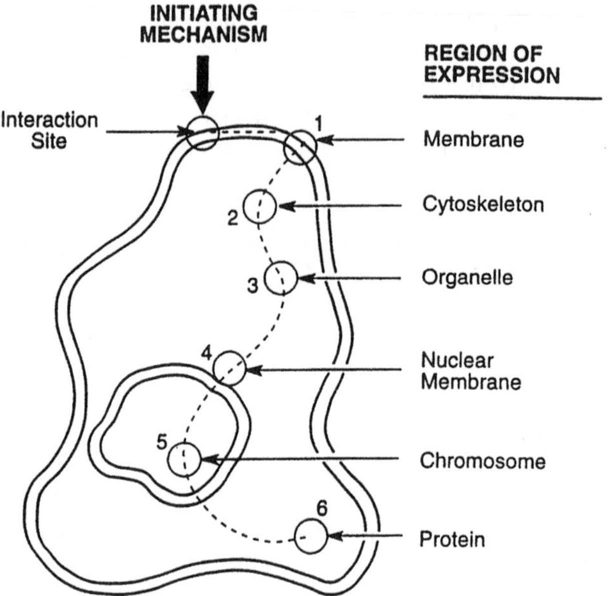

Figure 2 Cascade of responses initiated at cell membrane. (From Ref. 15.)

a membrane-bound particle could transfer enough energy to the membrane that an aqueous pathway (pore) is created. As in electroporation, metastable hydrophilic pores with a minimum radius of $r \approx 1$ nm and lifetimes of $t \approx 1-100$ s could occur, resulting in a significant influx of foreign molecules into the cytosol and alteration of cell biochemistry. Fields generated in biomagnetic stimulation (2 T) and other local sources in the human environment exceeded the minimum MF required to generate such pores.

A. Effect of MF on Ca^{2+} Transport Across Cellular Membranes

One of the most significant reported effects of a MF is the alteration of ion transport across cellular membranes. Though the transport of calcium ions has been widely studied, results have been contradictory. Several claims have been made about increased calcium uptake by lymphocytes when exposed to certain combinations of low-level AC and DC MF. In order to verify these claims and confirm an ICR effect for calcium ions, Coulton and Barker (18) monitored calcium uptake during a determined exposure period (up to 60 min) and found no change in intracellular calcium concentration when lymphocytes were exposed to combined AC and DC MF at 16 Hz (20.9 μT) and 50 Hz (65.3 μT).

The effect of combined AC/DC fields on ^{45}Ca uptake by human osteosarcoma cells was studied by Fitzsimmons et al. (9) by using field intensities of 40 μT for the AC MF and 20 μT for the DC static MF. These fields were calculated to couple with calcium according to the ion-resonance theory proposed by McLeod et al. (28). A 30-minute exposure of cells to the combined MF resulted in an increase in ^{45}Ca uptake. However, this effect was only observed when cells were exposed to the field, and returned to control level once the stimulus was suspended. ^{45}Ca uptake was observed to be frequency dependent with a peak at 16.3 Hz, which suggests the existence of a ''window.'' The physical mechanism of this frequency specificity remains unclear but has been observed by other researchers (7). EMF can increase ^{45}Ca uptake by lengthening the open time for a calcium channel or increasing the probability for opening a calcium channel. Since increased ^{45}Ca uptake was absent after MF exposure, increased calcium channel kinetics was not determined to be due to stable channel modifications.

Markov and Pilla (29) studied the effect of Ca^{2+}-calmodulin–dependent myosin phosphorylation and found that very weak static MF (200 μT) increased calcium-binding kinetics to calmodulin depending on calcium concentrations below saturation levels. However, when Ca^{2+} approached saturation levels, the MF effect diminished and eventually disappeared.

B. Effect on DNA and Gene Transcription

Alteration in cellular processes as a direct consequence of MF exposure has been reported elsewhere (9,29–31), though whether these changes are due to DNA modification remains unknown. Although the cell membrane has been proposed as a main interaction site, it has been observed that some cellular processes that occur far from it (i.e., protein synthesis) can also be altered by EMF exposure (32). EMF research on gene transcription is thus focused on the identification of some essential initial event in the interaction of an EMF with a biological system that could be useful to define a mechanism of interaction (33). Tuinstra et al. (27), using *Escherichia coli* and HeLa cell extracts, found that a MF of 1.1 mT and 45 Hz enhanced transcription. Based on these results it was hypothesized that alternative sites to membranes and/or interaction mechanisms may be capable of inducing a bioresponse in cells exposed to EMF. Belyaev et al. (31) suggest that changes in DNA-protein interaction due to the exposure to ELF MF involve electromagnetic as well as chemical processes.

Direct interaction with DNA as a result of an enzymatic cascade initiated at the cell membrane has been proposed by Blank and Goodman (30) as a possible mechanism of MF action. Research done on the Na K-ATPase (ion pump) enzyme in cell membrane showed that above a critical activation level, MF tend to stimulate enzymatic activity. Identification of a mechanism by which EMF exposure alters key processes is one of the greatest challenges in this area.

Altered protein synthesis of eukaryotic and prokaryotic cell cultures has been reported by Mittenzwey (7) and Blank (34). Protein p21 levels, which are argued to play a critical role at the membrane/cytoskeleton interface and regulation of some of the physical exchanges between the cell and its environment, showed a reduction of 70% compared to unexposed control cells after 16 hours of exposure to a 0.1 mT and 72 Hz pulsed MF. However, at longer exposure times, the protein level increased similarly to the control (26). The MF-induced increase in a given protein might be the result of more messages being available for translation, equivalent amounts of messages being differentially translated, or some combination of both (27).

Kropinski et al. (32), working with *E. coli* cells at midlog growth phase exposed for 2 hours to a 60 Hz MF of 3 mT at a constant temperature of 30°C, did not find any change in protein synthesis after a comparison with unexposed cells. In other experiments when the temperature was slowly increased (3°C over 70 min), protein synthesis was changed, suggesting that a gradual temperature rise with EMF exposure may profoundly affect the critical threshold for induction of protein synthesis in cells.

Cell size may also play an important role among variables in EMF exposure experiments. These variables include the characteristics of the signals, cell position within the coil, temperature regulation, exposure duration, field intensity, induced electric field, and rate of coil current change.

C. Effects on Microorganisms

Reported MF effects on bacteria have been varied and in general contradictory. All kinds of results can be found in the literature. Several studies have demonstrated that low-intensity MF are capable of altering growth in certain microorganisms by either stimulation or inhibition. Mehedintu and Berg (35) found that for a fixed frequency (50 Hz) and a field intensity of 0.2 mT, the cell number in *Saccharomyces cerevisiae* was decreased by 16 ± 5%, but when 0.5 mT was applied the cell number increased 25 ± 5%. They also studied the frequency dependence of cell growth, and at 0.5 mT obtained peaks at 15 and 50 Hz, both of which showed more than a 20% increase in cell number after 10 hours of exposure. Another suggestion was that temperature may play an important role since when this was decreased the stimulative effect of the 50 Hz and 0.5 mT almost disappeared (±3% stimulation), while for 15 Hz the stimulation was slightly increased (25 ± 5%).

The stimulative effect that MF cause on some microorganisms has been used to enhance the production of certain substances such as cellulases by *Trichoderma reesei*, nourseothricin by *Streptomyces noursei*, and nitrate reductase by *Pseudomonas stutzeri* as proved by Hönes et al. (36). For the last microorganism, biomass was increased 25–30% after 7 hours when the field intensity was 1.3 mT. However, at a field intensity of 0.6 mT only a 7% increase was observed after 8 hours of treatment.

Belyaev et al. (31) studied the genome conformational state response of *E. coli* cells (with densities varying from 5×10^5 to 10^9 cell/mL) exposed to a weak ELF (30 μT, 9 Hz) through anomalous viscosity time dependencies and observed a cell-to-cell interaction effect during the response to the MF. Saffer and Phillips (8) also identified cell density as an important factor in EMF experiments.

Mittenzwey et al. (7) reported that in experiments with MF of 1–10 mT and sinusoidal waves of 2–50 Hz with exposure times of 7.5 hours and temperatures from 25 to 37°C, a 3.8% difference in cell number between control and treated samples was observed. However, one must be very careful when attributing small differences to a MF effect, since many sources of variation could influence the results. No differences were observed in protein synthesis, and in other experiments no effect was found on *E. coli, Bacillus thuringiensis*, and *Proteus vulgaris*.

Saito et al. (37) reported no significant difference in the number of bacteria at each measured time due to exposure to a homogeneous static MF of intensities 150 and 450 mT incubated on normal broth for 12 hours at 37°C. However, when using a medium in which bacteria could not grow or only grow slowly (i.e., using saline solution to dilute the media), a statistically significant extension of the generation time was observed. The exposed culture generation time was extended four to five times compared to the control group. In the same work HeLa cells were incubated for 4 hours with different strains of adhesive *E. coli* under 450 mT to investigate the parasite-host relationship under the influence of a MF. The MF was observed to significantly increase the number of bacteria adhered to HeLa cells. However, when both bacteria and HeLa cells were individually exposed to the MF prior to the adhesive study, no changes were observed in the number of bacteria adhered. In this same paper it was reported that Ramon et al. (38) cultured *E. coli* cells at 0°C and exposed them to the ELF MF from 60 to 600 Hz to examine the growth. As a result they found that the bacteria exposed to this MF was inhibited by approximately 40% in comparison with the control after 60 hours of incubation and that the surface layer structure of the former had been destroyed.

VI. HIGH-INTENSITY MAGNETIC FIELDS

Reports on the effects of high MF exist in a variety of fields and with very different objectives. Distinct groups of investigators are working in areas such as fluid mechanics, environmental sanitation, foodborne pathogens, and metabolic studies to report the effects of high MF of different intensities and types on water, ion solutions, macromolecules such as proteins and nucleic acids, lipid bilayers, membranes, microorganisms, plants, and higher animals (1,5,39).

Another potential application of high MF is to enhance microbial degradation in environmental pollutants such as phenol biodegradation in sludge and bioremediation of hydrocarbon-contaminated soil. Ueno and Iwasaka (40) observed that water flowing through a plastic tube across a superconducting magnet reduced its flow as the MF increased and that it actually stopped at an intensity of 8 T. It was thus implied that when living systems are exposed to high gradient MF, diamagnetic liquids such as water are affected by magnetic forces, and it was hypothesized that exposure to such fields could affect the organisms due to their high water content.

It has also been suggested that high-intensity MF can affect membrane fluidity and other properties of cells (41). Magnetic orientation of diamagnetically anisotropic domains in artificial bilayers has been reported, and in studies with unilamelar liposome systems, a well-defined structural change in organization during phase transitions has been observed. In several cases a magnetically induced orientation of membrane phospholipids with temperatures on the phase transition of those molecules was found. In other experiments solute release by unilamelar membranes was enhanced during brief exposures to 7.5 T MF and again at temperatures near the characteristic phase transition (~41°C). In this way it was shown that MF-induced membrane permeability changes are possible in simple phospholipid bilayers and that these field interactions display a dependence on the structural organization of the membrane. However, changes in the curvature of the membrane bilayer due to MF near transition temperatures have also been observed. As in low-intensity MF, available binding or ligand sites have been proposed as a target site for high-intensity MF (41,42).

Studies on the effects of high MF on enzyme activities in vitro have revealed acceleration, inhibition, or no observable effect (43). Weissbluth et al., as cited in Frankel and Liburdy (41), reported no effect of up to 14 T MF on peroxidase and xanthine oxidase. Iwasaka and Ueno (44) noted than when applying 22 T MF to the enzymes, the reaction rate of hydrogen peroxide and catalase were 50–85% lower than controls when not treated with nitrogen gas bubbles, but when the reaction mixture was bubbled with nitrogen gas, no MF effect was observed. Since the dissolved oxygen concentration in the reaction mixture exposed to MF increased 20–25% compared to the controls, it was assumed that the MF affected the dynamic movement of the oxygen bubbles produced by the decomposition of hydrogen peroxide, but not the catalytic activity of the catalase itself (44).

The most important results of the application of MF to foods were those reported by Hoffman in his 1985 patent (45). As seen in Table 1, up to a 3 log cycle inactivation was achieved with no significant increase in temperature or appreciable changes in food quality as judged by consumer panelists.

Studies on various microorganisms have yielded different results: Malko et al. (46) observed no significant differences in a *Saccharomyces cerevisiae* count subjected to 1.5 T static MF for 15 hours at 30°C when compared to controls exposed to the geomagnetic field. However, Nakamura et al. (43) found that *Bacillus subtilis* cell numbers under a 5.2–6.1 T inhomogeneous MF were about

Table 1 Food Microorganism Inactivation

Microorganism	Product	Field intensity (T)	Number of pulses	Frequency (kHz)	Temperature		m.o./cm³		Differences detected by test panel
					Initial	Final	Before	After	
Streptococcus thermophilus	Milk	12	1	6	23	24	25,000	970	No
Saccharomyces cerevisiae	Yogurt	40	10	416	4	6	3,500	25	No
Saccharomyces cerevisiae	Orange juice	40	1	416	20	21	25,000	6	No
Mold spores	Dough	7.5	1	8.5	—	—	3,000	1	No

Source: Ref. 45.

Table 2 Effect of High-Intensity Magnetic Fields on Different Microorganisms

Microorganism	Type of MF	Field strength (Tesla)	Frequency of pulse (Hz)	Effect	Ref.
Wine yeast cells	Homogeneous SMF	1.1	0	No effect for 5, 10, 20, 40, or 80 min exposure	50
Serratia marcescens	Heterogeneous SMF	1.5	—	Growth rate equivalent to controls up to 6 h; decreases between 6 and 7 h; increases between 8 and 10 h. At 10 h, cell population equivalent to controls	51
Staphylococcus aureus	Heterogeneous SMF	1.5	0	Growth rate increases between 3 and 6 h; then decreases between 6 and 7 h. Cell population at 7 h is same as in controls	51
Staphylococcus aureus	Homogeneous SMF	1.4	—	Inhibition of growth after 16 h. No effect observed when exposure was interrupted hourly for 3 s	52
Saccharomyces cerevisiae	Homogeneous SMF	7.3	—	Increased survival to UV light for cultures grown in a MF prior to irradiation; application of MF after irradiation decreases survival	53

Saccharomyces cerevisiae	Homogeneous SMF	1.49	—	No statistical differences from controls	46
Saccharomyces ellipsoide	Homogeneous SMF	0.1–1.2	—	No observable differences	54
Claviceps purpurea	Homogeneous SMF	1.1–1.4	—	Possible slight influence of MF on germination and growth direction of germ tubes	54
E. coli	Homogeneous SMF	11.7	0	Neither mutagenic nor lethal effect of the MF was observed	55
E. coli	Homogeneous SMF	11.7	0	Super high MF accelerates or reduces the growth rate depending on the media When temperature was increased from 30 to 40°C, the growth rate was decreased in both media	56
E. coli	Homogeneous SMF	7	0	Comparing the ratio # of cells under HMF/# cells control:	57
E. coli	Heterogeneous SMF	5.2–6.1	0	before 6 h of treatment ratio <1 after 24 h of treatment ratio >1	
Bacillus subtilis	Homogeneous SMF	7	0	No difference in cell number between treated and untreated samples	43
Bacillus subtilis	Heterogeneous SMF	5.2–6.1	0	Twofold cell number with respect to the control due to a slower death rate of the vegetative cells	43

SMF = Static magnetic field.

twofold higher than that of the reference after 72 hours. The declining cell numbers were judged to mainly be due to autolysis of the vegetative cells, while the rate of their decline was reduced under the influence of the MF. The effect of the inhomogeneous MF was determined more important in the stationary phase, and it was further hypothesized that the constant changing of extracellular nutrients and chemical conditions during cultivation gave rise to metabolic differences between the logarithmic and stationary phase cells. This resulted in a greater concentration of components susceptible to high MF such as water (40), ion solutions, macromolecules such as proteins and nucleic acids (47), and the lipid bilayers of the membrane (48). Additionally, the enzyme activity produced during sporulation was reduced during the early stages under the inhomogeneous MF, but no differences were observed when the bacteria were subjected to a homogeneous 7 T MF. When genetically transformed, *B. subtilis* MI113 (pC112) was subjected to an inhomogeneous MF, the production of the antibiotic surfactin was increased due to higher cell numbers (49).

Tsuchiya et al. (39), working with homogeneous (7 T) and inhomogeneous (5.2–6.1 T and 3.2–6.7 T) MF, found a growth stage–dependent response of *E. coli* bacterial cultures. The ratio of cells under MF to cells under geomagnetic field was <1 during the first 6 hours of treatment and >1 after 24 hours. These authors also found that cell survival was more pronounced under inhomogeneous than homogeneous fields. Based on the assumption that MF could act as a stress factor, cells collected after 30 minutes of MF treatment or in the stationary phase after long-term treatment were heated to 54°C. No differences were observed between the treated and control samples.

Okuno et al. (14) investigated the influence of different bacteria strains, growth media, amino acids, and temperature on the growth of *E. coli* bacteria exposed to a homogeneous 11.7 T MF and concluded that the effect of the MF was dependent on neither the strain nor the media used. Variable results were obtained when different amino acids were added to the media, such as when the addition of cesamino acid (rich in amino acids) to an inorganic media shifted the growth rate of bacteria in comparison to the control. At lower temperatures (30°C) the growth was accelerated, while the reduced growth became apparent with an increase in temperature (45°C). These results confront the hypothesis that MF influence free radicals in enzyme reactions since growth inhibition should be diminished as temperature increases. In Table 2, the results obtained by different authors on high MF are summarized.

VII. CONCLUDING REMARKS

The use of MF as an alternative nonthermal technology for food preservation is still far from reach, as discrepancies in results and inability to replicate experiments reflect the complex nature of this technology.

Although microorganism inactivation may be the most desirable attribute of MF technology, its use should not be limited to this purpose. Some interesting alternative applications include low-intensity static MF as biomass promoters in the fermentation industry and biotechnology processes in either the food or pharmaceutical industries.

Finally, it should be stated that although this technology has not yielded consistent results on the inactivation of foodborne pathogens, it is far from being put aside. The interaction between MF and biological systems is not fully understood, and many factors such as intensity and frequency of MF, the physiological state of the organism, and other environmental variables such as temperature add complexity to an already intricate problem. On the other hand, development of new materials and technology makes possible overtaking previous maximum intensity MF boundaries.

REFERENCES

1. UR Pothakamury, GV Barbosa-Cánovas, BG Swanson. Magnetic-field inactivation of microorganisms and generation of biological changes. Food Technol 47(12):85–93, 1993.
2. W Grundler, F Kaiser, F Keilman, J Walleczek. Mechanism of electromagnetic interaction of cellular systems. Naturwissenschaften 79:551–559, 1992
3. M Blank. Biological effects of electromagnetic fields. Bioelectrochem Bioenerg 32: 203–210, 1993.
4. CB Grissom. Magnetic fields effect on biology: a survey of possible mechanisms with emphasis on radical pair recombination. Chem Rev 95:3–24, 1995.
5. PE Kovacs, RL Valentine, PJJ Alvarez. The effect of static magnetic fields on biological systems: implications for enhanced biodegradation. Crit Rev Environ Sci Technol 22(4):319–382, 1997.
6. MN Zhadin. Combined action of static and alternating magnetic fields on ion motion in a macromolecule: theoretical aspects. Bioelectromagnetics 19:279–292, 1998.
7. R Mittenzwey, R Sübmuth, W Mei. Effects of extremely low-frequency electromagnetic fields in bacteria—the question of a co-stressing factor. Bioelectromagnetics 40:21–27, 1996.
8. JD Saffer, JL Phillips. Evaluating the biological aspects of in vitro studies in bioelectromagnetics. Bioelectrochem Bioenerg 40:1–7, 1996.
9. RJ Fitzsimmons, JT Ryaby, FP Magee, DJ Baylink. Combined magnetic fields increased net calcium flux in bone cells. Calcif Tissue Int 55:376–380, 1994.
10. PA Valberg, R Kavet, CN Rafferty. Can Low-level 50/60 Hz electric and magnetic fields cause biological effects? Radiation Res 148:2–21, 1997.
11. FS Barnes. Mechanisms for biological changes resulting from electric and magnetic fields. Proceedings of the 1994 Japan-U.S. Science Seminar on Electromagnetic Field Effects Caused by High Voltage Systems. Preliminary Edition. Sapporo, Japan, 1994.

12. FE Van Nostran, RJ Reynolds, HG Hedrick. Effects of a high magnetic field at different osmotic pressures and temperatures on multiplication of *Saccharomyces cerevisiae*. Appl Microbiol 15:561–563, 1967.
13. TE Vaughan, JC Weaver. Molecular change due to biomagnetic simulation and transient magnetic fields: mechanical interference constrains on possible effects by cell membrane pore creation via magnetic particles. Bioelectrochem Bioenerg 46:121–128, 1998.
14. K Okuno, K Tuchiya, T Ano, M Shoda. Effect of super high magnetic field on the growth of *Escherichia coli* under various medium compositions and temperatures. J Ferment Bioeng 75(2):103–106, 1993.
15. AR Liboff. Cyclotron resonance in membrane transport. In: A Chiabrera, C Nicolini, HP Schwan, eds. Interactions Between Electromagnetic Fields and Cells. NATO ASI Series. Series A: Life Sciences 97, 1985, pp. 281–293.
16. VV Lednev. Possible mechanism for the influence of weak magnetic fields on biological systems. Bioelectromagnetics 12:71–75, 1991.
17. CF Blackman, JP Blanchard, SG Benane, DE House. Empirical test of an ion parametric resonance model for magnetic field interactions with PC-12 cells. Bioelectromagnetics 15:239–260, 1994.
18. LA Coulton, AT Barker. Magnetic fields and intracellular calcium: effects on lymphocytes exposed to conditions for 'cyclotron resonance.' Phys Med Biol 38:347–360, 1993.
19. S Galt, J Sandblom, Y Hamnerius, P Höjevick, E Saalman, B Norden. Experimental search for combined AC and DC magnetic field effects on ion channels. Bioelectromagnetics 14:315–327, 1993.
20. SD Smith, BR McLeod, AR Liboff, K Cooksey. Calcium cyclotron resonance and diatom mobility. Bioelectromagnetics 8:215–227, 1987.
21. SD Smith, BR McLeod, AR Liboff. Testing the ion cyclotron resonance theory of electromagnetic field interaction with odd and even harmonic tuning for cations. Bioelectromagnetics 38:161–167, 1995.
22. FS Smith. Lithium as a normal metabolite: some implications for cyclotron resonance of ions in magnetic fields. Bioelectromagnetics 9:387–391, 1988.
23. DT Edmonds. Larmor precession as a mechanism for the detection of static and alternating magnetic fields. Bioelectrochem Bioenerg 30:3–12, 1993.
24. R Adair. A physical analysis of the ion parametric resonance model. Bioelectromagnetics 19:181–191, 1998.
25. FL Cozens, JC Scaiano. A comparative study of magnetic field effects on the dynamics of geminate and random radical pair processes in micelles. J Am Chem 115: 5204–5211, 1993.
26. JL Phillips, W Haggren, WJ Thomas, T Ishida-Jones, WR Adey. Effect of 72 Hz pulsed magnetic field exposure in RAS p21 expression in CCRF-CEM cells. Cancer Biochem Biophys 13:187–193, 1993.
27. R Tuinstra, B Greenebaum, EM Goodman. Effects of magnetic fields in cell-free transcription in *E. coli* and HeLa extracts. Bioelectromagnetics 43:7–12, 1997.
28. BR McLeod, SD Smith, AR Liboff. Calcium and potassium cyclotron resonance curves and harmonics in diatoms (*A. coffeaeformis*). J Bioelect 6(2):153–168, 1987.
29. MS Markov, AA Pilla. Weak static magnetic field modulation of myosin phosphory-

lation in a cell-free preparation: calcium dependence. Bioelectrochem Bioenerg 43: 233–238, 1998.
30. M Blank, R Goodman. New and missing proteins in the electromagnetic and thermal stimulation of biosynthesis. Bioelectrochem Bioenerg 21:307–317, 1989.
31. IY Belyaev, YD Alipov, AY Matronchik. Cell density dependent response of *E. coli* cells to weak ELF magnetic fields. Bioelectromagnetics 19:300–309, 1998.
32. AM Kropinski, WC Morris, MR Szewczuk. Sinusoidal 60 Hz electromagnetic fields failed to induce changes in protein synthesis in *Escherichia coli*. Bioelectromagnetics 15:283–291, 1994.
33. JL Phillips. Effects of electromagnetic field exposure on gene transcription. J Cell Biochem 51:381–386, 1993.
34. M Blank. Biological Effects of electromagnetic fields. Bioelectrochem Bioenerg 32: 203–210, 1993.
35. M Mehedintu, H Berg. Proliferation response of yeast *Saccharomyces cerevisiae* on electromagnetic field parameters. Bioelectrochem Bioenerg 43:67–70, 1997.
36. I Hönes, A Pospischil, H Berg. Electrostimulation of proliferation of denitrifying bacterium *Pseudomonas sutzeri*. Bioelectrochem Bioenerg 44:275–277, 1998.
37. A Saito, J Arisawa, K Kimura, G Matsumoto. Effect of static magnetic field on bacterial growth and its adhesion to HeLa cells. Proceedings of the 1994 Japan–US Science Seminar on Electromagnetic Fields Effects Caused by High Voltage Systems. Sapporo, Japan 330–339, 1994.
38. C Ramon, M Ayaz, D Streeter. Inhibition of growth of *Escherichia coli* induced by extremely low-frequency weak magnetic fields. Bioelectromagnetics 2:285–289, 1981.
39. K Tsuchiya, K Nakamura, K Okuno, T Ano, M Shoda. Effect of homogeneous and inhomogeneous high magnetic fields on the growth of *Escherichia coli*. J Ferment Bioeng 81(4):343–346, 1996.
40. S Ueno, M Iwasaka. Properties of diamagnetic fluid in high gradient magnetic fields. J Appl Phys 75(10):7177–7179, 1994.
41. RB Frankel, RP Liburdy. Biological effects of static magnetic fields. In: C Polk, E Postow, eds. Handbook of Biological Effects of Electromagnetic Fields, 2nd ed. Boca Raton, FL: CRC Press, 1995.
42. H Aoki, H Yamazaki, T Yoshino, T Akagi. Effects of static magnetic fields on membrane permeability of a cultured cell line. Research Communications in Chemical Pathology and Pharmacology 69(1):103–106, 1990.
43. K Nakamura, K Okuno, T Ano, M Skoda. Effect of high magnetic field on the growth of Bacillus subtilis measured in a newly developed superconducting magnet biosystem. Bioelectrochemistry and Bioenergetics 43:123–128, 1997.
44. M Iwasaka, S Ueno. Influence of intense magnetic fields on enzymatic processes of SOD, peroxidase, xanthine oxidase and catalase. Second World Congress for Electricity and Magnetism in Biology and Medicine (abstract). 1997.
45. GA Hoffman. Deactivation of microorganisms by an oscillating magnetic field. US Patent 4,524,079, 1985.
46. JA Malko, I Constantinidis, D Dillejay, WA Fajman. Search for influence of 1.5 Tesla magnetic field in growth of yeast cells. Bioelectromagnetics 15:495–501, 1994.

47. G Maret, K Dransfeld. Macromolecules and membrane in high magnetic fields. Physica B 86–88, 1977.
48. JB Speyer, PK Sripada, SK Das Gupta, G Graham-Shipley, RG Griffin. Magnetic orientation of sphingomyelin-lecithin bilayers. Biophysical Journal 51:687–691, 1987.
49. M Shoda, K Nakamura, K Tsuchiya, K Okuno, T Ano. Bacterial growth under strong magnetic field. Second World Congress for Electricity and Magnetism in Biology and Medicine (abstract), 1997.
50. GC Kimball. The growth of yeast in a magnetic field. J Bacteriol 35:109–122, 1938.
51. VF Gerencser, MF Barnothy, JM Barnothy. Inhibition of bacterial growth by magnetic fields. Nature 196:539–541, 1962.
52. HG Hedrick. Inhibition of bacterial growth in homogeneous fields. In MF Barnothy, ed. Biological Effects of Magnetic Fields, Vol. 1. Plenum Press, New York:240–245, 1964.
53. B Schaarschmidt, I Lamprecht, K Muller. Influence of a magnetic field on the UV sensitivity in yeast. Z. Naturforsch., 29C, 447–448, 1974.
54. DJ Montgomery, AE Smith. A search for biological effects of magnetic fields. Biomed Sci Instrum 1:123–125, 1963.
55. K Okuno, T Ano, M Shoda. Effect of super high magnetic field on the growth of Escherichia coli. Biotech Letter 13(10):745–750, 1991.
56. K Okuno, K Tsuchiya, T Ano, M Shoda. Effect of super high magnetic field on the growth of Escherichia coli under various medium compositions and temperatures. Journal of Fermentation and Bioengineering 75(2):103–106, 1993.
57. K Tsuchiya, K Nakamura, K Okuno, T Ano, M Shoda. Effect of homogeneous and inhomogeneous high magnetic fields on the growth of Escherichia coli. Journal of Fermentation and Bioengineering 81(4):343–346, 1996.

16
Microbial Inactivation by High Pressure

Dallas G. Hoover
University of Delaware
Newark, Delaware

I. INTRODUCTION

A. Objective and Prospectus

The driving force for study and development of what can be called nonthermal food-processing methods is the quest for an effective preservative process that can be delivered to foods without substantial detrimental changes to sensory and nutritional qualities caused by exposure to cooking temperatures. Demand for additive-free and shelf-stable foods has also added impetus. Methods that by convention fall into this category include high hydrostatic pressure processing (HPP) as well as high-intensity pulsed electric fields, high-intensity pulsed light, oscillating magnetic fields, and submegahertz ultrasound (1). Irradiation can be included in this group because of its nonthermal nature, but since it has an established presence in food processing dating back to 1951, it is sometimes excluded from what is considered ''the new technologies.'' Other electromagnetic and electrothermal processing approaches that have been associated in this burgeoning field are pulsed x-ray or ultraviolet, combined ultraviolet light and low-concentration hydrogen peroxide, ozone, carbon dioxide/pressure, filtration, use of bacteriocins, and even mild ohmic, inductive, and microwave heating. Evidently the boundaries for this area are not definitive, and with use of multiple/combined treatments in food processing, distinctions are often blurred.

In the 1980s there was a seeming dearth of information regarding inactivation kinetics of HPP for important food microorganisms. Such is no longer the case; now the literature is relatively voluminous concerning the inactivation of microorganisms (and enzymes) in foods processed by HPP. Principles (or ''rules of thumb'') that have become established or reestablished are:

1. Increasing the pressure magnitude or time of pressure will increase the number of microorganisms inactivated (with bacterial endospores the exception).
2. Pressure inactivation rates will be enhanced by exposure to acidic pH or an increase in the pressure-treatment temperature above ambient.
3. Actual foods are more pressure-protective for microorganisms than buffers or pressure menstrua of microbiological media.
4. Gram-positive bacteria are usually more barotolerant than gram-negative bacteria (although there are notable exceptions).
5. Younger (an exponential phase) cells are more pressure-sensitive than old (stationary or death phase) cells.
6. Incomplete inactivation of microorganisms by pressure can result in injured cells capable of recovery under optimal growth conditions.
7. The more developed the life form, the more sensitive it is to pressure.

Regarding HPP as a food-processing technology, the greater the pressure level and time of application, the greater the changes to the sensory quality of the food. This is especially true for high-protein foods, where pressure-induced protein denaturation will be visually evident, and for physically fragile foods where high hydrostatic pressures will cause structure deformation. Usually these changes are undesirable because the food or beverage can then be easily determined to be processed and no longer fresh or raw; however, applications and food products do exist wherein the disadvantages of HPP are masked or nominal and its advantages are commercially viable. Food products brought to market that currently employ HPP in their manufacture includes fruit jellies and jams, fruit juices, raw squid, rice cakes, foie gras, and guacamole.

Primary factors that impact upon HPP inactivation rates are:

1. Type of microorganism
2. Culturing or growth conditions of the microorganism(s)
3. Composition, pH, and water activity of the food
4. Temperature, magnitude, and time of pressurization

Additional secondary factors influence the effectiveness of HPP. For example, the redox potential of the pressure menstruum may also play a role in the inactivation of some microorganisms (2).

The focus of this chapter is the use of HPP to pasteurize or sterilize foods; however, HPP can also be utilized to create unique textural properties in foods, such as in meat tenderization or in the coagulation and gelation of protein foods (3). Additional uses for HPP that have been explored or used are the regulation of enzymatic reactions, gelatinization of starch to tenderize legumes or grains, restructuring of foods (e.g., powder agglomeration), rapid formation of small ice crystals (reversible decrease in the ice melting point), reversible increase in the melting point of lipids, gas removal or solubilization with carbon dioxide, extraction of constituents (such as pectin) from food, and surface impregnation or coating of foods (e.g., adsorption of minerals or vitamins). Some of these food applications have been in use in some form for years.

Current commercial applications of HPP are limited to batch (pressure processing of prepackaged foods) and semi-batch (used with pumpable product, as in the case of fruit juices). Lack of a continuous processing system limits the throughput of product and therefore significantly affects the economics of HPP use in foods. A hold time of 7–10 minutes is common in the industry; however, since the effects of pressure are transmitted uniformly and instantly throughout the food regardless of shape or size, issues of energy transfer do not exist for HPP as they do for heat transfer in thermal processing.

B. Historical Perspective*

The French physiologist P. Regnard studied the effects of high hydrostatic pressure on invertebrates and fish under laboratory conditions of 10–100 MPa in 1884 (4). Regnard showed that pressures below 100 MPa reversibly affected enzymatic processes in bacteria. The first report of high hydrostatic pressure killing bacteria was by H. Royer in 1895 (4).

In food science and technology, the most important work (if not the most cited reference by authors of HPP papers) was that by Bert Hite, published in June 1899 (5). Hite originally experimented with the application of high hydrostatic pressure on foods and food microorganisms. He showed that the shelf life of raw milk could be extended by about 4 days after pressure treatment at 600 MPa for one hour at room temperature. Souring was delayed for about 24 hours after treatment at 200 MPa. In later work, Hite et al. (6) found that most pressure-treated fruits and vegetables remained commercially sterile for at least 5 years after processing at pressures ranging from 400 to 820 MPa. Hite's last contribution to the field was in 1929 (7) in which tobacco mosaic virus was treated at pressures more than 930 MPa.

At this time the effects of HPP on proteins were first being investigated. The Nobel Prize–winning physicist P. W. Bridgman (8) demonstrated that pressures of 800 MPa coagulated ovalbumin irreversibly. He first published on the

* In the literature a variety of pressure units have been used. I have taken the liberty of converting pressure magnitudes into megaPascals (MPa) to make comparison of cited work less cumbersome. The conversion is: 1 atmosphere = 14.696 pounds/in^2 = 1.013 bar = 0.1013 MPa.

shrinkage of water under pressure (i.e., 4% at 100 MPa, 7% at 200 MPa, 11.5% at 400 MPa, 15% at 600 MPa at 22°C).

In 1918, Larsen and coworkers (9) confirmed that HPP can inhibit microbial growth and cause cells to die. Vegetative types were killed after 14 hours at 607 MPa. It was recognized that spores of bacteria were extremely resistant to inactivation by pressure but could be killed at 1214 MPa.

In later years, Timson and Short (10) pressurized milk at 1034 MPa at 35°C for 90 minutes and learned that approximately 0.05% of the bacterial population was capable of surviving this pressure. Microbial analysis identified the survivors as spores of *Bacillus subtilis* and *Bacillus alvei*. It was suggested that the lethal effect of pressure was more evident in the solid phase than the liquid phase of water. That is, *B. subtilis* survived solid-phase transitions from Ice II, III, and V to Ice I. They found a neutral pH to be more protective to the spores than acid pH. Additionally, the presence of NaCl or glucose counteracted the damaging effect of pressure encountered at acid and alkaline pH. In their 1965 article, Timson and Short (10) noted a 1932 paper by J. Basset and M. A. Macheboeuf, who reported the survival of spores of *B. subtilis* exposed to more than 1724 MPa (250,000 psi) for 45 minutes!

At the IFT Annual Meeting in 1974, D. C. Wilson (11) presented a paper reestablishing the use of pressure and elevated temperatures as a food preservation method. Use of low pressures of 0.14 MPa with pasteurization temperatures of 82–103°C was effective for the sterilization of low-acid foods in sealed containers. The combination of heat with hydrostatic pressure realized a dramatic synergistic effect. At 0.35 MPa and 100°C the D value was 280 minutes for gram-positive spore-forming bacteria, while at 138 MPa and 100°C the D value was 2.2 minutes. Consequently, substantial reductions in microorganisms can be achieved when co-treatments of heat and pressure are utilized. Because of the instantaneous transmission of pressure, a 2–5 kg container can be processed as fast as a 30 g container if initial temperatures are in the pasteurization range (12).

II. FUNDAMENTALS OF PRESSURE APPLICATIONS

The various effects of high hydrostatic pressure are commonly organized into cell envelope–related effects, pressure-induced cellular changes, biochemical aspects, and effects on genetic mechanisms. It has been established that cellular morphology is altered by pressure and that cell division slows with the application of increasing pressures. Hydrostatic pressures of 100–300 MPa can induce spore germination and resultant vegetative cells are usually more sensitive to environmental conditions (13).

As a general rule (LeChatelier's principle), pressure enhances reactions that lead to volume decrease, and reactions involving increases in volume are generally inhibited or reduced by pressure application (14). Proteins will respond variably to pressure largely because hydrophobic interactions act in a peculiar manner under pressure. Up to pressures of 100 MPa hydrophobic interactions tend to result in a volume increase, but beyond this pressure range a volume decrease is associated with hydrophobic interactions and the pressure tends to stabilize these reactions (15). Consequently the extent of hydrophobicity of a protein will determine to a large degree the extent of protein denaturation at any given pressure (16). Additional factors for enzyme inactivation are the alteration of intramolecular structures and conformational changes at the active site (17). Enzyme inactivation under pressure is also affected by pH, substrate concentration, and subunit structure of the enzyme (18).

Pressurized membranes normally show altered permeabilities. A reduction in volume occurs along with a reduction in the cross-sectional area per phospholipid molecule. Pressure-induced membrane malfunctions cause inhibition of amino acid uptake, probably due to membrane protein denaturation, and it is generally felt that for microorganisms the primary site of pressure damage is the cell membrane (19). Numerous studies have shown loss of intracellular constituents from microorganisms during and after pressure treatment. Leakage of these components from the cells indicate damage to the cellular membrane and the higher amount lost from cells correlates with a greater degree of death and injury.

Bacteria with a relatively high content of diphosphatidylglycerol (shown to cause rigidity in membranes in the presence of calcium) are more susceptible to inactivation by HPP (20). Conversely, those compounds that enhance membrane fluidity tend to impart resistance of the organism to pressure (21).

Other important sites for pressure inactivation of microbial cells are enzymes, especially membrane-bound ATPases (22,23). Enzymes vary in their sensitivities to denaturation. It is assumed that in some organisms denaturation of key enzymes by pressure plays an important role in pressure-induced death and injury.

Numerous publications have presented electron micrographs of microbial cells that were exposed to different pressure levels. At pressures greater than 500 MPa it is not uncommon to view physical disruption to the surface of cells using scanning electron microscopy. At levels <500 MPa, it is possible to observe internal cellular damage using transmissible electron microscopy. Perrier-Cornet et al. (24) measured cell volume during high pressure application. An image analysis system was connected to a light microscope. For *Saccharomyces*, 250 MPa generated an observed compression rate of 25% with partial irreversibility of cell compression (10%) upon return to atmospheric pressure. The occurrence of mass transfer implied cell disruption or increase in membrane permeability.

III. BACTERIA

A. Genetic Implications

As noted earlier, the genetic predisposition of marine bacteria to hydrostatic pressure was realized in the nineteenth century. More recently, modern genetic analysis has identified genes used by bacteria for survival and proliferation in a high-pressure aqueous environment. *Methanococcus thermolyticus, Rhodotorula rubra*, and *E. coli* have been shown to synthesize proteins in response to hydrostatic pressure (25).

Bartlett et al. (26) cloned a gene, *ompH*, encoding for a pressure-inducible protein from a gram-negative eubacterium isolated from an ocean depth of 2.5 km (1.6 mi). *OmpH* was expressed at 28.4 MPa (optimal growth pressure for organism), but not at 0.1 MPa. It was known that in deep-sea bacteria, cultivation at high pressure results in changes to the degree of saturation of membrane lipids and production rates of several proteins. Data suggested that *ompH* functions in some manner in channeling of nutrients through the outer membrane under elevated pressures.

Kato et al. (27) isolated several barophilic and barotolerant bacteria from deep-sea sediments. The barophilic strains were generally too difficult to work with because they would only grow in the laboratory under high-pressure conditions. As a result, Kato et al. (27) studied the gene structure of one barotolerant isolate, identified as strain DSS12. Previous work from this laboratory indicated that pressure-regulated promotor regions are shared among many high-pressure–adapted marine bacteria. The genes of strain DSS12 that was found to be pressure-regulated, and therefore play a role in pressure adaptation, were sequences with high homology to proline dehydrogenase and ATP-dependent RNA helicase from *E. coli*.

Yano et al. (28) isolated two taxonomically unidentifiable bacteria, strains 16C1 (facultatively barophilic) and 2D2 (obligately barophilic), from the intestinal contents of deep-sea fish retrieved from depths of 3100 and 6100 m, respectively. In these bacteria there was a general trend from saturated to unsaturated fatty acids (especially docosahexaenoic acid [DHA], 22:6n-3) in the membrane with exposure to increasing magnitudes of pressure with growth. Their results suggested that DHA is an important factor in maintaining membrane fluidity under pressure. Furthermore, this same compositional change in the membrane was evident in strain 16C1 with growth at low temperatures.

B. Pure Culture Studies of Vegetative Bacteria

There does exist a trend where heat-resistant bacteria are usually more pressure-resistant than heat-sensitive types, but there are notable exceptions. *Salmonella* provides an example. *Salmonella* Senftenberg 775W is the most heat-resistant *Salmonella* known (29). In buffer, its decimal reduction time at 57.5°C is 15 minutes (30). Comparison with a heat-sensitive strain of *Salmonella* Typhimurium ($D^{57.5°C}$ = 3 min) showed *S.* Senftenberg 775W to be consistently more pressure-sensitive. *S.* Senftenberg was reduced by 5 $\log_{10}$ cycles within 30 minutes in buffer at 345 MPa and 23°C. It was also found that significant metabolic injury occurred in salmonellae that survived pressurization, and recovery of these cells was possible with incubation at 37°C in a nonselective enrichment medium. These data suggest that cells sublethally stressed by pressure may be more susceptible to a further means of inactivation if recovery (return of the capability to replicate) is prevented.

Vibrio parahaemolyticus, a marine bacterium that is also an important foodborne pathogen, is substantially more sensitive to the effects of high hydrostatic pressure than *Listeria monocytogenes*, an important gram-positive pathogen common in raw foods (31). A 10^6 CFU/mL population of *L. monocytogenes* is inactivated within 20 minutes by a 345 MPa pressurization in buffer at 23°C, while a similar concentration of *V. parahaemolyticus* is eliminated in half the time (10 min) at half the pressure (173 MPa) in clam juice. Milk, as compared to buffer, offers a protective effect for *L. monocytogenes*. This was similar to the protection afforded to pressurized *Salmonella* by strained-chicken baby food (30). Pressure in combination with low pH between 3.0 and 4.0 in citrate buffer destroyed *L. monocytogenes* populations of approximately 10^7 CFU/mL within 30 minutes (32). Treatment at pressures above 304 MPa at pH < 6.0 also resulted in no detectable survivors when tryptic soy agar plus 0.6% yeast extract was used as the plating medium; however, when *Listeria* recovery agar was used, approximately 10^2 CFU/mL were recoverable. These surviving cells represent an injured subpopulation that cannot recover at pH < 5.6.

Fujii et al. (33) evaluated several plating media to judge the effect of pressure-induced injury of *E. coli, V. parahaemolyticus*, and *L. monocytogenes*. These bacteria were pressure-treated to generate survival rates of 10–50% of the starting viable concentration. Their results showed that plating media such as Trypticase soy agar and nutrient agar were superior to brain heart infusion agar (BHI) and plate count agar (PCA) in the detection of sublethally treated cells exposed to pressure. Detection levels could be improved for BHI and PCA by the addition of horse blood. As anticipated, detection was relatively low for selective media and variable depending on the selective ingredients in these media.

Satomi et al. (34) observed a sharp drop in survivors and injury rate in *E. coli* above 182 MPa that corresponded to release of UV-absorbing substances. Barotolerance in *E. coli* was not affected by the type of growth media used to propagate the cells nor the presences of oxygen; however, barotolerance did increase with age of the culture and increase of osmotic pressure in the pressurizing menstruum. Barotolerance was reduced with a decline of pressure menstruum pH and pressure treatment at 44°C. Additional studies by Satomi et al. (35) assessed conditions of optimal recovery for *E. coli* and *V. parahaemolyticus* following exposure to debilitating levels of HPP. Parameters for most rapid recovery of *E. coli* were nutrient medium with <1.0% NaCl, pH 7.0 at 30–37°C, and aerobic incubation in nutrient medium of 0.5–3.0% NaCl, pH 7.0 at 37°C for *V. parahaemolyticus*.

Patterson et al. (36) examined several vegetative types of food-poisoning bacteria to HPP. *Yersinia enterocolitica* was the most sensitive bacterium in the study, reduced 5 $\log_{10}$ cycles with 275 MPa for 15 minutes in PBS. For comparable 5 $\log_{10}$ reductions using 15-minute treatments, *S.* Typhimurium required 350 MPa, *L. monocytogenes* required 375 MPa, *Salmonella* Enteritidis 450 MPa, *E. coli* O157:H7 required 700 MPa, and *Staphylococcus aureus* 700 MPa. The bacteria tended to be more barotolerant in UHT milk than meat or buffer. The authors remarked that the variability of pressure response in bacteria was dependent upon bacterial strain differences and different pressure substrates.

Patterson and Kilpatrick (37) used HPP against *E. coli* O157:H7 NCTC 12079 and *S. aureus* NCTC 10652 in milk and poultry. Their findings showed a practical necessity for combined use of pressure and elevated temperatures. Alone, neither treatment displayed effective inactivation of the pathogens. In UHT milk, 400 MPa/50°C/15 min reduced populations of *E. coli* approximately 5 $\log_{10}$ CFU/g, and 500 MPa/50°C/15 min delivered reductions of approximately 6 $\log_{10}$ CFU/g for *S. aureus*. In minced irradiation-sterilized poultry meat, *E. coli* was reduced by approximately 6 $\log_{10}$ CFU/g by 400 MPa/50°C/15 min, and *S. aureus* exposed to 500 MPa/50°C/15 min was reduced by approximately 5 $\log_{10}$ CFU/g. Also, polynomial expressions derived from the Gompertz equation were used to devise models to predict inactivation of each pathogen at different pressure-temperature combinations.

HPP of *L. monocytogenes* and *S.* Typhimurium in fresh pork loin was investigated by Ananth et al. (38), who found that at 25°C the D-values at 414 MPa were 2.17 minutes for *L. monocytogenes* and 1.48 minutes for *S.* Typhimurium. A treatment of 414 MPa/13 min/25°C inactivated either pathogen inoculated at levels of approximately 10^6 per chop, and there were also no detectable psychrotrophic plate counts from the pork loin after 7 days of storage at 4°C. After 7 days plate counts climbed, and at 33 days they nearly reached 10^6 CFU/g. Interestingly, sensory analysis (triangle test of difference) showed that samples cooked after pressurization were different ($p > 0.05$) from controls, but only for samples pressure-treated at 2°C, not at 25°C. It was determined that generally, pressure-treated meat was not significantly different from controls in sensory quality, and HPP did extend the shelf life of the product.

The effects of HPP on *L. monocytogenes* and pork chops were also studied by Mussa et al. (39) with pressurizations apparently conducted at ambient temperature. Strain Scott A was found to have a D-value at 400 MPa of 3.5 minutes while the indigenous microbiota of the pork was found to have a D-value at 400 MPa of 1.3 minutes.

Adaptation of HPP to ovine milk processing was investigated by Gervilla et al. (40) with study of samples of 6% fat inoculated with *E. coli* and *Pseudomonas fluorescens*. The test strain of *E. coli* was most resistant when pressure-treated at 10°C and was more resistant than the test strain of *P. fluorescens* that was most barotolerant when pressure-treated at 25°C. Inactivation in ovine milk of >6 $\log_{10}$ CFU/mL was attained for *E. coli* when treated at ≥450 MPa at 25°C for 5 minutes and for *P. fluorescens* treated at ≥400 MPa at 10°C for 5 minutes. Use of a pressure-treatment temperature of 50°C produced equivalent reductions of bacterial populations with use of 400 MPa for *E. coli* and 300 MPa for *P. fluorescens*.

In another study by Gervilla et al. (41), 6%-fat ovine milk was again used. *Listeria innocua* 910 CECT was investigated with special regard to pressure-treatment temperatures. Pressurizations at 2°C were superior to processing at ambient temperature (25°C) for *L. innocua* inactivation, while pressure treatment at 50°C was more effective than pressurization at 2°C. Complete elimination of starting inocula of 10^7–10^8 CFU/mL in ewes' milk was accomplished by the following conditions: 2°C/450 MPa/15 min; 10°C/450 MPa/15 min; 25°C/450 MPa/15 min; and 50°C/350 MPa/15 min. Five- and 10-minute treatment periods were also examined and found to require an additional 50 MPa increase in pressure magnitude with the exception that 5-minute treatments at 50°C required 450 MPa for complete destruction of *L. innocua*. The authors noted that the fat in ewes' milk has been shown to confer thermal protection for *L. monocytogenes* and *L. innocua* and voiced concern that this same character may increase the resistance of *Listeria* spp. and other detrimental bacteria treated with combinations of pressure and temperature in ovine milk.

Listeria innocua 910 CECT was examined in liquid whole egg by Ponce et al. (42); however, in this product, starting inocula of approximately 10^6 CFU/mL could not be totally inactivated with use of 300–450 MPa at −15 to 20°C for up to 15 minutes. The most effective treatment examined in this study (450 MPa/20°C/15 min) showed a reduction of about 5 $\log_{10}$ CFU/mL.

Three strains of *L. monocytogenes* showed a wide range of pressure sensitivities (43); Scott A was not eliminated by exposure to 450 MPa for 30 minutes at ambient temperature, another strain (Lm2) was eliminated at 400 MPa after 15 minutes (starting concentration for both 5×10^8 CFU/mL). These cultures were pressurized in phosphate-buffered saline (PBS) amended with bovine serum albumin (protein), glucose (carbohydrate), and olive oil (lipid); these components were found to protect the listeriae against pressure inactivation as compared to PBS alone.

Simpson and Gilmour (44) examined the barotolerance of 13 enzymes from three strains of *L. monocytogenes* that demonstrated a range of sensitivities to HPP. They found no evident trends between the pressure resistance of any specific enzyme and the strain from which it was derived, suggesting that none of the selected enzymes was the primary site of pressure inactivation in *L. monocytogenes*.

C. Spores

Unless extremely high hydrostatic pressures are used, mild heat is an additional requirement for effective elimination of bacterial endospores in low-acid foods. The articles described in this section present indications of the pressure magnitudes, treatment temperatures, and times of exposure necessary to deal with spores and describe the complex role spore germination plays in this process. Spores present the greatest challenge for inactivation by HPP.

Clouston and Wills (45) examined the effect of hydrostatic pressure up to 170 MPa at 25°C on the heat and radiation resistance of spores of *Bacillus pumilus*. Initiation of germination occurred at pressures exceeding 500 MPa and was the prerequisite for inactivation by compression. It was assumed that there was a net decrease in the volume of the system during initiation as a result of increased solvation of the spore components.

Butz et al. (46) investigated the effects of pressures between 150 and 400 MPa at temperatures of 25–40°C on bacterial spores and showed that pretreatment at relatively low pressures (60–100 MPa) led to accelerated inactivation of spores at high pressure. Several papers on use of HPP to inactivate spores have made similar suggestions for a two-exposure treatment with HPP to enhance the inactivation of spores; the first exposure to germinate or activate the spores, and the second exposure at a higher pressure to inactivate the germinated spores.

Effects of combined pressure (200–400 MPa) and temperature (20–90°C) on the reduction of *B. stearothermophilus* spores have been examined (47). Limited effects were found when spores were pressurized at 0.1 MPa (1 atmosphere) in conjunction with temperatures up to 90°C or 400 MPa and 20°C; however, marked effects on spore counts were observed when pressurized between 200 and 400 MPa at temperatures between 60 and 90°C. Initial counts of 3×10^6 CFU/mL were reduced to <10 at 90°C and 200 MPa, at 80°C and 350 MPa, or at 70°C and 400 MPa.

Kakugawa et al. (48) also examined heat and pressure effects on spore suspensions of *B. stearothermophilus*. Viable counts could be reduced from 10^6

to 10^2 spores/mL in 30 minutes by treatment at 110°C and 200 MPa and in 10 minutes by exposure to 100°C and 400 MPa. Attempts to reduce the viable spore counts below 10^2/mL could not be accomplished even after 50 minutes at 120°C and 400 MPa.

The effect of hydrostatic pressure on activation of *Bacillus* spp. spores as a preparatory state for synchronous germination was investigated by Nishi et al. (49). These workers found that activation of *B. subtilis* spores in milk by 200 MPa from 25 to 60°C resulted in a greater rate of spore germination than exposure to 80°C. They reported that most of the pressure-activated spores germinated within 1 hour of exposure to 37°C as indicated by loss of heat resistance.

Okazaki et al. (50) examined spores of *B. subtilis, B. coagulans*, and *C. sporogenes* PA3679 at pressures up to 400 MPa in combination with temperatures ranging from 25 to 110°C. In phosphate buffer, it was found that for the strains selected, spores of *B. subtilis* were more pressure-resistant than spores of *B. coagulans* and spores of *B. coagulans* were more pressure-resistant than spores of *C. sporogenes*. As a result, high treatment temperatures were required to eliminate spores suspensions of $\sim 10^7$/mL. At ambient temperature, Crawford et al. (51) was able to reduce *C. sporogenes* by 5 $\log_{10}$ cycles after 60 minutes at 680 MPa.

Rovere et al. (52) examined pressure-treatment parameters for inactivation of spores of *C. sporogenes* PA3679 starting with concentrations of approximately 10^5 spores/mL and pressure-hold times of 5 minutes. Elimination of these spore levels was possible with processes of 1400 MPa at 54°C and 800 MPa at 75°C in different model food systems. In a study involving spore suspensions of PA3679 in meat broth, Rovere et al. (53) noted that pressure acts as a complementary synergistic process to allow reduction of the thermal processing parameters necessary to eliminate problematic sporeformers in foods. Processing at 108°C and 800 MPa was found to be the most effective treatment with a calculated D-value of 0.695 minutes Heat treatment (110°C) alone generated a D of 13.3 minutes for spores of PA3679.

The pressure sensitivity of strains of several species of bacilli and *C. sporogenes* PA3679 were evaluated by Gola et al. (54). Pressure treatments of ~900 MPa for 10 minutes at 30°C were unable to completely destroy 8.4×10^2 *C. sporogenes* spores/mL in truffle cream. Total inactivations of *Bacillus cereus* (starting concentration 4×10^5 spores/mL), *B. licheniformis* ($\sim 6 \times 10^6$ spores/mL), and *B. stearothermophilus* ($\sim 4 \times 10^5$ spores/mL) were successful using a 20°C double-pulse treatment (200 MPa/1 min followed by 900 MPa/1min), 800 MPa for 3 minutes at 60°C, and 800 MPa for 3 minutes at 70°C, respectively, in phosphate buffer.

Ludwig et al. (55) found that the best conditions to germinate spores of *Bacillus* spp. were medium pressure, high temperature, and some additives such as salts, amino acids, and glucose. For spore suspensions of *B. stearothermophilus* exposed to 250 MPa and 60°C, a biphasic survivor curve was evident. It featured a rapid decrease in viability that represented the inactivation of vegetative cells, followed by a ''slow step'' that represented the spores. Similar results were obtained using a strain of *B. subtilis*. Ludwig et al. (55) noted that pressure only kills the germinated forms of the spores. Data were presented showing kinetics of germination as measured by the release of dipicolinic acid (DPA). Release was greatest at an ionic strength of 0.14 M NaCl and pressure between 100 and 250 MPa. Full germination (100% DPA release) was strongly dependent on treatment temperature; 40°C gave optimum germination within 30 minutes at 100 MPa.

More recent work by Ludwig et al. (56) with gram-positive sporeformers showed that *Clostridium sticklandii* ATCC 12662 was quite susceptible to HPP; cultures of 10^9 CFU/mL were eliminated from a 10-minute exposure to 300 MPa at 37°C, while treatment at 25°C and 300 MPa required 30 minutes for complete destruction. Pressure inactivation of this strain of *C. sticklandii* also showed biphasic kinetics, a large very sensitive population and a smaller, more resistant fraction in the ratio of $10^6:1$ representing vegetative cells to spores. In addition, Ludwig et al. (56) examined the release of dipicolinic acid (DPA) and amino acids by *B. subtilis* and found that optimal release of these components occurred at 110 MPa and 50–60°C. The authors recommended that for maximum inactivation of spores of *B. subtilis*, pressure cycles between low (0.1–60 MPa) and high pressures (500 MPa) and temperatures as high as possible would represent the best approach. They added that spore inactivation is best achieved by complex interplay between temperature and pressure effects on germination and inactivation processes.

Raso et al. (57,58) presented data indicating that the temperature of sporulation affected the pressure resistance of spores of the foodborne pathogen, *Bacillus cereus*. They found that at any water activity (0.92–0.99) or pH (3.5–7.8) of pressure treatment, *B. cereus* sporulated at a lower temperature (e.g., 20°C) was more pressure-resistant than *B. cereus* sporulated at a higher temperature (e.g., 37°C). Also affected by sporulation at lower temperatures was initiation of spore germination. High concentrations of sucrose were found to protect the spores from pressure inactivation. The basis for heightened resistance of the spores was stated to be due to the mechanism of pressure inactivation. That is, it occurs in two stages; exposure to pressure first germinates the spores, then pressure inactivates the germinated forms. Raso et al. (58) measured germination by plate counts of pressure-treated spore suspensions exposed to 65°C for 30 minutes.

Work by Wuytack et al. (59) added further to clarification of the mechanism of germination and induced pressure resistance in spores of *B. subtilis*. They found that germination can be initiated at low (100 MPa) and high (500 MPa) treatments of 30 minutes; however, germination is arrested by exposure to 500 MPa resulting in a significant portion of the spores becoming pressure-resistant as well as more resistant to hydrogen peroxide and UV light. Such findings indicate that exposure of bacterial spores to pressure can result in spores not only more recalcitrant to the pressure process itself, but also more resistant to other accompanying food preservation methods, which can worsen conditions for effective elimination or reduction of spores.

It appears that the high variability of heat resistance of spores of clostridia is caused by the immediate environmental history of the spores (60). For example, spores of the food pathogen *C. perfringens* vary dramatically in their resistance to heat (61). It is assumed that the wide range of heat resistances is due to the diversity of environments from which *C. perfringens* has evolved, as well as the inducible nature of spore heat resistance triggered by compounds in the environment (62). As such, the heat resistance of the spore of *C. perfringens* is chemically reversible between the resistant and sensitive states. Heredia et al. (63) demonstrated that not only will spores of *C. perfringens* show increased heat resistance by a sublethal heat shock of 55°C at 30 minutes, but the vegetative cells will become more heat-resistant as well (at least two- to threefold). Spores of *Clostridium botulinum* held in calcium acetate solutions (0.1–0.5 M) for 140 hours at 50°C raises heat resistance 5–10 times, while heat resistance can be lowered by holding the spores in 0.1 N HCI at 25°C for 16 hours (64). Such phenomena have been indicated by exposure of spores to the natural acid conditions of some foods. In this regard, the extent of variability of clostridial spores to pressure is largely unknown. This is important, not only because of the pathogenic nature of these two species of *Clostridium*, but because *C. botulinum* also contains some very pressure-resistant spores. Spores of *C. botulinum* 17B have shown little if any reduction in viability after exposure to 825 MPa at 70°C (J. Larkin, personal communication, 1999).

Hayakawa et al. (65,66) found that six cycles of oscillatory pressurization (5 min each) at 600 MPa and 70°C were required to eliminate 10^6 spores/mL of *B. stearothermophilus* IFO 12550. Continuous treatments at pressures up to 800 MPa and 70°C for 60 minutes showed that some spores survived (Fig. 1). Attempts were made to reduce the treatment temperature for inactivation of the spore suspensions (10^6/mL), but the need for an elevated treatment temperature could not be eliminated. The only treatment that resulted in complete destruction of the spores was the oscillatory approach with a treatment temperature of 70°C (Fig. 2). Additionally, it was found that a synergistic effect of spore bacteriostasis existed with a sucrose palmitic acid ester (<10 ppm) used in combination with 60°C for 60 minutes against spores of *B. stearothermophilus* IFO 12550 (66).

Use of pressure-pulsing or oscillatory pressurizations, as stated above, has been shown to be generally more effective than equivalent single pulses or continuous pressurization of equal times. This enhanced inactivation has been found not only with spores, but with vegetative bacteria as well. The difference in effectiveness varies, and the measure of improved inactivation by oscillatory pressurization must be weighed against the design capabilities of the pressure unit, added

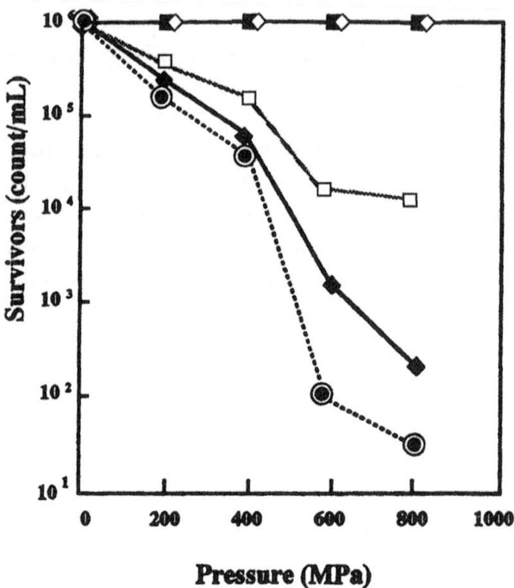

Figure 1 Effect of temperature and treatment time on the inactivation of spores of *Bacillus stearothermophilus* IFO 12550 by continuous pressurization: (■) 60 min at 20°C; (◇) 20 min at 20°C; (◆) 60 min at 60°C; (□) 20 min at 60°C; (◉) 60 min at 70°C. (From Ref. 66.)

wear on the pressure unit, possible detrimental effect to the sensory quality of the product, and additional time required for cycling. Besides pressure-pulsing, another modification to pressurization mechanics is the use of very rapid pressure release (measured in milliseconds). Rapid decompression can be attained in pressure units designed with a ''knuckle'' (a quick-release joint in the connecting rod linked to the piston applying pressure to the chamber) that permits a very rapid, but controlled release of high pressure. It is believed that rapid decompression invokes cavitations in the cells and spores that result in physical disruption and death. This approach is still quite novel, and further information is presently quite limited.

In the comparison of spore suspensions from six strains representing five different species of *Bacillus*, Nakayama et al. (67) found no correlation between pressure and heat resistances. Treatment conditions of 981 MPa at 5–10°C and neutral pH for 40 minutes demonstrated little, if any, reduction of spore viability. These findings indicated that use of pressure alone to eliminate spores at low temperature and neutral pH is not possible, requiring the use of elevated temperature with pressure treatment to inactivate spores (13,45). It was noted that the

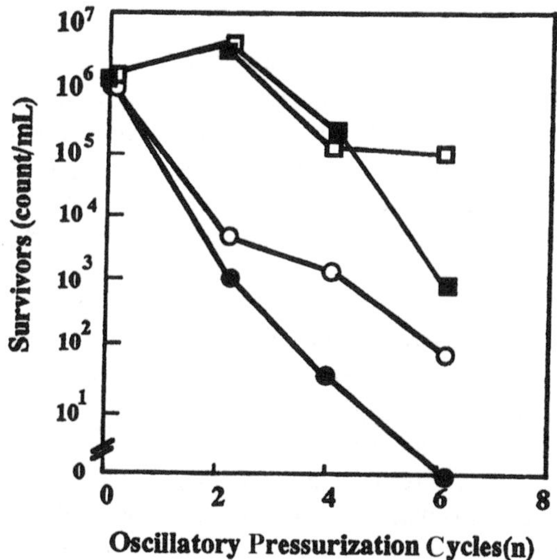

Figure 2 Effect of oscillatory pressurization cycles on spores of *Bacillus stearothermophilus* IFO 12550: (□) 400 MPa at 60°C; (○) 400 MPa at 70°C; (■) 600 MPa at 60°C; (●) 600 MPa at 70°C. (From Ref. 66.)

earlier work of Sale et al. (68) showed that exposure to elevated temperature germinated spores and made them susceptible to pressure inactivation, but unfortunately not all germinated spores are inactivated by pressure, making pressure sterilization of low-acid foods a fundamentally unreliable process.

For green infusion tea, Kinugasa et al. (69) found that 700 MPa at 70°C for 10 minutes produced commercial sterility, even in tea inoculated with spores of *Bacillus licheniformis, B. coagulans*, and *B. cereus* added at 10^6/mL. HPP was deemed superior to retort processing in that HPP had little or no effect on tea components, including catechins, vitamin C, and amino acids; the taste was unchanged.

IV. FUNGI

A. Yeasts

Except as in the case for ascospores of *Byssochlamys* and probably a few other species of fungi, eukaryotic microorganisms are normally more pressure-sensitive than prokaryotic microorganisms, and as a consequence fungi are easier to deal with than bacteria using HPP in a food preservation system. Of the fungi, yeasts have received the most attention in HPP research because of their importance in the spoilage of acid foods where bacterial endospores are of little concern. In low-pH foods, it is the yeasts and lactic acid bacteria that are the primary targets of preservative methods. It is no surprise that the first commercial HPP-treated products on the market were fruit products. In part, the slower-growing molds have been largely ignored in HPP research due to their pressure sensitivity and the availability of other preservative factors that can curtail mold growth using a hurdle approach in combination with HPP.

Examples of the lower levels of pressure that can be used to reduce spoilage caused by fungi include Hayashi (3), who found 200 MPa at 23°C effective for inactivation of yeasts and molds in freshly squeezed orange juice. Shimada et al. (70) measured extensive leakage of UV-absorbing substances released from cells of *Saccharomyces cerevisiae* after treatment at 400 MPa at room temperature that correlated with loss of viability. Freshly squeezed orange juice as well as juice inoculated with yeasts and molds when processed for 10 minutes at 400 MPa and 23°C showed no increase in total counts over 17 months of storage (71). Exposure of the orange juice to 45°C for 30 minutes allowed the lowering of pressure magnitude required for microbial inactivation (72); however, pressures in the 300–400 MPa range were unable to eliminate juice pectinesterase and peroxidase activities. Fungal spores could be inactivated at 25°C, whereas *Bacillus* spp. spores required 60°C and 608 MPa (73).

Oxen and Knorr (74) found that the viability of *Rhodotorula rubra* increased with a commensurate increase in sucrose concentration. Cultures of approximately 10^7 CFU/mL were inactivated by a 300 MPa treatment at 25°C in $<30\%$ sucrose and at 400 MPa in $<40\%$ sucrose. Without pressure, heat treatments of 70–80°C were necessary to generate similar levels of inactivation. From this work it was evident that HPP inactivation of yeasts in foods of a low water activity would be more challenging and require application of greater HPP parameters that foods and beverages with an a_w closer to 1.00.

Iwashasi et al. (75) suggested that the damage caused by HPP was essentially equivalent to the damage caused by high temperature and oxidative stress in yeast. The cellular membrane was noted as the primary lesion site. Their conclusion was based on observation of strains of *S. cerevisiae* by comparing tolerance under different applications of heat shock and recovery and different growth phases that also involved incorporation of HPP-resistant mutant strains. Comparable effects were found with HPP, heat treatment, and exposure to oxidative stress. It was suggested that plasma membrane ATPase may be the key component in tolerance of many environmental stresses in *Saccharomyces* spp.

For *S. cerevisiae* 2373 and *Zygosaccharomyces bailii* 36947, Pandya et al. (76) found that a population of approximately 10^7 CFU/mL was killed by a 10-minute exposure to 304 MPa at 25°C or 253 MPa at 45°C. These pressure parameters can be considered sufficient to inactivate the common types of yeast present in food. Under conditions where partial pressure inactivation of yeast was demonstrated, treatment at 45°C enhanced the lethality of the process. Pressure-induced sublethal stress or injury of yeast was apparent at exposure parameters that produced a 3–5 $\log_{10}$ CFU/mL differential between viable counts as measured on potato dextrose agar (PDA) and counts on PDA plus 35% glucose (a_w = 0.945; injured yeast unable to grow in a reduced water activity environment). Yeast injury was most pronounced in both yeast cultures when pressurized at 228 MPa for 30 minutes. Results showed that mild heat and acidity contributed to the effectiveness of death and injury of yeast by HPP and were worthy tools in the enhancement of yeast inactivation by HPP. Similar responses of yeasts to HPP were found by Chen and Tseng (77), with *S. cerevisiae* CCRC 20271 more heat- and pressure-resistant than *Zygosaccharomyces rouxii* CCRC 21873 using pressures up to 300 MPa and temperatures up to 55°C. The rate of inactivation of the yeast was found to be first-order, a finding normally encountered with other microorganisms. In most studies, the inactivation kinetics are linear for a time interval to allow determination of D-values. ''Tailing'' in survivor curves involving HPP is not uncommon.

Palou et al. (78) addressed the impact of water activity on the effectiveness of HPP inactivation of a strain of *Z. bailii* isolated from the heavy syrup of papaya. Using a treatment of 345 MPa at 21°C for 5 minutes, high water activities (>0.98) allowed complete elimination of added culture. When the water activity was lowered, pressure resistance of the yeast increased, as found by Oxen and Knorr (74). The protection was significant. At a water activity of 0.92, 345 MPa only reduced the population by 1 log cycle (starting inoculum ~10^7 CFU/mL).

Palou et al. (79) compared oscillatory application of HPP to single, continuous pressurizations using *Z. bailii*, finding that cyclic applications improved inactivation of the yeast in sucrose-amended (a_w = 0.98) Sabouraud glucose broth (pH 3.5); however, at least two 5-minute cycles were needed to detect a significant difference from a single pulse. Three cycles of 5 minutes each were necessary to generate a 1 $\log_{10}$ CFU/mL different in plate counts at the 276 MPa level as compared to a continuous application of 15 minutes. Come-up time (period necessary to reach treatment pressure) was approximately 2.7 minutes at 276 MPa; decompression was normally <15 s. It was assumed that the greater rate of inactivation of the yeast due to oscillatory HPP was due to greater injury to the cellular membrane from rapid changes in intracellular/extracellular differences at the membrane interface.

Important items of information not to be overlooked in HPP are the come-up times and pressure-release times. Obviously, long come-up times will add appreciably to the total process time and affect product throughout, but these periods will also affect inactivation kinetics of microorganisms, therefore consistency and awareness of these times are important in the process development of HPP.

B. Molds

Butz et al. (80) examined responses of the heat-resistant molds *Byssochlamys nivea, Byssochlamys fulva, Eurotium (Aspergillus fischeri), Eupenicillium* spp., and *Paecilomyces* spp. to HPP (300–800 MPa) used in combination with different treatment temperatures (10–70°C). All the vegetative forms were inactivated by exposure to 300 MPa/25°C with a few minutes; however, ascospores required treatment at higher pressures. A treatment of 600 MPa at 60°C eliminated all ascospores within 60 minutes except for the ascospores of *B. nivea* and *Eupenicillium. B. nivea* required 800 MPa and a processing temperature of 70°C to destroy a starting inoculum of $<10^6$/mL within 10 minutes, while 600 MPa at 10°C was adequate to eliminate 10^7 CFU/mL *Eupenicillium* within 10 minutes. In the range of 4.0–7.0, pH was found to have little effect on pressure inactivation of *Byssochlamys* sp. On the other hand, low water activities (from sucrose $a_w = 0.89$) increased pressure sensitivity of ascospores, as did treatment in grape juice (as compared to saline solution).

V. VIRUSES AND PARASITES

As stated earlier, the first attempt to estimate the pressure sensitivity of viruses was by Giddings et al. in 1929 with tobacco mosaic virus (7). Among viruses, there is a high degree of structural diversity and this is reflected in a wide range of pressure resistances (81). Human immunodeficiency viruses are reduced by 10^4–10^5 viable particles from exposure to 400–600 MPa for 10 minutes (82). Brauch et al. (83) showed that bacteriophages (DNA viruses) were significantly inactivated by exposures to 300–400 MPa, while Butz et al. (84) found Sindbis virus (a lipid-coated virus) relatively unaffected by pressures of 300–700 MPa at −20°C.

VI. APPLICATIONS OF HPP FOR FOOD PRESERVATION

A. Products and Microorganisms

Horie et al. (85) presented work on the development of pressure-processed jams from the Meidi-ya Food Factory Co. in Japan, whose jams and preservatives marketed in 1991 were the first commercial foods that incorporated HPP for preservative purposes. Elimination of yeasts (*S. cerevisiae* and *Z. rouxii*) was reported as well as bacteria (*Staphylococcus* spp., *Salmonella* spp., and a coliform) in jam processed at 294 MPa for 20 minutes (starting inocula were between 10^5 and 10^6 CFU/mL). Some heating of the product was required by Japanese law. Refrigeration of the jam was necessary due to browning and flavor changes caused by enzymatic activities and chemical reactions involving oxygen. Taste panels were reported to prefer the pressure-processed varieties to the jams prepared in the conventional manner. Nutritionally, the pressure-processed strawberry jam retained 95% of its vitamin C compared to the fresh product.

Parish (86) studied HPP applied to nonpasteurized Hamlin orange juice (pH 3.7). The target organism was *S. cerevisiae*. He calculated D-values of 4–76 s for ascospores treated at pressures between 500 and 350 MPa, respectively. For vegetative cells of *S. cerevisiae*, D-values were between 1 to 38 s; for the native microbiota of the orange juice, the D-values ranged from 3 to 74 s. Surviving organisms in the orange juice at 1–300 s of HPP were found to be yeasts and gram-positive and gram-negative rods. Figure 3 shows a schematic of a HPP system designed for the processing of fluid products, such as orange juice.

Shelf-life extension of fresh-cut pineapple was realized by application of 340 MPa for 15 minutes by Aleman et al. (87). D-values as determined on plate count agar (PCA) were 3.0 minutes for processing at 4°C, 3.1 minutes when treated at 21°C, and >2.5 minutes at 38°C. The posttreatment counts on PCA from pressure-treated pineapple were <50 CFU/g. As noted earlier, the inherent

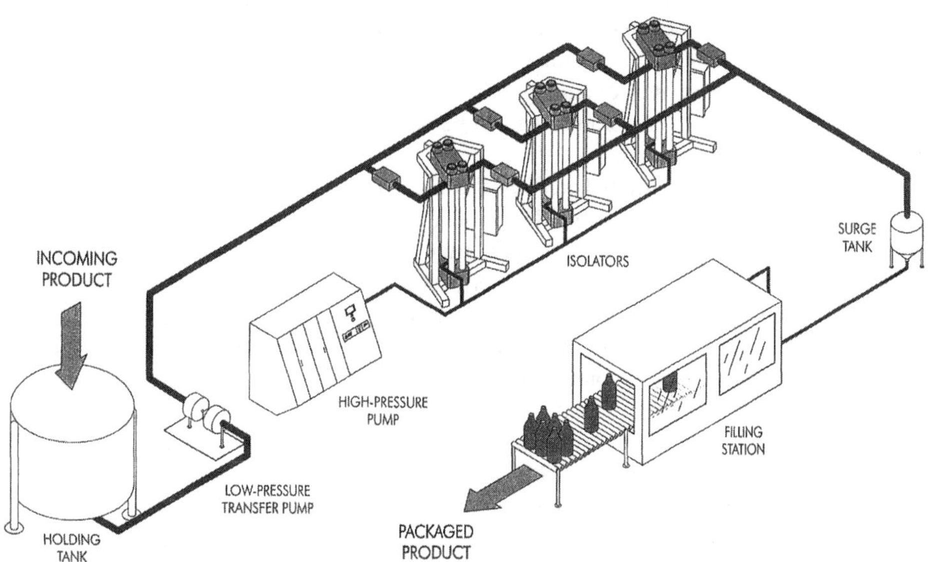

Figure 3 Schematic diagram of a pressure system designed for the processing of fluid products. (Courtesy of Flow International Corp., Kent, WA.)

acid pH of most fruit works very well with the ability of HPP to eliminate yeasts and vegetative bacteria, so much so that the limiting parameter of HPP in these products usually is the necessary inactivation of browning enzymes. As a result, blanching of the product is often required to eliminate this problem.

HPP application on nonpasteurized rice wine (Namazake) was examined by Hara et al. (88). All lactobacilli and yeasts were inactivated by a treatment at 294 MPa/10 min/25°C. Processing at 392 MPa resulted in a shelf-stable product with an equivalent taste profile as the control due to the inactivation of problematic enzymes and microorganisms.

Lettuce and tomatoes were inoculated by Arroyo et al. (89) and pressurized at 20°C for 10 minutes and 10°C for 20 minutes. Microorganisms were not significantly affected at 100 and 200 MPa, and gram-positive bacteria were not completely inactivated at 400 MPa (the highest pressure magnitude examined). Pressures of 300 and 350 MPa reduced populations of gram-negative bacteria, yeasts, and molds by at least one log cycle; however, in this range of pressures, skin loosened and peeled away in tomatoes, and lettuce underwent browning. The authors noted that hurdle technology would be necessary to maintain the desired organoleptic quality of these vegetables while using HPP as a preservation treatment to lower populations of undesired microorganisms.

In yogurt, Tanaka and Hatanaka (90) used HPP with the specific intent to reduce metabolic activity of the lactic acid bacteria used as starter culture so that the yogurt did not become over acidified with storage. This product defect was eliminated by a 200 MPa treatment for 10 minutes. Use of 300 MPa caused an undesirable drop in the remaining level of lactic acid bacteria with storage at 3°C. The treatment temperature had to be <20°C to avoid deterioration of the texture of the yogurt curd.

For the cheese strain *Lactobacillus helveticus* LHE-511, Miyakawa et al. (91) found that exposure at 400 MPa at 30°C for 10 minutes completely inhibited its acid-producing activity; however, two enzymes important in accelerated cheese-ripening (aminopeptidase and X-prolyl dipeptidyl aminopeptidase) were not affected by the treatment. In this case the potential was demonstrated for pressure treatment of cheese that would allow for expression of desirable flavor enzymes while excessive souring of the cheese is prevented at ripening temperatures.

HPP was adapted for preservation of spreadable smoked salmon cream (pH 5.95; a_w 0.95) by Carpi et al. (92). Pressure-treated products were superior to heat-treated creams with regard to sensory quality. A 3-minute exposure to 700 MPa extended shelf life at both 3 and 8°C from 60 to 180 days without changes in the sensory characteristics as compared to the product before treatment. In inoculated trials at 700 MPa at 3 minutes (starting inocula 10^3–10^4 CFU/g), *L. monocytogenes, S. aureus, S.* Typhimurium, and lactic acid bacteria were completely inactivated, while spores of sulfite-reducing clostridia were not affected and enterococci were only partially inactivated. Immediately after treatment the aerobic plate count was 3.0×10^3 CFU/g, but after 6 months of refrigerated storage the level was $<10^2$ CFU/g. Aerobic plate counts were mostly comprised of spores of *Bacillus*.

Carlez et al. (93) worked with freshly minced meat that was pressure-processed for 20 minutes at 20°C at 200–450 MPa and stored at 3°C in air and under vacuum for up to 22 days. They found treatment at 200 and 300 MPa was somewhat effective in that microbial growth was delayed 2–6 days. As one would expect, treatments at 400 and 450 MPa were more effective, reducing total counts of the meat by 3–5 $\log_{10}$ cycles. At the higher levels of pressure treatment, pseudomonads were the most problematic organisms in the meat. Data suggested that approximately 0.01% of the pseudomonads survived exposures to these pressures with subsequent growth at 3°C after a recovery period of 3–9 days. Lactobacilli also responded in a similar manner to such treatment. At the higher levels of pressure, there were evident changes in the color and texture of the minced meat.

Foie gras de canard (fatty duck liver) was produced with incorporation of an HPP preservative treatment by ElMoueffak et al. (94). Compared to classical thermal pasteurization of this product, 400 MPa at 50°C for 10 minutes stabilized the product as shown by reduction of the psychrotrophic microbiota, coliforms, and *S. aureus* below detectable levels with significant reduction of total mesophilic counts to approximately 10^2 CFU/g. Microbial analysis did not include an inoculated sample study, but instead used foie gras stored 13 days postslaughter to elevate the native microbial populations and allow estimation of the extent of inactivation. Treatment at 300 MPa was found to be ineffective for foie gras.

Fujii et al. (95) monitored changes in sensory quality and bacterial levels in minced mackerel pressure-treated at 203 MPa for 60 minutes and stored at 5°C. Growth of bacteria was delayed for approximately 4 days with populations of *Bacillus, Moraxella, Pseudomonas*, and *Flavobacterium* no longer evident after pressurization; however, coryneforms, *Staphylococcus*, and *Micrococcus* then dominated during refrigerated storage. It was noted that fat rancidity was enhanced in the pressurized mackerel, becoming a leading factor in deterioration of the product.

In surimi, Miyao et al. (96) found that levels between 300 and 400 MPa were adequate to kill most of the fungi, gram-negative bacteria, and gram-positive bacteria (in declining order). Notable pressure-resistant varieties were found and identified as *Moraxella* spp. (viable at 200 MPa), *Acinetobacter* spp. (viable at 300 MPa), *Streptococcus faecalis* (viable at 400 MPa), and *Corynebacterium* spp. (viable at 600 MPa). These pressure-treated isolates displayed significant lag time upon transfer to nutrient medium for batch culture. For example, following exposure to 400 MPa, growth of *S. faecalis* was delayed approximately 20 hours more as compared to control. The extracellular release of iron and magnesium ions, RNA, and carbohydrates was detected after pressurization, suggesting that damage to the membrane occurred and RNA degradation took place.

B. Enzymes

Although microorganisms are usually the top priority for inactivation or removal in food preservation systems, the deleterious effects of endogenous enzymes cannot be ignored. The problem of browning enzymes was noted earlier in this chapter. In the case of HPP-produced guacamole now being marketed in the United States, the browning enzymes determine the shelf life of the product, and processing conditions are understandably determined by successful inactivation of these proteins. Since enzymes are so highly variable in their response to high pressure (inactivation, no effect, or promotion of enzyme action), products need to be examined individually for overall effectiveness of processes employing HPP. Just a few examples are cited below as an addendum to the HPP work focused on inactivation of microorganisms.

Anese et al. (97) conducted pressure studies on crude enzymatic digests of carrots and apples to find that pressure (300–500 MPa for 1 min) caused a greatly enhanced activation of peroxidase and polyphenoloxidase (PPO) activities. These enzymes were completed inactivated after a 900 MPa treatment. The pH of the extracts did affect the degrees of enzymatic activation and inactivation; there was slight recovery of peroxidase activity 24 hours after pressure treatment (300–700 MPa). Castellari et al. (98) also found 900 MPa necessary to inactivate PPO in grape musts; 300–600 MPa resulted in only partial inactivation.

Papain required exposure to 800 MPa for 10 minutes for significant inactivation to occur (600 MPa had little effect), as reported by Gomes et al. (99). The extent of inactivation was increased when pressure treatment was conducted at 60°C (as compared to 20°C) and when oxygen concentration was very high (from an oxygen flush).

VII. HURDLE TECHNOLOGY EMPLOYING HPP IN FOOD PRESERVATION SYSTEMS

Hurdle technology is a term affixed to the evolved concept of using combinations of different preservation factors and techniques to achieve reliable preservation effects for foods and beverages (100). This recognition has given an identity to the approach and has instilled in food scientists an established manner for product and process development relying on cumulative effects to produce foods of enhanced quality, shelf life, and safety. In the case of HPP, a hurdle approach is almost automatic for any significant measure of widespread use in commercial

Table 1 Potential Hurdles for High-Pressure Use in Food Preservation

Physical hurdles: high temperature, low temperature, ultraviolet radiation, ionizing radiation, electromagnetic energy, photodynamic inactivation, ultra high pressure, ultra sonication, packaging film, modified-atmosphere packaging, aseptic packaging, and food microstructure
Physicochemical hurdles: low water activity, low pH, low redox potential, salt, nitrite, nitrate, carbon dioxide, oxygen, ozone, organic acids, lactic acid, lactate, acetic acid, acetate, ascorbic acid, sulfite, smoking, phosphates, glucono-δ-lactone, phenols, chelators, surface treatment agents, ethanol, propylene glycol, Maillard reaction products, spices, herbs, lactoperoxidase, and lysozyme
Microbially derived hurdles: competitive biota, protective cultures, bacteriocins, and antibiotics
Miscellaneous hurdles: monolaurin, free fatty acids, chitosan, and chlorine

Source: Ref. 100.

food processing. At the top of the difficulty list for broad application of HPP reside the inherent high resistances of bacterial endospores and food enzymes to hydrostatic pressure. Table 1 outlines the major hurdles that can be incorporated into food processing systems to preserve foods and beverages.

Covered elsewhere in this book is the subject of application of carbon dioxide under pressure. This method has been shown to harbor significant antimicrobial antagonism (101). This method, sometimes referred to as high-pressure carbon dioxide processing, does not employ the pressure levels associated with HPP. For example, Hong et al. (102) evaluated a CO_2 pressure process for the inactivation of lactobacilli in kimchi (fermented Korean vegetables, pH $\sim$ 4.2), and the optimal process parameters that decreased populations of lactobacilli by 5 $\log_{10}$ cycles were a 200-minute treatment at 30°C under a CO_2 pressure of 6.9 MPa. Ballestra et al. (103) examined pressures of 1.2, 2.5, and 5 MPa at 25, 35, and 45°C for the inactivation of *E. coli*; the higher treatment temperatures permitted a shortening of processing time to approximately 20 minutes for elimination of a cell suspension between 10^9 and 10^{10} CFU/mL in Ringer's solution. Again, the pressure magnitudes are modest by HPP standards but effective nonetheless, with carbon dioxide as the determinative antimicrobial force. The suggested lethal mechanism is a lowered intracellular pH caused by penetration of elevated levels of carbon dioxide into the cell, not by physical rupture of the cell walls or membrane due to the pressure of CO_2. Results were not as impressive for Wei et al. (104), who used 13.7 MPa for 2 hours at 35°C to kill inoculated *Salmonella* Typhimurium in chicken and egg yolk and inoculated *Listeria monocytogenes* in shrimp, orange juice, and egg yolk. Levels of microbial reduction varied considerably depending on the nature of the food and treatment conditions. Bacterial reductions ranged from limited effect to 9 log_{10} cycles. Results were poor for whole egg formulations. Enomoto et al. (105) reduced spores of *Bacillus megaterium* by 10^7 CFU/mL 30-hour exposures to 5.9 MPa and 60°C; above this pressure spore inactivation was lessened. An obvious commercial limitation for pressurized carbon dioxide is the lengthy processing times necessary to allow for diffusion of carbon dioxide into microbial cells.

Combination treatments of HPP and irradiation have been investigated by several laboratories. Paul et al. (106) targeted staphylococci in lamb meat. A population of approximately 10^4 staphylococci/g was reduced by only one log_{10} cycle by either treatment with gamma irradiation (1.0 kGy) or HPP (200 MPa for 30 min). When used in combination, no staphylococci were found immediately after completion of the tandem process. After 3 weeks of storage at 0–3°C, mannitol-negative staphylococci (presumably coagulase-negative as well) were detectable ($<10^3$ CFU/mL). Crawford et al. (51) were able to eliminate *C. sporogenes* in chicken breast using combinations of HPP and irradiation.

Many different antimicrobial compounds have been used in combination with HPP in a hurdle approach. Examples include HPP and lytic enzymes (lysozyme) (107), HPP and antimicrobial chitosans (108), and HPP and bacteriocins. Use of nisin with pressure has been addressed by several laboratories. Roberts and Hoover (109) examined the concurrent use of nisin with pressure treatment on *B. coagulans* 7050. While pressure alone (up to 400 MPa) had no effect in reducing the number of viable spores when treated at neutral pH and ambient temperature, the use of a 400 MPa/70°C/30 min pressure treatment at pH 4.0 and 0.8 IU/mL nisin resulted in the sterilization of spore crops containing 2.5×10^6 CFU/mL.

Kalchayanand et al. (110) examined the effectiveness of the pediocin AcH in combination with HPP. The goal of this work was to identify those HPP/AcH treatments capable of inactivating within 5 minutes 10^7–10^8 CFU/mL of *S. aureus, L. monocytogenes, S.* Typhimurium, *E. coli* O157:H7, *Lactobacillus sake, Leuconostoc mesenteroides, Serratia liquefaciens*, and *Pseudomonas fluorescens* in 0.1% peptone water. This could not be accomplished using HPP parameters of 345 MPa/50°C/5 min, unless 3000 AU/mL of pediocin AcH was included in the peptone water. Of the gram-negative bacteria, *E. coli* O157:H7 strain 932 was the most barotolerant, while for the gram-positive bacteria in the study, *L. sake* FM1 and *L. mesenteroides* Ly were the most pressure-resistant.

The monoterpenes were investigated by Adegoke et al. (111) in combination with HPP versus *S. cerevisiae*. Alone, *S. cerevisiae* IFO 10149 was found to be resistant to exposure to 300 and 600 μg/mL of α-terpinene but sensitive to a concentration of 1250 μg/mL. When 150 μg/mL of α-terpinene was combined with exposure to 177 MPa for 1 hour at 25°C, a reduction of 6 log_{10} cycles was found. A 3 log_{10} cycle reduction was found with similar pressure parameters but replacement of the α-terpinene with 200 μg/mL (+)-limonene.

Ishiguro et al. (112) examined the inactivation of *B. coagulans* in tomato juice with addition of the antimicrobial compounds, polylysine, protamine, and an extract of etiolated seedlings of adlay. Polylysine and protamine were ineffective processing aids; in fact, these compounds conferred protection to *B. coagulans* in the tomato juice treated at 400 MPa. The adlay extract did demonstrate enhanced destruction of *B. coagulans*, improving inactivation by approximately 1 $\log_{10}$ CFU/mL after 100 minutes. The treatment temperature was not specified, but regardless of that fact, treatment times of 100 minutes are not commercially practical.

VIII. HORIZONS FOR THE TWENTY-FIRST CENTURY

As shown by the presentation of just some of the literature presenting research in the field of HPP over the past 15 years, interest is expanding for the use of HPP in food production. Alternative methods to thermal processing have caught the eye of food scientists and engineers, entrepreneurs, regulatory officials, and the lay public. Design engineers will continue to improve pressure units to deal with current problems of reliability, capital cost, and the need for higher maximum pressures. Within the next 15 years it is realistic to expect many more new food products to emerge that incorporate HPP in their production. It will be interesting to see how successful these products will be.

REFERENCES

1. DG Hoover. Minimally processed fruits and vegetables: Reducing microbial load by nonthermal physical treatments. Food Technol 51(6):66–71, 1997.
2. DG Hoover. Pressure effects on biological systems. Food Technol 47(6):150–155, 1993.
3. R Hayashi. Application of high pressure to food processing and preservation philosophy and development. In: WE Spiess, H Schubert, eds. Engineering and Food, Vol. 2. New York: Elsevier, 1989, pp. 815–826.
4. K Heremans. Pressure effects of biochemical systems. H Kelm, ed. In: High Pressure Chemistry. Boston: D. Reidel Publishing Co., 1978, pp. 311–324.
5. BH Hite. The effect of pressure in the preservation of milk. Bull WV Univ Agric Exp Sta Morgantown 58:15–35, 1899.
6. BH Hite, NJ Giddings, CE Weakly. The effects of pressure on certain microorganisms encountered in the preservation of fruits and vegetables. Bull WV Univ Agric Exp Sta Morgantown 146:1–67, 1914.
7. NJ Giddings, HA Allard, BH Hite. Inactivation of the tobacco mosaic virus by high pressure. Phytopathology 19:749–750, 1929.

8. PW Bridgman. The coagulation of albumen by pressure. J Biol Chem 19(1):511–512, 1914.
9. WP Larsen, TB Hartzell, HS Diehl. The effects of high pressure on bacteria. J Infect Dis 22:271–279, 1918.
10. WJ Timson, AJ Short. Resistance of microorganisms to hydrostatic pressure. Biotechnol Bioeng VII:139–159, 1965.
11. DC Wilson. High pressure sterilization. 34th Annual Meeting of the Institute of Food Technologists, New Orleans, LA, May 12–15.
12. DG Hoover, C Metrick, AM Papineau, DF Farkas, D Knorr. Biological effects of high hydrostatic pressure on food microorganisms. Food Technol 43(3):99–107, 1989.
13. GW Gould, AJH Sale. Initiation of germination of bacterial spores by hydrostatic pressure. J Gen Microbiol 60:335–346, 1970.
14. FH Johnson, DH Campbell. The retardation of protein denaturation by hydrostatic pressure. J Cell Comp Physiol 26:43–46, 1945.
15. K Suzuki, Y Taniguchi. Effect of pressure on biopolymers and model systems. In: MA Sleigh, AG Macdonald, eds. The Effect of Pressure on Living Organisms. New York: Academic Press, Inc., 1972, p. 103.
16. R Jaenicke. Enzymes under extreme conditions. Ann Rev Biophys Bioeng 10:1–67, 1981.
17. C Suzuki, K Suzuki. The protein denaturation by high pressure. J Biochem 52:67–71, 1962.
18. KJ Laidler. The influence of pressure on rates of biological reaction. Arch Biochem 30:226–240, 1951.
19. KL Paul, RY Morita. Effects of hydrostatic pressure and temperature on the uptake and respiration of amino acids by a facultatively psychrophilic marine bacterium. J Bacteriol 108:835–843, 1971.
20. JPPM Smelt, AGF Rijke, A Hayhurst. Possible mechanism of high pressure inactivation of microorganisms. High Press Res 12:199–203, 1994.
21. NJ Russell, RI Evans, PF ter Steeg, J Hellemons, A. Verheul, T Abee. Membranes as a target for stress adaptation. Int J Food Microbiol 28:255–261, 1995.
22. BM Mackey, K Forestiere, N Isaacs. Factors affecting the resistance of *Listeria monocytogenes* to high hydrostatic pressure. Food Biotechnol 9:1–11, 1995.
23. RE Marquis, GR Bender. Barophysiology of prokaryotes and proton translocating ATPases. In: HW Jannisch, RE Marquis, AM Zimmerman, eds. Current Perspectives in High Pressure Biology. London: Academic Press, Ltd., 1987, pp. 65–73.
24. JM Perrier-Cornet, PA Marechal, P Gervais. A new design intended to relate high pressure treatment to yeast cell mass transfer. J Biotechnol 41:49–58, 1995.
25. DH Bartlett, C Kato, K Horikoshi. High pressure influences on gene and protein expression. Res Microbiol 146:697–706, 1995.
26. DH Bartlett, M Wright, AA Yayanos, M Silverman. Isolation of a gene regulated by hydrostatic pressure in a deep-sea bacterium. Nature 324(6249):572–574, 1989.
27. C Kato, A Ikegami, M Smorawinska, R Usami, K Horikoshi. Structure of genes in a pressure-regulated operon and adjacent regions from a barotolerant bacterium strain DSS12. J Mar Biotechnol 5:210–218, 1997.
28. Y Yano, A Nakayama, Kishihara, H Saito. Adaptive changes in membrane lipids

of barophilic bacteria in response to changes in growth pressure. Appl Environ Microbiol 64:479–485, 1998.

29. H Ng, GB Bayne, JA Garibaldi. Heat resistance of *Salmonella*, the uniqueness of *Salmonella senftenberg* 775W. Appl Microbiol 17:78–82, 1969.
30. C Metrick, DG Hoover, DF Farkas. Effects of high hydrostatic pressure on heat-resistant and heat-sensitive strains of *Salmonella*. J Food Sci 54:1547–1564, 1989.
31. MF Styles, DG Hoover, DF Farkas. Response of *Listeria monocytogenes* and *Vibrio parahaemolyticus* to high hydrostatic pressure. J Food Sci 56(5):1404–1407, 1991.
32. CM Stewart, FF Jewett Jr, CP Dunne, DG Hoover. Effect of concurrent high hydrostatic pressure, acidity and heat on the injury and destruction of *Listeria monocytogenes*. J Food Saf 17:23–36, 1997.
33. T Fujii, M Satomi, G Nakatsuka, T Yamaguchi. Effect of media on the detection rate of pressure-injured bacteria. Shokuhin Eiseigaku Zasshi (J Food Hyg Soc Jpn) 36(1):17–21, 1995.
34. M Satomi, T Yamaguchi, M Okuzumi, T Fujii. Effect of conditions on the barotolerance of *Escherichia coli*. Shokuhin Eiseigaku Zasshi (J Food Hyg Soc Jpn) 36(1): 29–34, 1995.
35. M Satomi, T Yamaguchi, M Okuzumi, T Fujii. Effect of several conditions on the recovery of pressure-injured bacteria. Shokuhin Eiseigaku Zasshi (J Food Hyg Soc Jpn) 36(3):344–351, 1995.
36. MF Patterson, M Quinn, R Simpson, A Gilmour. Sensitivity of vegetative pathogens to high hydrostatic pressure treatment in phosphate-buffered saline and foods. J Food Prot 58(5):524–529, 1995.
37. MF Patterson, DJ Kilpatrick. The combined effect of high hydrostatic pressure and mild heat on inactivation of pathogens in milk and poultry. J Food Prot 61(4):432–436, 1998.
38. V Ananth, JS Dickson, DG Olson, EA Murano. Shelf-life extension, safety and quality of fresh pork loin treated with high hydrostatic pressure. J Food Prot 61(12): 1649–1656, 1998.
39. DM Mussa, HS Ramaswamy, JP Smith. High-pressure destruction kinetics of *Listeria monocytogenes* on pork. J Food Prot 62(1):40–45, 1999.
40. R Gervilla, X Felipe, V Ferragut, B Guamis. Effect of high hydrostatic pressure on *Escherichia coli* and *Pseudomonas fluorescens* strains in ovine milk. J Dairy Sci 80:2297–2303, 1997.
41. R Gervilla, M Capellas, V Ferragut, B Guamis. Effect of high hydrostatic pressure on *Listeria innocua* 910 CECT inoculated into ewes' milk. J Food Prot 60(1):33–37, 1997.
42. E Ponce, R Pla, M Mor-Mur, R Gervilla, B Guamis. Inactivation of *Listeria innocua* inoculated in liquid whole egg by high hydrostatic pressure. J Food Prot 61(1): 119–122, 1998.
43. RK Simpson, A Gilmour. The effect of high hydrostatic pressure on *Listeria monocytogenes* in phosphate-buffered saline and model food systems. J Appl Microbiol 83:181–188, 1997.
44. RK Simpson, A Gilmour. The effect of high hydrostatic pressure on the activity of intracellular enzymes of *Listeria monocytogenes*. Lett Appl Microbiol 25:48–53, 1997.

45. JG Clouston, PA Wills. Initiation of germination and inactivation of *Bacillus pumilus* spores by hydrostatic pressure. J Bacteriol 97:684–690, 1969.
46. P Butz, V Trangott, H. Ludwig, J. Ries, H. Weber. The high pressure inactivation of bacteria and bacterial spores. Pharm Ind 52:487–491, 1990.
47. I Seyderhelm, D Knorr. Reduction of *Bacillus stearothermophilus* spores by combined high pressure and temperature treatments. ZFL Eur Food Sci 43:17, 1992.
48. K Kakugawa, T Okazaki, S Yamauchi, K Morimoto, T Yoneda, K Suzuki. Thermal inactivating behavior of *Bacillus stearothermophilus* under high pressure. In: R Hayashi, C Balny, eds. High Pressure Bioscience and Biotechnology. Amsterdam: Elsevier Science B.V., 1996, pp. 171–174.
49. K Nishi, R Kato, M Tomita. Activation of *Bacillus* spp. spores by hydrostatic pressure. Nippon Shokuhin Kogyo Gakkaishi 41(8):542–548, 1994.
50. T Okazaki, K Kakugawa, S Yamauchi, T Yoneda, K Suzuki. Combined effects of temperature and pressure on inactivation of heat-resistant bacteria. In: R Hayashi, C. Balny, eds. High Pressure Bioscience and Biotechnology. Amsterdam: Elsevier Science B.V., 1996, pp. 415–418.
51. YJ Crawford, EA Murano, DG Olson, K Shenoy. Use of high hydrostatic pressure and irradiation to eliminate *Clostridium sporogenes* in chicken breast. J Food Prot 59:711–715, 1996.
52. P Rovere, D Tosoratti, A Maggi. Prove di sterilizzazione a 15.000 bar per ottenere la stabilita microbiologica ed enzimatica. Indust Aliment XXXV:1062–1065, 1996.
53. P Rovere, A Maggi, N Scaramuzza, S Gola, L Miglioli, G Carpi, G Dall'Aglio. High-pressure heat treatments: evaluation of the sterilizing effect and of thermal damage. Indust Conserve 71:473–483, 1996.
54. S Gola, C Foman, G Carpi, A Maggi, A Cassara, P. Rovere. Inactivation of bacterial spores in phosphate buffer and in vegetable cream treated at high pressures. In: R Hayashi and C. Balny, eds. High Pressure Bioscience and Biotechnology. Amsterdam: Elsevier Science B.V., 1996, pp. 253–259.
55. H Ludwig, D Bieler, K Hallbauer, W Scigalla. Inactivation of microorganisms by hydrostatic pressure. In: C Balny, R Hayashi, K Heremans, P Masson, eds. High Pressure and Biotechnology. London: Colloque INSERM/John Libby Eurotext Ltd., Vol. 224, 1992, pp. 25–32.
56. H Ludwig, G Van Almsick, B Sojka. High pressure inactivation of microorganisms. In: R Hayashi, C Balny, ed. High Pressure Bioscience and Biotechnology. Amsterdam: Elsevier Science B.V., 1996, pp. 237–244.
57. J Raso, M Gongora, GV Barbosa-Canovas, BG Swanson. Effect of pH and water activity on the initiation of germination and inactivation of *Bacillus cereus* by high hydrostatic pressure. Institute of Food Technologists Annual Meeting Technical Program Abstracts, No. 59D-9, 1998, p. 153.
58. J Raso, GV Barbosa-Canovas, BG Swanson. Initiation of germination and inactivation by high hydrostatic pressure of *Bacillus cereus* sporulated at different temperatures. Institute of Food Technologists Annual Meeting Technical Program Abstracts, No. 59D-10, 1998, p. 154.
59. EY Wuytack, S Boven, CW Michiels. Comparative study of pressure-induced germination of *Bacillus subtilis* spores at low and high pressures. Appl Environ Microbiol 64:3220–3224, 1998.
60. JM Jay. Modern Food Microbiology. New York: Chapman & Hall, 1996, pp. 451–458.

61. KF Weiss, DH Strong. Some properties of heat-resistant and heat-sensitive strains of *Clostridium perfringens*. I. Heat resistance and toxigenicity. J Bacteriol 93:21–26, 1967.
62. G Alderton, N Snell. Bacterial spores: chemical sensitization to heat. Science 163: 1212, 1969.
63. NL Heredia, GA Garcia, R Luevanos, RG Labbe, JS Garcia-Alvarado. Elevation of the heat resistance of vegetative cells and spores of *Clostridium perfringens* type A by sublethal heat shock. J Food Prot 60(8):998–1000, 1997.
64. G Alderton, KA Ito, JK Chen. Chemical manipulation of the heat resistance of *Clostridium botulinum* spores. Appl Environ Microbiol 31:492–498, 1976.
65. I Hayakawa, T Kanno, M Tomita, Y Fujio. Application of high pressure for spore inactivation and protein denaturation. J Food Sci 59(1):159–163, 1994.
66. I Hayakawa, T Kanno, K Yoshiyama, Y Fujio. Oscillatory compared with continuous high pressure on *Bacillus stearothermophilus* spores. J Food Sci 59(1):164–167, 1994.
67. A Nakayama, Y Yano, S Kobayashi, M Ishikawa, K Sakai. Comparison of pressure resistances of spores of six *Bacillus* strains with their heat resistances. Appl Environ Microbiol 62(10):3897–3900, 1996.
68. AJH Sale, GW Gould, WA Hamilton. Inactivation of bacterial spores by hydrostatic pressure. J Gen Microbiol 60:323–334, 1970.
69. H Kinugasa, T Takeo, K Fukumoto, M Ishihara. Changes in tea components during processing and preservation of tea extract by hydrostatic pressure sterilization. Nippon Nogeikagaku Kaishi 66(4):707–712, 1992.
70. Shimada, Y Takada, T Denchi, R Hayashi, M Osumi. The structural damage and leakage of contents from *Saccharomyces cerevisiae* O-39 induced by hydrostatic pressure. In: R Hayashi, ed. Pressure Processed Food: Research and Development. Kyoto: San-ai Publ. Co., 1990.
71. H Ogawa, K Tukuhisha, K Sasai, Y Kubo, H Tukumoto. Effect of hydrostatic pressure on sterilization and preservation of various kinds of citrus juice. In: R Hayashi, ed. Pressure Processed Food: Research and Development. Kyoto: San-ai Publ. Co., 1990.
72. H Ogawa, K Fukuhisa, H Fukumoto, K Hori, R Hayashi. Effect of hydrostatic pressure on sterilization and preservation of freshly-squeezed, non-pasteurized citrus juice. Nippon Nogeikagaku Kaishi 63(6):1109–1114, 1989.
73. Y Taki, N Awao, N Mitsuura, Y Takagaki. Sterilization of Bacillus spores by hydrostatic pressure. In: R Hayashi, ed. Pressure Processed Food: Research and Development. Kyoto: San-ai Publ. Co., 1990.
74. P Oxen, D Knorr. Baroprotective effects of high solute concentrations against inactivation of *Rhodotorula rubra*. Lebensm Wissens Technol 26:220–223, 1993.
75. H Iwashashi, S Fujii, K Obuchi, SC Kaul, A Sato, Y Komatsu. Hydrostatic pressure is like high temperature and oxidative stress in the damage it causes to yeast. FEMS Microbiol Lett 108:53–58, 1993.
76. Y Pandya, FF Jewett Jr, DG Hoover. Concurrent effects of high hydrostatic pressure, acidity and heat on the destruction and injury of yeasts. J Food Prot 58:301–304, 1995.
77. C Chen, CW Tseng. Effect of high hydrostatic pressure on the temperature depen-

dence of *Saccharomyces cerevisiae* and *Zygosaccharomyces rouxii*. Proc Biochem 32(4):337–343, 1997.

78. E Palou, A Lopez-Malo, GV Barbosa-Canovas, J Welti-Chanes, BG Swanson. Effect of water activity on high hydrostatic pressure inhibition of *Zygosaccharomyces bailii*. Lett Appl Microbiol 24:417–420, 1997.
79. E Palou, A Lopez-Malo, GV Barbosa-Canovas, J Welti-Chanes, BG Swanson. Oscillatory high hydrostatic pressure inactivation of *Zygosaccharomyces bailii*. J Food Prot 61(9):1213–1215, 1998.
80. P Butz, S Funtenberger, T Haberditzl, B Tauscher. High pressure inactivation of *Byssochlamys nivea* ascospores and other heat-resistant moulds. Lebensm Wiss Technol 29:404–410, 1996.
81. JPPM Smelt. Recent advances in the microbiology of high pressure processing. Trends Food Sci Technol 9:152–158, 1998.
82. T Otake, H Mori, T Kawahata, Y Izumoto, H Nishimura, I Oishi, T Shigehisa, H Ohno. Effects of high hydrostatic pressure treatment of HIV infectivity. In: K Heremans, ed. High Pressure Research in the Biosciences and Biotechnology. Leuven: Leuven University Press, 1997, pp. 223–236.
83. G Brauch, U Haensler, H Ludwig. The effect of pressure on bacteriophages. High Pressure Res 5:767–769, 1990.
84. P Butz, G Habison, H Ludwig. Influence of high pressure on a lipid-coated virus. In: C Balny, R Hayashi, K Heremans, P Masson, eds. High Pressure and Biotechnology. London: John Libby & Co, Ltd., 1992, pp. 61–64.
85. Y Horie, K Kimura, M Ida, Y Yosida, K Ohki. Jam preparation by pressurization. Nippon Nogeikagaku Kaishi 65(6):975–980, 1991.
86. ME Parish. High pressure inactivation of *Saccharomyces cerevisiae*, endogenous microflora and pectinmethylesterase in orange juice. J Food Prot 18(1):57–65, 1998.
87. G Aleman, DF Farkas, JA Torres, E Wilhelmsen, S McIntyre. Ultra-high pressure pasteurization of fresh cut pineapple. J Food Prot 57(10):931–934, 1994.
88. A Hara, G Nagahama, A Ohbayashi, R Hayashi. Effects of high pressure on inactivation of enzymes and microorganisms in nonpasteurized rice wine (Namazake). Nippon Nogeikagaku Kaishi 64(5):1025–1030, 1990.
89. G Arroyo, PD Sanz, G Prestamo. Effect of high pressure on the reduction of microbial populations in vegetables. J Appl Microbiol 82:735–742, 1997.
90. T Tanaka, K Hatanaka. Application of hydrostatic pressure to yoghurt to prevent its after-acidification. Nippon Shokuhin Kogyo Gakkaishi 39(2):173–177, 1992.
91. H Miyakawa, K Anjitsu, N Ishibashi, S Shimamura. Effects of pressure on enzyme activities of *Lactobacillus helveticus* LHE-511. Biosci Biotechnol Biochem 58(3): 606–607, 1994.
92. G Carpi, S Gola, A Maggi, P Rovere, M Buzzoni. Microbial and chemical shelf life of high-pressure-treated salmon cream at refrigeration temperatures. Indus Conserve 70:386–397, 1995.
93. A Carlez, JP Rosec, N Richard, JC Cheftel. Bacterial growth during chilled storage of pressure-treated minced meat. Lebensm Wiss Technol 27:48–54, 1994.
94. EI Moueffak, C Cruz, M Antoine, M Montury, G Demazeau, A Largeteau, B Roy, F Zuber. High pressure and pasteurization effect on duck foie gras. Int J Food Sci Technol 30:737–743, 1995.

95. T Fujii, M Satomi, G Nakatsuka, T Yamaguchi, M Okuzumi. Changes in freshness indexes and bacterial flora during storage of pressurized mackerel. Shokuhin Eiseigaku Zasshi (J Food Hyg Soc Jpn) 35(2):195–200, 1994.
96. S Miyao, T Shindoh, K Miyamori, T Arita. Effects of high pressurization on the growth of bacteria derived from surimi (fish paste). Nippon Shokuhin Kogyo Gakkaishi 40(7):478–484, 1993.
97. M Anese, MC Nicoli, G Dall'Aglio, CR Lerici. Effect of high pressure treatments on peroxidase and polyphenoloxidase activities. J Food Biochem 18:285–293, 1995.
98. M Castellari, L Matricardi, G Arfelli, P Rovere, A Amati. Effects of high pressure processing on polyphenoloxidase enzyme activity of grape musts. Food Chem 60(4):647–649, 1997.
99. MRA Gomes, IG Sumner, DA Ledward. Effects of high pressure on papain activity and structure. J Sci Food Agric 75:67–72, 1997.
100. L Leistner, LGM Gorris. Food preservation by hurdle technology. Trends Food Sci Technol 6(2):41–46, 1995.
101. GJ Haas, HE Prescott, E Dudley, R Dik, C Hintlan, L Keane. Inactivation of microorganisms by carbon dioxide under pressure. J Food Saf 9:253–265, 1989.
102. SI Hong, WS Park, YR Pyun. Inactivation of *Lactobacillus* sp. from kimchi by high pressure carbon dioxide. Lebensm Wiss Technol 30:681–685, 1997.
103. P Ballestra, AA Da Silva, JL Cuq. Inactivation of *Escherichia coli* by carbon dioxide under pressure. J Food Sci 61(4):829–836, 1996.
104. CI Wei, MO Balaban, SY Fernando, AJ Peplow. Bacterial effect of high pressure CO_2 treatment of foods spiked with *Listeria* or *Salmonella*. J Food Prot 54(3):189–193, 1991.
105. A Enomoto, K Nakamura, M Hakoda, N Amaya. Lethal effect of high-pressure carbon dioxide on a bacterial spore. J Ferment Bioeng 83(3):305–307, 1997.
106. P Paul, SP Chawla, P Thomas, PC Kesavan. Effect of high hydrostatic pressure, gamma-irradiation and combination treatments on the microbiological quality of lamb meat during chilled storage. J Food Saf 16(4):263–271, 1997.
107. L Popper, D Knorr. Applications of high-pressure homogenization for food preservation. Food Technol 44:84–89, 1990.
108. AM Papineau, DG Hoover, D Knorr, DF Farkas. Antimicrobial effect of water-soluble chitosans with high hydrostatic pressure. Food Biotechnol 5:45–47, 1991.
109. CM Roberts, DG Hoover. Sensitivity of *Bacillus coagulans* spores to combinations of high hydrostatic pressure, heat, acidity and nisin. J Appl Bacteriol 81:363–368, 1996.
110. N Kalchayanand, A Sikes, CP Dunne, B Ray. Interaction of hydrostatic pressure, time and temperature of pressurization and pediocin AcH on inactivation of food-borne bacteria. J Food Prot 61(4):425–431, 1998.
111. GO Adegoke, H Iwahashi, Y Komatsu. Inhibition of *Saccharomyces cerevisiae* by combination of hydrostatic pressure and monoterpenes. J Food Sci 62(2):404–405, 1997.
112. Y Ishiguro, T Sato, T Okamoto, K Okamoto, H Sakamoto, T Inakuma, Y Sonoda. Effects of hydrostatic pressure and antimicrobial substances on the sterilization of tomato juice. Nippon Nogeikagaku Kaishi 67(12):1707–1711, 1993.

17
Ohmic Heating

Sevugan Palaniappan
The Minute Maid Company
Houston, Texas

Sudhir K. Sastry
The Ohio State University
Columbus, Ohio

I. INTRODUCTION

Interest in rapid methods of heating and nonthermal microbial inactivation resulted in revived attention to electrical and electromagnetic treatments in the 1980s. Ohmic or electrical resistance heating, induction heating, microwave heating, radio-frequency heating, pulsed electric field, electric arc discharge, and oscillating magnetic field treatments are some of the technologies investigated for microbial inactivation. The electrical heating treatments are rapid heating methods, and the others are low-temperature or nonthermal treatments. The electrical heating methods generate heat within food products without the use of an external heating medium. Consequently these methods hold promise for microbial inactivation through thermal means. The question is regarding the nature of interaction of microorganisms with electric fields and the nonthermal effects, which could act in synergy with internal heat generation. This chapter reviews ohmic heating in relation to mechanisms and kinetics contributing to microbial death.

II. PRINCIPLE OF OHMIC HEATING

Ohmic or electrical resistance heating involves an application of a low-voltage alternating current to a continuously flowing food product. Heat is generated within the food product during this process due to the electrical conductivity of food. The first commercial application of electricity in food processing was introduced in the early twentieth century for milk pasteurization (1). This method of processing was called the ''electropure process'' and involved application of 220 V alternating current using parallel carbon electrodes in a vertical rectangular tube consisting of glass insulators between them (2). However, these units virtually disappeared from the dairy industry in succeeding years (3). The reasons for discontinued usage of this technology could be due to lack of suitable inert electrodes, instrumentation, and controls.

APV Baker (APV, Crawley, West Sussex, United Kingdom) licensed a technology for manufacturing continuous flow ohmic heaters from the Electricity Council of Great Britain and reintroduced a commercial ohmic system for food processing in the late 1980s (4). The manufacturer reported rapid and uniform heating of the food products in the ohmic heater to microbiologically lethal temperatures. APV does not make any claim of lethal effects due to electric current except from pure heating.

III. APPLICATIONS OF OHMIC HEATING IN FOOD PROCESSING

Ohmic heating has potential applications for processing of liquid-particle mixtures, highly viscous, and heat-sensitive food products. The particular interest in this technology stemmed from the food industry interest in aseptic processing of particulate foods. Conventional aseptic process systems rely on heating of the liquid phase to transfer heat to the solid phase. Process design in such cases is conservative to ensure product safety, thereby compromising product quality. Ohmic heating offers an attractive alternative, because it heats materials by internal heat generation, but without some of the nonuniformities commonly associated with microwave heating. The heating rates of the particles can be manipulated by formulation, and in fact the particles can be heated faster than the carrier fluid. Ohmic process can deliver high quality food particles with good integrity and microbiological safety.

Ohmic heating could also produce better quality highly viscous and heat-sensitive (e.g., liquid egg) food products than conventional heating for the same reason of internal heat generation. Production runtimes can potentially be increased because of less fouling (no heat transfer surface) of chamber walls. There are currently a few commercial systems operating around the world for pet foods, particulate foods, and liquid eggs.

IV. REVIEW OF EARLY MICROBIOLOGICAL STUDIES

Tracy (5) obtained a reduction of 2–3 log cycles of yeast cells due to low-voltage alternating current in a chamber with carbon electrodes. The chamber was externally cooled, and the highest temperatures attained were recorded by a thermocouple. The inactivation obtained was attributed purely to the passage of alternating current through the inoculated grape juice sample. It was speculated that the formation of temporary toxic substances like free chlorine might be the reason for the death of yeast cells. Since the temperature in these studies was not controlled and measured accurately (increasing heating rate with increasing electrical conductivity) with fast responding sensors, even a few seconds above the lethal temperature might have caused the microbial death resulting in an experimental artifact.

Horrall (6) reported that the temperature required to kill *Escherichia coli* and *Microbacterium tuberculosis* using the electropure process was lower compared to conventional heating. However, no data showing the temperature histories of both processes were presented to substantiate these findings. Gelpi and Devereux (7) reported a 2 log reduction of bacterial spores during electropure process compared to none with batch pasteurization of milk. However, the temperature during electric pasteurization was >71°C for at least 15 seconds compared to 62.8°C for 30 minutes during batch heating. Hall and Trout (3) stated that some studies had concluded that there was no additional effect due to electricity on the microbial kill other than the effect of heat generated. Rosenberg et al. (8) reported an inhibition of cell division during a low-voltage alternating current 1000 Hz frequency. They concluded that this inhibition effect on cell division was due to the production of certain group VIII transition metal compounds of 1–10 ppm during the treatment. Pareilleux and Sicard (9) studied the effects of low-voltage alternating current (50 Hz), ranging from 10 to 200 mA, on *E. coli* at temperatures below 40°C. There was no decrease in the number of viable cells immediately after the treatment, but the number decreased with the holding time. They reported that the bactericidal effect depended on the current level, presence of chloride-containing compounds, and the hold time after the treatment. These results seemed to indicate that the mechanisms influencing bacterial death are complex, involving a number of interactions between the organisms, medium, electrode material, etc.

Shimada and Shimahara (10,11) conducted experiments with *E. coli* B cells suspended in a phosphate buffer solution. A cell suspension of 8.5 mL (10^5 cells/mL) was exposed to low-voltage alternating current (50 Hz) of 0–300 mA/cm^2. Although the method of temperature measurement was not presented, it was reported that the temperature was kept below 32°C during the treatment. The inactivation of cells was attributed to the toxicity of hydrogen peroxide produced during the treatment, and it was confirmed that a definite current density, the amount of hydrogen peroxide produced increased with increasing exposure time under aerobic conditions. It was concluded that hydrogen peroxide was formed on the surface of carbon electrodes by electrolytic reduction of dissolved oxygen.

Shimada and Shimahara (12) suggested that the permeability of the cell membrane was modified after the alternating current treatment since the stainability of exposed cells with crystal violet was low. Shimada and Shimahara (13) presented results showing higher concentration of UV-absorbing materials in the supernatant fractions of the cell suspensions exposed to a current density of 600 ± 60 mA/cm^2 and temperature of <37°C. Based on an increased absorbance at 260 nm with increased exposure time, they suggested that the intracellular materials (including DNA) were released from the cells during alternating current exposure. Electron micrographs of thin sections showed more disorganized materials in the central areas within exposed (5 h) cells than in unexposed (only shaken for 5 h) or fresh (untreated) cells. They concluded that alternating current exposure enhanced the aggregation of DNA related materials within cells following the leakage of cellular contents from the cells.

V. KINETIC STUDIES ON MICROBIAL INACTIVATION

Microbial inactivation kinetics during ohmic heating was not investigated in the early studies. Palaniappan et al. (14) reviewed early literature on the effects of electricity on microorganisms and found that most studies either did not measure sample temperatures accurately or failed to recognize it as a variable during the treatment. It is essential that any study comparing conventional and ohmic heating be conducted under temperature histories that are identical as possible. Kinetic studies comparing D- and z-values or rate constant (k) and activation energy (E_a) would show any additional nonthermal lethal effect due to ohmic heating.

Table 1 D-Values and Kinetic Reaction Rate Constants (k) for *Zygosaccharomyces baillii* Under Conventional and Ohmic Heating

Temperature (°C)	D-value for conventional heating (min)	k for conventional heating (s^{-1})	D-value for ohmic heating (min)	k for ohmic heating (s^{-1})
49.8	4.91	0.008	4.57	0.009
52.3	2.50	0.016	1.88	0.021
55.8	0.79	0.049	0.72	0.054
58.8	0.28	0.137	0.30	0.130
z-values (°C) or activation energy (E_a) (kcal/mol)	7.19[a]	29.63[b]	7.68[a]	27.77[b]

[a] z-value.
[b] Activation energy.
Source: Ref. 15.

Table 2 D-Values and Kinetic Reaction Rate Constants (k) for *E. coli* Under Conventional Heating and Conventional Heating with Electrical Pretreatment

Temperature (°C)	D-value for conventional heating (min)	k for conventional heating (s^{-1})	D-value for electrical pretreatment (min)	k for electrical pretreatment (s^{-1})
60.0	2.75	0.014	3.00	0.013
64.5	0.80	0.048	0.85	0.045
68.5	0.38[a]	0.100[a]	0.35[a]	0.109[a]
71.0	0.20	0.194	0.20	0.195
z-values (°C) or activation energy (E_a) (kcal/mol)	9.87[b]	23.08[c]	9.36[b]	24.30[c]

[a] Significantly different at 95% confidence level.
[b] z-value.
[c] Activation energy.
Source: Ref. 15.

Table 3 D-Values and Kinetic Reaction Rate Constants (k) for *B. subtilis* Spores Under Conventional and Ohmic Heating

Temperature (°C)	D-value for conventional heating (min)	k for conventional heating (s^{-1})	D-value for ohmic heating (min)	k for ohmic heating (s^{-1})
88	32.8	0.00117	30.2	0.001271
92.3	9.87	0.003889	8.55	0.004489
95	5.06	0.007586		
95.5			4.38	0.008763
97	3.05	0.012585		
99.1			1.76	0.021809
z-values (°C) or activation energy (E_a) (kcal/mol)	8.74[a]	70.0[b]	9.16[a]	67.5[b]

[a] z-value.
[b] Activation energy.
Source: Ref. 16.

Table 4 D-Values and Reaction Rate Constants (k) for Inactivation of *B. subtilis* Spores During Single and Double Stage Conventional and Ohmic Heating at 90°C

Stage no.	D-value for conventional heating (min)	k for conventional heating (s^{-1})	D-value for ohmic heating (min^{-1})	k for ohmic heating (s^{-1})
1	17.1	0.002245	14.2	0.002703
2	9.2	0.004172	8.5	0.004516

Source: Ref. 16.

Palaniappan et al. (15) attempted to compare ohmic and conventional heat treatments on the death kinetics of yeast cells (*Zygosaccharomyces baillii*) in a phosphate buffer sollution, under identical histories, and found no difference (Table 1). However, a mild electrical pretreatment of *E. coli* in a buffer solution decreased the subsequent inactivation requirement in certain cases (Table 2). In this study, the temperature of the samples was maintained below 36°C during the electrical pretreatment (current density of 2300 A/m^2) for 30 minutes. An insignificant reduction of 0.2–0.3 log of *E. coli* was reported. Unlike Shimada and Shimahara (11), Palaniappan et al. (15) found no hydrogen peroxide in the samples immediately after the treatment. They concluded that the microbial inactivation during ohmic heating was due primarily to thermal effects. Cho et al. (16) conducted a study under near-identical temperature conditions and reported that the kinetics of inactivation of *Bacillus subtilis* spores can be accelerated by an ohmic treatment (Table 3). A two-stage ohmic treatment (ohmic treatment, followed by a holding time prior to a second heat treatment) was found to further accelerate death rates (Table 4). A recent study (17) indicated that leakage of intracellular constituents of *Saccharomyces cerevisiae* was enhanced under ohmic heating, as compared to conventional heating in boiling water.

VI. MECHANISMS OF INACTIVATION

Different explanations have been offered for the microbicidal property of electric current, but substantiating data are not available for all the claims. Mechanisms of microbial inactivation reported in the literature can be categorized into thermal, chemical, and mechanical effects.

A. Thermal Effects

It is well established that lethal temperatures can rapidly be attained in an electrically conducting medium during ohmic heating (4). The heat that is internally generated in the treatment medium is the major cause for bacterial inactivation.

B. Chemical Effects

Presence of chloride-containing compounds (9) or hydrogen peroxide (10) in the samples after the low-voltage alernating current treatment was found to cause bacterial death. Production of chemical compounds in the treatment medium depends on electrode reactions depending upon the type of electrodes, current density, frequency, and the medium used.

C. Mechanical Effects

Although there is no shock wave formation during ohmic heating, membrane damage causing permeability modification and leakage of cellular contents was reported by Shimada and Shimahara in 1985 (13,14). More recent studies also suggest that a mild electrical permeation mechanism may be at play during ohmic heating. For example, studies on fermentation of *Lactobacillus acidophilus* under the presence of a mild electric field (18) have indicated that the lag phase can be significantly reduced under these conditions, but that the ultimate productivity of the fermentation is reduced by the presence of the electric field. It is hypothesized that this may be due to the presence of mild electroporation, which improves nutrient transport at the early stages of fermentation, thereby accelerating it. At later stages, the electroporation effect improves the transport of metabolites into the cell, thereby causing an inhibitory effect. The presence of pore-forming mechanisms on cellular tissue has been confirmed by a few recent studies (19–21).

The additional nonthermal lethal effect of ohmic treatment may be due to the low frequency (50–60 Hz) of ohmic heating, which allows cell walls to build up charges and form pores. This is in contrast to high-frequency methods such as radio-frequency or microwave heating, where the electric field is reversed before sufficient charge build-up. Some contrary evidence exists; in particular, the work of Lee and Yoon (17) has indicated that greater leakage of *S. cerevisiae* constituents occurs under high frequencies. However, the details of temperature control within this study are not available at this time; thus, it is not known if these researchers have adequately eliminated temperature effects.

VII. MECHANISM OF ELECTROPERMEABILIZATION

Electropermeabilization of cell membranes has received considerable attention in the biophysics literature, principally with reference to electroporation and electrofusion. The cell membrane breakdown mechanism has been shown to consist of three parts:

1. The occurrence and growth of membrane shape fluctuations. This process has been shown to last on the order of microseconds and can be described by the thin film model of Dimitrov (22).

2. The increasing amplitude of the shape fluctuations results in decreasing membrane thickness at local points. This causes a rapid, nonlinear increase in the driving force due to the electric field. Then, molecular rearrangements, leading to the discontinuity of the membrane, can occur. This process is extremely short (on the order of nanoseconds).
3. The further growth of pores, resulting in irreversible mechanical breakdown of the membrane. This is a slow process, lasting on the order of milliseconds and more. This process has been studied by Chernomordik and Abidor (23), among others. Sugar and Neumann (24) have also modeled this phenomenon as a stochastic process.

VIII. MODELS FOR CELL MEMBRANE BREAKDOWN

The cell membrane is typically regarded as a capacitor with a dielectric material of low dielectric constant compared to water (25). Crowley (26) modeled the irreversible breakdown of a lipid bilayer membrane by considering it to be an isotropic elastic layer between two electrically conducting liquids. The calculation of the resulting electrical and elastic stresses showed the existence of a critical membrane potential at which rupture occurred. Subsequently, a number of models have been developed, including those of Zimmermann et al. (27,28), which included mechanical external forces in the analysis. Further literature has included the consideration of surface tension effects, as well as the viscoelastic character of the cell membrane (22,29).

IX. REVERSIBLITY AND REPAIR OF PORES

Evidence for pore formation in lipid membranes has been presented by Benz et al. (30), who noted a reversible change in lipid bilayers, with resealing times of about 2–20 seconds. However, cell membranes were found to reseal over much longer times (up to 10 min). It is believed (25) that conformational changes of proteins are responsible for the long-duration permeabilization in biological membranes.

X. SUMMARY

Although much of the early literature has been inconclusive, recent studies on ohmic heating suggest that, in addition to the thermal effects, some mild permeabilization of the microbial cells might contribute to lethal effects. While this effect has been observed with certain bacteria, sufficient data does not exist for a wide range of microorganisms. Since it is likely that this technology may be used principally for rapid and uniform heating, it may be unnecessary or insignificant to claim additional nonthermal lethal effects in most cases.

REFERENCES

1. AK Anderson, R Finkelstein. A study of the electropure process of treating milk. J Dairy Sci 2:374–406, 1919.
2. BE Getchell. Electric pasteurization of milk. Agric Eng 16:408–410, 1935.
3. CW Hall, GM Trout. Milk Pasteurization. New York: Van Nostrand Reinhold/AVI, 1968.
4. CH Biss, SA Coombes, PJ Skudder. The development and application of ohmic heating for the condinuous heating of particulate foodstuffs. In: RW Field, JA Howell, eds. Process Engineering in the Food Industry. Essex, England: Elsevier Applied Science Publishers, 1987, pp. 17–27.
5. RL Tracy. Lethal effect of alternating current on yeast cells. J Bacteriol 24:423–438, 1932.
6. BE Horrall. A study of the lecithin content of milk and its products. Indiana Agr. Expt. Station. Bulleting 401, 1935.
7. AJ Gelpi Jr., ED Devereux. Effect of the electropure process and of the holding method of treating milk upon bacterial endospores. Paper No. 16, Michigan State College Agricultural Experimental Station, 1930.
8. B Rosenberg, LV Camp, T Krigas. Inhibition of cell division in *Escherichia coli* by electrolysis products from a platinum electrode. Nature 205:698–699, 1965.
9. A Pareilleux, N Sicard. Lethal effects of electric current on *Escherichia coli.* Appl Microbiol 19(3):421–424, 1970.
10. K Shimada, K Shimahara. Factors affecting the surviving fractions of resting *Escherichia coli* B and K-12 cells exposed to alternating current. Agric Biol Chem 45: 1589–1595, 1981.
11. K Shimada, K Shimahara. Responsibility of hydrogen peroxide for the lethality of resting *Escherichia coli* B cells exposed to alternating current in phosphate buffer solution. Agric Biol Chem 46:1329–1337, 1982.
12. K Shimada, K Shimahara. Changes in surface charge, respiratory rate and stainability with crystal violet of resting *Escherichia coli* B cells anaerobically exposed to an alternating current. Agric Biol Chem 49:405–411, 1985.
13. K Shimada K Shimahara. Leakage of cellular contents and morphological changes in resting *Escherichia coli* B cells exposed to an alternating current. Agric Biol Chem 49:3603–3607, 1985.
14. S Palaniappan, SK Sastry, ER Richter. Effects of electricity on microorganisms: a review. J Food Proc Pres 14:393–414, 1990.
15. S Palaniappan, SK Sastry, ER Richter. Effects of electroconductive heat treatment and electrical pretreatment on thermal death kinetics of selected microorganisms. Biotech Bioeng 39:225–232, 1992.
16. H-Y Cho, AE Yousef, SK Sastry. Kinetics of inactivation of *Bacillus subtilis* spores

by continuous or intermittent ohmic and conventional heating. Biotech Bioeng 62(3): 368–372, 1999.

17. CH Lee, SW Yoon. Effect of ohmic heating on the structure and permeability of the cell membrane of *Saccharomyces cerevisiae.* Abstract No. 79 B-6, presented at the 1999 Annual IFT Meeting, Chicago, IL, July 24–28, 1999.
18. H-Y Cho, AE Yousef, SK Sastry. Growth kinetics of *Lactobacillus acidophilus* under ohmic heating. Biotech Bioeng 49:334–340, 1996.
19. T Imai, K Uemura, N Ishida, S Yoshizaki, A Noguchi. Ohmic heating of Japanese white radish *Rhaphanus sativus* L. Int J Food Sci Technol 30:461–172, 1995.
20. W-C Wang. Ohmic heating of foods: physical properties and applications. PhD dissertation, The Ohio State University, Columbus, OH, 1995.
21. SA Kulshrestha, SK Sastry. Low-frequency dielectric changes in vegetable tissue from ohmic heating. Abstract No. 79 B-3, presented at the 1999 Annual IFT Meeting, Chicago, IL, July 24–28, 1999.
22. DS Dimitrov. Electric field-induced breakdown of lipid bilayers and cell membranes: a thin viscoelastic film model. J Membrane Biol 78:53–60, 1984.
23. LV Chernomordik, IG Abidor. The voltage-induced local defects in unmodified BLM. Bioelectrochem Bioeng 7:617–623, 1980.
24. IP Sugar, E Neumann. Stochastic model for electric field-induced membrane pores electroporation. Biophys Chem 19:211–225, 1984.
25. U Zimmermann. Electrical breakdown, electropermeabilization and electrofusion. Rev Physiol Biochem Pharmacol 105:176–256, 1986.
26. JM Crowley. Electrical breakdown of bimolecular lipid membranes as an electromechanical instability. Biophys J 13:711–724, 1973.
27. U Zimmermann, G Pilwat, M Riemann. Dielectric breakdown of cell membranes. Biophys J 14:881–899, 1974.
28. U Zimmermann, F Beckers, HGL Coster. The effect of pressure on the electrical breakdown of membranes of *Valonia utricularis.* Biochim Biophys Acta 464:399–416, 1977.
29. RK Jain, C Maldarelli. Stability of thin viscoelastic films with application to biological membrane deformation. Ann NY Acad Sci 404:89–102, 1983.
30. R Benz, F Beckers, U Zimmermann. Reversible electrical breakdown of lipid bilayer membranes: A charge-pulse relaxation study. J Membrane Biol 48:191–204, 1979.

18
Modeling Acid Inactivation of Foodborne Microorganisms

Robert L. Buchanan and Richard C. Whiting
U.S. Food and Drug Administration
Washington, D.C.

Marsha H. Golden
U.S. Department of Agriculture
Wyndmoor, Pennsylvania

I. INTRODUCTION

Humans have struggled since ancient times to safeguard their food supply from spoilage and foodborne diseases. Traditional food preservation techniques have evolved over millennia to prolong shelf stability. One such approach is the control of foodborne microorganisms by acidification. The inherent acidity of certain foods (e.g., fruits), acidification by fermentation (e.g., pepperoni, yogurt, sauerkraut), and the direct addition of organic acids (e.g., vinegar-based marinades) is the primary means of controlling pathogenic bacteria in a wide range of ready-to-eat foods that do not receive a thermal process. The application of these techniques is largely based on historical information related to individual products. As public health or microbial spoilage concerns associated with these products have arisen, scientific efforts have been directed largely to solving the immediate problem. Until recently, there has been little sustained effort to examine acid-mediated inactivation of foodborne microorganisms in a systematic manner. However, there is an increasing need to be able to rely on acid inactivation processes with the same degree of confidence that we have currently for thermal processing.

Mention of brand or firm name does not constitute an endorsement by the U.S. Department of Agriculture over others of a similar nature not mentioned.

This current need reflects the emergence of foodborne pathogens that combine high acid tolerance with low infectious doses, which has resulted in an increase in disease outbreaks associated with products previously considered inherently safe because of their acidity. For example, there have been recent outbreaks of *Escherichia coli* O157:H7 in fermented salami (1), yogurt (2), and unpasteurized apple cider (3); *Salmonella* in Lebanon bologna and unpasteurized orange juice; *Cryptosporidium parvum* in unpasteurized apple cider; and *E. coli* O111:H^- in mettwurst (4). These outbreaks established the need to better validate the efficacy of food preservation techniques that rely on acid-mediated inactivation of pathogenic microorganisms. However, this means of microbial control is very complex and requires a firm understanding of how the various factors that influence acid inactivation interact.

The development of relatively simple mathematical models that describe the response of microorganisms to thermal stresses was one of the key scientific developments that was instrumental in the widespread, systematic use of thermal processing as a preservation technique. These models allow food processors to specify treatments that could be relied on to deliver the desired degree of microbiological inactivation. Understanding and using D- and Z-values, two simple models, has become an integral part of food microbiology. As a means of addressing increasingly pressing food safety concerns related to nonthermally processed foods that rely on acidification as their primary means of microbial control, several years ago scientists at the U.S. Department of Agriculture (USDA) Agricultural Research Service, Eastern Regional Research Center (ERRC), began a major effort to acquire sufficient inactivation kinetics data and supporting physiological studies to be able to model mathematically acid-mediated inactivation of selected foodborne pathogens. The purpose of this chapter is to provide an overview of how this project and related basic research on bacterial responses to acid stress are providing a scientific basis for how to use these ancient food preservation techniques more effectively.

II. CHARACTERISTICS OF MICROBIAL SURVIVAL IN ACID ENVIRONMENTS

When placed in an environment or condition that does not support growth, vegetative bacterial cells begin to die. With strongly acidic conditions (pH 1.5–3.5), inactivation typically follows first-order kinetics, i.e., a linear response is observed when the logarithm of survivor counts is plotted against time. However, when the environment is less harsh, non–first-order kinetic survivor curves that have shoulders or tails are often observed. Examples of different survivor curves observed with *Listeria monocytogenes* are depicted in Fig. 1. The survival of foodborne pathogens under acidic conditions is a highly complex integration of

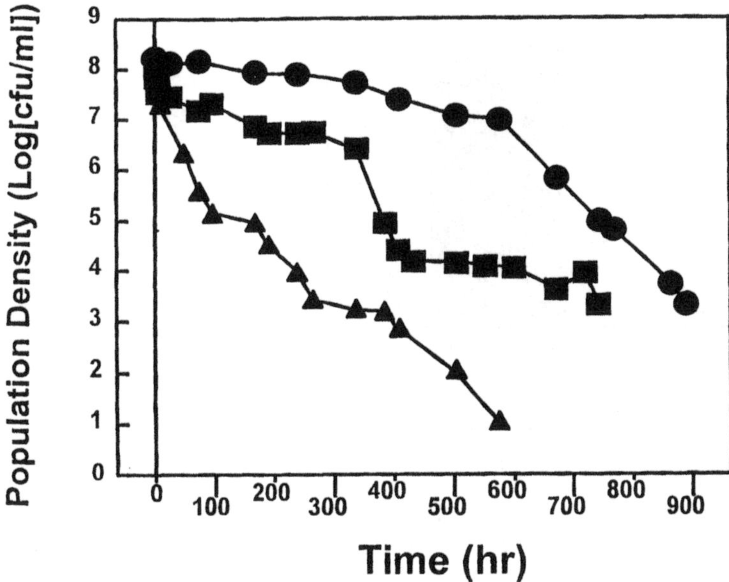

Figure 1 Examples of survivor curves for a three-strain mixture of *Listeria monocytogenes*: (■) 12°C/13.1% NaCl/0.5% lactic acid/pH 4.0/15μg/mL $NaNO_2$; (●) 28°C/13.1% NaCl/0.0% lactic acid/pH 7.3/5 μg/mL $NaNO_2$; (▲) 12°C/6.3% NaCl/0.5% lactic acid/pH 4.5/200 μg/mL $NaNO_2$. (From Ref. 12.)

several physiological processes, and a variety of factors influence inactivation kinetics (Table 1). Several rules of thumb can be generally relied on in relation to the survival of foodborne pathogens:

Exponential phase cells are less resistant than stationary phase cells (Fig. 2a).
Inactivation rates increase as storage temperatures are raised (Fig. 2b).
Inactivation rates are inversely related to pH (Fig. 2c).
Salt may be protective until high levels are added.

The identity of the acidulants used to modify the pH of a food system also affects survival rates. Hydrochloric acid is generally assumed to be the ''gentlest'' acid and has been used extensively as the control acid to assess the impact of pH alone. In this instance, the rate of *L. monocytogenes* inactivation has been shown to be linearly related to pH at values ≤5.5 (Fig. 3) (5–8). The relationship is somewhat more complex with weak organic acids due to their ''anion effects,'' which are related to acids' impact on bacterial metabolism (Fig. 2c) (5,6,8,9). In many instances, the antimicrobial activities of organic acids are related to their

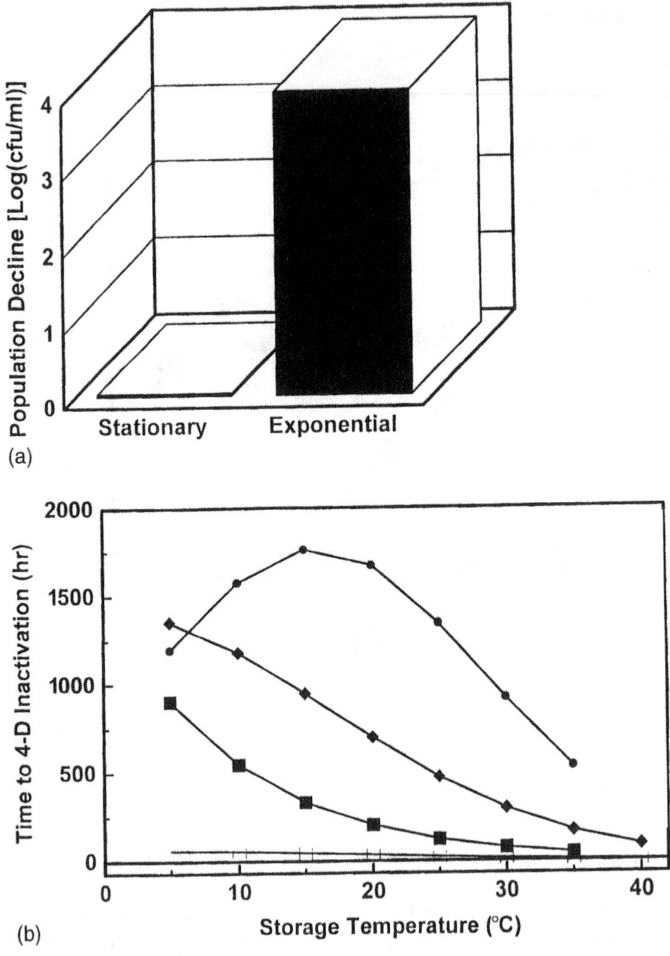

Table 1 Factors That Influence the Survival of Foodborne Pathogens in Acidic Environments

pH
Acid identity
Acid concentration
Temperature
Water activity/Sodium chloride content
"Age" of cells
Nutrient content
Antimicrobials
Atmosphere
Preexposure conditions

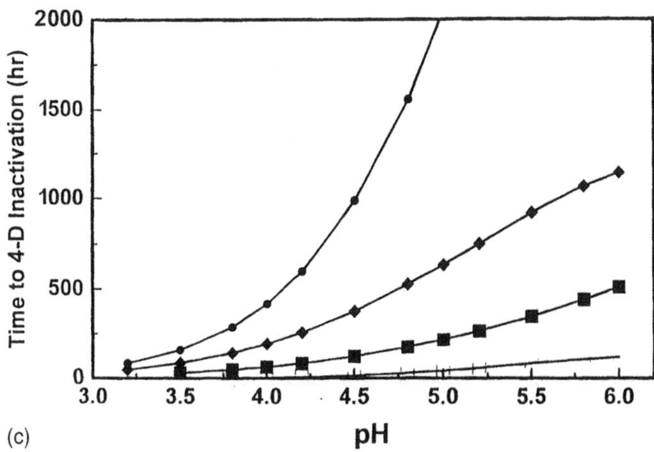

Figure 2 Examples of general characteristics of acid-mediated inactivation of foodborne pathogens. (a) Decline in *Escherichia coli* levels when exponential and stationary phase cells were exposed to pH 3.0 at 37°C for 2 h. (From Ref. 10.) (b) Effect of storage temperature on the predicted time (from Ref. 18) to achieve a 4-D inactivation of *Escherichia coli* (■), *Listeria monocytogenes* (◆), *Salmonella* spp. (□), and *Staphylococcus aureus* (●) under conditions of pH 4.7/0.3% lactic acid/4.5% NaCl, and 25 μg/mL $NaNO_2$. (c) Effect of pH on the predicted time (from Ref. 18) to achieve a 4-D inactivation of *Escherichia coli* (■), *Listeria monocytogenes* (◆), *Salmonella* spp. (□), and *Staphylococcus aureus* (●) under conditions of 25°C/0.3% lactic acid/4.5% NaCl, and 25 μg/mL $NaNO_2$.

pKa values because they must be in a completely undissociated acid form to cross the cell membrane. However, for other acids (e.g., citric acid) there may be specific transport systems that mediate their uptake. There can be substantial differences in the ability of foodborne pathogens to tolerate different acids, even when the pH is the same (Fig. 4). Since the primary acid associated with fermented foods is lactic acid, it has been the focus of most of the modeling research. As will be discussed later, this has an impact when using and interpreting nonthermal inactivation models.

The concentration of the organic acid used as an acidulant also influences survival rates. For monocarboxylic acids, increasing the acid concentration increases inactivation rates (8). However, at higher pH levels (5.0–7.0) there are indications that low levels of certain dicarboxylic and tricarboxylic acids may actually provide some degree of protection (6,7). In both cases antimicrobial activity appears to be related to the concentration of undissociated acid. There has been little research examining combinations of acids, but it appears that their impact is additive, not synergistic (10).

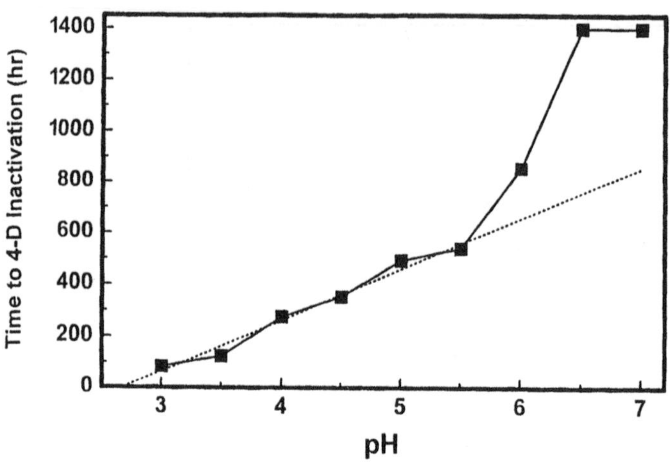

Figure 3 The effect of pH on the time for a 4-D inactivation of *Listeria monocytogenes* in brain heart infusion broth adjusted with hydrochloric acid. (From Ref. 8.)

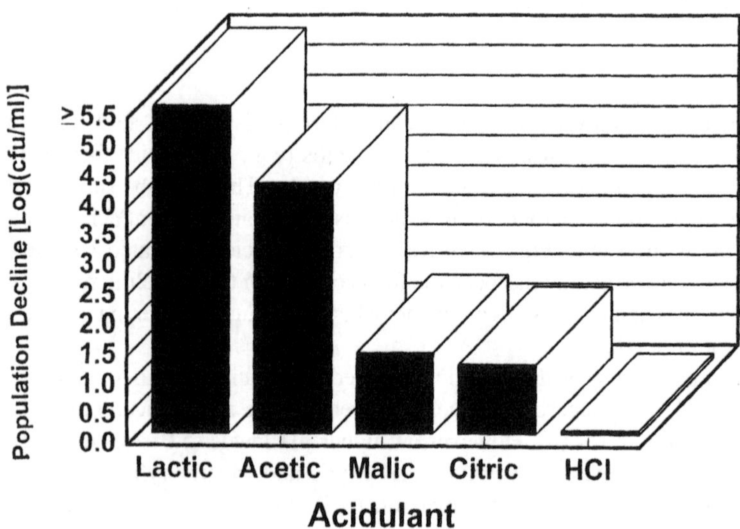

Figure 4 Effect of acidulant identity on the extent of *Escherichia coli* O157:H7 inactivation. Conditions: pH 3.0, 37°C, organic acid concentration 0.5% (w/v). (From Ref. 10.)

III. DEVELOPMENT OF INACTIVATION MODELS

Four microorganisms were selected for the development of survival inactivation models, *L. monocytogenes* (9,11,12), *E. coli* O157:H7 (13), *Staphylococcus aureus* (14), and *Salmonella* spp. (15). Since the particular interest was to predict pathogen inactivion in fermented meat products, the independent variables selected included temperature, pH, sodium chloride content/water activity, sodium nitrite, and lactic acid concentrations. The general approach for all of the species was similar. A mixture of three to four strains of a species was inoculated into brain heart infusion broth that had been adjusted to contain the designated pH, NaCl, lactic acid, and $NaNO_2$ levels. The initial population density of the test organisms was 10^8–10^9 cfu/mL. A high level was used to better estimate inactivation kinetics after preliminary studies indicated that inactivation rates were independent of inoculum size (11). Once inoculated, the cultures were stored at temperatures varying from 4 to 42°C for up to 6 months, with samples being plated and assayed for viable counts periodically. In selecting variable ranges, combinations of cultural and environmental conditions that supported growth prior to the initiation of inactivation were included to better characterize the often hazy boundary between growth and survival. In these instances, the small degree of growth observed when starting with a large inoculum was included as a shoulder in the inactivation curve.

Once the laboratory data were generated, model development proceeded in two steps. First, the data for each individual survivor curve were fitted to a primary model (16). Then the parameters for the primary model were fitted to a secondary model so that the survival of the pathogen could be described as a function of the different environmental and cultural conditions investigated. A number of different primary models were used, including linear, linear with a lag period, and logistic (Table 2). The selection of a primary model was, in large part, based on the number of survivor curves that had "shoulders," "tails," and other deviations from first-order inactivation kinetics. Due to the large number of variables being considered, polynomial response surface modeling has been the major approach to secondary models employed, though more mechanistic models based on a consideration of the concentrations of undissociated acids have been explored (6–9,14). Where linear primary models have been used, the secondary models have been used to mathematically describe the effects of the environmental variables on D-values, the times for one $\log_{10}$ decrease. However, where lag periods prior to initiation of first-order inactivation (i.e., shoulders) have had to be considered, the models have been calculated based on an inactivation end point. A 4 log decrease in population density was selected as the target inactivation modeled. More recently, these models have been recalculated to consider the number of "Ds" of inactivation as a variable and thus provide a means for estimating a range of inactivation endpoints (17). In all cases, the natural

Table 2 Two-Phase Linear and Logistics Primary Models Used to Mathematically Depict Individual Survivor Curves

A. Two-Phase Linear Inactivation Model

$$Y = Y_0 \quad [t < t_L]$$
$$Y = Y_0 + s(t - t_L) \quad [t \geq t_L]$$

where:

Y = $\log_{10}$ count of bacteria at time t, log(cfu/mL)
Y_0 = $\log_{10}$ count of bacteria at time t = 0, log(cfu/mL)
s = slope of survivor curve, log(cfu/mL])/h
t = time, h
t_L = duration of lag period prior to commencement of inactivation, h

D-values were calculated by taking the negative reciprocal of the "s" term, and "time to a 4-D(99.99%) inactivation" was calculated using the equation:

$$t_{4-D} = t_L + (4*D)$$

B. Logistics Inactivation Model

$$\log_{10}(Y/Y_0) = \log_{10}\left[F_1 \frac{(1 + e^{-b_1 t_L})}{(1 + e^{b_1(t-t_L)})} + \frac{(1 - F_1)(1 + e^{-b_2 t_L})}{(1 + e^{b_2(t-t_L)})}\right]$$

where:

b_1 = $2.3/D_1$ = inactivation of major population group
b_2 = $2.3/D_2$ = inactivation of minor population group
F_1 = fraction of population in major group
$(1 - F_1)$ = fraction of population in minor group
t_L = duration of lag period

Time to a 4-D (99.99%) inactivation was calculated based on the D_1 and t_L values using the following equation:

$$t_{4-D} = \frac{LN[((1 + e^{-b_1 t_L})/0.0001) - 1] + b_1 t_L}{b_1}$$

logarithm of the inactivation kinetic was the dependent variable actually modeled; this transformation is used to help stabilize the variance.

While it was not uncommon to observe "tailing" of survivor curves, and the logistic primary model was developed to describe resistant subpopulations, we have made little effort to model this attribute. In part, this reflects uncertainty about the importance and meaning of this phenomenon. When cells that have survived under conditions producing substantial tailing are regrown and rechal-

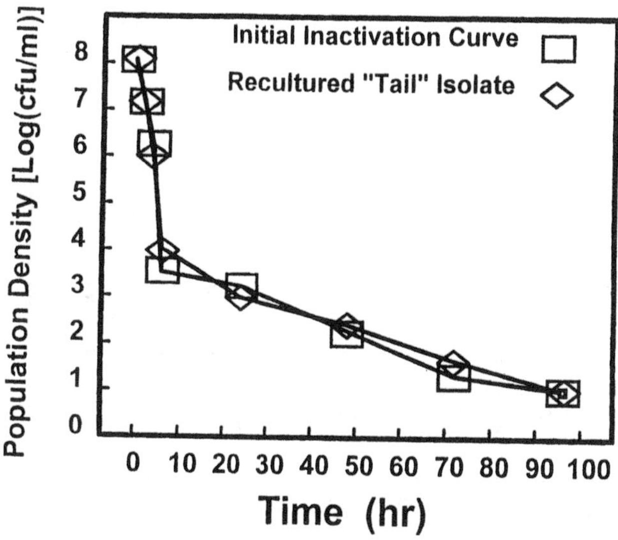

Figure 5 Effect of reculturing a *Listeria monocytogenes* isolate from the "tail" of survivor curve on the microorganism's survival when subsequently rechallenged with the same conditions. Conditions: 19°C/6.3% NaCl/1.0% lactic acid/100 μg/mL $NaNO_2$. (From Ref. 12.)

lenged, the survivor curves are virtually identical (Fig. 5). This suggests that tailing is not the result of a genetically stable, resistant subpopulation.

IV. USING PREDICTIVE MODELS

A key assumption made in developing models for pathogen survival was that the microorganisms' behavior in microbiological medium provides reasonable estimates of their survival in foods. The goal was to provide predictions that are as accurate as possible, but with the model biased toward safety. If there were differences between the survivor rates predicted by the model and those observed in foods, the model should "fail" on the "fail-safe" side (i.e., provide a prediction that was longer than the actual time to achieve the inactivation).

Comparison of models against survivor data with various foods has indicated that the models provide good "first-estimate" predictions of pathogen survivors for a range of products, and where there are discrepancies, the models are conservative. In part, this reflects the fact that foods often contain additional

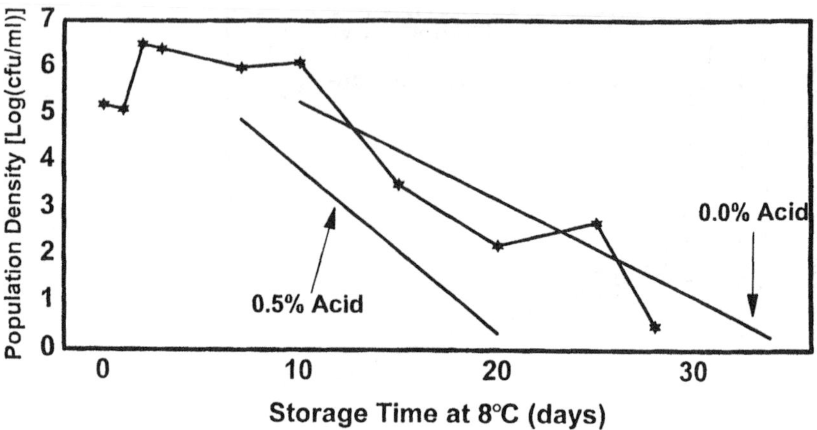

Figure 6 Comparison of the survivor curve for *Escherichia coli* O157:H7 in apple cider (from Ref. 19) with those predicted by the USDA Pathogen Modeling Program 5.1 using 0.0 and 0.5% acid as input values.

factors (e.g., antimicrobials) that further accelerate bacterial inactivation. Therefore, any use of the models for other than initial estimates should be predicated on their validation for the specific product of interest.

Predictive models are used most effectively when the user has knowledge of both the food and the survival characteristics of the pathogen. For example, Figure 6 depicts a comparison of the inactivation of *E. coli* O157:H7 in apple cider (18) versus the survivor curve predicted for two acid values using the USDA Pathogen Modeling Program (13,19). It is apparent that assuming 0% acid provided a more accurate prediction than 0.5%, the approximate level of acid in apple juice. As discussed earlier, this reflects the fact that the malic acid in apple cider has less of an anion effect than the lactic acid used to develop the model (Fig. 4). In such instances, the appropriately conservative use of the model would be to assume that inactivation was only due to the acid's impact on pH (i.e., % acid = 0) until data could be acquired relating the anion effects of the two acids.

The direct use of mathematical models can be cumbersome. However, with the widespread availability of personal computers, software applications that present the models under a user-friendly format can be designed readily. Models developed by ERRC scientists have been incorporated into the USDA Pathogen Modeling Program (17). In addition to having survival models, this user-friendly software package also contains growth models for nine foodborne pathogens and probabilistic models for proteolytic and nonproteolytic *Clostridium botulinum*. Provided without charge, the software has been distributed and used worldwide.

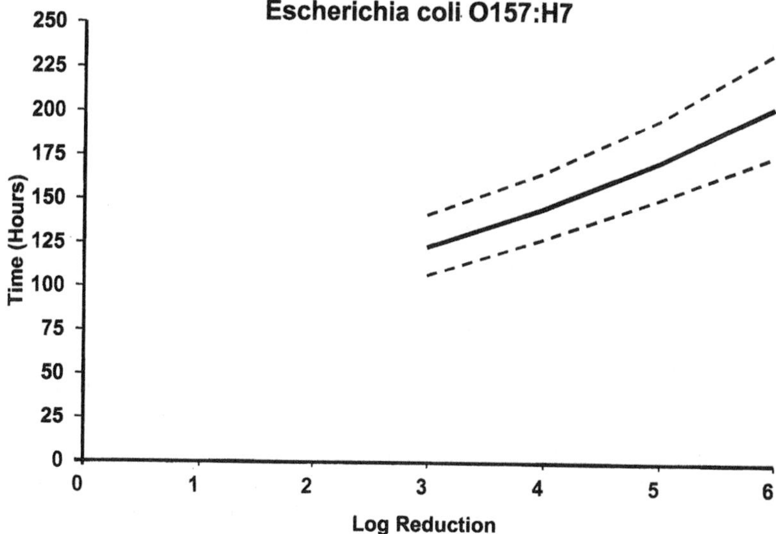

Figure 7 Computer screen from USDA Pathogen Modeling Program 5.1 for survival model for *Escherichia coli* O157:H7. (From Ref. 18.)

Copies of the software can be obtained either via the Internet (www.arserrc.gov/mfs/) or by e-mailing the developers (apickard@arserrc.gov).

The inactivation models in PMP5.1 include two features that are important for the proper use of microbial models (Fig. 7). The first is that predictions are statistical estimates; each is made with a certain degree of confidence. To adequately interpret predictions, this variability must be considered. The PMP5.1 automatically provides 95% confidence intervals for all survival predictions. The second feature is the selection of permissible variable ranges. As with any regression-based model, extrapolations made beyond the limits of the experimental data upon which the model is based are not advisable. The PMP5.1 restricts the user to variable input values that correspond to the laboratory studies that underlay each model.

V. EFFECTS OF BIOVARIABILITY AND PHYSIOLOGICAL STRESS RESPONSE ON ACID RESISTANCE

In developing models for the inactivation of foodborne pathogens, several decisions were made in regard to the strains employed and the means by which they were grown prior to being challenged. An underlying goal was to ensure that the models were conservative, i.e., a predicted value would represent the pathogen in its most resistant state. These decisions have proved to be fortuitous as research since the inception of the inactivation modeling project continues to redefine our concepts about acid resistance and its role as a factor influencing foodborne disease. In addition to allowing microorganisms to survive longer in acidic foods (19,20), acid resistance is increasingly recognized as an important virulence determinant. Constitutive and inducible acid resistance increases the likelihood that infectious bacteria will survive their passage through the hostile environment of the stomach. Further, recent studies have indicated that enhanced acid resistance also increases infectivity once the pathogen reaches the intestinal tract (21,22).

Acid resistance can vary among strains of the same species or even within the same serotype. For example, *E. coli* O157:H7 isolates vary substantially in their ability to survive an acid challenge (23–25) (Fig. 8). A set of the three or four representative strains per species was selected to develop the survival models. By using a mixture of resistant strains, the resulting survivor curves reflect the survival of the strain that was most resistant for each specific condition being examined. Data acquired in this manner have again helped ensure that the models developed are appropriately conservative.

A number of foodborne pathogens, including *L. monocytogenes, Salmonella, E. coli*, and *Shigella*, possess constitutive and inducible systems that in-

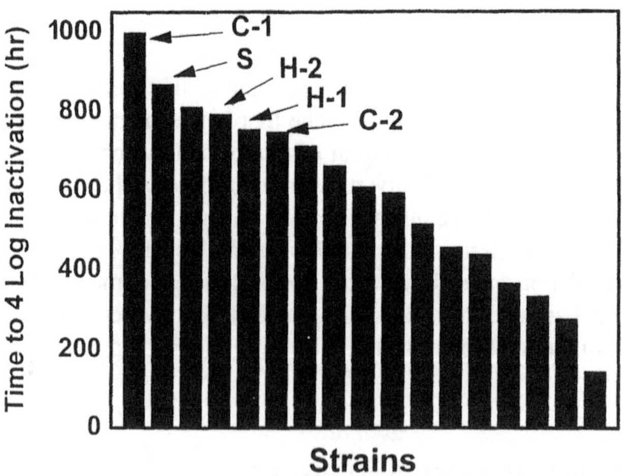

Figure 8 Variability in the acid resistance of seventeen strains of *Escherichia coli* O157:H7 when challenged with conditions of 12°C/5.0% NaCl/pH 4.8/0.3% lactic acid/25 ppm $NaNO_2$. Letter-designated strains are those from large outbreaks (C-1 and C-2: cider associated; H-1 and H-2: hamburger associated; S: fermented sausage associated.) (From Ref. 26.)

crease their ability to survive in acidic environments (11,24,26–28). While the specific mechanisms for acid tolerance/resistance vary among species, the general pattern of response is similar: (a) preexposure to moderately acidic conditions enhances survival when bacteria are exposed subsequently to harsher conditions, (b) acid resistance is increased when cells are grown in a glucose-containing, acidogenic medium, and (c) stationary phase cells are more resistant than exponential phase cells. The acid resistance of the strains used to develop the nonthermal inactivation models was maximized by growing the cells to stationary phase in an acidogenic medium. Again, this was done to ensure that the models developed were based on the pathogens being in their most resistant state.

Recent studies have demonstrated that understanding bacterial responses to acid stress is important in relation to other aspects of microbiological modeling. In particular, the induction of acid resistance can significantly impact the ability of pathogenic bacteria to withstand other stresses. Acid tolerance–associated cross-protection can influence resistance to heating, salt, antimicrobials, and ionizing and nonionizing irradiation (Fig. 9) (29–32). Alternatively, prior exposure to acids can sensitize bacteria to alkali conditions and alkali antimicrobials (27). The effects of acid resistance would have to be taken into account to accurately model these approaches to controlling foodborne pathogens.

VI. CONCLUDING REMARKS AND THE FUTURE

Before foods are eaten by a consumer, they may be processed, stored, and prepared by the manufacturer, wholesaler, retailer, restaurateur, and consumer. Each of these steps has an impact on both the frequency and extent of contamination by foodborne pathogens and, as such, has an impact on the risks faced by the consumer. The expanding use of Hazard Analysis Critical Control Point (HACCP) programs to manage microbial food safety risks and the accompanying emergence of quantitative microbial risk assessment as a means of linking the stringency of HACCP programs to public health impacts are two areas where there is an increasing need to deal with the behavior of foodborne pathogens quantitatively (33). Likewise, the emergence of microbiological performance criteria and safety objectives as means for setting "food safety/public health" targets while allowing industry the flexibility to devise innovative solutions is another area where an increase in our understanding of the growth and survival characteristics of foodborne pathogens is imperative (34).

It has become apparent in the last several years that predictive microbiology will play an important role in realizing these advances. This will be particularly true with ready-to-eat products that rely on nonthermal inactivation techniques such as fermented or acidic foods. This inactivation of foodborne pathogens as a result of exposure to an acidic pH is a highly complex process, requiring knowl-

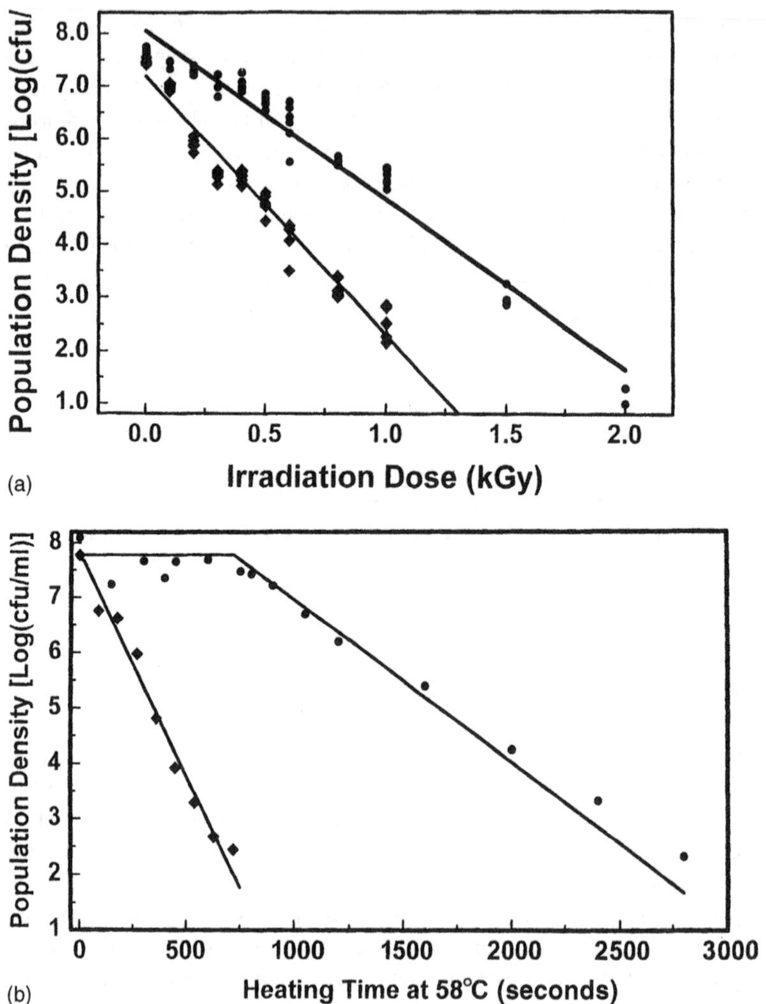

Figure 9 Examples of cross-protection effects associated with the induction of acid resistance in *Escherichia coli* O157:H7. (a) Cells grown to stationary phase in an acidogenic medium (TSB + 1% dextrose) or a nonacidogenic medium (TSB w/o dextrose) and then exposed to ionizing radiation (from Ref. 33). (b) Cells grown as in (a) and then resuspended in brain heart infusion (pH 6.0) and heated at 58°C. (From Ref. 35.)

edge of the interaction of pH with temperature, water activity, antimicrobials, and a variety of other factors. Without the understanding achieved through the systematic acquisition and analysis of data of the type used in developing validated microbiological models, producers will have to continue to rely exclusively on expensive, time-consuming ''inoculated pack'' studies to validate the efficacy of their processes. As continuing research addresses the complexities of acid tolerance and biovariability, there will be further increases in the sophistication of available models. This, coupled with increased experience in model use, should provide increased confidence that this traditional food preservation technology can be used effectively and predictably to produce foods that are simultaneously organoleptically unique and microbiologically safe.

REFERENCES

1. J Tilden, Jr., W Young, A-M McNamara, C Custer, B Boesel, MA Lambert-Fair, J Majkowski, D Vugia, SB Werner, J Hollingsworth, JG Morris, Jr. A new route of transmission for *Escherichia coli*: Infection from dry fermented salami. Am J Public Health 86:1142–1145, 1996.
2. D Morgan, CP Nawman, DN Hutchinson, AM Walkes, B Rowe, F Jaijd. Verotoxin-producing *Escherichia coli* O157:H7 infections associated with the consumption of yoghurt. Epidemiol Infect 111:181–187, 1993.
3. RE Besser, SM Lett, JT Weber, MP Doyle, TJ Barrett, GJ Wells, PM Griffin. An outbreak of diarrhea and hemolytic uremic syndrome from *Escherichia coli* O157:H7 in fresh-pressed apple cider. J Am Med Assoc 269:2217–2220, 1993.
4. AW Paton, RM Ratcliff, RM Doyle, J Seymour-Murray, D Davos, JA Lanser, JC Paton. Molecular microbiological investigation of an outbreak of hemolytic-uremic syndrome caused by dry fermented sausage contaminated with Shiga-like toxin-producing *Escherichia coli*. J Clin Microbiol 34:1622–1627, 1996.
5. RL Buchanan, SG Edelson. PH-dependent stationary-phase acid resistance response of enterohemorrhagic *Escherichia coli* in the presence of various acidulants. J Food Prot 62:211–218, 1999.
6. RL Buchanan, MH Golden, Interaction of citric acid concentration and pH on the kinetics of *Listeria monocytogenes* inactivation. J Food Prot 57:367–370, 1994.
7. RL Buchanan, MH Golden. Interactions between pH and malic acid concentrations on the inactivation of *Listeria*. J Food Safety 18:37–48, 1998.
8. RL Buchanan, MH Golden, RC Whiting. Differentiation of the effects of pH and lactic or acetic acid concentration on the kinetics of *Listeria monocytogenes* inactivation. J Food Prot 56:474–478, 1993.
9. RL Buchanan, MH Golden, JG Phillips. Expanded models for the non-thermal inactivation of *Listeria monocytogenes*. J Appl Microbiol 82:567–577, 1997.
10. MH Golden, RL Buchanan, RC Whiting. Effect of sodium acetate or sodium propionate with EDTA and ascorbic acid on the inactivation of *Listeria monocytogenes*. J Food Safety 15:53–65, 1995.

11. RL Buchanan, MH Golden, RC Whiting, JG Phillips, JL Smith. Non-thermal inactivation models for *Listeria monocytogenes*. J Food Sci 59:179–188, 1994.
12. RL Buchanan, MH Golden. Model for the non-thermal inactivation of *Listeria monocytogenes* in a reduced oxygen environment. Food Microbiol 12:203–212, 1995.
13. RC Whiting, MH Golden. *E. coli* survival model, USDA Pathogen Modeling Program, Version 5.1. USDA ARS Eastern Regional Research Center, Wyndmoor, PA, *www.arserrc.gov/mfs/*. 1997.
14. RC Whiting, S Sackitey, S Calderone, K Morely, JG Phillips. Model for the survival of *Staphylococcus aureus* in nongrowth environments. Int J Food Microbiol 31:231–243, 1996.
15. RC Whiting. Modeling bacterial survival in unfavorable environments. J Ind Microbiol 12:240–246, 1993.
16. RC Whiting, RL Buchanan. Letter to the editor: A classification of models in predictive microbiology—a reply to KR Davey. Food Microbiol 10:175–177, 1993.
17. RL Buchanan, RC Whiting, A Pickard. USDA Pathogen Modeling Program 5.1. USDA ARS Eastern Regional Research Center, Wyndmoor, PA, *www.arserrc.gov*. 1997.
18. T Zhao, MP Doyle, RE Besser. Fate of enterohemorrhagic *Escherichia coli* O157: H7 in apple cider with and without preservatives. Appl Environ Microbiol 59:2526–2530, 1993.
19. LG Miller, CW Kaspar. *Escherichia coli* O157:H7 acid tolerance and survival in apple cider. J Food Prot 57:460–464, 1994.
20. CGM Gahan, B O'Driscoll, C Hill. Acid adaptation of *Listeria monocytogenes* can enhance survival in acidic foods and during milk fermentation. Appl Environ Microbiol 62:3128–3132, 1996.
21. FG del Portillo, JW Foster, BB Finlay. Role of acid tolerance response genes in *Salmonella typhimurium* virulence. Infect Immun 61:4489–4492, 1993.
22. B O'Driscoll, CGM Gahan, C Hill. Adaptive acid tolerance response in *Listeria monocytogenes*: Isolation of an acid-tolerant mutant which demonstrates increased virulence. Appl Environ Microbiol 62:1693–1698, 1996.
23. MM Benjamin, AR Datta. Acid tolerance of enterohemorrhagic *Escherichia coli*. Appl Environ Microbiol 61:1669–1672, 1995.
24. RL Buchanan, SG Edelson. Culturing enterohemorrhagic *Escherichia coli* in the presence and absence of glucose as a simple means of evaluating the acid tolerance of stationary phase cells. Appl Environ Microbiol 62:4009–4013, 1996.
25. RC Whiting. Overview of *E. coli* O157:H7 growth and decline modeling. USDA, FSIS Public Meeting on the Risk Assessment for *E. coli* O157:H7 in Beef. Arlington, VA, Oct. 28, 1998.
26. JW Foster. Low pH adaptation and the acid tolerance response of *Salmonella typhimurium*. Crit Rev Microbiol 21:215–237, 1995.
27. RJ Rowbury. An assessment of environmental factors influencing acid tolerance and sensitivity in *Escherichia coli, Salmonella* spp. and other enterobacteria. Lett Appl Microbiol 20:333–337, 1995.
28. S Bearson, B Bearson, JW Foster. Acid stress responses in enterobacteria. FEMS Microbiol Lett 147:173–180, 1997.

29. JM Farber, F Pagotto. The effect of acid shock on the heat resistance of *Listeria monocytogenes*. Lett Appl Microbiol 15:197–201, 1992.
30. GJ Leyer, EA Johnson. Acid adaptation induces cross-protection against environmental stresses in *Salmonella typhimurium*. Appl Environ Microbiol 59:1842–1847, 1993.
31. Y Lou, AE Yousef. Adaptation to sublethal environmental stresses protects *Listeria monocytogenes* against lethal preservation factors. Appl Environ Microbiol 63: 1252–1255, 1997.
32. RL Buchanan, SG Edelson, G Boyd. Effects of pH and acid resistance on the radiation resistance of enterohemorrhagic *Escherichia coli*. J Food Prot 62:219–228, 1999.
33. RL Buchanan, RC Whiting. Risk assessment and predictive modeling. J Food Prot 59(suppl):31–36, 1996.
34. Intern. Commission Microbiol. Specifications for Foods Working Group on Microbial Risk Assessment. Potential application of risk assessment techniques to microbial issues related to international trade in food and food products. J Food Prot 61: 1075–1086, 1998.
35. RL Buchanan, SG Edelson. Effect of pH-dependent, stationary phase acid resistance on the thermal tolerance of *Escherichia coli* 0157:H7. Food Microbiol. (In press.)

19

Management of Microbial Control in HACCP Systems

O. Peter Snyder, Jr.
Hospitality Institute of Technology and Management
St. Paul, Minnesota

Vijay K. Juneja
U.S. Department of Agriculture
Wyndmoor, Pennsylvania

An effective management program must be established for the control of pathogens in food processing and production, or the food operation is vulnerable to significant liability. Microbial control processes are only as good as management control of the stability of the processes. When the safety parameters of the process have been established through research and development, the key to evaluating the safety of a process is to measure its stability.

I. HACCP AS A BASIS FOR SYSTEM CONTROL

The National Advisory Committee on Microbiological Criteria for Foods (NACMCF) (1) describes the Hazard Analysis Critical Control Point (HACCP) program and how it should be documented. The two important components of the NACMCF-HACCP program are prerequisite program controls and food process controls. Prerequisite program controls are listed in Table 1.

For food process control, the NACMCF program lists seven principles of HACCP (see Table 2). These principles are actually a seven-step sequence for

---

Mention of brand or firm name does not constitute an endorsement by the U.S. Department of Agriculture above others of a similar nature not mentioned.

Table 1 Prerequisite Controls for a HACCP Program

Listing of HACCP management team and responsibilities
Description of the food and its distribution
Description of the intended use and consumers of the food
Facility construction and maintenance
Linear product flow
Supplier controls: GMPs, supplier guarantees, HACCP
Specifications written for ingredients, products, and packaging
Production equipment construction, installation, preventive maintenance, calibration
Cleaning and sanitizing procedures; master sanitation schedule
Personal hygiene; all employees and visitors
Training documentation covering personal hygiene, GMPs, cleaning safety, HACCP, food and ingredient handling
Chemical control and segregation
Receiving, storage, shipping procedures, temperature, humidity
Traceability and recall; lot coding
Pest control
Quality Assurance (QA) procedures
Process and recipes
Product formulation
Labeling
Glass control

Source: Ref. 1.

Table 2 NACMCF Seven Principles of HACCP for Process Control

1. Conduct the hazard analysis. Identify steps in the process with hazards.
2. Apply HACCP decision tree to each step with hazards. Determine which steps are critical control points.
3. Establish critical limits for preventive measures associated with each CCP (critical control point).
4. Establish CCP monitoring requirements. Establish procedures for using the results of monitoring to adjust the process and maintain control.
5. Establish corrective action to be taken when monitoring indicates that there is a deviation from an established critical limit.
6. Establish procedures for verification that the HACCP system is working correctly.
7. Establish effective record keeping procedures that document the HACCP system.

Source: Ref. 1.

developing a validated, safe process. These principles are used only after the system and all prerequisite programs are specified by a process authority and a HACCP team (a group consisting of management and line employees).

II. THE INTEGRATION OF NACMCF-HACCP WITH TOTAL QUALITY MANAGEMENT FOR FULLY FUNCTIONAL MICROBIOLOGICAL INACTIVATION ASSURANCE

While the NACMCF-HACCP (1) document provides a listing of some necessary elements of a HACCP program, it does not provide a method or approach for documentation of a complete HACCP program. A number of management components are not specified. There is no emphasis on continuous improvement of the program. It is implied that monitoring and record keeping are just for government review. However, it is even more important that operators use the monitoring data to continually improve the processes and reduce variability.

Table 3 lists 12 sections of a fully functional HACCP-TQM continuous improvement program as developed by the Hospitality Institute of Technology and Management (2) for foodservice operations. This program incorporates NACMCF-HACCP components together with management elements necessary for system control.

III. ESTABLISHING CRITICAL LIMITS FOR A PROCESS: HACCP DOCUMENTATION OF THE RECIPE/FOOD PROCESS

In order to simplify the documentation of the NACMCF-HACCP food process chart, it is convenient to combine a flow chart and HACCP analysis into one form, as shown in Table 4. When all prerequisite programs are functioning, the product can be produced in accordance with this HACCP procedure.

The Process Step column lists the steps for the production line preparation/processing of the product sequentially. Typically, this is written just as a computer flow chart is written. The first steps are the ''get-ready'' steps (e.g., turn on equipment and make sure that the process line is functioning correctly). Then, each process step is listed, beginning with preparation. The inactivation process follows, and finally, cooling (if necessary), packaging, labeling, and final storage for shipment of the food.

Table 3 HACCP-TQM Retail Food Operations Manual: Table of Contents

I. Food Safety Policy
II. Organization for HACCP-based TQM
    A. Organization chart
    B. HACCP-TQM team
III. System Description
IV. Good Manufacturing Practices
    A. Management
        1. Senior management commitment and involvement for food safety
            a. Policies and procedures
            b. PIC (Person in Charge)
        2. Hazard analysis and control
        3. Manager communication and employee training
        4. Handling emergencies
        5. Facility improvement
    B. Personnel
        1. Employee responsibility
        2. Disease control
        3. Cleanliness
    C. Environment
        1. Area around facility or grounds
        2. Water
        3. Sewage
        4. Pest control and materials used
        5. Poisonous (toxic) materials
    D. Facilities
        1. Facility design
        2. Floors, walls, and ceilings
        3. Lighting
        4. Plumbing
    E. Equipment
        1. Equipment and utensil construcuon
        2. Equipment operation
        3. Food contact surface equipment
        4. Non-food contact surface equipment
    F. Supplies and materials
        1. Purchasing
        2. Supplier certification
        3. Ingredient specification
    G. Food production and service
        1. General production policy
        2. Home-prepared food
        3. Milk
        4. Receiving
            a. Inspection of incoming products
            b. Substandard products
            c. Food containers
            d. Container disposal
            e. Labeling
            f. Use-by date
            g. Food storage areas
        5. Pre-preparation
            a. Ingredient inspection and control
            b. Food thawing
            c. Chemical additives
            d. Raw food handling

Table 3 Continued

- 6. Preparation
  - a. Potentially hazardous food
  - b. Hard foreign objects
  - c. Food pasteurization
- 7. Holding, serving, and transporting
  - a. Food holding temperatures
  - b. Conveyors
  - c. Serving, packaging, and transporting
- 8. Storing prepared food
  - a. Food cooling time
  - b. Storage to prevent cross-contamination
  - c. Storage time
  - d. Storage containers
  - e. Leftovers

H. Consumer
- 1. Consumer information
- 2. Returned food
- 3. Food sabotage

V. Supplier HACCP
- A. Supplier HACCP Qualification Standards
- B. Supplier HACCP/QA Qualification List by Ingredients Purchased
- C. Ingredient Specifications

VI. Recipe HACCP
- A. Quality-assured Product/Recipe Procedures
- B. Product Specifications

VII. Cleaning and Sanitizing Schedule and Instructions
- A. Cleaning and Sanitizmg Schedule
- B. Sanitation Procedures and Standards

VIII. Maintenance Schedule and Instructions
- A. Maintenance Schedule
- B. Maintenance Procedures and Standards

IX. Pest Control Schedule and Instructions
- A. Pest Control Schedule

X. HACCP-TQM Employee Training Program and Record
- A. Employee Training Record
  - 1. New Employee Training Record
  - 2. Continuing Education Training Record

XI. Self-inspection, Continuous Quality Improvement
- A. New Product/Process Development
- B. Product Process Monitoring/Sampling Plan
- C. Daily Self-inspection
- D. Weekly, Yearly Self-inspection

XII. Food Safety Program Verification and Certification
- A. Performance Verification and Capability Certification

Source: Ref. 2.

Table 4 HACCP Analysis and Flow Chart

Process step, control measures, and linear process flow chart	Hazard and critical limit	Monitoring procedure frequency, record, and person responsible	Corrective action, record, and person responsible	Verification procedure, record, and person responsible
1. Get ready. Start the production line. Verify that process controls are working. ↓	Not a CCP			
2. Get raw ingredients. Unpackage, cut, and make ready to process. ↓	Not a CCP			
3. Process. Using an inactivation method, reduce pathogens to a safe level. ↓	CCP 1	Cook takes 5 random product temperatures.	If product is not greater than ____, continue to process. Record on ____.	QC (Quality Control) check and record on ____.
4. Cool if necessary. Do not let the product come in contact with contaminated surfaces. ↓	CCP 2			
5. Package. Avoid cross-contamination. Label with lot number and consumer handling. ↓	CCP 3			
6. Store.	Not a CCP			

Source: Adapted from Ref. 1.

IV. DECIDING ON THE PROCESS PERFORMANCE STANDARDS

Each country has its own microbiological inactivation process (lethality) performance standards. For example, in the United States the commercial sterilization process standard for low-acid canned food is processing at temperatures for times that assure a 10^{12} reduction in *Clostridium botulinum* (proteolytic types A and B) spores.

A key starting point for the design of pasteurized food processes is to use U.S. Department of Agriculture (USDA) requirements for the reduction of pathogens in food. Decimal reduction values (D-values) are used to indicate the time required to destroy populations of microorganisms. One D-value is the time necessary to destroy 90% of a population of cells (or spores) at a given temperature. The Code of Federal Regulations, Title 9 and Title 21 (3,4), provides current requirements for pathogen reduction in food processes. The USDA requires that large cuts of red meat (e.g., beef roasts) receive a 6.5D *Salmonella* reduction. Poultry requires a 7.0D *Salmonella* reduction. For fruits, the U.S. Food and Drug Administration (FDA) has tentatively established a requirement of 5D *Salmonella* reduction. There is no established thermal reduction for fish, but the FDA 1999 Retail Food Code (5) requirement for a heat treatment of 63°C (145°F) for 15 seconds would only be sufficient for inactivation of parasitic worms that may be present in the center of fish.

It is essential to understand that these government process recommendations are ''safe harbors.'' They assume that the processor is starting with normally contaminated ingredients. If a processor uses HACCP, these critical limits are not required if the processor can show that the production line is starting with a product that has a much lower level of contamination than the government-assumed contamination level used to establish government process performance standards.

V. DEVELOPING THE MANAGEMENT PROGRAM

When management decides to use a new technology for processing food and understands government critical limits for surviving microorganisms (especially *Salmonella* in the food), management can implement a hazard control program to assure that, ''from farm to fork,'' the food will be safe when consumed. The input is the specification of ingredient contamination level from the farm and slaughtering operations. For example, the deep muscle tissue of meat and poultry from healthy animals is basically sterile; however, contamination can occur during and after slaughter. The output is the product that the consumer receives, which may either require some preparation or be ready-to-eat food. Thus, the operator's process must produce a product that meets each product's specifications with zero defects.

The more an operator can work with a supplier to reduce the numbers of bacteria on incoming supplies by using microbial specifications, the less processing will be required to achieve the government specification for pasteurized food of no detectable salmonellae, *Escherichia coli* O157:H7, or *Listeria monocy-*

togenes. Current government recommended process reductions are based on the assumption that there are approximately 1000 *Salmonella* per gram of raw poultry and meat. Within the limits of statistical variation, these pathogenic bacteria must be reduced to approximately 1 *Salmonella* spp. per 250 of food. Processing to achieve this microbial standard is actually more than adequate to assure food safety.

VI. IMPLEMENTATION

After the processes are planned and validated, management designs and checks the processing facility, selects the equipment, and provides for cleaning, maintenance, and employee training (all prerequisite programs), so that the processes will reliably deliver food that is safe when consumed. Six steps used in the management cycle of a process operation are listed in Table 5.

VII. WHAT SIGNIFIES A SAFE PROCESS?

The best way to assure a safe process is to verify the capability of the process. When the critical process is designed, process variables with target values are determined that, when met, assure adequate pathogen reduction. These process variables may include achieving a designated center temperature for a specified period of time, use of ionizing radiation, or heating the product in steam under pressure for a specified period of time. As the process runs, the critical process

Table 5 Steps in the Management Cycle of Process Operation

1. Management commitment of time and resources so that an effective program is developed
2. Hazard analysis of process
3. Documentation of procedures and standards and validation that the operating plan is effective
4. Training of employees and calibration of the processes so that they are as stable as possible for the production of the product
5. Operation of the process that includes monitoring and using corrective action to keep the process in control
6. Analyzing the data in order to revise operating procedures and the operating manual to achieve better stability; repeating the cycle the following day (i.e., begin with #1)

control target values are monitored and charted, and process capability (Cp) is determined (6). The process capability can be calculated with the following formula.

$$\text{Process capability (Cp)} = \frac{\text{upper control} - \text{lower control limit}}{\text{6 sigma deviation in the process}}$$

This formula is widely used in manufacturing facilities to judge the capability of a process to produce a standard-performing product. When the process capability is 1.0, it can be expected that in a two-tailed statistical analysis, there will be 28 excursions in 10,000 beyond the upper or lower control limit. If the Cp is >1.3, the excursion will be only 96 times in 1,000,000. Therefore, the objective of management is to determine the common and special causes of process fluctuation and to minimize those causes with continuous improvement. Thus, process stability and capability are improved. The improvement in process stability becomes a standard goal of a company's quality control department.

A. Risk Management and HACCP

To establish any national and international standards for microbiological, chemical, or physical hazards in food, it is essential to understand that there will always be a risk factor associated with some contamination of ingredients. Food grown in the ground may be contaminated with pathogens such as *Bacillus cereus*, *C. botulinum*, and *L. monocytogenes*, as well as molds, parasites, and viruses. Rodents, insects, and wild birds are known to carry pathogenic microorganisms to farm animals, poultry, and fish. Hence, meat and poultry products produced in a typical farm environment may be contaminated with *Salmonella* spp., *Campylobacter jejuni*, *E. coli*, and other pathogens. Fish and shellfish are likely to be contaminated with *Vibrio* spp. Using methods of farming and harvesting that reduce or minimize contamination can reduce microbial levels in raw food items.

One method of controlling pathogens in meat and poultry items in the future may be the development of vaccines to be used for meat animals and poultry so that they are not colonized with pathogens. Another method of control is to raise animals and poultry in pathogen-reduced environments. One way to do this is to use potable water (e.g., water treated with chlorine) to raise animals and to wash field soil off of vegetables. However, raising animals without some degree of environmental contamination is very costly. This means that the processor must reduce pathogenic microorganisms in foods to low-risk levels. This can be accomplished by pasteurization, cooking food, addition of acid, fermentation, or by washing food items. When these controls are effectively applied in a stable process, the risk of consumers becoming ill is dependent on the functioning immune system of each person and the extent to which consumers follow safe food-handling procedures.

Risk, as related to food consumption, is the likelihood that individuals or a population will incur an increased incidence of adverse effects such as foodborne illness, disease, or death as a result of consuming a food (7). The risk can be further defined by defining the specific cause of illness or disease (e.g., the risk of salmonellosis due to consumption of underheated shell eggs or the risk of foodborne illness due to *E. coli* O157:H7 in ground beef).

In order to estimate how much illness or injury can be expected from exposure(s) to a given risk agent, and to assist in judging whether these consequences are great enough to require increased management or regulation, a risk assessment or analysis should be conducted as a part of a HACCP program. Risk assessment can be used as a tool to determine sources of the most serious hazards and apply action to reduce the presence of these hazards. A risk assessment can also be used to ensure that operational risk management decisions are rational and are based on the best available science (8).

After a risk assessment is completed, risk can be expressed in quantitative probability terms or in qualitative terms. An example of the use of quantitative probability terms is the number of illnesses over a lifetime in a population of 1 million exposed people. A risk of 1 illness in 10,000 is described as "10^{-4} risk," while 1 illness in 1 million is described as a "10^{-6} risk." Historically, risks of less than 10^{-6} in magnitude have not been the object of concern. Risk can also be expressed in qualitative terms of "low," "medium," and "high." These terms are used when quantification is either not feasible or unnecessary.

Table 6 shows an organizational chart for integrated risk management. Integrated risk management can be divided into risk analysis, management risk control, and risk communication.

B. Components of Risk Analysis

When the process is specified, the following components of risk analysis can be used:

1. Hazard analysis is based on quantitative epidemiology, expert knowledge, data, or research evidence that shows that an agent associated with the consumption of a particular food may cause human illness or injury.
2. Risk assessment evaluates the chance that a hazard will be in the food. This type of assessment includes consumers, what they eat, and in what type of environments. For example, "Is the product take-out food? How do consumers handle the food? Is there consumer abuse? What is the amount, frequency, and source of the hazard in food?" It is also

Table 6 Integrated Risk Management

1. Operations process step description	2. Risk analysis			3. Management risk control	4. Risk communication
	A. Hazard analysis	B. Risk assessment	C. Hazard control assessment		
Employee procedures and controls Get ready Do ______ Until ______ Check ______ If ______ Then ______ Else ______ Record if ____ Clean up Put away	Hazard identification Hazard quantification Critical limits	Exposure assessment Dose-response assessment Risk characterization (dollars)	Analysis of effectiveness of current unit controls Analysis of consumer control Failure mode effect analysis Expected cost per year from failure of this step	If the risk is acceptable, certify the step If the risk is not acceptable, improve control	Consumer is informed how to control remaining risk

Source: Ref. 8.

important to assess the effectiveness of the controls of the hazard from food production to consumption.

A part of the assessment is the "dose response," or the probability of consumers becoming ill at various dose intakes. For example, the level that normally healthy people can consume and still remain healthy and possibly even develop some immunity must be known, as well as the levels known to cause probable illness or injury.

3. Risk characterization is the severity and cost of a hazard. For example, the development of listeriosis in pregnant women and their fetuses is a severe and costly hazard, while the incidence illness due to consumption of *C. perfringens* in improperly cooled food will only cause an illness of short duration and consequence and virtually no cost.
4. Hazard control assessment is the fourth component of risk analysis. This type of assessment is generated by the HACCP team through evaluation of the effectiveness of the current process, consumer controls, and estimates of the probability of failure. The HACCP team also estimates the probable cost to the business in the event of possible incidents related to an expected failure of control at this step.

C. Management Risk Control

If the risk of financial loss and human suffering is unacceptable after the risk has been assessed, management can take measures to control or reduce the risk to an acceptable level. This can be accomplished through application of Good Manufacturing Practices (GMPs) and the use of a HACCP-based Total Quality Management (TQM) program in the improvement of prerequisite programs and the production of food products.

D. Risk Communication

Risk is never "zero." For example, microbial spores in food survive food preparation processes in retail operations. Consumers must be aware of this risk in foods that they take out of the food operation to consume at a later time. Also, many people desire undercooked/raw food such as beef, eggs, and fish. While the operators can buy these kinds of items from highly reliable suppliers with HACCP programs, there is still risk. Producers and operators of retail food establishments should communicate the amount of residual risk associated with food items to consumers through labels on containers or other various media presentations. Communicating the risk and consumer control responsibilities after the food leaves food operations control does not relieve management of all liability, but it does help in a "due diligence" defense.

VIII. FOOD OPERATIONS RISK MANAGEMENT BASED ON COMBINED HAZARD ANALYSIS RISK ASSESSMENT AND FAILURE MODE EFFECT ANALYSIS

Risk management of food operation procedures begins with assuring that prerequisite programs are effective. These programs are the basis for assuring that the process controls will be effective. Prerequisite programs assure that there is a control system. Every employee, in addition to HACCP process control, must be able to perform systems control at his or her work place or site.

When a process authority or government official certifies a process as having an acceptable risk, the process should be reviewed, using the principles of (failure mode effect analysis) (FMEA) (6). This is a systematic method for identifying and preventing process and product problems before they occur. FMEAs are focused on preventing process deviations and improving stability. Hence, safety is enhanced. Ideally, FMEAs are conducted in the food product (recipe) design or process development stages. However, conducting an FMEA on existing products and processes can also yield benefits. The objective of a food safety FMEA is to look at a HACCP plan for all of the ways that a process can fail and thus produce an unsafe product. Even the simplest foods have many opportunities for failure in terms of microbiological, chemical, and particulate hazard risks. For example, spores can outgrow in retail food; the growth spoilage microorganisms can produce histamine in scombroid fish; chemical sanitizers can contaminate food; and metal or plastic package clips can fall into open containers of food. The production process must be tested to find out how the controls can fail. In this way, the hazards, standards, and controls at each step can be set to achieve an acceptable risk, and the process can be certified as acceptably safe.

IX. SUMMARY

As scientists continue to develop new, innovative methods for microbial inactivation and making food safe, HACCP will be used to assess and document the safety of processes. Processes have been evolving that make food safe with the least possible change in the quality characteristics of the food. Canning, for instance, creates a very safe product, but it changes the characteristics of the product considerably. Perishable products such as lettuce and other leafy vegetables, which cannot withstand the canning process, must be subjected to other processing, packaging, and storage methods to increase their safety and stability. Many minimally processed foods on dining tables today are the result of new innovative methods for extending the shelf life of food and making it safe. If the processor uses a stable HACCP-TQM program to develop quality-assured products, the risk of liability can be controlled, and new inactivation processes will continually evolve.

REFERENCES

1. National Advisory Committee on Microbiological Criteria for Foods (NACMCF). Hazard analysis and critical control point principles and application guidelines. J Food Prot 61(6):762–775, 1998.
2. HACCP-TQM Retail Food Operations Manual. St. Paul, MN: Hospitality Institute of Technology and Management. 1999.
3. Code of Federal Regulations (CFR). Title 9. Animal and Animal Products. Part 200 to end. Superintendent of Documents. Washington, DC: U.S. Government Printing Office., 1999.
4. Code of Federal Regulations (CFR). Title 21. Food and Drugs. Part 100 to 169. Superintendent of Documents. Washington, DC: U.S. Government Printing Office, 1999.
5. FDA Food Code. U.S. Public Health Service, U.S. Department of Health and Human Services. Pub. No. PB99-115925. Washington, DC: Food and Drug Administration, 1999.
6. VE Kane. Defect Prevention: Use of Simple Statistical Tools. New York: Marcel Dekker, Inc., 1989.
7. National Advisory Committee on Microbiological Criteria for Foods (NACMCF). Principles of risk assessment for illness caused by foodborne biological agents. J Food Prot 61(8):1071–1074, 1998.
8. National Advisory Committee on Microbiological Criteria for Foods (NACMCF). Potential application of risk assessment techniques to microbiological issues related to international trade in food and food products. J Food Prot 61(8):1075–1086, 1998.

20
Hurdle Technology

Lothar Leistner
International Food Consultant
Kulmbach, Germany

I. INTRODUCTION

The microbial stability and safety as well as the sensory and nutritive quality of most preserved foods are based on a combination of several empirically applied preservative factors called hurdles and more recently on knowingly employed hurdle technology. Deliberate and intelligent application of hurdle technology allows a gentle but efficient preservation of foods and is advancing worldwide. Various expressions are used for the same concept in different languages (e.g., Hürden Technologie in German, hurdle technology in English, technologie des barrières in French, barjernaja technologija in Russian, tecnologia degli ostacoli in Italian, métodos combinados in Spanish, shogai gijutsu in Japanese, and shan lan ji shu in Chinese) (1). At present, the term "hurdle technology" is most often used.

The theme of this volume is the inactivation of foodborne microorganisms, and, therefore, the significance of hurdle technology will be discussed in this context. Proper application of hurdle technology will inhibit the growth of microorganisms in foods, and under certain conditions their survival will be shortened. In order to achieve the desired results, the principles and basic aspects of hurdle technology have to be understood, and they will be discussed first. Furthermore, examples of recent applications of hurdle technology to foods in industrialized and developing countries will be mentioned, as well as the use of hurdle technology in food design.

II. PRINCIPLES OF HURDLE TECHNOLOGY

Many traditional preservation methods are used to make foods stable and safe, e.g., heating, chilling, freezing, drying, salting, curing, sugar addition, acidification, fermentation, oxygen removal, and smoking. However, these processes are based on relatively few parameters or hurdles, e.g., high temperature (F-value), low temperature (t-value), water activity (a_w), acidity (pH), redox potential (Eh), preservatives, and competitive flora. In some of the preservation methods mentioned, these parameters are of major importance; in others they are only secondary hurdles (2,3). The use of inhibiting factors in combination is advantageous by allowing the less extreme use of any single treatment (4). Also in modern food preservation techniques using bacteriolytic enzymes, bacteriocins, irradiation, high pressure, or pulsed technologies, secondary hurdles are employed to achieve the desired preservation. For instance, pulsed electric fields can be combined with other hurdles such as pH, a_w, temperature, or preservatives. The effect of high pressure in food preservation is also substantially improved if combined with heat, antimicrobials, ultrasound, or ionizing radiation. Therefore, the use of combined methods is most likely the future of food preservation (5).

The critical values of many preservative factors for the death, survival, or growth of microorganisms in foods have been determined in recent decades and are now the basis of food preservation. However, the critical value of a particular parameter changes if additional preservative factors are present in the food. It is, for instance, well known that the heat resistance of bacteria increases at low a_w and decreases at low pH or in the presence of preservatives, whereas a low Eh increases the inhibition of microorganisms caused by reduced a_w. The simultaneous effect of different preservative factors (hurdles) could be additive or even synergistic. In food preservation the combined effect of preservative factors must be taken into account, which is illustrated by the hurdle effect.

A. Hurdle Effect

For each stable and safe food a certain set of hurdles is inherent, differing in quality and intensity depending on the particular product. In any case the hurdles must keep the "normal" population of microorganisms in this food under control. The microorganisms present (at the start) in a food should not be able to overcome ("leap over") the hurdles present, otherwise the food will spoil or even cause food poisoning. This situation is illustrated by the so-called hurdle effect, first introduced in 1978 (6), which is of fundamental importance for the preservation of intermediate-moisture foods (7) as well as high-moisture foods (2). Leistner and coworkers acknowledged that the hurdle effect illustrates only the well-known fact that complex interactions of temperature, water activity, acidity, redox potential, preservatives, etc., are significant for the microbial stability and safety of most foods.

In previous publications (3,6,8) figures with several examples illustrating the hurdle effect in foods have been presented, which will not be repeated here. However, it should be mentioned that the hurdle effect is, for instance, important for the ultraclean or aseptic packaging of foods, because if there are only few microorganisms present at the start, then a few or low hurdles are sufficient for the stability of the product. The same proves true if the initial microbial load of a food (e.g., on carcass meat or in high-moisture fruits) is substantially reduced (e.g., by the application of steam), because after such decontamination procedures fewer microorganisms are present, which are then more easily inhibited. The number and intensity of hurdles needed for microbial stability is also lower if the microorganisms present are sublethally injured, because then they lack ''vitality'' and thus are easier to inhibit. On the other hand, a food rich in nutrients and vitamins will foster the growth of microorganisms (the ''booster'' or ''trampoline'' effect), and thus the hurdles in such a product must be enhanced, otherwise they will be overcome. The latter also happens if due to bad hygiene too many undesirable microorganisms are initially present, because then the usual hurdles inherent to a product may be unable to prevent spoilage or food poisoning. In fermented foods (salami, cheese, pickled vegetables, etc.) a sequence of hurdles is active, arising in different stages of the ripening process and leading to a microbiologically stable and safe finished product. Further research in this area is promising and should clear and optimize the sequence of hurdles in various fermented foods.

The most important hurdles commonly used in food preservation are temperature (high or low), a_w, acidity, Eh, preservatives (e.g., nitrite, sorbate, sulfite), and competitive microorganisms (e.g., lactic acid bacteria). More than 60 potential hurdles for foods of animal or plant origin, which improve the stability and/or the quality of these products, have been described, and the list of possible hurdles for food preservation is by no means complete (9). At present, physical, nonthermal processes (high hydrostatic pressure, mano-thermo-sonication, oscillating magnetic fields, pulsed electric fields, light pulses, etc.) are receiving considerable attention, since in combination with other secondary hurdles they are of potential use for the microbial stabilization of fresh-like food products with little degradation of nutritional and sensory properties (5). With these novel processes often not a sterile product but only a reduction in the microbial load is intended, and growth of the residual microorganisms is inhibited by additional, secondary hurdles. Another group of hurdles of special interest in industrialized as well as in developing countries is ''natural preservatives''—spices and their extracts, hop extracts, lysozyme, chitosan, protamine, pectine hydrolysate, etc. In most countries these ''green preservatives'' are preferred because they are not synthetic chemicals, whereas in some African countries short of foreign currency, they are given preference because spices are more available and cheaper than imported chemicals.

B. Hurdle Technology

A better understanding of the impact and interaction of different preservative factors (hurdles) in foods is the basis for improvements in food preservation. If the hurdles in a food are known and their interaction visualized, the microbial stability and safety of the food might be optimized by changing the intensity or quality of these hurdles (9). An understanding of the hurdle effect is the key for an understanding of the effectiveness of traditional preservation methods for foods. The next step was the optimization of traditional foods as well as the development of novel products by intelligent combination of hurdles. Thus, from an understanding of the hurdle effect, hurdle technology has been derived (10), which means that hurdles are deliberately and intentionally combined in the preservation of traditional and novel foods. Using an intelligent mix of hurdles, it is possible to improve not only the microbial stability and safety but also the sensory and nutritive quality as well as the economic aspects of a food. For the economy of a food, water content is important. However, it is essential that the water content in the product be compatible with its microbial stability, and if an increased a_w is compensated by other hurdles (pH, Eh, etc.), the food becomes more economical.

Hurdle technology is increasingly used in industrialized and developing countries for optimizing traditional foods and for making new products according to needs. For instance, if energy preservation is the goal, then energy-consuming hurdles such as refrigeration are replaced by other hurdles (a_w, pH, or Eh) that do not demand energy but still ensure a stable and safe food (6). Furthermore, if we want to reduce or replace preservatives (e.g., nitrite) in meats, we could emphasize other hurdles (e.g., a_w, pH, refrigeration, or competitive flora, which would stabilize the product) (11). More recent examples related to the application of hurdle technology will be given in the Sec.IV.

C. Total Quality

Stanley (12) proposed that the hurdle technology approach could be applicable to a wider concept of food preservation than just microbial stability, but that in order for it to work a precise knowledge of the effectiveness of each hurdle for a given food would be required. Furthermore, he suggested distinguishing between positive and negative hurdles for the quality of foods. Certainly hurdle technology is applicable not only to safety, but also to quality aspects of foods, although this area of knowledge has been much less explored than the safety aspect. McKenna (13) emphasized that while hurdle technology is appropriate for securing the microbial stability and safety of foods, the total quality of foods is a much broader field and encompasses a wide range of physical, biological, and chemical attributes. The concept of combined processes should work towards the total quality of foods rather than the narrow but important aspects of microbial stability and safety. But at present the tools for applying hurdle technology to total food quality are still not adequate, and this is equally true for predicting food quality by modeling. However, researchers should appreciate the wider power of the hurdle technology concept, and the food industry should use the available tools of combined processes for as many quality enhancements as possible (13).

Some hurdles, e.g., Maillard reaction products, influence the safety as well as the quality of foods, because they have antimicrobial properties and at the same time improve the flavor of the products. This also applies to nitrites used in the curing of meat. The possible hurdles in foods can influence the stability and safety, as well as the sensory, nutritive, technological, and economic properties of a product, and they may be negative or positive in securing the desired total quality of a food. Moreover, the same hurdle could have positive or negative effects on foods, depending on its intensity. For instance, chilling to an unsuitably low temperature is detrimental to fruit quality (''chilling injury''), whereas moderate chilling is beneficial. Another example is the pH of fermented sausages, which should be low enough to inhibit pathogenic bacteria, but not so low as to impair taste. If the intensity of a particular hurdle in a food is too small, it should be strengthened; on the other hand, if it is detrimental to the total food quality, it should be lowered. By this adjustment, the hurdles in foods should be kept in the optimal range, considering safety as well as quality (8,14).

III. BASIC ASPECTS OF HURDLE TECHNOLOGY

Food preservation implies exposing microorganisms to a hostile environment in order to inhibit their growth, shorten their survival, or cause their death. The feasible responses of the microorganisms to such a hostile environment determine whether they grow or die. More basic research is needed in this area, because a better understanding of the physiological basis for the growth, survival, and death of microorganisms in food products could open new dimensions for food preservation (8). Furthermore, such an understanding would be the scientific basis for an efficient application of hurdle technology in the presevation of foods. Recent advances have been made by considering the homeostasis, metabolic exhaustion, and stress reactions of microorganisms, as well as by introducing the concept of multitarget preservation for gentle yet effective preservation of foods (15,16).

A. Homeostasis

A key phenomenon that deserves more attention in food preservation is the interference by the food with the homeostasis of microorganisms (17). Homeostasis is the strong tendency of organisms to maintain their internal environment stable and balanced. For instance, the maintenance of a defined pH within narrow limits is a feature and prerequisite of living organisms (18); this applies to higher organisms as well as to microorganisms. Much is already known about homeostasis in higher organisms in the fields of molecular biology, biochemistry, physiology, pharmacology, and medicine (18). This knowledge should be transferred to microorganisms important in the toxicity and spoilage of foods. If the homeostasis of microorganisms, i.e., their internal equilibrium, is disturbed by preservative factors (hurdles) in foods, they will not multiply but will remain in the lag phase or even die before their homeostasis is reestablished. Thus, food preservation is achieved by disturbing the homeostasis of microorganisms in a food temporarily or permanently (8).

Gould (17) has pointed out that during evolution a wide range of more or less rapidly acting mechanisms (e.g., osmoregulation to counterbalance a hostile water activity in food) developed in microorganisms that act to keep important physiological systems operating, in balance, and unperturbed even when the environment around them is greatly perturbed (19). In most foods microorganisms are operating homeostatically in order to react to environmental stresses imposed by the preservation procedures applied. The most useful procedures employed to preserve foods are effective in overcoming the various homeostatic mechanisms the microorganisms have evolved in order to survive extreme environmental stresses (19). The repair of a disturbed homeostasis demands much energy, and thus the restriction of energy supply inhibits repair mechanisms in microbial cells and leads to a synergistic effect of preservative factors (hurdles). Energy restrictions for microorganisms are, for example, caused by anaerobic conditions, such as in vacuum or modified-atmosphere packaging of foods. Therefore, low a_w (and/or low pH) and low redox potential act synergistically (19). Such interference with the homeostasis of microorganisms or entire microbial populations provides an attractive and logical focus for improvements in food-preservation techniques (19).

B. Metabolic Exhaustion

Another phenomenon of practical importance is the metabolic exhaustion of microorganisms, which could lead to "autosterilization" of foods. This was first observed by us in 1970 (20). Mildly heated (95°C core temperature) liver sausage was adjusted to different water activities by the addition of salt and fat, and the product was inoculated with *Clostridium sporogenes* PA 3679 and stored at 37°C.

Clostridial spores that survived the heat treatment vanished in the sausage during storage if the products were stable (i.e., did not allow growth of *C. sporogenes*). Later this type of behavior for both *Clostridium* and *Bacillus* spores was regularly observed during storage of shelf-stable meat products (SSP), especially F-SSP (21). The most likely explanation is that bacterial spores that survive the heat treatment are able to germinate in these foods under less favorable conditions than those under which vegetative bacteria are able to multiply (3). Therefore, during storage of these products some viable spores germinate, but the germinated spores or vegetative cells derived from these spores die. Thus, the spore counts in stable hurdle-technology foods actually decrease during storage, especially in unrefrigerated foods. During studies in our laboratory of Chinese dried meat products, we observed the same behavior of microorganisms (22). If these meats were contaminated after processing with staphylococci, salmonellae, or yeasts, the counts of these microorganisms on stable products decreased quite fast during unrefrigerated storage, especially on meats with a water activity close to the threshold for microbial growth. The same phenomenon was observed by Latin American researchers (23–26) in studies of high-moisture fruit products (HMFP) because the counts of a variety of bacteria, yeasts, and molds that survived the mild heat treatment decreased quite fast in the products during unrefrigerated storage because the hurdles applied (pH, a_w, sorbate, sulfite) did not allow growth.

A general explanation for this behavior might be that vegetative microorganisms that cannot grow will die, and they die more quickly if the stability of the food is close to the threshold for growth, storage temperature is elevated, antimicrobial substances are present, and the organisms are sublethally injured (e.g., by heat) (8). Apparently, microorganisms in stable hurdle-technology foods strain every possible repair mechanism for their homeostasis in order to overcome the hostile environment. By doing this they completely use up their energy and die. This leads eventually to an autosterilization of such foods (16).

Thus, due to autosterilization, hurdle-technology foods, which are microbiologically stable, become even more safe during storage, especially at ambient temperature. So, for example, salmonellae that survive the ripening process in fermented sausages will vanish more quickly if the product is stored at ambient temperature, and they will survive longer and possibly cause foodborne illness if the products are stored under refrigeration (8). It is also well known that salmonellae survive in mayonnaise at chill temperatures much better than at ambient temperatures. Unilever laboratories in Vlaardingen have confirmed metabolic exhaustion in water-in-oil emulsions (resembling margarine) inoculated with *Listeria innocua*. In these products listeria vanished faster at ambient (25°C) than at chill (7°C) temperature, at pH 4.25 > pH 4.3 > pH 6.0, in fine emulsions more quickly than in coarse emulsions, and under anaerobic conditions more quickly than under aerobic conditions. From these experiments it was concluded that metabolic exhaustion is accelerated if more hurdles are present, and this might be caused by increasing energy demands to maintain internal homeostasis under stress conditions (P. F. ter Steeg, Unilever Vlaardingen, personal communication).

C. Stress Reactions

A limitation to the success of hurdle-technology foods could be adaptation or stress reactions of microorganisms. Some bacteria become more resistant or even more virulent under stress, when they generate stress shock proteins. Synthesis of protective stress shock proteins is induced by heat, pH, a_w, ethanol, etc. as well as by starvation. These stress reactions cause a higher tolerance toward the stress that induced them, i.e., sublethal heat treatment creates microorganisms that tolerate higher heat treatment. However, stress reactions also might have a nonspecific effect, which means that actually exposing bacteria to certain sublethal stresses may significantly improve their responses to other, unrelated types of stresses to which they are exposed at a later stage; such behavior is called "cross-tolerance" (27). The responses of microorganisms under stress might hamper food preservation and could turn out to be problematic for the application of hurdle technology. On the other hand, the activation of genes for the synthesis of stress shock proteins, which help organisms to cope with stress situations, should be more difficult if different stresses are received at the same time. Simultaneous exposure to several stresses will require energy-consuming synthesis of different or at least much more protective stress shock proteins, which in turn may cause the microorganisms to become metabolically exhausted (15). Therefore, multitarget preservation of foods could be the way to avoid synthesis of such stress shock proteins, which otherwise could jeopardize the microbial stability and safety of hurdle-technology foods (16).

D. Multitarget Preservation

The multitarget preservation of foods should be the ultimate goal for a gentle but most effective preservation of foods (15). It has been suspected for some time that different hurdles in a food might not have just an additive effect on microbial stability, but might act synergistically (6). A synergistic effect could be achieved if the hurdles in a food hit, at the same time, different targets (e.g., cell membrane, DNA, enzyme systems, pH, a_w, Eh) within the microbial cell and thus disturb the homeostasis of the microorganisms present in several respects. If so, the repair of homeostasis as well as the activation of stress shock proteins would become more difficult (8). Therefore, employing different hurdles simultaneously in the preservation of a particular food should achieve optimal microbial stability. In practical terms, this means that it may be more effective to use a combination of different preservative factors with low intensities when these hit different targets or act synergistically, rather than a single preservative factor with a high intensity (14,27).

It is anticipated that the targets in microorganisms of different preservative factors (hurdles) for foods will be elucidated and that hurdles could then be grouped in classes according to their targets. A mild and effective preservation of foods, i.e., a synergistic effect of hurdles, is likely if the preservation measures are based on intelligent selection and combination of hurdles taken from different target classes (8). This approach seems valid not only for traditional food-preservation procedures, but for modern processes (e.g., ultra-high pressure, mano-thermo-sonication, pulsed technologies) as well. An example of a multitarget novel process is the application of nisin, which damages the cell membrane, in combination with lysozyme and citrate, which are then able to easily penetrate the cell and disturb the homeostasis with different targets (16).

Food microbiologists could learn in this respect from pharmacologists, because the mechanisms of action of biocides have been studied extensively in the medical field. At least 12 classes of biocides are already known, which often have more than one target within the microbial cell. Frequently the cell membrane is the primary target, becoming leaky and disrupting the organism, but biocides also impair the synthesis of enzymes, proteins, and DNA (28). Multidrug attack has proved successful in the medical field to fight bacterial infections (e.g., tuberculosis) as well as viral infections (e.g., AIDS), and thus a multitarget attack of microorganisms should be a promising approach in food microbiology, too (16).

IV. APPLICATION OF HURDLE TECHNOLOGY

Foods based on combined preservation methods (hurdle technology) are prevalent in industrialized as well as in developing countries. In the past and often still today hurdle technology was applied empirically without knowing the governing principles in the preservation of a particular food. But with a better understanding of these principles and improved monitoring devices, the deliberate application of hurdle technology has advanced.

A. Industrialized Countries

Deliberate and intelligent hurdle technology for food preservation started in the early 1980s in Germany with meat products, but now is progressing in several countries with a variety of foods.

Hurdle technology was first used for the gentle preservation of mildly heated, fresh-like meats storable without refrigeration (10). In the meantime, four categories (F-SSP, a_w-SSP, pH-SSP, Combi-SSP) of these shelf-stable products (SSP) have evolved, which are present in large quantities on the German market and have caused no problems related to spoilage or food poisoning (3). In the manufacturing plants processing these shelf-stable meats, no microbiological tests have to be carried out, but other parameters must be strictly controlled—time, temperature, pH, and a_w (21).

Better understanding of the sequence of hurdles that leads to microbial stability of fermented sausages (salami) has improved the safety and quality of these products (29). In fermented sausages the microstructure of the products, studied by electron microscopy, turned out to be an important hurdle related to the behavior of pathogens as well as starter cultures in salami (8,30).

More recent is the application of hurdle technology for microbial stabilization of novel healthful foods, derived from meat, poultry, or fish, which contain less fat and/or salt and therefore are more prone to spoil or cause food poisoning. The reduction of salt and fat as well as the substitutes and replacers of these traditional ingredients for muscle foods diminish the microbial stability, since several hurdles (a_w, pH, preservatives, and possibly Eh and microstructure) will change. Compensation could be achieved by an intelligent application of hurdle technology (31).

The advantages of hurdle technology are most obvious in high-moisture foods which are shelf stable at ambient temperature due to an intelligent application of combined methods. However, the use of hurdle technology is approriate for chilled foods too, because in the case of temperature abuse, which can easily happen during food distribution, the stability and safety of chilled foods break down, especially if low-temperature storage is the only hurdle. Therefore, it is advisable to incorporate into chilled foods (e.g., sous vide dishes, salads, fresh-cut vegetables) some additional hurdles (e.g., modified-atmosphere packaging) that will act as a back-up in case of temperature abuse. This type of safety precaution for chilled foods is called ''invisible technology,'' implying that additional hurdles act as safeguards in chilled foods ensuring that they remain microbiologically stable and safe during storage in retail outlets as well as in the home (9).

Packaging is an important hurdle for most foods, since it supports the microbial stability and safety as well as the sensory quality of food products. Industrialized countries have the tendency to overpackage foods, and this is especially true for Japan, where, e.g., ''active'' packaging (using scavengers, absorbers, emmiters, antimicrobial or antioxidative packaging materials, etc.) has been developed to perfection. These ''smart'' packaging systems are very sophisticated, but also wasteful. Therefore, Japanese experts are aiming now for less packaging of foods (32). Future packaging shall provide only necessary information and some convenience to the consumer, but the required shelf life of the product should depend not on the packaging, but should be based on superclean packaging procedures combined with just-in-time delivery or on the development of hurdle-technology foods that are stable and safe in spite of minimal packaging (K. Ono, Snow Brand Tokyo, Japan, personal communication, 1996).

B. Developing Countries

Most of the food in developing countries, which preferably must be storable without refrigeration, since electricity is expensive and not continuously available, is preserved based on empiric use of hurdle technology. However, such foods have been optimized by intentional application of hurdles. Relevant examples are meat products in China and Taiwan as well as dairy and meat products of India. In several countries of Latin America (especially Argentina, Mexico, and Venezuela) using hurdle technology high-moisture fruit products (HMFP) have been developed that, in spite of a high water activity (a_w 0.98–0.93), are storable in fresh-like condition for several months at ambient temperatures and become even sterile during storage due to metabolic exhaustion of the bacteria, yeasts, and molds originally present in these products. There is a general trend in developing countries to move gradually away from the traditional intermediate-moisture foods because they are often too salty or too sweet and have a less appealing texture and appearance than high-moisture foods, and this goal can be achieved by the application of intentional hurdle technology. The progress made in application of intelligent hurdle technology in Latin America, China, India, and Africa has recently been reviewed (33).

C. Food Design

Hurdle technology as a concept proved useful in the optimization of traditional foods as well as in the development of novel products. There are similarities to the concepts of predictive microbiology and hazard analysis critical control point (HACCP). The three concepts have related but different goals: hurdle technology is primarily used in food design, predictive microbiology for process refinement, and HACCP for process control. Nevertheless, in product development these three concepts should be combined.

We have suggested a 10-step procedure for the optimization of traditional foods or the design of new hurdle-technology foods encompassing hurdle technology, predictive microbiology, and HACCP (34). This approach proved suitable when solving real product development tasks in the food industry (21).

Predictive microbiology is a promising concept that involves computer-based and quantitative predictions of microbial growth, survival, and death in foods, and thus should be an integral part of advanced food design. However, the predictive models constructed so far are primarily applicable to pathogenic bacteria, and models to predict the behavior of the spoilage flora are just emerging. Furthermore, the available predictive models handle only up to four different factors (hurdles) simultaneously. There are numerous hurdles to be considered that are important for the stability, safety, and quality of a particular food; more than 60 different hurdles have been described. It is unlikely that all or even a majority of these hurdles could be covered by predictive modeling. Thus, predictive microbiology cannot be a quantitative approach to the totality of hurdle technology. However, it does allow quite reliable predictions of the fate of microorganisms in food systems, while considering the most important hurdles. Because several hurdles are not taken into account, the predicted results are often on the safe side, i.e., the limits indicated for growth of pathogens in foods by the models available are generally more conservative than the limits in real foods. Nevertheless, predictive microbiology will be an important tool for advanced food design, because it can considerably narrow the range over which challenge tests with relevant microorganisms need to be performed. Although predictive microbiology will not render challenge testing obsolete, it may greatly reduce both the time for and the costs of product development (8,9,21).

After a food has been properly designed, its manufacturing process must be effectively controlled, for which purpose the application of HACCP might be suitable. In a strict sense the HACCP concept only controls the hazards of foods, not their stability or quality, even though in commercial practice safety and quality issues will often overlap if HACCP is applied (21). Since for hurdle-technology foods microbial safety and stability as well as quality, i.e., the total quality of the food, is essential, the HACCP concept might be too narrow for this purpose since it relates only to biological, chemical, and physical hazards. Therefore, the HACCP concept should be broadened in order to cover the microbial safety (food poisoning) and stability (spoilage) of foods as well as their sensory quality. If this is not acceptable, the production process should be controlled by good manufacturing practice (GMP), and rules or guidelines for the production of each food item must be defined (35). For hurdle-technology foods in developing countries, GMP guidelines are often more acceptable because the application of HACCP poses practical difficulties where many small producers exist (9,33).

In food design different types of researchers, including microbiologists and technologists, must work together. The microbiologist should determine which types and intensity of hurdles are needed for the necessary safety and stability of a particular food product, and the technologist should determine which ingredients or processes are proper for establishing these hurdles in a food, taking into account the legal, technological, sensory, and nutritive limitations. Because the engineering, economic, and marketing aspects must also be considered, food design is indeed a multidisciplinary endeavor (8,21).

V. CONCLUSIONS AND PROSPECTS

The total quality of a food, which includes microbial safety and stability as well as sensory quality, is decisive for consumer satisfaction. However, the microbial safety and sensory quality of foods are often contradictory. Mildly processed foods taste and look good but have unsatisfactory microbiological stability, whereas severely processed foods are certainly safe and stable, but have shortcomings related to their flavor, texture, and appearance. The application of advanced hurdle technology enables food technologists to create minimally processed food products with the desired sensory and nutritional properties which at the same time are microbiologically stable and safe. It is anticipated that deliberate, educated hurdle technology will be increasingly applied to food products.

Hurdle-technology foods are in general less robust than traditional foods, which are often overprocessed and thus possess a large margin of safety. Therefore, if hurdle-technology foods are produced, the applied processes must be exactly defined and controlled. In the design of advanced hurdle-technology foods, a 10-step procedure, which comprises hurdle technology, predictive microbiology, and HACCP (or GMP guidelines), has proved useful.

Hurdle technology should reduce the amount of additives used even if their number might increase. It is of paramount importance that the additional hurdles be introduced into food products only after careful consideration of the necessity and in essential amounts, otherwise an undesirable chemical overloading of the food might result.

At present, in industrialized countries the hurdle-technology approach is of most interest for minimally processed, fresh-like foods which are mildly heated or fermented, and for underpinning the microbial stability and safety of foods coming from future lines, such as healthful foods with less fat and/or salt or advanced hurdle-technology foods that require less packaging. For refrigerated foods chill temperatures are the major and sometimes the only hurdle. But if exposed to temperature abuse during distribution this hurdle breaks down, spoilage or even food poisoning could happen. In the future, additional hurdles might be incorporated as safeguards into chilled foods using an approach called "invisible technology."

In developing countries the intentional application of hurdle technology for foods that remain stable, safe, and flavorful even if stored without refrigeration has recently made impressive strides, especially in Latin America with the development of novel high-moisture fruit products, which are ambient stable for several months. Interest in intentional hurdle technology is also emerging for meat products in China as well as for dairy and meat products of India. Such applications will require a thorough understanding of the principles involved as well as more back-up of their production by guidelines based on good manufacturing practice (GMP) or, whenever appropriate, by application of HACCP.

The physiology of microorganisms will be taken more into account in the design of hurdle-technology foods, particularly relating to the homeostasis and metabolic exhaustion of microorganisms and to their stress reactions. Metabolic exhaustion of microorganisms is already known to be instrumental in the microbial stability and safety of minimally processed meats and fruits stored without refrigeration. However, the possibilities and limitations of metabolic exhaustion of microorganisms during preservation and storage of food should be explored in detail, because better knowledge of this phenomenon would be beneficial for many minimally processed foods stored at ambient temperature. For consumers and even for food scientists it may sound strange that refrigeration is not always good for the safety and quality of foods. However, there are well-documented examples that in stable hurdle-technology foods, in which microorganisms are not able to grow, they will die off and vanish if such foods are stored without refrigeration. Therefore, the conditions that accelerate metabolic exhaustion of microorganisms in foods and the limitations of this approach for food preservation should be studied systematically.

Another future challenge for improving food preservation is the intelligent application of multitarget preservation of foods, i.e., the selection and application of preservative hurdles according to their targets within the microbial cells. Today the preservative factors (hurdles) for microbial stabilization of food items are often chosen at random; even empirical knowledge is used as a guide. However, the targets within the microbial cells of the different possible hurdles for food preservation should be elucidated in detailed studies based on the physiological responses of microorganisms. For mild and most efficient food preservation, hurdles from different target classes should be chosen and applied, because these hurdles will most likely act synergistically and thus be effective even if used at low strength/intensity.

Metabolic exhaustion of microorganisms and multitarget preservation of foods should become major research priorities in food preservation. Better knowledge about these phenomena could improve the future application of hurdle technology and food preservation in general.

REFERENCES

1. L Leistner. Hurdle technology. In: FJ Francis, ed. The Wiley Encyclopedia of Food Science & Technology, 2nd ed. New York: John Wiley & Sons, 1999, pp. 1302–1307.
2. L Leistner, W Rödel, K Krispien. Microbiology of meat and meat products in high- and intermediate-moisture ranges. In: LB Rockland, GF Stewart, eds. Water Activity: Influences on Food Quality. New York: Academic Press, 1981, pp. 855–916.
3. L Leistner. Food preservation by combined methods. Food Res Int 25:151–158, 1992.
4. GW Gould, MV Jones. Combination and synergistic effects. In: GW Gould, ed. Mechanisms of Action of Food Preservation Procedures. London: Elsevier, 1989, pp. 401–422.

5. GV Barbosa-Cánovas, UR Pothakamury, E Palou, BG Swanson. Nonthermal Preservation of Foods. New York: Marcel Dekker, 1998, pp. 235–268.
6. L Leistner. Hurdle effect and energy saving. In: WK Downey, ed. Food Quality and Nutrition. London: Applied Science Publishers, 1978, pp. 553–557.
7. L Leistner, W Rödel. The stability of intermediate moisture foods with respect to micro-organisms. In: R Davies, GG Birch, KJ Parker, eds. Intermediate Moisture Foods. London: Applied Science Publishers, 1976, pp. 120–134.
8. L Leistner. Principles and applications of hurdle technology. In: GW Gould, ed. New Methods of Food Preservation. London: Blackie Academic & Professional, 1995, pp. 1–21.
9. L Leistner. Combined methods for food preservation. In: M Shafiur Rahman, ed. Food Preservation Handbook. New York: Marcel Dekker, 1999, pp. 457–485.
10. L Leistner. Hurdle technology applied to meat products of the shelf stable product and intermediate moisture food types. In: D Simatos, JL Multon, eds. Properties of Water in Foods in Relation to Quality and Stability. Dordrecht: Martinus Nijhoff Publishers, 1985, pp. 309–329.
11. L Leistner, I Vuković, J Dresel. SSP: meat products with minimal nitrite addition, storable without refrigeration. Proceedings of 26th European Meeting of Meat Research Workers, Colorado Springs, 1980, Vol. II, pp. 230–233.
12. DW Stanley. Biological membrane deterioration and associated quality losses in food tissue. Crit Rev Food Sci Nutr 30:487–553, 1991.
13. BM McKenna. Combined processes and total quality management. In: L Leistner, LGM Gorris, eds. Food Preservation by Combined Processes. FLAIR Final Report Concerted Action no. 7, Subgroup B. Brussels: European Commission, EUR 15776 EN, 1994, pp. 99–100.
14. L Leistner. Further developments in the utilization of hurdle technology for food preservation. J Food Eng 22:421–432, 1994.
15. L Leistner. Food protection by hurdle technology. Bull Japan Soc Res Food Prot 2: 2–27, 1996.
16. L Leistner. Basic aspects of food preservation by hurdle technology. Int J Food Microbiol 55:181–186, 2000.
17. GW Gould. Interference with homeostasis—food. In: R Whittenbury, GW Gould, JG Banks, RG Board, eds. Homeostatic Mechanisms in Micro-organisms. Bath: Bath University Press, 1988, pp. 220–228.
18. D Häussinger. pH Homeostasis—Mechanisms and Control. London: Academic Press, 1988.
19. GW Gould. Homeostatic mechanisms during food preservation by combined methods. In:GV Barbosa-Cánovas, J Welti-Chanes, eds. Food Preservation by Moisture Control, Fundamentals and Applications. Lancaster: Technomic, 1995, pp. 397–410.
20. L Leistner, S Karan-Djurdjić. Beeinflussung der Stabilität von Fleischkonserven durch Steuerung der Wasseraktivität. Fleischwirtschaft 50:1547–1549, 1970.
21. L Leistner. Food Design by Hurdle Technology and HACCP. Kulmbach: Adalbert-Raps-Foundation, 1994.
22. HK Shin. Energiesparende Konservierungsmethoden für Fleischerzeugnisse, abgeleitet von traditionellen Intermediate Moisture Foods. PhD dissertation, Universität Hohenheim, Stuttgart-Hohenheim, 1984.

23. S Sajur. Preconservación de Duraznos por Métodos Combinados. MS dissertation, Universidad Nacional de Mar del Plata, Argentina, 1985.
24. SM Alzamora, MS Tapia, A Argaiz, J Welti. Application of combined methods technology in minimally processed fruits. Food Res Int 26:125–130, 1993.
25. SM Alzamora, P Cerrutti, S Guerrero, A López-Malo. Minimally processed fruits by combined methods. In: GV Barbosa-Cánovas, J Welti-Chanes, eds. Food Preservation by Moisture Control, Fundamentals and Applications. Lancaster: Technomic, 1995, pp. 463–492.
26. MS Tapia de Daza, A Argaiz, A López-Malo, RV Díaz. Microbial stability assessment in high and intermediate moisture foods: special emphasis on fruit products. In: GV Barbosa-Cánovas, J Welti-Chanes, eds. Food Preservation by Moisture Control, Fundamentals and Applications. Lancaster: Technomic, 1995, pp. 575–601.
27. LMG Gorris. Hurdle technology, a concept for safe, minimally processed foods. In: RK Robinson, CA Batt, PD Patel, eds. Encyclopedia of Food Microbiology. London: Academic Press, 1999, pp. 1071–1076.
28. SP Denyer, WB Hugo, eds. Mechanisms of Action of Chemical Biocides: Their Study and Exploitation. London: Blackwell Scientific Publications, 1991.
29. L Leistner. Stable and safe fermented sausages world-wide. In: G Campbell-Platt, PE Cook, eds. Fermented Meats. London: Blackie Academic & Professional, 1995, pp. 160–175.
30. K Katsaras, L Leistner. Distribution and development of bacterial colonies in fermented sausages. Biofouling 5:115–124, 1991.
31. L Leistner. Microbial stability and safety of healthy meat, poultry and fish products. In: AM Pearson, TR Dutson, eds. Production and Processing of Healthy Meat, Poultry and Fish Products. London: Blackie Academic & Professional, 1997, pp. 347–360.
32. K Ono. Packaging design and innovation. Booklet for a regional training course in food packaging, Snow Brand Tokyo, Japan, 1994.
33. L Leistner. Use of combined preservative factors in foods of developing countries. In: BM Lund, AC Baird-Parker, GW Gould, eds. The Microbiological Safety and Quality of Food. Gaithersburg: Aspen Publishers, 2000, Vol. I, pp. 294–314.
34. L Leistner. User guide to food design. In: L Leistner, LMG Gorris, eds. Food Preservation by Combined Processes. FLAIR Final Report Concerted Action no. 7, Subgroup B. Brussels: European Commission, EUR 15776 EN, 1994, pp. 25–28.
35. L Leistner. Shelf-stable products and intermediate moisture foods based on meat. In: LB Rockland, LR Beuchat, eds. Water Activity: Theory and Applications to Food. New York: Marcel Dekker, 1987, pp. 295–327.

21
Summary and Future Prospects

Martin B. Cole
Food Science Australia
North Ryde, New South Wales, Australia

Terry A. Roberts
Consultant
Reading, England

I. INTRODUCTION

As outlined in Chapter 1, not only is the incidence of food-borne illness increasing globally but the nature of food poisoning is also changing, influenced by demographics, industrialization and centralization of food production and supply, travel and trade, and microbial evolution and adaptation. Although traditional foodborne pathogens such as Salmonellae, *Staphylococcus aureus,* and *Clostridium perfringens* are still important in terms of the number of cases of food poisoning they cause, the emergence of pathogens such as *Campylobacter jejuni/coli*, *Escherichia coli* O157:H7, *Listeria monocytogenes*, and *Cyclospora cayetanensis* that were not recognized 20 years ago has changed the nature of foodborne illness. Twenty years ago most food poisoning was associated with an outbreak where the organism could usually be readily isolated from the implicated food vehicle, with high levels being needed to cause illness. In addition, a low mortality rate was associated with illness and the types of foods traditionally associated with food poisoning were well known. With the emergence of infectious pathogens such as pathogenic *E. coli* with a low infective dose, food poisoning cases are now often of a sporadic nature and only detectable through surveillance. With the increased trade in food between countries there is often an international aspect, and because of the low infective dose, the risk is often undetected by traditional sampling and testing regimes. Of particular concern is the fact that infection is frequently associated with a higher mortality rate and may now be associated with the consumption of a wider range of foods, including fresh produce that does not receive a pathogen reduction step prior to consumption.

These trends, along with an overall increased concern for the safety of foods, has prompted industry, government, and academia to consider new techniques to both assess and manage microbiological risks. The International Commission on Microbiological Specifications for Foods (ICMSF) has proposed a comprehensive ''preventive'' system for the management of microbiological hazards in foods in international trade. Initial steps in the scheme include risk assessment and the establishment of a tolerable level of risk in terms of a public health outcome or goal. Because industry cannot address public health goals (e.g., ''reduce the number of cases per 100,000 population''), a Food Safety Objective (FSO) is proposed as a concept for translating public health goals into risk management control measures.

An FSO is a statement of the frequency or maximum concentration of a microbiological hazard in a food acceptable for consumer protection (1) and therefore provides a functional link between risk assessment and risk management. Following the initial development of a proposed FSO, risk managers in industry and government must confirm that the FSO is technically achievable through the implementation of available control measures such as good hygienic practice (GHP) and the hazard analysis and critical control point (HACCP) system.

Control measures generally fall into three categories: controlling the initial level of a hazard, preventing an increase in the level of a hazard, and reducing the level of a hazard. One or more control measures may be applied at different steps along the food chain. The effectiveness of such control measures can be assessed by adherence to performance and process criteria set by industry, trade associations, governments, and regulatory bodies with the aim of meeting an FSO. This chapter considers the control measures outlined in the previous chapters within this framework and discusses the use of predictive modeling as a means of validating control measures as well as future prospects.

II. FOOD SAFETY OBJECTIVES

Traditionally, risk assessment in international trade has been defined in terms of having the chemical or microbial risk ''as low as reasonable.'' This has caused great difficulties for a number of reasons. Both the technological capabilities and the idea of what is considered reasonable differ from country to country and even within countries from one company to another. Managing tolerable risk implies balancing public health considerations with other factors such as economic cost and public acceptability, which also has culturally different aspects. Developments in quantitative risk assessments in microbiology (2,3) have, by making a range of assumptions, begun to link the exposure assessment of a pathogen to likely public health outcomes. The ICMSF scheme for managing microbiological risks for foods in international trade proposes that the FSO be a functional link between risk assessment and risk management. The FSO is defined as "a statement of the frequency or maximum concentration of a microbiological hazard in a food considered acceptable for consumer protection" (1) and allows the equivalence of different control measures to be established.

III. CONTROL MEASURES

Control measures are the actions and activities used to prevent, eliminate, or reduce a food safety hazard to a tolerable level. One or more control measures may be necessary at different steps along the food chain to prevent, eliminate, or reduce a hazard to an acceptable level at the time the food is consumed and meet a given FSO. Each person along the food chain, from primary production to the consumer, has a responsibility to contribute to the provision of safe foods. The control measures outlined in the previous chapters of this book can be considered within this framework. However, depending on conditions, some control measures may be involved in both preventing an increase of a hazard as well as causing a reduction in the level of hazard. An example is the use of chemical preservatives that may prevent growth at low concentrations but may also cause a reduction in a hazard through inactivation at higher concentrations.

1. Controlling initial levels

Avoiding foods with a history of contamination or toxicity (e.g., raw milk, raw molluscan shellfish harvested under certain conditions)
Selecting ingredients (e.g., pasteurized liquid eggs or milk)
Using microbiological testing and criteria to reject unacceptable ingredients or products

2. Preventing increase of levels

Preventing contamination (e.g., adoption of GHPs that minimize contamination during slaughter, separating raw from cooked ready-to-eat foods, implementing employee practices and the use of packaging)
Preventing growth of pathogens (e.g., the use of cold temperatures, chemicals, natural antimicrobials, formulation of products to prevent growth of *Clostridium botulinum*)

3. Reducing levels

Destroying pathogens (e.g., thermal treatments such as traditional heating, ohmic heating and microwave heating, nonthermal treatments such as irradiation, pulsed electric fields, ultra-high-pressure and magnetic fields, and finally chemical treatments for inactivation such as the use of acids and bacteriocins)
Removing pathogens (e.g., washing, ultra-filtration, centrifugation)

IV. PERFORMANCE CRITERIA

To achieve a defined FSO it is necessary to implement one or more control measures at one or different steps in the food chain. At these steps, hazards can either be controlled or reduced, or they may, if not controlled, increase. The outcome of these control measure(s) is defined as ''performance criteria'': the required outcome of a step or a combination of steps that can be applied to assure a FSO is met. A step is a point, procedure, operation, or stage in the food chain including raw materials from primary production to final consumption. An example of performance criteria would be a 6D reduction of *L. monocytogenes* in ready-to-eat chilled foods (4).

When establishing performance criteria, consideration must be given to the initial level of a hazard and changes occurring during production, distribution, storage, preparation, and use of a product. A performance criterion is preferably less but at least equal to the FSO and can be expressed by the following equation:

$$H_0 - \Sigma R + \Sigma I \leq FSO \quad (1)$$

where FSO = Food Safety Objective, H_0 = initial level of the hazard, ΣR = total (cumulative) reduction of the hazard, and ΣI = total (cumulative) increase of the hazard, all of which are expressed in $\log_{10}$ units.

These criteria are usually not established for control measures designed to avoid certain foods, although they may be applied to ensure that the initial levels of hazards in ingredients are not excessive. Microbiological testing may thus be used to select ingredients or to obtain information on the initial level of a hazard.

This systematic approach can provide the framework for validation of control measures and the establishment of verification activities for use by food operators and control authorities. Application of processes validated to achieve specified performance criteria is more reliable for ensuring food safety than attempting to rely upon microbiological testing of foods to separate ''safe'' from ''unsafe'' foods.

V. PROCESS AND PRODUCT CRITERIA

A performance criterion is met by implementing process criteria such as time and temperature of a heat treatment and/or product criteria such as the water activity of the product, alone or in combination, to achieve control over a specific hazard. The definition of a ''process criterion'' can be given as the control parameters of a step or combination of steps that can be applied to achieve the performance criterion. An example of a process criterion could be heating for 2 minutes at 70°C or higher in order to achieve a 6D reduction in *L. monocytogenes.*

Predictive microbial models are invaluable tools for estimating the cumulative increase or decrease of hazard and hence serve as one measure of validation of the likely effectiveness of a given combination of control measures.

VI. PRODUCT AND PROCESS VALIDATION

Although a considerable amount of information is available in the literature and other sources for traditional processes, such as those based on thermal inactivation, for new, novel processes it may be necessary to develop information to verify the efficacy of the control measures. Validation can include the use of laboratory data in the form of predictive microbial models and challenge tests, the use of data collected during normal processing in the food operation, comparison with similar processes/products, as well as the use of other expert knowledge.

A. Predictive Modeling

Modeling microbial responses has become popular with the availability of powerful personal computers and user-friendly software. In some instances, however, the urge to model has overcome researchers before they have considered fully the controls necessary on experimental design (i.e., the range of conditions in two or three dimensions to be covered), the precision with which the rate to be modeled (growth or death) can be estimated, and the nature of the modeling procedure that (often) converts a multitude of estimates of rate into a response surface describing the effect of the two- or three-dimension controlling factors on the change in rate. Few publications have considered the reproducibility of the data generated (i.e., when rates are measured again in the same lab). Checking that predictions describe the observed biological responses is also often overlooked. Comparison of the resulting model against the available literature is also an important aspect of validation (5,6). The different types of predictive microbial models and their relevance to the validation of different control measures is summarized below.

Growth models estimate responses where at least part of the range of conditions permits growth to occur and can describe the increase in numbers with time (kinetic), the conditions allowing growth or no growth (boundary), or the chance of growth (probabilistic).

Kinetic models are usually based on growth over a range of conditions (e.g., pH, brine concentration, temperature, and sometimes a fourth factor such as CO_2 or nitrite) and can allow growth rates to be predicted. These models can also predict the time needed to go from the initial condition to the final condition of interest (e.g., hours for 3-log increase in numbers). If desired, they can predict a complete growth curve for the conditions of interest.

Boundary models describe the limits of combinations of conditions that permit or do not permit a response, for example, growth of the organism. This type of model predicts a ''time to growth'' based on data collected as a qualitative response, growth or no growth, at intervals of time over a specified time period. A feature of the datasets used to build these models is that the no-growth observations, known as ''censored'' data, tell only that growth was not observed during the experiment. Due to the censoring, these data cannot be handled by standard regression modeling approaches. Rather, statistical techniques designed for ''survival'' analysis must be used.

Probability or probabilistic models describe the likelihood of a particular response being observed or the time until an event occurs. Such models are appropriate when a binary response (i.e., either ''growth'' or ''no growth'') is available and a model for the probability of growth at time *t* is desired.

In the context of the previous section, growth models are used to validate product and process criteria that will prevent or limit the increase in a hazard.

Death or inactivation models are designed to predict for conditions in which a lethal process is deliberately applied. As a consequence, microbial death is relatively rapid. These are most commonly encountered in thermal processing but have been presented for other deliberately lethal conditions such as irradiation treatment. They are usually kinetic in nature and describe inactivation with time of lethal treatment.

Survival models relate to transitional conditions between growth and death. Typically death occurs relatively slowly in conditions where growth cannot occur but no deliberately lethal treatment is applied, for example, rate of death at ambient temperature when the a_w (water activity) is too low to permit growth.

Death and survival models are used to validate product or process criteria that will allow a reduction in the level of a hazard.

B. Model Validation

Before predictive models can be used to validate process and product criteria, they must themselves be validated for the particular food and situation that they will be used for. In the predictive modeling, validation has two components:

1. Model validation—to confirm that the model meets certain statistical/mathematical criteria and follows biological experience
2. Food validation—to confirm that the outputs from the model (the predictions) mimic the behavior of the microbe in question in real foods

Having developed a model from sufficient laboratory data and confirmed that it meets statistical/mathematical criteria, critics would still comment that the model may have little relevance to the microbiology of food products. Ideally, the model should be evaluated by storing the microbe in a range of foods under conditions representative of the range of conditions in the model, including extremes of conditions. However, that approach is slow and costly, and a less expensive alternative is to compare the outputs of the model with independent data already in existence (e.g., the scientific literature). When the literature does not contain sufficient information/data to make comparisons, targeted challenge tests can be performed.

Data developed through laboratory challenge tests can involve the food, culture media, or other material that may be appropriate. Challenge studies in a food processing environment can provide a higher degree of assurance concerning the ability to meet performance criteria; however, this requires the use of surrogate test microorganisms. Pathogenic microorganisms should never be introduced into the food production or processing environment for the purpose of process validation. In some cases it may be possible to follow changes in the population of naturally occurring pathogens throughout a process. Such studies could, for example, be conducted during the preparation and processing of raw agricultural commodities into ready-to-eat foods. Ideally, validation could involve laboratory challenge tests with pathogens in the laboratory and then revalidation after the control measures have been implemented. This may be impractical in situations where the prevalence of a pathogen is very low and large numbers of samples are necessary to develop meaningful data.

C. Challenge Tests

When conducting laboratory challenge studies, factors such as the intrinsic resistance of the pathogen, composition of the food, and the conditions of storage, distribution, and use should be taken into account.

The inocula should be prepared under conditions that yield resistance of the pathogen appropriate to the process. Vegetative cells of Salmonellae and pathogenic *E. coli*, for example, may reach a maximum resistance to heat and acidic conditions when in the stationary phase after having been grown at elevated temperatures. Sufficient numbers of the pathogen (e.g., cells, spores, viral particles, oocysts) should be used to eliminate biovariability effects (see below).

Strains to be tested should not include isolates with unrealistically extreme resistances or growth characteristics when these are not associated with public health concerns. For example, *Salmonella* Senftenberg 775W should not be used as the basis for thermal processing requirements for liquid egg products because of its unusually high heat resistance and rare occurrence but is appropriate to evaluate survival in chocolate and similar products (7). Validation of control measures in a food operation can be accomplished through the use of nonpathogenic microorganisms if they have been shown to have the same growth pattern or resistance as the pathogen of concern.

Composition of the food can affect inactivation, survival, and/or growth of pathogens and therefore must be known and taken into account. Factors such as pH, a_w, Eh, humectants, acidulants, solutes, antimicrobials, substrates, and competing microflora can affect the chemical and physical properties of the food and subsequently the pathogen of concern. Normal variation in the concentration and distribution of food constituents and microorganisms also must be known and understood.

Factors affecting the safety of a food during storage, distribution, and preparation for use must be identified and controlled. Information on the intended use and an estimate of likely misuse of the product may be necessary. Examples of parameters that often have a significant effect include time and temperature, the potential for contamination, and faulty preparation before consumption.

D. Process Variability

The variability that occurs in a food operation must be considered when establishing the critical limits associated with control measures. Examples of factors that can influence the variability of a process include equipment performance and reliability, integrity of container seals, processing times and temperatures, pH, humidity, flow rates, and turbulence.

It is essential that the "typical" microbial load, variability of process parameters, and product formulation be taken in account when setting critical limits. In general terms, the critical limits at a critical control point (CCP) for a process operating under a high degree of control (low variability) can be closer to the conditions necessary for control of a hazard. Conversely, the critical limits for a less controlled process (high variability) must be more conservative and more restrictive. In other words, critical limits must be based on the capability of the process to achieve a given criterion under normal operating conditions taking into account variability. Monitoring and verification procedures specified in a HACCP plan should be designed to determine when the process is operating outside this normal variability so that appropriate corrective actions can be taken.

E. Monitoring and Verification

After effective control measures have been established, it is necessary to establish procedures to monitor each CCP in HACCP plans and verify that the control measures are being implemented as planned. Monitoring and verification can consist of a variety of measurements, such as sensory, chemical, physical, time, or even microbial counts. For further information and examples of FSOs and the establishment of control measures to meet them, see Ref. 8.

VII. FUTURE PROSPECTS

A. Biovariability

An important consideration for both growth and inactivation modeling is better estimates of inherent within- and between-strain variability. This will be essential for the application of predictive models to very low pathogen levels. This appreciation of natural biovariability has been termed quantum or quantal microbiology (9) and compares our current understanding of microbiology to the level of understanding in physics during the Newtonian years. For the last 120 years the study of microorganisms as brought to us by Koch and others has assumed that a pure subculture of a microorganism will contain identical cells. Future mathematical models will have to account for natural biovariability as described above for inactivation kinetics and in the area of modeling the fate of small numbers of cells. Single cell techniques such as image analysis (10), flow cytometry (11), and automated turbidometry (12) will be invaluable for this.

In the case of inactivation models the traditional first-order approach to describing bacterial inactivation assumes that all of the cells or spores in a population are identical in their sensitivity to the lethal agents. It is now becoming apparent that this assumption is incorrect. New modeling approaches based on an understanding of biovariability and an assumption that, in any population of cells or spores, there will be a distribution of sensitivities, etc., are now being employed for heat and ultra–high pressure to describe the inactivation kinetics of key microorganisms (13–16). The new approach to determining microbial inactivation kinetics has a number of implications including the need to use sufficiently high numbers to achieve the required decimal reduction without the need for extrapolation as well as an increased importance of the ''come-up'' or equilibration time.

B. Prospects for Emerging Preservation Technologies

In the last few years there has been growing research and commercial interest, especially in Europe and the United States, in a number of the nonthermal or cold pasteurization techniques described in earlier chapters, such as ultra–high pressure, pulsed electric field treatments, and ionizing radiation. Interest in such technologies has been fueled by a continuing consumer desire for foods that are more fresh-like but convenient and safe. A decontamination step that does not significantly alter the organoleptic qualities of the food would have obvious advantages. Cold pasteurization technologies offer the promise of foods that approximate the freshness, flavor, color, texture, and nutritional value of nonheated products while at the same time exhibiting enhanced microbiological safety. Despite this, all of the preservation technologies covered in this book have some disadvantages as well as advantages. For example, most nonthermal treatments cannot currently be relied upon to inactivate bacterial spores, which means that they must be combined with an additional preservation hurdle to prevent spore outgrowth such as refrigeration or acid formulation. The advantages and limitations of a number of the preservation treatments described in earlier chapters are summarized in Table 1. Taking ultra–high pressure (UHP) as an example, UHP cannot currently be relied upon to kill spores, but it can inactivate vegetative cells. It has no residual preservative effect to contribute to product shelf life once the product is open. The effect of UHP on enzymes is variable, with the activity of some enzymes being enhanced with pressure treatment while the activity of others is reduced under UHP. UHP can be used for solids as well as liquids and can be applied as an in-pack treatment but is currently a batch not an in-line process. Finally, UHP treatment is not just a surface treatment. Because each preservation technology has its own advantages and disadvantages, it is likely that different technologies will be commercialized for different applications. It is also likely that for some applications it will continue to be necessary to combine two or more preservation regimes in order to achieve the required product safety and/or stability.

In the case of thermal processing there is around 80 years of experience in establishing safe and reliable processes. If the true potential of some of the emerging technologies outlined in this book is to be realized, it will be important to develop systematic, kinetic data describing their efficacy against key target microorganisms. The use of FSOs and associated performance and product/process criteria offers a useful framework in which to establish the equivalency of emerging processes as control measures.

Table 1 Uses and Limitations of New and Emerging Preservation Technologies

Use	Ionizing irradiation	Microwave heating	Ohmic heating	High pressure	Pulsed E-field	Pulsed light	Natural antimicrobials
Killing spores (ambient shelf life)	+	+	+	−	−	−	(−/+)
Killing veg. cells (closed shelf life)	+	+	+	+	+	−	(−/+)
Preventing growth (open shelf life)	−	−	−	−	−	−	+
Enzyme inactivation	−	+	+	(−/+)	−	−	−
Solids	+	+	+	+	−	+	+
Liquids	+	+	+	+	+	+	+
In-pack treatment	+	+	−	+	−	+	+
In-line treatment	+	+	+	−	+	+	+
Surface only	−	−	−	−	−	+	−

REFERENCES

1. M van Schothorst. Principles for establishment of microbiological food safety objectives and related control measures. Food Control 9(6):379–384, 1998.
2. RL Buchanan, WG Damert, RC Whiting, M van Schothorst. The use of epidemiologic and food survey data to estimate a purposefully conservative relationship for *Listeria monocytogenes* levels and incidence of listeriosis. J Food Prot 60:918–922, 1997.
3. RC Whiting, RL Buchanan. Development of a quantitative risk assessment model for *Salmonella enteritidis* in pasteurized liquid eggs. Int J Food Microbiol 36(2/3): 111–125, 1997.
4. BM Lund, MR Knox, MB Cole. Destruction of *Listeria monocytogenes* during microwave cooking. Lancet (Jan. 28):218, 1989.
5. J Baranyi, TA Roberts. A dynamic approach to predicting bacterial growth in food. Int J Food Microbiol 23:277–294, 1994.
6. J Baranyi, TA Roberts. Mathematics of predictive microbiology. Int J Food Microbiol 26:199–218, 1995.
7. ICMSF (International Commission on Microbiological Specifications for Foods). Microorganisms in Foods 6: Microbial Ecology of Food Commodities. Blackie Academic and Professional, London, 1998.
8. ICMSF (International Commission on Microbiological Specifications for Foods). Microorganisms in Foods 7: Microbiological Testing in Food Safety Management. Aspen Publishers (in press).
9. EY Bridson, GW Gould. Quantal microbiology. Lett Appl Microbiol 30:95–98, 2000.
10. PJ Coote, CM-P Billon, S Pennell, PJ McClure, DP Ferdinando, MB Cole. The use of confocal scanning laser microscopy (CSLM) to study the germination of individual spores of *Bacillus cereus*. J Microbiol Methods 21:193–208, 1995.
11. G Nebe-von-Caron, A Balley. Bacterial characterisation by flow cytometry. In: M Al-Rubeai, AN Emery, eds. Flow Cytometry Applications in Cell Culture. Marcel Dekker, New York, 1996.
12. PJ Stephens, JA Joynson, KW Davies, R Holbrook, HM Lappin-Scott, TJ Humprey. The use of an automated growth analyser to measure recovery of single heat-injured Salmonella cells. J Appl Microbiol 83:445–455, 1997.
13. MB Cole, KW Davies, G Munro, CD Holyoak, DC Kilsby. A vitalistic model to describe thermal inactivation of *L. monocytogenes*. J Indust Microbiol 12:232, 1993.
14. CL Little, MR Adams, WA Anderson, MB Cole. Application of a log-logistic model to describe the survival of *Yersinia enterocolitica* at sub-optimal pH and temperature. Int J Food Microbiol 22:63, 1994.
15. WF Anderson, PJ McClure, AC Baird-Parker, MB Cole. The application of a log-logistic model to describe the thermal inactivation of *C. botulinum* 213B at temperatures below 121.1°C. J Appl Bacteriol 80:283, 1996.
16. M Peleg, MB Cole. Reinterpretation of microbial survival curves. Crit Rev Food Sci 38(5):353–380, 1998.

Index

Milton Keynes UK
Ingram Content Group UK Ltd.
UKHW020006071024
449327UK00031B/2672